Thermodynamics

in

Materials
Science

Other Materials Science Textbooks

Physical Metallurgy, by William F. Hosford

Selection of Engineering Materials and Adhesives, by Lawrence W.
Fisher, PE

Advanced Mechanics of Materials and Applied Elasticity, by Anthony
Armenàkas, Ph.D.

Engineering Designs with Polymers and Composites, James C. Gerdeen,
Harold W. Lord, and Ronald A.L. Rorrer

Modern Ceramic Engineering, Third Edition, by David Richerson

COMING SOON

Introduction to Chemical Polymer Science: A Problem Solving Guide,
by Manas Chanda

Second Edition

Thermodynamics
in
Materials
Science

Robert DeHoff

Taylor & Francis
Taylor & Francis Group

Boca Raton   London   New York

A CRC title, part of the Taylor & Francis imprint, a member of the
Taylor & Francis Group, the academic division of T&F Informa plc.

CRC Press
Taylor & Francis Group
6000 Broken Sound Parkway NW, Suite 300
Boca Raton, FL 33487-2742

© 2006 by Taylor & Francis Group, LLC
CRC Press is an imprint of Taylor & Francis Group, an Informa business

No claim to original U.S. Government works

Printed in the United States of America on acid-free paper
Version Date: 20110713

International Standard Book Number: 978-0-8493-4065-9 (Hardback)

Library of Congress Cataloging-in-Publication Data

DeHoff, Robert T.
  Thermodynamics in materials science / Robert DeHoff.-- 2nd ed.
    p. cm.
  Includes bibliographical references and index.
  ISBN-13: 978-0-8493-4065-9 (alk. paper)
  ISBN-10: 0-8493-4065-9 (alk. paper)

  1. Materials science. 2. Thermodynamics. I. Title.
  TA403.6.D44 2006   2005036364

Visit the Taylor & Francis Web site at
http://www.taylorandfrancis.com

and the CRC Press Web site at
http://www.crcpress.com

Preface to the Second Edition

The presentation of the principles and strategies at the heart of thermodynamics has been retained from the first edition. The principles and laws, the definitions, the criterion for equilibrium, and the strategies for deriving relationships among variables and for finding the conditions for equilibrium are all intact. There is a new emphasis on the structure of thermodynamics, introduced in a new Chapter 1, which provides a visualization of how all of these components integrate to solve problems. There is a new emphasis on the main goal of thermodynamics in materials science, which is the use of thermochemical databases to generate maps of equilibrium states, such as phase diagrams, predominance diagrams, and Pourbaix corrosion diagrams. There are many other useful applications of thermodynamic information, but the equilibrium maps are clearly the most widely used tools in the field.

Although computer software to convert database information into equilibrium maps was available at the writing of the first edition, such software now comes with more comprehensive databases and breadth of application and, perhaps most importantly, user-friendliness. It is also more widely available for student use as materials science programs have acquired it for use in their research or teaching. The CALPHAD origins of these programs is dealt with in Chapters 8 to 10.

There is a danger that applications in this field may achieve a "black box" status in which the results of all this software, in the form of equilibrium maps and other information, come to be used without an understanding of their origins. In industry, this kind of information may be a component in a decision-making process that has millions of dollars on the line. It is crucial that the connections between the results and the fundamentals provided by this kind of text be maintained.

Preface to the First Edition

In his classic paper in 1883, J. Willard Gibbs completed the apparatus called phenomenological thermodynamics, which is used in engineering and science to describe and understand what determines how matter behaves. This work is all the more remarkable in the light of the enormous expansion of our knowledge in science and technology in the 20th century. During the last century, hundreds of books have been written on thermodynamics. In most cases, these texts were directed at students in a particular field. Thermodynamics plays a key role in biology, chemistry, physics, chemical and mechanical engineering, and materials science. Each presentation offered its own slant to its intended audience. Several of these texts are classics that have endured for decades, experiencing many revisions and many printings.

An author who undertakes an introductory text in thermodynamics in the face of this history had better be sure of his subject, and have something unique to say. After teaching introductory thermodynamics to materials scientists for nearly three decades at both the graduate and undergraduate levels, I am convinced that the approach used in this text is unique and in many ways better than that available elsewhere.

Thermodynamics in Materials Science is an introductory text intended primarily for use in a first course in thermodynamics in materials science curricula. However, the treatment is sufficiently general so that the text has potential applications in chemistry, chemical engineering, and physics, as well as materials science. The treatment is sufficiently rigorous and the content sufficiently broad to provide a basis for a second course either for the advanced undergraduate or graduate level.

Thermodynamics is a discipline that supplies science with a broad array of relationships between the properties that matter exhibits as it changes its condition. All of these relationships derive from a very few, very general and pervasive principles (the laws of thermodynamics) and the repetitive application of a very few, very general strategies. It is not a collection of independent equations conjured out of misty vapors by an all-knowing mystic for each new application. There is a structure to thermodynamics that is elegant and, once contemplated, reasonably simple.

The approach that undergirds the presentation in this text emphasizes the connections between the foundations and the working relations that permit the solution of practical problems. In this emphasis, and in its execution, it is unique among its competitors. The difference is crucial to the student seeing the subject for the first time.

Most texts spend a significant amount of print and the student's time in presenting the laws of thermodynamics and in laying out arguments that justify the laws and lend intuitive interpretation to them. This presentation is based on the recognition that such diversions are a significant waste of time and effort for the student and, what is worse, are usually confusing to the uninitiated. Worse still, students may be left with an inadequate intuition that merely serves to mislead them when they attempt to apply it to complex systems. Thermodynamics is fundamentally a rational subject, rich with deductions and derivations. Intuition in thermodynamics is not for the uninitiated.

Thus, the laws are presented as *fait accompli*: "great accomplishments of the 19th century" that distilled a broad range of scientific observation and experience into succinct statements that reflect how the world works. It is best at this beginning stage that these laws be presented with clear statements of their content, without the perpetual motion arguments, Carnot cycle, and other intuitive trappings.

The most significant departure of this text from other works lies in the treatment of the concept of equilibrium in complex physical systems, and in the presentation of a general strategy for finding the conditions for equilibrium in such systems. A general criterion for equilibrium is developed directly from the second law of thermodynamics. The mathematical procedure for deriving the equations that describe the internal condition when it is at equilibrium is then presented with rigor. It is the central viewpoint of this text that, since all of the "working equations" of thermodynamics are mathematical statements of these internal conditions for equilibrium, establishment of the connection between these conditions and first principles is crucial to a working understanding of thermodynamics. Indeed, the remainder of the text is a series of applications of this general strategy to the derivation of the condition for equilibrium in systems of increasing complexity, together with strategies for applying these equations to solve problems of practical interest to the student. With each increment in the level of sophistication being treated, new parts of the apparatus of thermodynamics are introduced and developed as they are needed. The general strategy for getting to the working equations is the same for all of these applications. Thus, the connection to the fundamental principles is visible for each new development. Furthermore, this connection can be maintained without introducing any mathematical or conceptual shortcuts. Repetition builds confidence; rigor builds competence.

One early chapter introduces the concepts of statistical thermodynamics. This subject is treated as an algorithm for converting an atomic model for the behavior of the system, formulated as a list of the possible states that each atom may exhibit, into values of all of the thermodynamic properties of the system. The strategy for deriving the conditions for equilibrium in this case applies to the derivation of the Boltzmann distribution function, which reports how the atoms are distributed over the energy levels when the system attains equilibrium. The algorithm is then illustrated for the ideal gas model and the Einstein model for a crystal. Statistical thermodynamics is used very little in subsequent chapters because the classes of systems that are the domain of materials science tend to be too complex for tractable treatment, much less for presentation to first-time students of the subject.

Most chapters contain several illustrative examples, designed to emphasize the strategies that connect principles to hard numerical answers. Each chapter ends with a summary that reviews the important concepts, strategies, and relationships that it contains. Each chapter also ends with a collection of homework problems, many of which are designed so that they are best solved using a personal computer: the astute student may find it useful to write some more general programs that can be used repeatedly as the level of sophistication increases. Examples of homework problems will be drawn more or less uniformly from the major classes of materials: ceramics, metals, polymers, electronic materials, and composites. This approach serves to illustrate the power of the concepts, laws, and strategies of phenomenological thermodynamics by demonstrating that they can be applied to all states of matter.

The experience gained in 25 years of teaching an undergraduate course in thermodynamics in materials science, together with more than 15 years of teaching a graduate course in the same area, has resulted in an approach to the topic that is unique. The approach accents rigor, generality, and structure in developing the concepts and strategies that make up thermodynamics because the connections between first principles and practical problem solutions are sharply illuminated; the first-time student can hope not only to apply thermodynamics to the sophisticated end of systems that are the bread and butter of materials science, but to understand their application.

It is a pleasure to acknowledge the help of Heather Klugerman, who provided advice in the more sophisticated aspects of word processing involved in putting together this text. Pamela Howell proofread the manuscript with remarkable skill before it was submitted to the publisher. David C. Martin, University of Michigan, and Monte Poole, University of Cincinnati, offered many helpful comments and suggestions while reviewing the manuscript. My thanks to the many students, both graduate and undergraduate, who for many years encouraged me to undertake this text. Finally, I am grateful to my wife, Marjorie, who sacrificed many evenings, weekends, and vacations as I disappeared into the den to work on the project.

About the Author

Now Professor Emeritus of the Department of Materials Science and Engineering at the University of Florida, **Robert T. DeHoff** was one of the founding fathers of that program. For more than four decades he has developed and taught graduate and undergraduate courses that relate to microstructures in materials science and engineering, including courses in the geometry of microstructures and the kinetics of their evolution, diffusion, phase diagrams, quantitative characterization of microstructures (stereology), and undergraduate and graduate courses in thermodynamics. Because of his longevity, it is very likely that Professor DeHoff has taught classes in thermodynamics more often than anyone else on the planet. He has also been involved in the development and evolution of the curriculum in the materials science and engineering program. His research and publications have also centered around the evolution of microstructure in most of the areas cited above as his teaching experience. He has received a number of awards based on his research and teaching from a variety of professional societies, most recently the Educator Award from The Minerals, Metals & Materials Society (2005).

Table of Contents

APPENDICES

1 Why Study Thermodynamics?

CONTENTS

1.1 THE POWER AND BREADTH OF THERMODYNAMICS

A survey of undergraduate curricula in materials science and engineering showed that every program requires a core course in thermodynamics. In more than 90% of those programs, the course is taught within the department. Most graduate programs have one or more courses in the subject. Evidently there is widespread agreement that this subject is a central one in materials science and engineering. The same statement can be made for programs in chemical engineering and chemistry and, perhaps to a lesser extent, physics.

Why?

Five primary reasons:

1. Thermodynamics is pervasive.
2. Thermodynamics is comprehensive.
3. Thermodynamics is established.
4. Thermodynamics provides the basis for organizing information about how matter behaves.
5. Thermodynamics enables the generation of maps of equilibrium states that can be used to answer a prodigious range of questions of practical importance in science and industry.

Thermodynamics is pervasive; it applies to every volume element in every system at every instant in time. How pervasive can you get?

1

Thermodynamics is comprehensive. The apparatus is capable of handling the most complex kinds of:

Systems: metals, ceramics, polymers, composites, solids, liquids, gases, solutions, crystals with defects

Applications: structural materials, electronic materials, corrosion-resistant materials, nuclear materials, biomaterials, nanomaterials

Influences: thermal, mechanical, chemical, interfacial, electrical, magnetic

Thermodynamics is established. J. Willard Gibbs essentially completed the apparatus of phenomenological thermodynamics in 1883 in his classic paper, *On the Equilibrium of Heterogeneous Substances*. The scientific and technological explosion of more than a century has not required a significant modification of Gibbs' apparatus.

Thermodynamics provides the basis for organizing information about how matter behaves. Thermodynamics identifies the properties of systems that are scientifically and technologically important in a wide range of applications. It identifies the subset of these properties that are sufficient to compute all the others. These are the properties that are measured in the thermochemistry laboratories of the world and have been accumulating in the databases of the world since Gibbs' time. The apparatus then provides relationships between these database properties and the functions that are crucial in predicting the behavior of matter.

Thermodynamics enables the generation of maps of equilibrium states for this broad spectrum of systems and influences. A variety of species of such maps are widely used in science and industry to answer real-world questions about the behavior of matter.

Will cadmium melt at 545°C?

If the temperature of the air outside drops eight more degrees, will it get foggy?

If I heat this Nb–Ti–Al alloy in air to 1100°C, will it oxidize?

Can this polymer solvent dissolve 25% PMMA at room temperature without phase separating?

How can I prevent the oxidation of silicon carbide when I hot press it at 1350°C?

How can I control the defect concentration in this fuel cell membrane?

What source temperatures should I use to codeposit a 40 to 60 Ge–Si thin film from the vapor phase?

Will silicon carbide fibers be stable in an aluminum nitride matrix at 1300°C?

Will titanium corrode in seawater?

Finally, a warning: thermodynamics is a very rational subject. The logic is largely linear. C follows from B, which follows from A. Nonetheless the predictions of thermodynamics are full of surprises. Accordingly, intuition applied to thermodynamics can be dangerous and misleading, particularly for the uninitiated. Students for which intuition plays an important role in their learning processes may have difficulty with thermodynamics. It is important to understand the laws and strategies of thermodynamics and let the logic lead where it will.

1.2 THE GENERIC QUESTION ADDRESSED BY THERMODYNAMICS

The questions in the preceding paragraph are all forms of a generic question which thermodynamics addresses (see Figure 1.1): "if I take System A in Surroundings I and put it into Surroundings II, what will happen?" Rudimentary thermodynamics concepts implicit in this question include:

System, which is the collection of matter whose behavior is the focus of the question.
Surroundings, which is the matter in the vicinity of the system that is altered because it interacts with it.
Boundary, implicit in the concept of a system and its surroundings, which may limit the kinds of exchanges that can occur between the two.
Properties, required in the definition of the condition of a system and its surroundings.

As an example of this generic scenario, consider a block of cadmium sitting on a laboratory bench so that its initial condition is ambient pressure and temperature (nominally 1 atm and 25°C). It is picked up with a pair of tongs and placed in a furnace, which has its temperature controlled at 545°C.

In Figure 1.1, System A is the piece of solid cadmium. Surroundings I is the ambient pressure and temperature of the laboratory. Surroundings II is the atmosphere in the furnace also at ambient pressure but a temperature of 545°C. System A experiences a change in surroundings when it is placed in the furnace. As a result, System A begins to change its condition toward a final state B, which is in equilibrium with this new Surroundings II. It is necessary to consult a thermo-dynamics database that has information about cadmium to determine that the melting point of cadmium is 321°C and the vaporization temperature is 767°C. Thus, the final equilibrium state in its new surroundings is liquid cadmium.

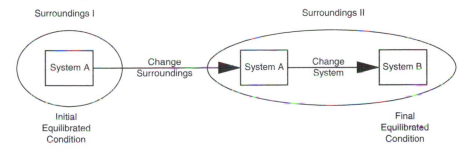

FIGURE 1.1 The generic question addressed in thermodynamics.

What happens? The cadmium melts. The process involved in this change is melting, a phase transformation in which the crystalline structure of solid cadmium is converted to a structure that is liquid. Pockets of the liquid phase nucleate, forming a solid/liquid interface. The motion of this interface toward the solid phase increases the amount of liquid at the expense of the solid phase until there is no solid phase left.

Practically speaking, a number of issues that are ignored in this simple description of the process also need to be addressed. Since the cadmium will melt, it has to be placed in a container, e.g., a crucible, before it is put in the furnace. Will the container react chemically with the cadmium? Also, cadmium vapor will form over the liquid. How high will the vapor pressure become? Cadmium vapor is toxic, so significant precautions will have to be taken to contain the sample and its vapor. If the ambient atmosphere is air, will cadmium oxide (or other compounds) form? Evidently a comprehensive answer to the question, "what will happen?" requires answers to all of these questions. Thermodynamics has the power to address all of these issues.

The scenario shown in Figure 1.1 can be used to frame a variety of rearrangements of the generic question:

> What Surroundings II must be provided to convert System A into a specific version of System B? (For example, what range of temperatures can I use to convert BCC iron to FCC iron?)
>
> What Surroundings II must be used to prevent the conversion of System A into a specific System B? What surroundings must be avoided? (For example, what range of furnace atmosphere compositions must be used to avoid the oxidation of a set of turbine blades during heat treatment?)

The apparatus of thermodynamics provides the answer to these kinds of questions by providing the basis for determining the equilibrium state of any system in any surroundings.

1.3   THERMODYNAMICS IS LIMITED TO SYSTEMS IN EQUILIBRIUM

Thermodynamics is limited to the description of systems that are in equilibrium with their surroundings. It provides the basis for predicting what the properties of an equilibrated system will be as a function of the content of the system and the characteristics of its surroundings. Thermodynamics does not permit the prediction of the step-by-step time-dependent evolution of a system toward equilibrium. That level of description of the behavior of matter, which is contained in a formalism called the thermodynamics of irreversible processes, requires the solution of sets of simultaneous partial differential equations. Implementation of this time-dependent description requires a great deal more information about the system than does the description of its equilibrium state in given surroundings. Irreversible thermodynamics is beyond the scope of this text. However, equilibrium thermodynamics, which is the subject of this text, provides the context within which these time-dependent processes occur.

How then can equilibrium thermodynamics be usefully applied in answering the question, "what will happen?"

System A has some set of properties when it was in Surroundings I. These are the initial properties of the system when it is placed in Surroundings II. Thermodynamics predicts what the state of this system will be when it comes to equilibrium with its new Surroundings, II. This provides a basis for deducing what processes must occur to change the system from its initial condition, inherited from Surroundings I to its equilibrium state in Surroundings II.

Thermodynamics provides not only the equilibrium state in such cases, but also some measure of how far the system is from the equilibrium state. These thermodynamic measures, perhaps misleadingly labeled "driving forces" in kinetic descriptions of processes, play a central role in the more sophisticated attempts to describe the sequence of states through which the system passes as it moves toward equilibrium and its rate of progress through that sequence.

The real utility of thermodynamics lies in its ability to predict whole patterns of behavior for a range of systems in a range of surroundings. These patterns are conveniently presented in the form of maps of equilibrium states. Thermodynamics produces a variety of such maps for different classes of systems operating in appropriate types of surroundings. Generation of these maps is a main topic of this text. Such a map provides an ability to answer the question, "what will happen?" for any combinations of systems and surroundings encompassed by the map.

Figure 1.2 is a sketch of such a map for a familiar substance, water. The system under consideration (System A) is some fixed quantity of the molecular specie H_2O. It is known that this specie can exhibit a number of structures (phases), depending upon its surroundings: solid, liquid and vapor (ice, water and steam or water vapor). (Ice can exist in a number of different crystalline forms but these occur outside the window of temperature and pressure represented here.) The condition of the surroundings that may be considered as imposed upon the system is specified

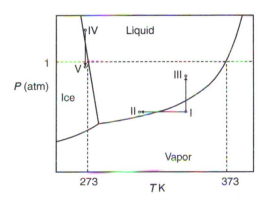

FIGURE 1.2 Sketch of a unary phase diagram for water. The negative slope of the solid–liquid line is real, but exaggerated to illustrate a point in the text.

by two variables: pressure, P and temperature, T. The map, called a phase diagram, is a display of the equilibrium state of this system for any selected state of the surroundings. The domain labeled solid is the range of surrounding conditions, temperatures and pressures, for which the final equilibrium state of the structure is solid water, i.e., ice. The other two areas labeled liquid and vapor are the ranges of combinations of P and T in the surroundings for which the equilibrium state is, respectively, liquid and vapor.

To illustrate how this diagram addresses the "what happens?" question posed above, suppose that a quantity of specie H_2O (System A) is initially at a temperature of 70°C with a vapor pressure of 0.62 atm, point I in Figure 1.2. The map reports that in these "Surroundings I" System A is in the vapor state, i.e., the H_2O exists as water vapor. Suppose the temperature drops to 30°C without changing the pressure in the surroundings. This "Surroundings II" is represented by the point II in Figure 1.2. Point II lies within the domain for which the equilibrium state is liquid water. What happens? Droplets of liquid begin to form and grow (dew perhaps evolving into rain, depending upon a host of other conditions). There is a nucleation process in which collisions of water molecules form tiny clusters that eventually attain a critical size of the liquid phase for growth. These grow by further collisions with molecules in the vapor and drop to the bottom of the container to form the liquid phase in bulk. Sophisticated theories for each of these kinetic processes (nucleation and growth) predict the rate at which they may happen, the dispersion of droplet sizes, and identify the kinetic and thermodynamic variables that control these processes.

A similar sequence of events would occur if System A in Surroundings I were contained in a cylinder with a piston and a force applied to move the piston to increase the vapor pressure to create a new Surroundings III, point III in Figure 1.2. This state also lies in the domain of liquid water on the map. Again, water droplets will nucleate and grow and eventually coalesce to form the bulk liquid. The resulting liquid occupies a small fraction of the precursor vapor phase so that the piston will rapidly drop to nearly the bottom of the cylinder when the vapor condenses.

A container of liquid water initially in Surroundings IV (point IV in Figure 1.2) would contain liquid water at this temperature and pressure. If the pressure is relieved so that it drops to 1 atm at the same temperature (point V), what happens? The equilibrium state in Surroundings V is solid H_2O. You may have observed that, when the top is popped off a bottle of liquid cola that has been in the freezer for a while (and which is under pressure from the dissolved gases that make it effervesce) the liquid cola may suddenly freeze completely. This situation leads to the consideration of the process of nucleation of ice crystals followed by their growth.

Figure 1.3 is a sketch of the map of the domains of stability of the phases for the element molybdenum. Like H_2O, molybdenum exhibits three phase forms, solid, liquid and vapor. The maps in Figure 1.2 and Figure 1.3 are qualitatively similar, but the quantitative differences are spectacular. At 1 atm pressure liquid water is stable between 273 and 373 K. Under 1 atmosphere pressure the range of stability of liquid molybdenum is from 2980 to 4912 K. This huge difference in behavior reflects the nature of the bonds that hold water molecules together in comparison to those that act between molybdenum atoms.

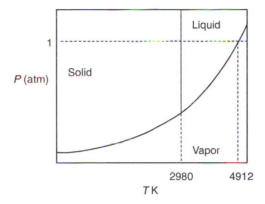

FIGURE 1.3 Sketch of a unary phase diagram for molybdenum. The diagram is qualitatively similar to that for water, but the quantitative differences are enormous.

The thermodynamic equations underlying the calculation of these two maps are identical in form. Thermodynamics provides the basis for defining and identifying the properties of each phase that must be determined in the laboratory in order to calculate these two maps. A database, laboriously developed over time, that collects and summarizes values for these properties for the elements in the periodic table and for many compounds, provides the experimental information specific to each specie that must be used in the computation of its map. Strategies, also derived from thermodynamic principles, are then applied to compute the map from its database.

Thus thermodynamics provides the definitions of the properties that must be measured to form the database for a phase diagram map for systems that contain one chemical component. The discipline also provides the principles and strategies needed to produce quantitative maps of equilibrium states from this database information.

1.4 THE THERMODYNAMIC BASIS FOR EQUILIBRIUM MAPS

Figure 1.4 provides a summary of the component parts of thermodynamics and how they fit together to produce maps that are ultimately used to answer practical questions. The concepts and connections in this figure provide a very useful basis for understanding the rudiments of how thermodynamics works and will be referred to frequently as the arguments develop in the text.

The content of the field is contained in a few principles that are applied through a few strategies.

1.4.1 THE PRINCIPLES

In phenomenological thermodynamics each system is a structureless glop that is endowed with properties. As a first step, properties that make thermodynamics work must be identified and defined, such as temperature, pressure, composition,

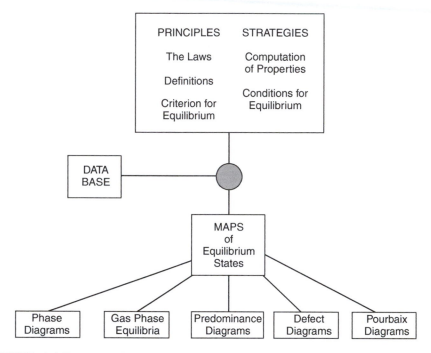

FIGURE 1.4 Representation of the structure of thermodynamics illustrating how the component parts of thermodynamics join together to generate maps of equilibrium states.

heat capacity, coefficient of thermal expansion and compressibility, entropy, and various measures of the energy of the system. Definitions of properties are introduced throughout the text as new system and surroundings variables are introduced.

The central principles of thermodynamics are the three laws that are described in Chapter 3.

The general criterion for equilibrium, deduced from the second law, is developed in Chapter 5. Since a primary goal of the text is the exposition of equilibrium states that matter will exhibit, there must be a basis for determining when a system is in equilibrium.

1.4.2 THE STRATEGIES

A general strategy for calculating all of the thermodynamic properties of a system from a minimum list of database properties is developed in Chapter 4.

The conditions for equilibrium are a set of equations between properties of the system that must be satisfied when the system is in equilibrium. These equations are the basis for calculating maps of equilibrium states from database information. The list of equilibrium equations expands as the nature of the system under study grows more complex, requiring additional variables in the description of its state. This strategy is first applied to a simple system in Chapter 5. The same strategy for deriving conditions for equilibrium is applied repeatedly in the remaining chapters of the text as systems of increasing variability and complexity are treated.

1.4.3 DATABASES

Thermodynamics identifies the minimum information set that must be obtained to compute the properties of a system. The list of properties in this minimum set expands as the system under consideration exhibits more variables. For example, there are no composition variables in a one-component (unary) system; heat capacity, coefficients of thermal expansion and compressibility are sufficient to compute the rest of the properties of such a simple system (Chapter 4). Additional information is required to treat systems that exhibit more than one phase, e.g., solid plus liquid, (Chapter 7). Treatment of multicomponent system requires additional chemical variables with associated required database properties (Chapter 8). Further information is required to treat additional phases that a multicomponent system may exhibit (Chapters 9 and 10). Systems capable of chemical reactions (Chapter 11) require yet another kind of data.

The vast scientific literature continues to expand, and information gleaned from the work of thousands of experimenters over decades continues to accumulate, be analyzed and assessed as it passes into the thermochemical databases of the world. The principles and strategies of thermodynamics render that data into the practical form of equilibrium maps. The map provides answers to "what will happen?" questions.

1.4.4 MAPS OF EQUILIBRIUM STATES

Examples of the maps shown in the bottom row of Figure 1.4 are described briefly below.

1. The phase diagram shown in Figure 1.5 is for the silver–magnesium system at 1 atm pressure. A point in the diagram represents the equilibrium structure of a particular Ag–Mg composition at a particular temperature. The vapor phase is stable above the range of temperature presented in this diagram. This map may be used to answer "what will happen?" questions similar to those illustrated in Figure 1.2 and Figure 1.3.
2. The equilibrium gas composition map shown in Figure 1.6 displays the equilibrium composition (expressed as its partial pressure in the mixture) of one of the components in a mixture of gases, in this case, the molecule O_2, as a function of the overall chemistry of the system reported by the relative quantities of the elements carbon, hydrogen and oxygen that make up the system. The lines on the diagram are calculated from a database reporting properties of the chemical reactions involved in forming the molecular components that can be made by combining these three elements. These "iso-oxygen contours" report the locus of points that will provide a fixed partial pressure of oxygen for the temperature

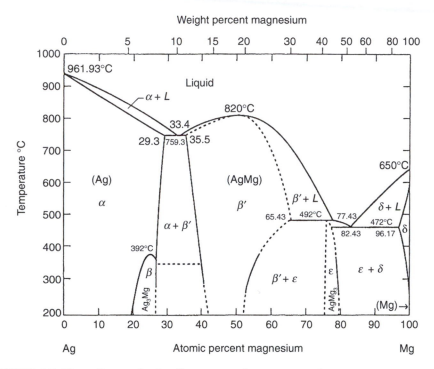

FIGURE 1.5 Phase diagram for the silver–magnesium system at 1 atm pressure.

of the diagram. This kind of map provides the basis for the design of furnace atmospheres with controlled chemistry for heat treatment, coating formation, vapor deposition and stoichiometry control.

3. The predominance diagram shown in Figure 1.7 is computed from a thermodynamic database that provides information about the chemical reactions that form the compounds displayed on the diagram. Regions on this diagram represent domains of predominance of each compound considered in the database relative to all the others in the system. The information is presented at a fixed temperature as a function of the chemistry of the gas atmosphere that forms its surroundings. Predominance diagrams provide a reasonable approximation to the phase diagram in such complex systems and require significantly less data. These maps may be used in conjunction with a gas composition map to determine the range of elemental composition of the atmosphere necessary to produce each compound.

4. An example of an equilibrium crystal defect diagram is shown in Figure 1.8. A database providing thermodynamic property changes associated with the formation of defects (vacancies, interstitials, anti-site defects, etc.) in compound crystals permits calculation of the concentration of each class of defect as a function of the departure of the composition of the compound crystal from its stoichiometric formula.

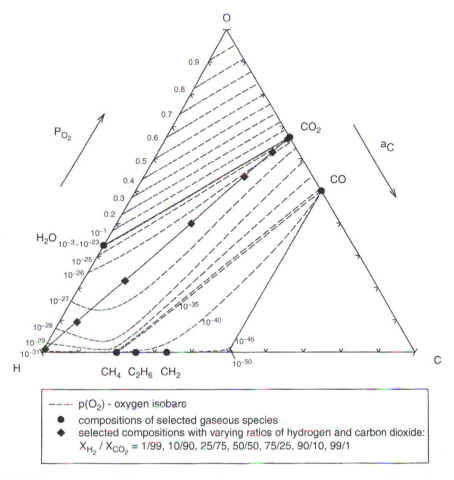

FIGURE 1.6 An equilibrium gas composition map for a gas mixture containing carbon, hydrogen and oxygen displays contours of constant partial pressure of one of the components (in this case the oxygen molecule, P_{O_2}) as a function of the elemental content of the gas mixture.

This departure can be controlled by manipulating the chemistry of the gas atmosphere with which the compound is equilibrated. Manipulation of this defect chemistry has important applications in microelectronic devices, nuclear fuel pellets, sensors, fuel cells, batteries and other devices, which lie at the interface between chemistry and electronic behavior.

5. Figure 1.9 illustrates a Pourbaix electrochemical diagram named after M. Pourbaix who devised the diagram and popularized its use in the analysis of corrosion behavior. The diagram shown illustrates the behavior of copper in aqueous solutions in the presence of an applied electromotive force. This form of predominance diagram is computed with procedures that mimic those needed in evaluating the more

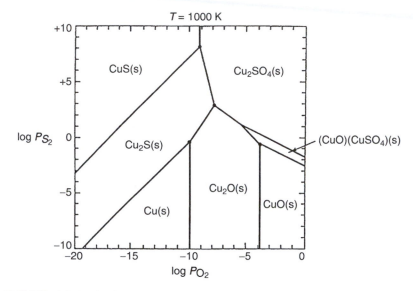

FIGURE 1.7 Predominance diagrams display domains of predominance of chemical compounds (in this example, copper compounds in an atmosphere containing sulfur and oxygen) as a function of chemistry of the gas atmosphere of the system.

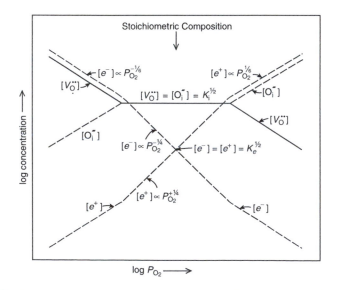

FIGURE 1.8 Sketch of the variation of the concentration of crystal defects of an oxide with departure from the stoichiometric composition of the compound, here represented by the equilibrium partial pressure of oxygen in the atmosphere, which controls such departures.

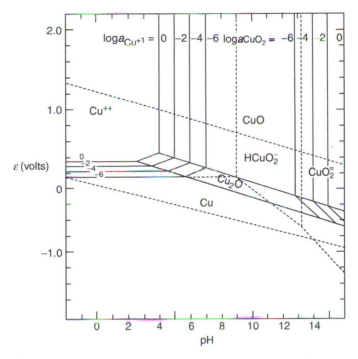

FIGURE 1.9 Pourbaix diagram displays domains of predominance of ionic and nonionic components in aqueous solutions containing copper as a function of the impressed emf and the hydrogen potential of the mixture.

traditional predominance diagrams illustrated in Figure 1.7. In this kind of application the variables subject to control to determine the surroundings are the pH of the aqueous environment and the magnitude of an electromotive force applied to the system. In the domain where Cu is predominant the metal is said to be immune to corrosion; copper corrodes in domains where ionic forms predominate in the aqueous environment. In other regions a compound predominates; it may form on the surface of the copper and protect the metal from corrosion resulting in passivation of the material.

This sampling of the kinds of maps that can be generated from thermodynamic data is by no means exhaustive. New forms can be created by choosing other variables to form the axes of the map. Additional information (e.g., iso-activity lines on a phase diagram) can be superimposed on these maps. These maps all share the attribute that they represent competition for stability or predominance among the various forms that the atoms of the elements in the system may exhibit. Each has an appropriate set of database information required to generate the map, and each has an underlying principle that determines the competition and an underlying strategy for connecting database to map. Each has its realm of applications that address appropriate versions of the question, "what happens?"

1.5   THREE LEVELS OF THE THERMODYNAMIC APPARATUS

What determines how matter behaves? There are three levels of sophistication that are used in answering this question.

1. The phenomenological description, in which matter is treated like a structureless glop that possesses a set of properties like temperature, pressure, chemical composition, heat capacity, etc. The behavior of an evolving system is described solely in terms of changes in its properties. This level of description contains no information about the underlying structure of matter that determines these properties. Relationships between properties are described, but there is no attempt to explain the source of values of these properties.
2. The statistical description, which recognizes that matter is composed of atoms that exist in different structural arrangements like gas, liquid, crystal, molecule. The individual atoms have properties (size, mass, electronegativity, energy, etc.); pairs of atoms (bonds) have properties associated with the interactions between the atoms (bond strength, bond energy). The behavior of a system is derived by statistical strategies that connect the properties of these units to the properties of the huge collection of atoms that make up a system. This level of description provides an explanation of the phenomenological properties of the system in terms of the behavior of the atoms that compose it.
3. The quantum description, which recognizes that single atoms and groups of atoms have an internal structure mostly residing in the distribution of electrons in space and in associated energy in the system. This level of description explains the behavior of atoms and ensembles of atoms, and hence that of systems, at the most fundamental level.

Phenomenological thermodynamics has been the most widely used level of description in the practical application of thermodynamics. The question, "what happens?" can be answered at the phenomenological level for very complicated, practical systems. A description of this level of detail is sufficient for many (if not most) practical applications. The more profound question, "why does this happen this way?" which seeks a level of explanation of observed phenomenon requires the statistical approach, and, even more fundamentally, the quantum mechanical approach. These fundamental descriptions rely heavily on computational materials science in which the properties of large collections of atoms are computed from sophisticated models for the interactions between atoms in the structure. Explanations provided by these first principle calculations have the potential for provoking new insights that ultimately may permit the prediction of database properties that have not yet been measured, or the extrapolation of results for simple systems to more complex ones. Because thermodynamics identifies the properties needed to predict the behavior of matter, thermodynamics is the basis for establishing the focus of these more fundamental scientific endeavors.

1.6   SUMMARY

- Thermodynamics provides the basis for answering the question, "if I take System A in Surroundings I and put it into Surroundings II, what will happen?"
- Thermodynamics is important because it is pervasive, comprehensive, established, the basis for organizing information about the behavior of physical systems and for developing maps of equilibrium states.
- Maps of equilibrium states are used to supply answers to the "what happens?" question for a wide variety of systems in a wide variety of surroundings.
- The behavior of matter has been described at three different levels of sophistication: (1) phenomenological thermodynamics, (2) statistical thermodynamics, (3) quantum statistical thermodynamics.

2 The Structure of Thermodynamics

CONTENTS

Figure 1.4 provides a visualization of the structure of thermodynamics. At its apex are a very few very general, and therefore very powerful, principles: the laws of thermodynamics. From these few principles can be deduced predictions about the behavior of matter in a very broad range of human experience, frequently expressed in the form of the equilibrium maps that descend from the apex. An understanding of how matter behaves in every situation rests directly upon these laws.

In their simplest and most general form the laws apply to the universe as a whole:

1. There exists a property of the universe, called its energy, which cannot change no matter what processes occur in the universe.
2. There exists a property of the universe, called its entropy, which can only change in one direction no matter what processes occur in the universe.
3. A universal absolute temperature scale exists and has a minimum value, defined to be absolute zero, and the entropy of all substances is the same at that temperature.

More precise, mathematically formulated statements of the laws are developed in Chapter 3.

In practice, the focus of thermodynamics is on a subset of the universe, called a system, (Figure 2.1). In order to apply thermodynamics, the first step is to identify the subset of the universe that encompasses the problem at hand. It is necessary to be explicit about the nature of the contents of the system, and the specific location and character of its boundary.

The condition of the system at the time of observation is described in terms of its properties, quantities that report aspects of the state of the system such as its temperature, T, its pressure, P, its volume, V, its chemical composition, and so on. As the system is caused to pass through a process its properties experience changes (Figure 2.1). A very common application of thermodynamics is a calculation of the changes that occur in the properties of some specified system as it is taken through some specified process. Thus, an important aspect of the development of thermodynamics is the deduction of relationships between the properties of a system, so that changes in some properties of interest, e.g., the entropy of the system, may be computed from information given or determined about changes in other properties of the system, e.g., temperature and pressure.

An understanding of the structure of thermodynamics is aided greatly by deliberately organizing the presentation on the basis of a series of classifications, which compartmentalize these characteristics of the field, and thus permit a focus upon the subset of the thermodynamic apparatus that is appropriate to a specific problem. Accordingly, presented in this chapter are classifications of:

1. Thermodynamic systems.
2. Thermodynamic properties.
3. Thermodynamic relationships.

Mastery of the strategy for developing relationships among the properties of thermodynamic systems paves the way for deducing the sets of working equations that are of most practical importance in thermodynamics: the conditions for equilibrium. These equations are the basis for developing the maps of equilibrium states discussed in Chapter 1. A general criterion that may be used to determine when a system has attained thermodynamic equilibrium is introduced in Section 2.4. This concept is developed precisely and mathematically in Chapter 5, along with a general strategy for finding the conditions for equilibrium in the most complex kind of thermodynamic system.

2.1   A CLASSIFICATION OF THERMODYNAMIC SYSTEMS

The complete thermodynamic apparatus is capable of evaluating the equilibrium conditions of the most complex kind of system as it experiences the full range of potential influences that have been identified that may affect its condition. Most practical problems in thermodynamics do not require the invocation of the whole thermodynamic structure for their solution. In order to pinpoint the part of the apparatus that must be used to handle a given case, it is useful to devise a classification of thermodynamics systems. Use of such a classification at the beginning of consideration of any problem serves to focus attention on the specific set of influences that may operate, and, perhaps more important, those that may be excluded from consideration. This classification also serves as a basis for laying out the sequence of presentation in this text.

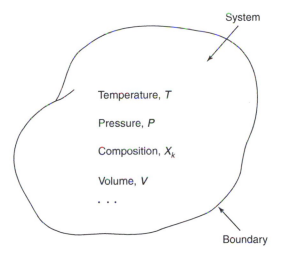

FIGURE 2.1 The subset of the universe in focus in a particular application of thermodynamics is usually called the system. At any given instant of observation the condition of the system is described by an appropriate set of properties. Limitations on changes in these properties are set by the nature of its boundary.

At the outset of consideration of any problem, classify the system under study according to each of the following five categories:

1. Unary vs. multicomponent.
2. Homogeneous vs. heterogeneous.
3. Closed vs. open.
4. Nonreacting vs. reacting.
5. Otherwise simple vs. complex.

Each of these descriptive words has explicit meaning in thermodynamics.

Category 1 identifies the complexity of the chemistry of the system. Systems with the simplest chemical composition are unary, which means: one chemical

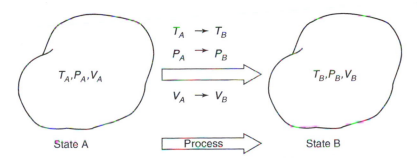

FIGURE 2.2 A process is a change in the condition or state of the system. Properties change from their values in some initial state A to some final state B.

component. If a system has more than one chemical component, i.e., is multicomponent, additional apparatus must be devised to describe its behavior; its composition may vary.

The word homogeneous in category 2 has a specific thermodynamic meaning: single phase. If a system is composed of more than one phase (e.g., a mixture of water and ice), it is heterogeneous. Treatment of heterogeneous systems adds to the thermodynamic apparatus.

In category 3, closed has a specific thermodynamic meaning: closed describes a system that makes no exchanges of matter with its surroundings for the processes being considered. If matter is transferred across the boundary, the system is an open system, and terms must be added to allow for changes in condition associated with the addition of matter to the system.

Category 4 brings into consideration systems that can exhibit chemical reactions and focuses upon the additional apparatus required to describe chemical reactions.

The last category lumps all other influences into a single listing. If a system is capable of exhibiting kinds of energy exchange other than those arising from thermal, mechanical or chemical changes, e.g., if in the problem at hand there may be involved gravitational, electrical, magnetic or surface influences, then it is classified as complex in this category. If none of these special kinds of influences operates in the problem at hand, it is an otherwise simple system.

Figure 2.3 is a cross-section through a thin film device. From the point of view of thermodynamics the system consists of a large number of chemical components, some as impurities added to control the electronic properties, distributed through several phases. Chemical reactions may occur at the gas/solid interface and between the solid phases. Thus, during processing this system may be classified as a multicomponent, multiphase, closed, reacting, otherwise simple system.

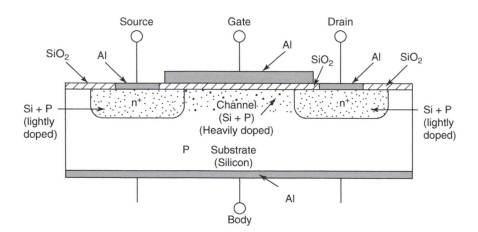

FIGURE 2.3 Cross-section through a MOSFET (metal oxide semiconductor field effect transistor) thin film device shows it to be a multicomponent, multiphase system in which chemical reactions and the influence of an electric field are important.

The most rudimentary kind of system that may be encountered is classified as an unary, homogeneous, closed, nonreacting, otherwise simple system. This classification of simplest of systems is the focus of Chapter 4. Chapter 7 introduces unary heterogeneous systems. Chapter 8 presents the apparatus for handling multicomponent, homogeneous systems. Multicomponent, heterogeneous systems are dealt with in Chapters 9 and 10. The apparatus for handling reacting systems is contained in Chapter 11. Complex systems are dealt with in Chapters 12 to 15. The text progresses through a sequence of classes of systems of increasing complexity until, at the end, the tools for describing the behavior of matter in the most complicated kind of system: multicomponent, heterogeneous, open, reacting, complex are in hand.

2.2 CLASSIFICATION OF THERMODYNAMIC VARIABLES

The internal condition of a thermodynamic system, the changes in its condition, and the exchanges in matter and energy, which it may experience, are quantified by assigning values to variables that have been defined for that purpose. These variables are the mathematical stuff of thermodynamics; their evaluation is the justification for inventing the apparatus in the first place. They fall into two major classes: state functions and process variables.

2.2.1 STATE FUNCTIONS

A state function is a property of a system that has a value that depends upon the current condition of the system and not upon how the system arrived at that condition. The temperature of the air in the room has a certain value at the moment which does not depend upon whether the room heated up to that temperature or cooled down to it. Other familiar properties that have this attribute are: pressure, volume and chemical composition. Figure 2.4 shows the mathematical nature of a state function.

One of the great accomplishments of thermodynamics is the identification of these and other properties of systems, perhaps not so familiar, which are also functions only of the current condition of the system. These include various measures of the energy of the system, its entropy, a variety of properties associated with components in solutions, and properties associated with complex aspects of the system. Complete definitions of these properties are first developed in Chapters 3 and 4; the list is expanded as the apparatus in later chapters requires.

The fact that such state functions exist gives rise to one of the most important strategies for the thermodynamic analysis of the complicated processes that are likely to be encountered in the real world of science and technology. A process converts the condition of a system from some initial state, A, to some final state, B. Precisely because this class of properties, state functions, depends only upon the state of the system, the change in any state function for any process is always simply its value for the final state minus its value for the initial state. Thus, the change in any state function must be the same for every process that converts the system from the

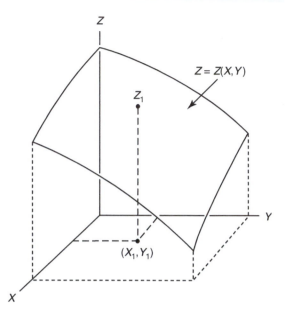

FIGURE 2.4 If the value of the variable Z depends only upon the current values of the variables X and Y, then all three variables are state functions. The functional relationship among these variables, written $Z = Z(X, Y)$, is represented by a surface in X, Y, Z space. For any given values (X_1, Y_1) there is a corresponding value $Z = Z_1$.

same initial state A to the same final state B. The value for the change in any state function is independent of the path (sequence of intermediate states) or process by which the system is converted from state A to state B (Figure 2.5). As a consequence, the change in any state function accompanying a very complicated real-world process, which alters the system from state A to state B, may be computed by concocting or imagining the simplest process that connects the same two end states. A computation of the changes in state functions for this simple process will yield the same result as would be obtained for the very complex process.

The importance of this strategy, its application, and other consequences of the existence of state functions is developed in Chapter 4.

2.2.2 PROCESS VARIABLES

In contrast to the notion of state functions, process variables are quantities that only have meaning for changing systems. Their values for a process depend explicitly upon the path, i.e., the specific sequence of states traversed, that takes the system from state A to state B. Change is inherent to the very concept of these quantities. There are two primary subcategories of process variables: work done on the system as it changes, and heat absorbed by the system.

The concept of work is developed in classical mechanics in physics. A force acts upon a body. If the point of application of the force moves, then the force does work. Let the vector $\mathbf{F}$ denote the force, and the vector $d\mathbf{x}$ denote an increment of its

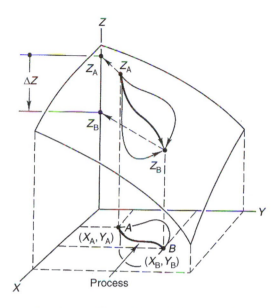

FIGURE 2.5 A process that changes the condition of the system from state A to state B may (if it is simple enough) be represented by a curve in the $(X-Y)$ plane: this represents the sequence of states through which the system passes in changing from state A to state B. Evidently, since Z is a state function, the change in Z, written $\Delta Z = Z_B - Z_A$, will be the same for all paths connecting A and B.

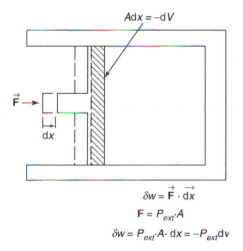

FIGURE 2.6 The generic concept of work is illustrated for mechanical work due to the external pressure acting on the system; the displacement of the force **F** through the distance dx does work.

displacement (Figure 2.6). The increment of work done by this displacement is defined to be:

$$\delta w = \mathbf{F} \cdot \mathrm{d}x \qquad (2.1)$$

where the notation represents the dot product of the two vectors. For a finite process the force is moved along some path through space; the value of the force and the direction of force and displacement may change as it moves. The work done is defined to be:

$$w = \int \mathbf{F}(x) \cdot \mathrm{d}x \qquad (2.2)$$

where $\mathbf{F}(x)$ describes how the force varies with position, x, and the integration is a line integral along the path traversed. The mathematical details are not important in the present context. It is important to note that the displacement of the force is an inherent component in the concept of work. Work cannot be associated with a system at rest; it is a process variable.

It is possible for work to be done through a variety of influences that may act upon the system. Each of the forces that have been identified in physics:

1. The force exerted by the pressure on a system.
2. Force due to gravity.
3. Body forces in a rotating system.
4. The force acting on a charged particle in an electrical field.
5. Force on a magnetic dipole in a magnetic field.
6. The force associated with surface tension.

may be displaced to do work. The early chapters in the text limit consideration primarily to work done by the mechanical force exerted by the external pressure on the system. The remaining forces listed are in the complex system category in the classification of systems; each has its own set of thermodynamic apparatus, as developed in Chapter 12 to Chapter 15.

If the boundary of a system is rigid and impermeable so that no matter can cross it and no force acting upon it may move, the internal condition of that system can still be caused to change. There thus exists a kind of influence, which can alter the condition of a system, that is not a form of work or matter transfer. During the past three centuries concepts and methods have gradually developed that quantified this thermal influence, beginning with the development of the thermometer. A temperature scale was devised and evolved into a tool of general application. The calorimeter provided a means for determining relative quantities of this thermal energy transferred in different processes. The quantity of heat that flows into or out of a system during a process can now be determined with accuracy, at least under carefully controlled experimental conditions. Heat always carries with it a change in the condition of a system; it is thus also a process variable. Just as is true for work, it is meaningless to visualize a quantity of heat associated with a system that is not changing; the notion of the heat content of a system is meaningless. Flow and change are inherent aspects of the nature of heat.

2.2.3 EXTENSIVE AND INTENSIVE PROPERTIES

State functions may be further classified as extensive or intensive properties of the system.

If the value of the property is reported for the system as a whole, then it is called an extensive property of the system. For example, the volume V of a system is an extensive property. The number of moles of a given chemical component n_k in the system is extensive, as are the internal energy U and entropy S, to be defined in Chapter 3. In general, extensive properties depend upon the size or extent of the system. The most direct measure of size of a system is the quantity of matter that it contains. In a comparison of two systems, which have identical intensive properties, doubling the quantity of matter doubles all of the extensive properties.

A property of the system is intensive if it may be defined to have a value at a point in the system. For example, temperature T is an intensive thermodynamic property; the temperature has a value at each point in the system and may indeed vary from point to point. Pressure P may also be defined at each point in the system; in the Earth's atmosphere, pressure varies with height as well as horizontally. Maps of this variation are routinely presented in weather reports.

It is possible to derive an intensive property for each of the extensive properties defined in thermodynamics. Such a definition visualizes a limit of the ratio of two extensive properties in a small region of the system. For example, the molar concentration of a component k, c_k (moles of component k/liter), may be defined at a point by visualizing a small volume element neighboring the point (ΔV) and the number of moles of component k (Δn_k) in that element. The concentration is the limit of the ratio

$$c_k \equiv \lim_{\Delta V \to 0} \frac{\Delta n_k}{\Delta V} \tag{2.3}$$

as ΔV approaches zero. This strategy may be used to define densities of internal energy, entropy, and any of the other extensive properties defined in Chapter 4. A rigorous development of these concepts is given in Chapter 14.

Similar definitions may be developed by reporting extensive properties per mole of matter in the system. Thus, the entropy per mole or volume per mole of the system may be defined for volume elements in the system and may vary from point to point. The most familiar and widely used example of molar properties is the mole fraction of component k, X_k (see Chapter 8), used to describe chemical composition. The mole fraction is the limit of the ratio of the number of atoms (or molecules) of component k to the total number of atoms or molecules in the volume element as the total number of moles goes to zero. This measure of composition may vary from point to point in the system.

It may be confusing to find some intensive properties treated as if they were properties of the system as a whole. For example, a value for the temperature, pressure, or the atom fraction of CO_2 in a gas mixture may be reported for the system. However, this is only possible in the special case, which is frequently encountered in introductory thermodynamics, in which these intensive properties do not vary with position for the system under consideration. It is possible to quote a value of the temperature for a system if, and only if, the temperature is uniform in that system. Intensive properties, by concept, may be defined at each point in a system, and have the potential to vary from point to point. If in simple systems they do not happen to vary then they have a single value that describes that property for the system. However, this does not convert them into extensive properties.

By their nature, intensive properties can only depend upon the values of other intensive properties. It is clear that the value of a property, which may be defined at a point in the system, cannot depend upon the value of another property that is a characteristic of the whole system. Extensive properties may be expressed as integrals of intensive properties over the extent of the system (see Chapter 14).

2.3   CLASSIFICATION OF RELATIONSHIPS

The apparatus of thermodynamics provides connections between the properties in a system. These connections are frequently unexpected: they may not be intuitively evident. For example, how the entropy of a system varies with pressure is determined by the coefficient of thermal expansion of the system, as will be demonstrated in Chapter 4. The apparatus also provides equations for computing changes in these properties when a specified system is taken through a specified process. Also, thermodynamics introduces a variety of variables that are defined in terms of previously defined quantities. As a result, the formalism of thermodynamics generates a large number of relationships between the quantities with which it deals. The potential confusion that might result may, to some extent, be relieved by organizing the presentation of the relationships in thermodynamics.

This section introduces a classification of thermodynamic relationships so that this formidable array of equations may be subdivided into categories that reflect their origin:

1. The laws of thermodynamics are the fundamental equations that form the physical basis for all of these relations.
2. Definitions present new measures of the energy of systems expressed in terms of previously formulated variables. These defined quantities are generally introduced because they simplify the description of some particular class of system or process that may be commonly encountered. Another set of definitions introduces the quantities that are commonly measured in the laboratory and reported in thermodynamic databases.
3. Coefficient relations emerge from the description of changes in state functions during an infinitesimal step in a process. The differential relation

$$dZ = MdX + NdY + \cdots \qquad (2.4)$$

describes in a formal way how the state function Z (the dependent variable in this relation) changes as a result of changes in the functions $(X, Y, \ldots)$

(the independent variables) that describe the state of the system. The coefficients in this equation are related to specific partial derivatives of Z:

$$M = \left(\frac{\partial Z}{\partial X}\right)_Y \quad \text{and} \quad N = \left(\frac{\partial Z}{\partial Y}\right)_X \cdots \quad (2.5)$$

This differential equation is illustrated for two independent variables in Figure 2.7. These coefficient relations frequently establish connections between variables that are not intuitively obvious.

4. Maxwell relations also derive from the mathematical properties of state functions. In order for Equation 2.4 to be mathematically valid, it is necessary and sufficient that:

$$\left(\frac{\partial M}{\partial Y}\right)_{X,\dots} = \left(\frac{\partial N}{\partial X}\right)_{Y,\dots} \quad (2.6)$$

Since every state function may be expressed as functions of all of the possible combinations of the others (see Chapter 4), equations of the form Equation 2.5 abound in thermodynamics. Maxwell relations hold for all of these equations.

5. Conditions for equilibrium are sets of equations that describe the relationships between state functions, which must exist within a system, when it attains its equilibrium state. These are the working equations for calculating the maps of equilibrium states discussed in Chapter 1.

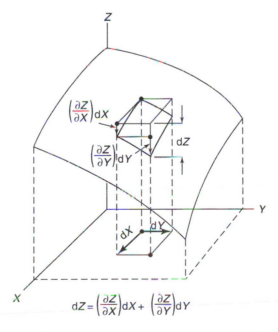

$$dZ = \left(\frac{\partial Z}{\partial X}\right)dX + \left(\frac{\partial Z}{\partial Y}\right)dY$$

FIGURE 2.7 Geometrically, the change dZ associated with the changes dX and dY is given by the slope of the surface in the X direction times dX plus the slope of the surface in the Y direction times dY.

The laws are developed with precision and generality in Chapter 3. In Chapter 4, definitions and the mathematical basis for coefficient and Maxwell relations are presented in detail, together with examples for their application. Derivations of the conditions for equilibrium and systematic application of these equations to generate equilibrium maps for systems of increasing complexity occupy most of the remainder of the text.

2.4 CRITERION FOR EQUILIBRIUM

A system placed in a new set of surroundings will change spontaneously until it has exhausted its capacity for change. When a system attains this final state of rest it is described as being in equilibrium with itself and its surroundings. The prediction and description of this equilibrium state for any given system is a problem of central importance in thermodynamics. This description is expressed in terms of a set of equations, called the conditions for equilibrium, which are relationships among the internal properties that must obtain in order for the system to be at rest. As a simple and familiar example of such equations, consider a unary, two phase, closed, nonreacting otherwise simple system consisting of ice floating in water. The condition for thermal equilibrium in this system is evidently

$$T^{\text{ice}} = T^{\text{water}} \tag{2.7}$$

Additional conditions must apply in order for the system to be in complete equilibrium.

The conditions for equilibrium are extremely important in thermodynamics. For the most complex kind of system these equations may be deduced from a single principle: the general criterion for equilibrium. This powerful principle, which follows directly from the second law of thermodynamics, is the subject of Chapter 5. The remainder of the text deals chiefly with the repeated application of this general criterion to the deduction of equations describing the equilibrium state for each category in the hierarchy of the classification of thermodynamic systems introduced at the beginning of this chapter, and the application of the resulting equations to computing equilibrium maps and solving practical problems in thermodynamics.

2.5 SUMMARY

Thermodynamics provides the basis for describing the behavior of arbitrarily complex physical systems. This description flows from the laws of thermodynamics:

- Energy is conserved.
- Entropy is created.
- Temperature has a zero.

The part of the apparatus of thermodynamics required to handle a given problem may be brought into focus by classifying the system under consideration according to:

- Its number of components.
- Its number of phases.
- The permeability of its boundary.
- Its ability to exhibit chemical reactions.
- Its interactions with nonmechanical forces.

The strategy of presentation in this text is based upon a progressive development of systems of increasing complexity in this hierarchical classification.

Changes in the condition experienced by such a system are described in terms of state variables, with values that depend only upon the current condition of the system; and process variables, which only have meaning for processes, i.e., changes in state of the system.

Changes in state functions are independent of the process by which the system passes from its initial to its final condition. Thus, these changes may be computed by visualizing the simplest process connecting the end states, and making the calculation for that process.

Properties of thermodynamic systems are either intensive, if in concept they are defined at each point in the system or extensive, if they report an attribute of the system as a whole.

Changes in properties may be evaluated for any system taken through any process by applying the relationships of thermodynamics if an appropriate database exists for that system. Knowledgeable manipulation of this potentially bewildering array of relationships is aided by classifying them as

- The laws of thermodynamics.
- Definitions.
- Coefficient relations.
- Maxwell relations.
- Conditions for equilibrium.

A general criterion for equilibrium forms the basis for deducing these conditions for equilibrium. The conditions for equilibrium lead directly to the working equations that yield maps of equilibrium states for that system. The maps in turn permit the prediction of the behavior of matter in the most complex kinds of systems.

HOMEWORK PROBLEMS

Problem 2.1.   Classify the following thermodynamic systems in the five categories defined in Section 2.1:
   a. A solid bar of copper.
   b. A glass of ice water.

c. A yttrium stabilized zirconia furnace tube.

d. A Styrofoam coffee cup.

e. A eutectic alloy turbine blade rotating at 20,000 r/min.

You may find it necessary to qualify your answer by defining the system more precisely; state your assumptions.

Problem 2.2. It is not an overstatement to say that without state functions thermodynamics would be useless. Discuss this assertion.

Problem 2.3. Determine which of the following properties of a thermodynamic system are extensive properties and which are intensive:

a. The mass density.

b. The molar density.

c. The number of gram atoms of aluminum in a chunk of alumina.

d. The potential energy of a system in a gravitational field.

e. The molar concentration of NaCl in a salt solution.

f. The heat absorbed by the gas in a cylinder when it is compressed.

Problem 2.4. Why is heat a process variable?

Problem 2.5. Write the total differential of the function:

$$z = 12u^3 v \cos(x)$$

a. Identify the coefficients of the three differentials in this expression as appropriate partial derivatives.

b. Show that three Maxwell relations hold among these coefficients.

Problem 2.6. Describe what the notion of equilibrium means to you. List as many attributes as you can think of that would be exhibited by a system that has come to equilibrium. Why do you think these characteristics of a system in equilibrium are important in thermodynamics?

3 The Laws of Thermodynamics

CONTENTS

The laws of thermodynamics, at the apex of Figure 1.4, are highly condensed expressions of a broad body of experimental evidence. The formulation of these succinct empirical statements about the behavior of matter was essentially completed by the end of the 19th century and has not required significant alteration in the light of scientific experience since that time. This is a remarkable fact in view of the enormous scientific progress that has been achieved in the 20th century. The laws of thermodynamics are thus soundly based and broad in their application.

The laws are empirical: derived from experimental observations of how matter behaves. No claim is made that they may be deduced from any fundamental philosophical principles.

The laws of thermodynamics have a status in science that is similar to Newton's laws of motion in mechanics and are similarly subject to potential revision in the light of new information. When it was found in physics that new evidence could only be explained by modifying Newton's laws with Einstein's relativistic concepts, the laws of motion were generalized. However, this generalization was in such a form that the new equations simplify to Newton's laws when the velocity of the system is not a significant fraction of the velocity of light. Newton's laws were not abandoned; they were expanded to include newly discovered phenomena. Classical mechanics could be viewed as a special case of the new relativistic mechanics. This strategy was necessary because classical mechanics successfully described a great body of scientific observations with plausibility and precision.

It is possible that new experimental evidence could require a reformulation of the laws of thermodynamics. Up to now this has not been necessary, although the discovery of nuclear energy has had to be accommodated by expanding the framework established in the 19th century.

The laws of thermodynamics are:

1. There exists a property of the universe, called its energy, which cannot change no matter what processes occur.
2. There exists a property of the universe, called its entropy, which always changes in the same direction no matter what processes occur.
3. There exists a lower limit to the temperature that can be attained by matter, called the absolute zero of temperature, and the entropy of all substances is the same at that temperature.

A "zeroth law of thermodynamics" is frequently cited, which acknowledges that a temperature scale exists for all substances in nature and provides an absolute measure of their tendencies to exchange heat.

This chapter presents each of the laws in turn, formulating the concepts involved with precision and ultimately stating the laws mathematically for the most complex class of thermodynamic system.

3.1 THE FIRST LAW OF THERMODYNAMICS

Energy has achieved the status of a household word, in everyday use. It is thought to be a common sense concept with attendant intuitive meaning. A careful examination of the concept of energy in all of its diverse aspects reveals that such an intuition is superficial and potentially misleading. For example, try to visualize the intuitive meaning of kinetic energy, $mv^2/2$, a quantity which implicitly contains the "square of the rate of displacement" of the matter in a system. Fortunately, from a practical point of view, an intuitive understanding of what energy is, is not crucial. Energy can be defined both physically and mathematically and can be measured with precision. The description of the state of a system and the prediction of its behavior rests upon the mathematical and physical formulation of the concept, not on an intuitive understanding of it.

Three broad categories of energy have been identified in scientific experience:

1. Kinetic energy, which is associated with the motion, translation or rotation, of a particle or body and nothing else.
2. Potential energy, which is associated with the position of a particle or body in a potential field and nothing else.
3. Internal energy, which is associated with the internal condition of the body and does not otherwise depend upon its motion or position in space.

Thermodynamics at first focuses upon the influences that change the internal condition of a system at rest. The apparatus ultimately is extended to include potential and kinetic energy as well as internal energy (Chapter 14).

In its most pretentious form the first law may be stated for the behavior of the universe. In practical applications the focus of attention is on some small subset of the universe, called "the system." Except for the case of a system which is isolated from its surroundings, changes that occur inside the system are always accompanied by changes in the condition of the matter in the vicinity of the system. The part of the universe that is external to the system but is also affected by changes that are caused to occur in the system is called "the surroundings" (Figure 3.1). Thus, from the standpoint of any particular process that may occur in practice, the sum of the changes that occur in the system and the surroundings includes all of the changes in the universe associated with that process (Figure 3.1).

By the first law the total energy of the universe cannot change for any process. Energy can be transported, or converted from one form to another but cannot be created or destroyed. Since conversion of energy from one form to another does not change the total quantity of energy, the only way that the internal energy of a system can change is by transferring energy across its boundary. A mathematical statement of the first law for a system can thus be formulated from the statement that the change in internal energy of a system for a process must be equal to the sum of all energy transfers across the boundary of the system during the process. It is only necessary to enumerate all of the possible kinds of energy transfers that may occur for the class of system being considered and set the sum of these energy transfers equal to the change in internal energy for the system for the process.

Let U be a thermodynamic state function called the internal energy of the system. For any process, define ΔU to be the increase in the internal energy of the system. Enumerate all of the kinds of energy transfers that may occur:

Q is the quantity of heat that flows into the system during the process.

W is defined to be the mechanical work done on the system by the force exerted by the external pressure in the surroundings.

W' is defined to be all other kinds of work done on the system during the process.

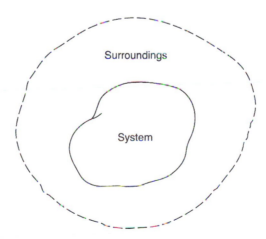

FIGURE 3.1 The part of the universe that is *outside* the system and experiences changes as a result of the changes that occur in the system is called the *surroundings*.

Note that in introducing these definitions it was necessary to establish conventions for the sign of each quantity. By this choice of convention, if Q is positive, heat (and thus energy) flows into the system, increasing its internal energy; if for a given process heat flows out of the system, Q is negative. Similarly, if W (or W') is positive, the surroundings does work on the system and energy flows into the system. If for the process under consideration work is done by the system on the surroundings, W (or W') is negative and energy is transferred out of the system. The statement that the increase in the internal energy of the system during a process is the sum of the transfers across the boundary may be written:

$$\Delta U = Q + W + W' \qquad (3.1)$$

This is a mathematical statement of the first law of thermodynamics. In order to apply it in practice, it will be necessary to devise an apparatus for evaluating the process variables Q, W and W' for each kind of process that may be encountered by each class of system.

In order to follow the course of a process in detail it is useful to consider it as a succession of incremental steps, which produce infinitesimal changes in the internal condition of the system and result from infinitesimal transfers of heat and work across the boundary. The first law is pervasive; it applies not only to the overall finite changes that a system may experience but also to each incremental step along the way. Energy must be conserved during each small incremental step in a process (Figure 3.2). Thus, for an infinitesimal change in the condition of the system,

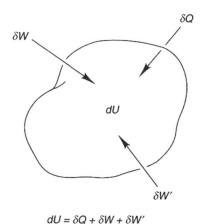

$$dU = \delta Q + \delta W + \delta W'$$

FIGURE 3.2 Illustration of the first law of thermodynamics for an infinitesimally small step in a process: the change in the internal energy of a system must equal the algebraic sum of all of the energies transferred across its boundary.

Equation 3.1 may be written:

$$dU = \delta Q + \delta W + \delta W' \tag{3.2}$$

where dU is the change in the state function U for an infinitesimal step in the process, and δQ, δW and $\delta W'$ are incremental quantities of heat and work that are transferred across the boundary of the system. It is widely accepted in thermodynamics that an infinitesimal quantity designated by a d, as in dU, represents the change in a state function and thus has the mathematical properties of a differential of the state function. Prefixing a quantity with δ, as in δQ above, denotes in infinitesimal quantity of heat or work; however, these infinitesimal quantities do not have the mathematical properties of a differential of a function implied by the operator, d. There is no mathematical function of the state of the system, Q, of which δQ is the differential, because Q is not a state function; it has meaning only for a process.

Thus U is a state function, while Q, W and W' are process variables. Q, W and W' have values that depend explicitly upon the path, i.e., the sequence of thermodynamic states, traversed by the system during the process; ΔU depends only on the initial and final states and is independent of the sequence of states that connect them.

There are no restrictions involved in making the mathematical statement of the first law of thermodynamics contained in Equation 3.1 and Equation 3.2. They apply to any system taken through any process. This very general form of statement is useful for emphasizing the concept that is the first law; however, it is not very useful for applications to practical systems and processes because it does not address the key question: "How can Q, W and W' be evaluated?" This question will be addressed in detail at the beginning of Chapter 4.

3.2   THE SECOND LAW OF THERMODYNAMICS

It is a common observation in human experience that processes observed in nature have a natural direction of change. If a motion picture or videotape is run backwards, it will be immediately evident to any viewer that something is wrong with the motions perceived. A pebble dropped into a pond produces a splash and a set of ripples that dissipate at the pond's edges. Viewed in reverse, ripples of waves spontaneously generate at the edges of the pond, converge to the middle, and amid an eruption of water, toss a pebble into the air. Energy is conserved in both processes, but the second one will never occur; there is something about it which is obviously contrary to experience.

When processes in nature are examined, either macroscopically or in detail, it is found that this principle of a proper direction for change is pervasive. A wilted flower does not revive to full bloom and then curl into a bud. Time flows in one direction. The second law of thermodynamics distills this aspect of experience and states it succinctly and quantitatively, albeit abstractly. It was a great accomplishment of the 19th century to identify a state function, called for obscure reasons the entropy, which when summed for both system and surroundings for any process, always changes in the same direction. By convention, this function is defined so that it always increases.[1]

Similar to the first law, the second law of thermodynamics is general and pervasive. No step in any process is exempt from its application. It applies microscopically to every volume element in a system that is experiencing change, as well as macroscopically to the system as a whole. It applies during each infinitesimal increment of time in the process, as well as to the process as a whole. Specifically stated, in every volume element of any system and surroundings that may be experiencing change, at every instant in time, the entropy production is positive. It will now be demonstrated that this does not imply that the entropy of a system can only increase.

Similar to energy, entropy changes arise from influences that operate to change the condition of a system. However, the relationship between the entropy changes and the influences that act is not the same as it is for energy. Just as with energy, entropy can be transferred across the boundary of a system in association with heat, work and mass transfers. Unlike energy, the change in entropy of a system is not restricted to entropy transferred across its boundaries. There is an additional contribution; namely, the production of entropy inside the system. The first law is a conservation law; the second law is not.

It is always possible to decompose the change in entropy experienced by a system undergoing change into two parts:

ΔS_t, the entropy transferred across the boundary during the process.
ΔS_p, the entropy production inside the system.

The change in entropy for the system is the sum:

$$\Delta S_{sys} = \Delta S_t + \Delta S_p \tag{3.3}$$

The transfer term may be positive or negative depending upon the nature of the process and the flows at the boundary of the system. The second law states that, for all processes for all systems,

$$\Delta S_p \geq 0 \tag{3.4}$$

Whether ΔS_{sys} is positive or negative depends upon the sign of ΔS_t as well as its relative magnitude in comparison with ΔS_p. Thus, it is not unusual for the entropy of a system to decrease for a given process. Indeed, many of the practical processes that make technology work are explicitly set up to produce a negative change in entropy

---

[1] In retrospect, this may have been a bad choice for the convention since the production of entropy is associated with dissipation or consumption of the capacity for spontaneous change when a process occurs. Every change that occurs for every process for every system uses up some of the finite capacity for change that the universe possesses. If the opposite sign had been chosen in this convention, then entropy would be destroyed in the unverise for every real process that occurs rather than being produced.

of a system, i.e., to make a system change in a direction which is opposite to that which would occur "naturally." In order to accomplish this it is necessary to place the system in an environment — a surroundings — that is capable of causing the transfers necessary to produce the "unnatural" change.

Consider the entropy change experienced by the surroundings to a system during a process. Treat the surroundings as a system. Denote by a prime the properties of the surroundings. Application of Equation 3.3 gives:

$$\Delta S_{sur} = \Delta S'_t + \Delta S'_p \tag{3.5}$$

Compute now the change in entropy for the universe, i.e., system plus surroundings:

$$\Delta S_{un} = \Delta S_{sys} + \Delta S_{sur} = [\Delta S'_t + \Delta S'_p] + [\Delta S_t + \Delta S_p]$$

The entropy transferred across the boundary into the system is the negative of the entropy transferred into the surroundings. Thus, these two terms cancel and

$$\Delta S_{un} = \Delta S'_p + \Delta S_p \tag{3.6}$$

The second law states that both of these production terms are positive. Thus, while the entropy of a system may increase or decrease during a process, the entropy of the universe, taken as system plus whatever surroundings are involved in producing the changes within the system, can only increase.

These statements about entropy have been presented for a finite change in the condition of a system. They may also be applied to an infinitesimal step in the course of this change, for which the entropy changes are infinitesimals:

$$dS_{sys} = dS_t + dS_p \tag{3.7}$$

in which dS_t may be positive or negative but, by the second law, $dS_p \geq 0$. The most sophisticated formulation of this concept applies it locally, to an element of volume in an arbitrary system, and states that the rate of change of the local density of entropy is due to the divergence of the local entropy flows (the transfer term) plus the local rate of entropy production σ (reported as entropy per unit volume per unit time). In this local formulation (the details of which are beyond the scope of this text), the second law states that

$$\sigma \geq 0 \tag{3.8}$$

See Refs. [1] and [2] for the sophisticated formulation of the concept of the local rate of entropy production.

3.3   INTUITIVE MEANING OF ENTROPY PRODUCTION

It is a tautology to state that all changes that occur in the universe are spontaneous. To state that a given system placed in a given surroundings will experience a specific process spontaneously merely acknowledges that for that system in that surroundings that process will happen and not its reverse, or some other process. Left in air iron will oxidize. Water placed in a 300°C oven will vaporize. A balloon kept in a refrigerator will expand when taken out into the kitchen air. All of these changes can be reversed by placing the system in the proper surroundings. Iron oxide can be reduced to iron by putting it in the presence of carbon in the blast furnace. Steam condenses to water below 100°C (at 1 atmosphere pressure). The balloon collapses if it is returned to the refrigerator. A given system, in a given surroundings, will always change in the same direction.

These ideas are inextricably intertwined with the notion of time. Time has a forward direction. If for a given familiar sequence of conditions that a system experiences the timescale is reversed, it will be perceived that the changes observed are "improper."

It has been said that "entropy is time's arrow."[3] Entropy is the property of the universe which monitors the pervasive experience that there is a direction of change that is proper or spontaneous for a given system and surroundings. The total entropy of the system plus the surroundings always increases with increasing time; entropy is constantly produced. If a process or change is imagined for which entropy is destroyed, then the direction of time is reversed for that process: it will not happen.

Changes in the real world are always accompanied by dissipations. When the pebble is dropped into the pond its kinetic energy at impact is dispersed by the wave motion throughout the pond; it is dissipated in the absorbing water of the pond. Eventually, frictional effects associated with the viscosity of the water dissipate this dispersed kinetic energy as heat, which is further dispersed into the surroundings. In this dispersed state, it would require an unbelievable cooperation among the atoms of the pond to generate wavelets at its perimeter, which amplify as they flow toward the center, concentrating kinetic energy underneath the pebble to launch it into the air. This is the key to the concept of dissipation: in experiencing spontaneous change the system exhibits a spatial redistribution of matter, energy or both that cannot be undone simply by reversing the direction of the influences that caused the change in the first place. The entropy production in a system is a quantitative measure of this dissipation.

There exists a qualitative correlation between the rate of entropy production, or dissipation, occurring in a system during a process and two related aspects of the behavior of the system:

1. How far the process has to go before the system will achieve equilibrium.
2. The rate of the process.

Qualitatively, the further a system is from its equilibrium state, the faster it will tend to change, the greater the frictional or dissipative effects, and the larger its rate of entropy production. Quantitatively, the correlation is very complex. These questions are developed in a sophisticated formalism for the description of rates of change of complex thermodynamic systems called the thermodynamics of irreversible processes mentioned in Chapter 1, and are beyond the scope of this text[1,2]. To provide an idea of the nature of the level of detail of information required to compute the dissipation function and the rate of entropy production, an expression borrowed from that theory is presented in Table 3.1, along with definitions of the factors involved.

Processes that are caused to occur slowly, by setting up conditions so that they are never very far removed from equilibrium, experience less dissipation and a correspondingly small entropy production. It is possible to visualize a process that is carried out so slowly that its internal state is never more than infinitesimally removed from equilibrium with itself and its surroundings. For example, one may visualize adding heat to a system under conditions in which the temperature in the system is only infinitesimally smaller than the temperature of the surrounding heat source. Similarly, work may be imparted from an expanding gas by reducing the external pressure applied to the system in very small increments, allowing the system to expand to come to equilibrium with the new value of pressure before the next incremental change. In the limiting case, in which the process is imagined to proceed infinitely slowly, and the incremental changes applied are infinitesimal, the dissipative effects approach zero. For such a limiting process, the entropy production is zero. The entropy change experienced by any system undergoing such a process is thus entirely due to entropy transferred across the boundary of the system.

Such processes, carried out infinitely slowly by infinitesimal changes in the influences that operate to produce change, are called reversible processes, for reasons which will shortly become evident. All real processes have a finite rate in response to finite influences; real processes are called irreversible to emphasize the contrast with the conceptual reversible processes. Irreversible processes suffer dissipations that result in the production of entropy and thus a permanent change in the universe.

TABLE 3.1

Expression Derived in the Thermodynamics of Irreversible Processes for the Local Rate of Entropy Production[1]

$$\sigma = -\frac{1}{T^2} J_Q \text{grad } T + \frac{1}{T} \sum_{k=1}^{c} J_k [K_Q - (\text{grad } \mu_k)_T] + \frac{1}{T} \sum_{v=1}^{r} b_v A_v - \frac{1}{T} \sum_{i=1}^{3} P_{ij} \left(\frac{\partial v_i}{\partial z_j} \right)$$

σ, is the local rate of entropy production; T, is the absolute temperature; grad T, is the temperature gradient; J_Q, is the heat flux; J_k, is the diffusional flux of component k; K_k, is the body force acting on the mass of component k; grad μ_k, is the gradient in the chemical potential of component k; b_v, is specific rate of chemical reaction v; A_v, is the affinity for chemical reaction v; P_{ij}, is the ij component of the local stress tensor; v_i, is the ith component of the local velocity vector; z_j, is the jth component of the local position vector.

Consider a piston containing a gas at some temperature and pressure, Figure 3.3a. For a given temperature and number of molecules, the pressure in the gas when it is in equilibrium with its surroundings is determined by the pressure and temperature in the surrounding atmosphere and the weight of the piston, the pan and its contents sitting on the top of the piston. Suppose the majority of the weight is supplied by the contents of the pan, which is lead powder.

Now suppose the gas is allowed to expand through two different processes which will be compared to illustrate the difference between reversible and irreversible processes. For process I (Figure 3.3b), the pan and its contents are lifted off the top of the piston. The system (the gas in the chamber) experiences a sudden drop in the force applied to it at the top and expands rapidly with time. Intermediate conditions of the gas are complicated, described by two three-variable functions: $T(x, y, z)$ and $P(x, y, z)$. These functions change with time. Local variations in these functions with position (gradients in T and P) induce finite flows of mass and heat. Eventually, the piston will come to rest at some new position, and the gas again comes to equilibrium with its surroundings. Its temperature will be that of its surroundings and its pressure will be lower, determined by the atmospheric pressure and the weight of the piston. It is even possible that during this expansion the piston may overshoot its ultimate position, then drop below it, and after some oscillations settle down to its final position. (Whether this actually happens in a given case

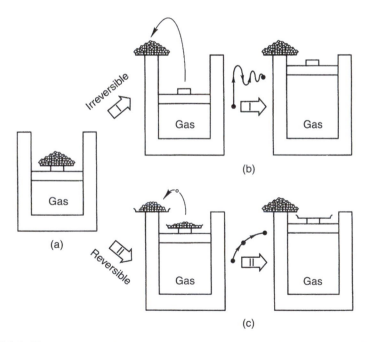

FIGURE 3.3 The pressure in the gas in the piston (a) is determined by the weight of lead powder in the pan. If the tray is removed the gas expands suddenly and irreversible and settles to a final state (b). If the gas expands slowly, by removing one particle of lead powder at a time, the process is reversible (c).

depends upon the viscosity of the gas and the details of the process.) The heat and mass flows gradually dissipate as the system approaches its new equilibrium state.

If the pan and its contents are then placed back on the piston, a second complex process ensues as the system ultimately returns to its initial state. The added force at the top immediately increases the gas pressure in that region, again producing gradients in pressure and temperature; these evolve and ultimately disappear as the final resting state is attained for the system. This state will be identical with that which originally existed for the system because the content of the system is the same and the surroundings are the same as the initial state.

The key point in this demonstration is: when change in the external influence on the system (force derived from the pan of lead shot) is reversed, the complicated sequence of conditions $T(x, y, z, t)$ and $P(x, y, z, t)$ experienced by the gas is not simply the reverse of that traversed when the pan was first removed. Indeed, it is not possible to devise a process that will retrace the sequence of conditions passed through during the expansion of the gas. The process is said to be irreversible.

Now consider a second process, II (Figure 3.3c), in which the gas is caused to expand by removing the lead powder with a pair of tweezers one particle at a time and letting the system come to rest before the next particle is removed. At each step during this expansion the pressure in the gas is only very slightly different from that exerted by its surroundings. Gradients in pressure and temperature within the system are negligible. The description of the sequence of states that the system traverses is very simple since each state is specified by only two numbers: its current pressure and temperature, which are both uniform in the system.

If the change in external influence is reversed, by individually replacing the particles of lead powder just removed, the sequence of states that the system traverses will be identical to that which it experienced as those particles were being slowly removed, traversed in the opposite in direction. Thus, reversal of the change in external influence simply reverses the direction of the trajectory of the system along the path connecting the initial and final states. This argument applies to any segment of the path in the process or to the whole path: the process is said to be carried out reversibly.

For a process carried out reversibly, there are no dissipations, no entropy production and no permanent changes in the universe (system plus surroundings). Irreversible processes are accompanied by dissipation, finite entropy production and permanent changes in the universe.

Calculation of the process variables, heat absorbed and work done on the system is relatively easy for reversible processes, primarily because each intermediate state is described by just a few numbers, e.g., temperature and pressure (T, P). In contrast, the calculation of these process variables for irreversible processes is extremely complicated because the description of evolution of the system requires specification of the variation with time of the positional variation within the volume of the system of temperature and pressure functions $[T(x, y, z, t), P(x, y, z, t)]$. This calculation involves the simultaneous solution of two partial differential equations with time-dependent boundary conditions.

On the other hand, calculation of changes in state functions can be straightforward for both classes of processes because such changes depend only upon the initial and final states and are therefore independent of the path connecting these two states. Thus, given a complex irreversible process connecting states A and B, changes in state functions for this process may be computed by imagining the simplest process that connects the states A and B. This process will, of course, be a reversible process. The changes in state functions for that simple process will be identical to the changes in the state functions that would be determined for the specified irreversible process. This strategy points up the importance of the fact that state functions have been identified in nature, and at the same time, justifies the central role played by the analysis of reversible processes in thermodynamics even though such processes are, practically speaking, unrealistic.

3.4   RELATION BETWEEN ENTROPY TRANSFER AND HEAT ABSORBED

The previous section developed the qualitative relation between dissipation occurring in irreversible processes and entropy production. This section focuses upon the other contribution to the change in entropy of a system: entropy flow across the boundary.

A quantitative treatment of entropy transfer for irreversible processes requires the formalism of the thermodynamics of irreversible processes and is beyond the scope of this text. A quantitative treatment of entropy transfer for reversible processes establishes a connection between the reversible heat flow across the boundary of the system and its change in entropy, and forms the basis for the evaluation of the entropy from database information. The argument that demonstrates this connection is somewhat convoluted and is presented in Appendix H. The interested reader will find it useful to delay examining that argument until Chapter 4 has been studied since some of the relations used are developed in that chapter. However, experience has shown that it is not essential to understand this argument in order to apply its conclusion. Therefore, only an outline of the sequence of logic involved is presented here.

The argument applies to any system that is taken through any reversible process. The nature of the system or the process is otherwise unrestricted, so that the result is general for reversible processes. Focus on an infinitesimal step in this reversible process. Let δQ_{rev} be the heat absorbed by the system during that step. The current temperature of the system is T. The argument focuses on the differential form $\delta Q_{rev}/T$, which has units of J/K. It sums the values of this differential form as the sequence of states is traversed for the process. This sum has the mathematical form of a line integral along the path:

$$\oint \frac{\delta Q_{rev}}{T} \tag{3.9}$$

The argument uses this function to demonstrate that, although δQ_{rev} is a process variable, $\delta Q_{rev}/T$ is the differential of a state function. This is demonstrated by considering a general cyclic process, i.e., one that returns the system to its original state. Since the system ends up where it started, changes in all state functions for any cyclic process must be zero. The development, based upon the Carnot cycle, demonstrates that the function defined in expression 3.9 is zero for any reversible cyclic path; it is thus a state function. It follows that the change in this function for any increment along the path is a state function, and ultimately, for each infinitesimal step along any path, $\delta Q_{rev}/T$ is a state function. This state function is then defined to be the entropy of the system. The practical form of this result is

$$\delta Q_{rev} = T\,dS \tag{3.10}$$

$$Q_{rev} = \oint T\,dS \tag{3.11}$$

which permits computation of the heat absorbed for any reversible process by integration of the combination of state functions, $T\,dS$. This is the most frequently used consequence of the argument. Note that, while T and S are state functions, the differential form $T\,dS$ represents a process variable: the value of Q_{rev} in Equation 3.11 is different for different paths.

One result of these developments is an important inequality that applies to irreversible (real) processes. For any reversible process that changes the condition of the system from state A to state B the change in entropy is completely due to transfer of heat across the boundary:

$$\Delta S_{rev}[A \rightarrow B] = \oint \frac{dQ_{rev}}{T} \tag{3.12}$$

Consider now any irreversible process that connects the same two end states A and B. Since entropy is a state function, the change in entropy for this process will be the same as for the reversible process:

$$\Delta S_{rev}[A \rightarrow B] = \Delta S_{irr}[A \rightarrow B]$$

However, the entropy change for the irreversible process must include some production of entropy inside the system. Therefore, the entropy transferred across the boundary will be smaller (algebraically) than in the reversible case:

$$\Delta S_{irr,t} + \Delta S_{irr,p} = \Delta S_{rev,t} + 0$$

$$\Delta S_{irr,t} = \Delta S_{rev,t} - \Delta S_{irr,p}$$

Insofar as the transfer of entropy is associated with heat transfer across the boundary,

$$\Delta S_{irr,t} = \oint \frac{\delta Q_{irr}}{T}$$

It follows that

$$\oint \frac{\delta Q_{irr}}{T} < \oint \frac{\delta Q_{rev}}{T} \tag{3.13}$$

For an isothermal process (i.e., one carried out in a system with temperature control so that T is held constant in the experiment), this result further simplifies to

$$[Q_{irr}]_T < [Q_{rev}]_T \tag{3.14}$$

In a comparison of all possible isothermal processes that convert a system from state A to state B, the heat absorbed by a reversible process is larger than for any irreversible process. Under these conditions, Q_{rev} is the maximum heat absorbed for these processes

3.5 COMBINED STATEMENT OF THE FIRST AND SECOND LAWS

An expression for the mechanical work done during any arbitrary reversible process may also be derived. Recall the general definition of work, Equation 2.1:

$$\delta W = F dx \tag{2.1}$$

The force acting in simple mechanical work is derived from the external pressure P_{ext} on the system as shown in Figure 2.6. This force vector is directed toward the system, in the $+x$ direction in Figure 2.6. Work results from the displacement of this applied force. The force is displaced if the boundary of the system on which it is acting moves. Consider a displacement of the boundary, dx, defined to be positive if it is in the $+x$ direction. Multiply and divide by the area A in Equation 2.1:

$$\delta W = \frac{F}{A} A dx \tag{3.15}$$

(F/A) is P_{ext}, the external pressure on the system. The combination $A dx$ is related to the change in volume of the system due to the motion of its boundary. The volume change of the system is given by

$$dV = -A dx$$

The negative sign is necessary because if dx is positive in Figure 6.2 (the boundary moves into the system) the volume of the system decreases so that dV is negative. If dx is negative, $-A dx$ is positive, which describes the accompanying volume increase. Insert these results into Equation 3.15:

$$\delta W = P_{ext}(-dV) \tag{3.16}$$

The goal of this development is to obtain an expression for the mechanical work done in terms of functions of the internal state of the system. If (and only if) the process is carried out reversibly, then the internal pressure in the system, P, will be essentially equal to the external pressure, $P = P_{ext}$ at each infinitesimal step in the process, and Equation 3.16 becomes:

$$\delta W_{rev} = -PdV \tag{3.17}$$

If the process is carried out irreversibly, Equation 3.16 may still be used to evaluate the work done on the system; however, in this case $P_{ext} \neq P$, i.e., the external pressure is not equal to the internal pressure since the internal pressure is not expected to be uniform.

Equation 3.10 and Equation 3.17 permit evaluation of the process variables, heat and mechanical work for any arbitrary reversible process. Substitution of these expressions into the general statement of the first law, Equation 3.2 yields the most frequently used equation in the practical application of thermodynamics to simple systems, called the combined statement of the first and second laws:

$$dU = TdS - PdV + \delta W' \tag{3.18}$$

The central role played by this equation in deriving relationships among the variables of thermodynamics is developed in detail in Chapter 4. It is also at the base of derivations of the conditions for equilibrium in thermodynamic systems, which are in turn the bases for calculating maps of equilibrium states which play a central role in the application of thermodynamics in materials science and engineering.

3.6   THE THIRD LAW OF THERMODYNAMICS

Similar to energy, temperature is taken to be a familiar concept with daily practical application in household technology, as well as science and industry. Weather reports give high, low and current temperatures along with predictions about tomorrow's conditions. Thermostats control the temperature of the house, office, oven or refrigerator. Temperature is understood to be a property of matter that provides a universal measure of the tendency of systems to exchange heat.

The earliest attempts to quantify this aspect of the behavior of matter date from Fahrenheit in the mid-seventeenth century. He monitored this property by recording changes in volume of a liquid in a tube, now familiar as a thermometer. In the Fahrenheit temperature scale that he devised he chose his own benchmarks, defining a zero point and a point for 100°F. It is a familiar fact that water freezes at 32°F and boils at one atmosphere at 212°F on this scale. More than a century later, the scientific community agreed that very pure water should be used to define the benchmarks on the scale and adopted the centigrade, now called the Celsius, scale in which water freezes at 0°C and boils at 100°C.

As refrigeration devices came into use in laboratories, it became clear that matter could exist at temperatures significantly below the defined zero point for the

Celsius scale. Near the turn of the last century, experimenters studying the behavior of matter at very low temperatures established that there is a lower limit to the temperature that matter can exhibit. In the light of this observation it made sense to define this lower limit to be the zero point on a new, absolute scale of temperature, now called the Kelvin scale. This zero point, written 0 K, has been shown to correspond to $-273.150°C$ on the Celsius scale. The interval between temperatures was chosen to agree with the Celsius scale so that the temperature difference between the freezing and boiling water at one atmosphere pressure is also 100 units on the Kelvin scale. This observation, that there exists an ultimate lower limit to the temperature that matter can exhibit, is one component of the third law of thermodynamics.

The same cryogenic studies that established the existence of the absolute zero in temperature explored other aspects of the behavior of matter in the range between absolute zero and room temperature. To illustrate the principle that evolved from these studies, consider the cyclic process shown in Figure 3.4. A system consisting of one mole of pure silicon and one mole of carbon initially at absolute zero is heated from 0 to 1500 K. The entropy change for this process (step I) can be computed from the heat capacities of silicon and carbon, which have been measured in this temperature range (see Chapter 4 and Appendix B). At 1500 K, the silicon and carbon are reacted to form the compound, silicon carbide, SiC. The entropy change for this reaction (II) can be experimentally determined by measuring the heat absorbed in a calorimeter during this reaction at the known temperature. The compound, SiC, is then cooled to absolute zero; the change in entropy for this process (III) can be computed from measured values of the heat capacity of the compound, SiC, which is, of course, different from those of pure silicon and carbon. The entropy difference between the compound, SiC and pure Si and C at 0 K, which would be characteristic of reaction (IV) in Figure 3.4 if it could be made to occur at that very low temperature, may be computed by invoking the fact that entropy is a state function.

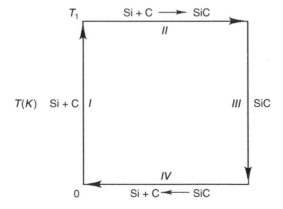

FIGURE 3.4 Cyclic process involving the reaction that forms silicon carbide from its components illustrates the third law of thermodynamics.

If the system is taken through all four processes, it is returned to its original state: i.e., the total process is a cycle. Since entropy is a state function, the change in entropy for the cycle must be zero:

$$\Delta S_{cyc} = \Delta S_I + \Delta S_{II} + \Delta S_{III} + \Delta S_{IV} = 0 \qquad (3.19)$$

However, when the individual entropy changes for steps I, II and III are calculated from experimental measurements it is found that, within the experimental error,

$$\Delta S_I + \Delta S_{II} + \Delta S_{III} = 0 \qquad (3.20)$$

It is thus concluded that the entropy change for step IV is 0:

$$\Delta S_{IV} = 0 \qquad (3.21)$$

That is, the entropy of the compound, SiC, is the same as that of the pure silicon and carbon at 0 K.

Repetition of this kind of computation for a large number of substances established that it is a pervasive observation. The entropy of all substances is the same at 0 K. (Where deviations from this observation have been found, they could be explained quantitatively by establishing that the substances involved had not achieved total internal equilibrium at these low temperatures.) Since all substances have the same value of entropy at absolute zero, it makes sense to choose this condition as a zero point for the definition and measurement of entropy. Thus, these empirical observations have established the third law of thermodynamics:

> There exists a lower limit to the temperature that can be attained by matter, called the absolute zero of temperature, and the entropy of all substances is the same at that temperature.

The primary practical application of this law (in the context of this text) lies in the evaluation of entropy changes for chemical reactions from database information. In the cyclic process just described the entropy change for the reaction (step II)

$$Si + C = SiC$$

may be evaluated without requiring that the reaction actually be experimentally studied. Rearrangement of Equation 3.20 gives:

$$\Delta S_{II} = -(\Delta S_I + \Delta S_{III}) \qquad (3.22)$$

The entropy changes on the right side of this equation (which are associated with heating the reactants from 0 K and cooling the products to 0 K) can be computed from heat capacity information about these substances. Information about the chemical reaction thus is not required to compute this entropy change for step II. It is not necessary to set up and carry out an experiment to measure the change in entropy for the reaction; this entropy change can be computed as the absolute entropy of the products minus that of the reactants.

Computations of entropy changes for chemical reactions are made more convenient with the aid of tabulations of values of absolute entropies of elements and compounds at room temperature, taken to be 25°C or 298 K (see Appendices B and E). To illustrate this application, consider the reaction that represents the formation of alumina (Al_2O_3) from aluminum and oxygen at 298 K. Refer to Appendices B and E to obtain the following values for the absolute entropies of these reactants at 298 K:

$$S^\circ_{Al,298\ K} = 28.3\ \text{J/mol K}$$

$$S^\circ_{O_2,298\ K} = 205.03\ \text{J/mol K}$$

$$S^\circ_{Al_2O_3,298\ K} = 51.1\ \text{J/mol K}$$

A stoichiometrically balanced statement of the reaction is

$$2Al + \frac{3}{2}O_2 = Al_2O_3$$

The entropy change for this reaction at 298 K is

$$\Delta S^\circ_{298\ K} = S^\circ_{Al_2O_3,298\ K} - \left[2S^\circ_{Al,298\ K} + \frac{3}{2}S^\circ_{O_2,298\ K}\right]$$

$$\Delta S^\circ_{298\ K} = 51.00 - \left[2(28.32) + \frac{3}{2}(205.03)\right] = -313.19\ \text{J/mol K}$$

3.7 SUMMARY

In all processes energy is conserved. Therefore, the change in internal energy of any system in any process is the sum of the energy transfers across its boundary. Mathematically, for each infinitesimal step in any process,

$$dU = \delta Q + \delta W + \delta W' \tag{3.2}$$

In every volume element at every instant in time, in all processes, entropy is created. This entropy production is largest for irreversible processes that are occurring rapidly and far from equilibrium; it approaches zero in the limiting case of reversible processes for which each step is but infinitesimally removed from equilibrium.

Entropy is a state function. For reversible processes the change in entropy is related to the heat absorbed by the system at each step:

$$\delta Q_{rev} = T dS \tag{3.10}$$

where T is the absolute temperature.

The most frequently used equation in treating the thermodynamics of simple systems is the combined statement of the first and second laws:

$$dU = T dS - P dV + \delta W' \tag{3.18}$$

Both entropy and temperature have absolute values because both have an empirically observed zero point.

HOMEWORK PROBLEMS

Problem 3.1.   The laws of thermodynamics are "pervasive." Explain in detail the meaning of this important statement.

Problem 3.2.   List the kinds of energy conversions involved in propelling an automobile.

Problem 3.3.   List the kinds of energy conversions involved in operating a hand calculator.

Problem 3.4.   List the kinds of energy conversions involved in using your right arm to turn the page in this text.

Problem 3.5.   Suppose the convention were adopted that defines W and W' in the first law of thermodynamics to be the "work done on the system by the surroundings."
    a. Write the first law with this alternate convention.
    b. Why do the signs change?

Problem 3.6.   Give five examples of the operation of the second law of thermodynamics in your daily experience; they must be different from those given in the text.

Problem 3.7.   Biological systems — organelles, cells, organs, plants and animals — are highly ordered, yet form spontaneously. Does the formation and growth of biological systems violate the second law of thermodynamics? Explain your answer.

Problem 3.8.   "Irreversible" is an awkward adjective. Why is this term so appropriate in its application to the description of processes in thermodynamics?

Suggest two or three alternate words or phrases that might be used to replace "irreversible" in these contexts.

Problem 3.9.   Contrast the relative magnitudes of the entropy transfer vs. entropy production in the following processes:
   a. A thermally insulated container has two compartments of equal size. Initially one side is filled with a gas and the other is evacuated. A valve is opened and the gas expands to fill both compartments.
   b. A gas contained in a steel cylinder is slowly expanded to twice its volume.

Problem 3.10.   Consider an isolated system (no heat, matter or work may be exchanged with the surroundings) consisting of three internal compartments A, B and C, of equal volumes. The compartments are separated by partitions; each partition has a valve which may be opened remotely. Initially the central volume B is filled with a gas at 298 K (25°C) and the outer two are evacuated. Consider the following two processes:
   a. The valve to the A side is opened, the gas expands freely into compartment A, and the system comes to equilibrium. Then, the valve to the C side is opened, and the system again comes to equilibrium.
   b. Both valves are opened simultaneously, the gas expands freely into both compartments, and the system comes to its equilibrium.

Which of these processes produces more entropy?

Problem 3.11.   It will be shown in Chapter 4 that the change in entropy ΔS associated with process A in Problem 3.8 is 4.60 (J/mol K) and that the initial and final states will be at the same temperature. Application of Equation 3.10 suggests that the heat absorbed by the system during this process is

$$Q = T\Delta S = (298 \text{ K})(4.60 \text{ J/mol K}) = 1370 \text{ J/mol}$$

Yet the description of the system says it is isolated from its surroundings so that $Q = 0$. Explain this apparent contradiction.

Problem 3.12.   Give three examples of processes, which are important in materials science, that are thermodynamically "irreversible." Speculate briefly about the nature of the dissipations in these processes that contribute to the production of entropy.

Problem 3.13.   The notion of a "reversible" process is a fiction in the real world. What makes this concept, which at first glance would appear to be only of academic interest, so useful in applying thermodynamics to real world "irreversible" processes?

Problem 3.14. The combined statement of the first and second laws of thermodynamics, Equation 3.15, evaluates the heat absorbed and mechanical work done on a system with relationships that are only valid for reversible processes. Since reversible processes do not occur in the real world, how is it possible for the combined statement to play an important role in the analysis of practical "irreversible" processes encountered in nature and in technology?

Problem 3.15. Describe the kinds of experimental observations that have been invoked to support the hypothesis that the entropy of all substances is the same at absolute zero.

Problem 3.16. Use the following values of absolute entropies of elements and compounds at 298 K to compute the entropy changes associated with as many chemical reactions as you can generate from this list.

Element	$S^\circ_{298\,K}$ (J/mol K)	Compound	$S^\circ_{298\,K}$ (J/mol K)
Al	28.3	CO	197.9
C(gr)	5.69	CO_2	213.64
Si	18.83	Al_2O_3	51.1
O_2	205.03	SiC	16.54
		SiO_2	41.5

REFERENCES

1. Haase, R., *Thermodynamics of Irreversible Processes*, Addison-Wesley, Reading, MA, p. 243, 1969, Also reprinted by Dover, New York, 1990.
2. DeGroot, S.R. and Mazur, P., *Non-Equilibrium Thermodynamics*, North Holland, Amsterdam, 1962.
3. Eddington, Sir Arthur, *The Nature of the Physical World*, University of Michigan Press, Ann Arbor, MI, 1958.

4 Thermodynamic Variables and Relations

CONTENTS

System A in Surroundings I is placed in Surroundings II. What will happen?

This chapter supplies answers to that question for the simplest class of thermodynamic system: a unary, homogeneous, closed, nonreacting, otherwise simple system. A part of the answer to the question, "What will happen?" is the more specific question, "What changes will occur in its properties?" Strategies for computing thermodynamic properties, referred to in Figure 1.4, begin with methods developed for such simple systems.

So far we have defined the following variables:

Process variables (values depend on the path specified for the process):
    Q — heat absorbed (joules).
    W — mechanical work done on the system (joules).
    W' — all other kinds of work done on the system (joules) (equal to 0 for the simplest systems in this chapter).
State variables (values depend only upon the current state):
    T — temperature (K).
    P — pressure (atmospheres).

U — internal energy (joules).
S — entropy (joules/K).
V — volume (cubic meters).

Additional state variables will be defined in terms of these variables in this chapter.

It is established that, in such a simple system in equilibrium with itself and its surroundings, the state is completely specified by two variables.[1] For example, if the temperature and pressure are specified then the state of such a system is fixed. This statement implies that the values of all other state properties (so far, V, U and S) of this system are determined given values of the variables (T, P). Functions exist connecting these properties to T and P: e.g., $V(T, P)$ implies that for each value of (T, P) for a given substance there is a corresponding value of its molar volume. This relationship may be represented graphically as a surface over the (T, P) plane. Each substance has its own surface $V(T, P)$ describing how its molar volume varies with temperature and pressure. For example, the molar volume (volume per mole) of liquid water is a different function of T and P than that for water vapor. At ordinary temperatures and pressures 1 mol of water vapor is about 20,000 times larger than 1 mol of liquid water. The molar volume of lead is roughly twice that of aluminum.

Since U and S are also state functions it follows that for any given substance they also have functional relations to T and P. Thus, for example, for solid copper there exist surfaces over the (T, P) plane for $U(T, P)$ and $S(T, P)$. The additional state functions defined in this chapter each have their own functional relationships to T and P, and their own surfaces over the (T, P) plane.

This chapter also introduces a set of experimental variables: properties of materials that are measured in laboratories, assessed, and gathered into databases for application in thermodynamic calculations. These properties are familiar from introductory chemistry and physics courses:

α — the volume coefficient of thermal expansion (with units K^{-1}).
β — the volume coefficient of compressibility (with units atm^{-1}).
C_P — the heat capacity at constant pressure (with units J/K).
C_V — the heat capacity at constant volume (with units J/K).

These properties are also functions of temperature and pressure for each substance. This chapter introduces procedures for computing all of the state functions for a simple system from these database properties.

Since all of the state functions that will be defined are related to temperature and pressure, it may be concluded that they are all related to each other. These relationships may also be derived from database information.

---

[1] To evaluate extensive properties of the system one needs in addition the total number of moles n_T of components contained in the system. A common strategy evaluates extensive properties per mole of system. Results per mole are easily converted to values for the system by multiplying by n_T.

The simplest class of problems that may be solved thermodynamically is those in which some change in the condition or properties of a system is brought about by controlled or measured changes in the influences that may operate on the system. A few representative examples are:

(a) Hydrogen gas in a 20-l steel cylinder at ambient temperature (18°C) is found to be at 10 atmospheres of pressure. Is the pressure significantly affected if the room temperature heats up to 25°C?

(b) Calculate the change in entropy of 50 g of nickel when it is heated from 0 to 500°C at one atmosphere pressure.

(c) A 40-g sample of the elastomer, polyisoprene, is stretched so that its length increases by 50%. Estimate the change in temperature of the sample if it is initially at 20°C.

(d) Estimate the amount of heat required to raise the temperature of an alumina (Al_2O_3) charge weighing 50 kg from room temperature to 1350°C at one atmosphere pressure.

Solutions to these problems are based upon relationships between the properties given and required; the strategy for deriving such relationships is the subject of this chapter.

The first step in applying this strategy is the translation of the problem encountered in the real-world (or in the text) from English into thermodynamics. This exercise in translation has the following elements:

1. Identify the properties of the system about which information is given. For example, in problem b these properties are:

   Temperature (T) (changes from 0 to 500°C).

   Pressure (P) (remains fixed at one atmosphere).

   These variables are called the *independent variables* in the problem because they are subject to experimental control in the context of the problem and thus may be changed independently.

   In simple systems there are two independent variables.

2. Identify the property in the system about which you are seeking information. In problem b, this property is the entropy of the system.

   This property is called the *dependent variable* in the problem because its value is determined by the changes in the controlled (independent) variables.

3. Find or derive a relationship between the sought (dependent) variable and the given (independent) variables. In problem b, this relation has the generic form: $S = S(T, P)$, read "entropy as a function of temperature and pressure."

   This relationship will necessarily contain quantities that are properties of the material that makes up the system, such as heat capacity, thermal expansion coefficient, and the like.

4. Obtain values for these quantities, either from tabulations such as those reviewed in Appendices B and E, from the literature, compiled databases or, if necessary, by direct experimental evaluation.
5. Substitute these values into the relationship and carry out the mathematical operations necessary to obtain a numerical value for the dependent variable in the problem.

This strategy will be applied repeatedly in this and subsequent chapters to yield numerical answers to practical problems.

The crucial application of the principles of thermodynamics in this strategy is evidently contained in step 3, which requires the derivation or deduction of the relationship that connects the dependent variable in the problem to the independent variable through database information. A general procedure for deriving such relationships is the focus of this chapter.

4.1 CLASSIFICATION OF THERMODYNAMIC RELATIONSHIPS

Thermodynamics abounds with relationships. This potentially confusing aspect of the structure of thermodynamics exists because a variety of state functions are defined in terms of other state functions. This potentially bewildering array of equations may be tamed by organizing them into five classifications introduced in Section 2.3 in Chapter 2:

I. The laws of thermodynamics.
II. Definitions in thermodynamics.
III. Coefficient relations.
IV. Maxwell relations.
V. Conditions for equilibrium.

The first two categories are self-explanatory; categories III and IV are derived from mathematical properties of state functions. Relationships in category V are derived from the general criterion for equilibrium developed in Chapter 5; much of the remainder of the text focuses upon these relationships.

4.1.1 THE LAWS OF THERMODYNAMICS

The relationships in this category have been presented in Chapter 3. Succinctly reviewed, they are:
The first law:

$$dU = \delta Q + \delta W + \delta W'$$
(4.1)

Reversible mechanical work:

$$\delta W_{\text{rev}} = -P \, dV$$
(4.2)

Reversible heat absorbed:

$$\delta Q_{rev} = T\,dS \tag{4.3}$$

Combined statement:

$$dU = T\,dS - P\,dV + \delta W' \tag{4.4}$$

Equation 4.4 is a centerpiece in the mathematical framework of thermodynamics and will be applied repeatedly in subsequent developments.

4.1.2 DEFINITIONS IN THERMODYNAMICS

Relationships presented in this section are purely and simply *definitions*. There are two subcategories in this classification: the *energy functions* and the *experimental (database) variables*.

The *energy functions* are measures of the energy of a system that differ from the internal energy U originally introduced in the first law. It will be demonstrated that use of these functions simplifies the description of systems that may be subjected to certain classes of processes; they are simply more convenient measures of energy in such applications than is the function, U.

4.1.2.1 Enthalpy, H

$$H \equiv U + PV \tag{4.5}$$

a state function, since it is defined in terms of other state functions (U, P, V). Consider a small change in the state of any system; the change in enthalpy is obtained in general by taking the differential of its definition:

$$dH = dU + P\,dV + V\,dP$$

Use the combined statement of the first and second laws, Equation 4.4, to substitute for dU in this equation:

$$dH = [T\,dS - P\,dV + \delta W'] + P\,dV + V\,dP$$

Simplify:

$$dH = T\,dS + V\,dP + \delta W' \tag{4.6}$$

This equation is an alternate form of the combined statement of the first and second laws of thermodynamics. It has the same level of generality and pervasiveness as Equation 4.4.

Historically, the enthalpy function was introduced because it was convenient in the description of heat engines taken through cycles. Only mechanical work is done in this case ($\delta W' = 0$) and cycles occur at atmospheric pressure ($dP = 0$). Thus, for such *isobaric* processes in simple systems,

$$dH_P = T\, dS_P = \delta Q_{rev,P}$$

For this class of process (and *only* for this class) the state function defined as the enthalpy provides a direct measure of the reversible heat exchanges of the engine with its surroundings. For this reason, the enthalpy is sometimes called the *heat content* of the system. This notion is potentially very misleading. A system at rest does not contain a "quantity of heat"; heat is a process variable and only has a value for a system that is changing its state. The *change* in the state function H is related to the process variable, Q, but only for reversible, isobaric processes.

4.1.2.2   Helmholtz Free Energy, F

$$F \equiv U - TS \tag{4.7}$$

is also a state function. Again consider a small arbitrary change in the state of any system. The change in Helmholtz free energy is

$$dF = dU - T\, dS - S\, dT$$

Substitute for dU from Equation 4.4:

$$dF = [T\, dS - P\, dV + \delta W'] - T\, dS - S\, dT$$

Simplify:

$$dF = -S\, dT - P\, dV + \delta W' \tag{4.8}$$

This is yet another alternate form of the combined statement of the first and second laws which may be used to relate property changes in any system taken through any process. This function was devised because it simplifies the description of systems subject to temperature control, which is a common experimental scenario in the laboratory. If the process is carried out at constant temperature (*isothermally*) then $dT = 0$ at each step along the way. For isothermal processes,

$$dF_T = -P\, dV_T + \delta W'_T = \delta W_T + \delta W'_T = \delta W_{T,tot}$$

That is, for isothermal processes, the Helmholtz free energy function reports the *total (reversible) work done on the system*. For this reason this property is sometimes called the *work function*. Indeed, in Helmholtz's original presentation, he used the symbol A for this function to stand for arbeiten, the German word for work. This expression, i.e., the work function, may be misleading in the same way as the term heat content is misleading for the enthalpy function. A system at rest does not contain a given "quantity of work"; work can only be associated with a change in state along a specified path. The *change* in the state function F is related to the total work done on the system only for isothermal, reversible processes.

4.1.2.3 Gibbs Free Energy, G

$$G \equiv U + PV - TS = H - TS \tag{4.9}$$

is evidently also a state function since it is also defined in terms of other state functions. Following the strategy used for the other two functions, for an arbitrary infinitesimal change in state, the change in Gibbs free energy is

$$dG = dU + P\,dV + V\,dP - T\,dS - S\,dT$$

Again substitute Equation 4.4:

$$dG = [T\,dS - P\,dV + \delta W'] + P\,dV + V\,dP - T\,dS - S\,dT$$

Simplify:

$$dG = -S\,dT + V\,dP + \delta W' \tag{4.10}$$

which is yet another equivalent but alternate statement of the combined first and second laws. The Gibbs free energy function was introduced because it simplifies the description of systems that are controlled in the laboratory so that *both* temperature and pressure remain constant. For processes carried out under these conditions [isothermal ($dT = 0$) and isobaric ($dP = 0$)]

$$dG_{T,P} = \delta W'_{T,P}$$

Thus, for such processes, which include, for example, phase transformations and chemical reactions, this function reports the total work done on the system *other than mechanical work*.

Four combined statements of the first and second laws have been presented, Equation 4.4, Equation 4.6, Equation 4.8, and Equation 4.10. All represent valid ways of describing the thermodynamic changes for any system taken through any process. Each of the defined energy functions provides a simple measure of some aspect of some process variable for specifically defined classes of processes. All of these equations will be used frequently in subsequent developments. Mnemonic devices have been concocted to aid in memorizing these equations. However, it is recommended rather that the *definitions* of H, F and G be committed to memory; then the simple two-line strategy presented here derives these combined statements of the first and second laws.

A set of *experimental variables* provides the core of practical information about a specific material essential to solving thermodynamic problems involving that material. These quantities are commonly measured in the laboratory and published in extensive tables or databases, such as those contained in the Appendices B and E and in Refs. [2–7] at the end of this chapter. A significant amount of this information is available online without charge as noted in the references. Development of this kind of information is expensive, and assessed databases for specific classes of materials (e.g., nickel-based alloys for aerospace applications) are available for purchase from companies that compile and assess the data.

The definitions of these experimental variables directly reflect the measurements involved.

The *coefficient of thermal expansion*, α, is obtained from a measurement of the volume change of the material when its temperature is increased, with the system constrained to a constant pressure. The definition:

$$\alpha \equiv \frac{1}{V}\left(\frac{\partial V}{\partial T}\right)_P (K^{-1}) \tag{4.11}$$

reports the fractional change in molar volume of a substance with temperature. The coefficient of expansion of any material substance varies with temperature, pressure, and composition of the system. Typical values of α are listed in Table 4.1 and Appendix B.

The *coefficient of compressibility*, β, is determined by measuring the volume change of the substance as the pressure applied to it is increased while the temperature is held constant. Its definition,

$$\beta = -\frac{1}{V}\left(\frac{\partial V}{\partial P}\right)_T (atm)^{-1} \tag{4.12}$$

is normalized to report the fractional change in volume with increasing pressure. The derivative in this definition is inherently negative (as P is increased, V decreases); thus, inclusion of the minus sign guarantees that tabulated values of β will be positive numbers. Note that β also varies with the temperature and pressure at which it is measured and is different for different materials. Table 4.2 and Appendix B list typical values for β.

Experimental information about the thermal behavior of substances is contained in the concept of *heat capacity*. This quantity is determined experimentally by precisely measuring the rise in temperature when a small measured quantity of heat is caused to flow reversibly into the system. Since heat is a process variable, it is necessary to specify the process path that was used in a particular heat capacity measurement. Virtually all heat capacity data are measured and compiled for either of two simple processes.

If the temperature rise, dT, is measured in a system that is set up so that the pressure is held constant and the heat, δQ, is transferred reversibly, then the heat capacity at constant pressure, C_P, is obtained from the definition

$$\delta Q_{rev,P} \equiv C_P \, dT_P \tag{4.13}$$

Measured heat capacities are normalized to obtain the value per mole of substance; thus the units of C_P in databases are (J/mol K). This experimental variable also changes with temperature, pressure, and composition. Almost all heat capacity

TABLE 4.1
Typical Values of Coefficients of Thermal Expansion α for Common Materials[a]

Material	$\alpha_L \times 10^6 \ K^{-1}$	$\alpha \times 10^6 \ K^{-1}$
Aluminum	23.5	70.5
Chromium	6.5	19.5
Copper	17.0	51.0
Lead	29.0	87.0
Potassium	83.0	250
Sodium	71.0	213
Alumina (Al_2O_3)	7.6	23
Silica (SiO_2)	22.2	66
Silicon Carbide (SiC)	4.6	14

[a] Reported values are for linear expansion coefficients, α_L; for isotropic systems the volume coefficient $\alpha = 3\alpha_L$.

Source: Brandes, E.A., Ed., Smithells Metals Reference Book, 6th ed., Butterworths, London, 1983.

TABLE 4.2
Typical Values of Coefficients of Compressibility May Be Estimated from DataBase Value for the Modulus of Elasticity, E[a]

Material	$\beta \ (atm)^{-1} \times 10^7$
Aluminum	12
Carbon (Graphite)	340
Copper	6.6
Iron	5.9
Tungsten	2.9
Alumina (Al_2O_3)	8.3
Boron Nitride (BN)	37
Silicon Carbide (SiC)	6.5
Silica Glass	42

[a] For isotropic materials $\beta = 3/E$ with properly converted units.

Sources: Reynolds, C.L., Jr., Faugham, K.A., and Baker, R.E., Metals, J. Chem. Phys., Vol. 59, p. 2934, 1973; Kingery, W.D., Bowman, H.K., and Uhlman, D.R., Ceramics, Introduction to Ceramics, John Wiley & Sons, New York, p. 777, 1976.

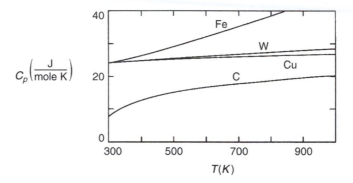

FIGURE 4.1 Variation of heat capacity with temperature for a variety of common materials.

measurements are made at one atmosphere; the pressure dependence of this quantity may be computed theoretically and can be shown to be small; the dependence upon pressure has not been widely investigated.

Figure 4.1 illustrates the temperature dependence of C_P values for a variety of substances at one atmosphere pressure and above room temperature. The variation in the temperature range below room temperature is significantly more complicated. Above room temperature, where most practical applications of thermodynamics arise, it is found that the temperature variation of C_P follows a relatively simple empirical relation:

$$C_P(T) = a + bT + \frac{c}{T^2} + dT^2 \tag{4.14}$$

Tables of heat capacity data, such as Table 4.3 and Appendix B, provide values of a, b and c for a variety of elements and chemical compounds. For precise thermodynamic computations some databases apply equations that have up to nine terms.

Heat capacity may also be measured in a system contained in a rigid enclosure so that, as the heat flows into the system, the volume of the system is constrained to remain constant. Precise measurement of the temperature rise accompanying the influx of a known quantity of heat gives the *heat capacity at constant volume*, defined in the relation

$$\delta Q_{rev,V} \equiv C_V \, dT_V \tag{4.15}$$

After normalization, the units for C_V are also (J/mol K); C_V also varies with temperature, pressure, and composition.

For the system at constant pressure, absorption of heat results in both a temperature rise and an expansion of the volume of the system; in the case of the system constrained to constant volume, all of the heat acts to raise the temperature. Thus, more heat is required to raise the temperature of a substance one degree at constant pressure than at constant volume. It is concluded that, in general, $C_P > C_V$, and this is generally observed. It will be demonstrated in a later section that these

TABLE 4.3
Experimental Values of the Variation of Heat Capacity at Constant Pressure (C_P) for Common Materials above Room Temperature[a]

Material	a	$b \times 10^3$	$c \times 10^{-5}$
Aluminum	20.7	12.3	—
Carbon (Diamond)	9.12	13.2	—
Copper	22.6	5.6	—
Gold	23.7	5.19	—
Iron (α)	37.12	6.17	—
Nickel	17.0	29.5	—
Silver	21.3	8.5	1.5
Silicon	23.9	2.5	−4.1
Tungsten	24.0	3.2	—
Alumina (Al_2O_3)	21.9	3.7	—
Silica (SiO_2)	15.6	11.4	—

[a] Emperical expression for the heat capacity for this table is

$$C_P(T) = a + bT + \frac{c}{T^2}$$

Values are reported in Joules per mole.

Source: Brandes, E.A., Ed., *Smithells Metals Reference Book*, 6th ed., Butterworths, London, 1983. (Compare to Appendix D.)

two experimental variables are related to each other (see Equation 4.48); given information about α and β for the system, if one heat capacity is known the other may be computed.

It will be further demonstrated later in this chapter that if α, β and C_P are known for any simple system, i.e., one for which $\delta W' = 0$, then changes in *all* of the state functions may be computed for any arbitrary process through which that system may be taken. No additional information is required. Thus, these experimental variables are at the center of the strategy for solving practical problems that require thermodynamics. The ranges of values that these quantities may have in solids, liquids, and gases are summarized in Tables 4.1 to 4.3.

4.1.3 COEFFICIENT RELATIONS

Coefficient relations derive from the mathematical properties of functions of several variables. Since the state variables of thermodynamics have been shown to satisfy the properties of ordinary mathematical functions, these coefficient relations apply to equations among all of the state variables that have been identified and defined.

Suppose that X, Y and Z are state variables, and that there exists a relationship among these variables such that, given values for two of them, say X and Y, the third,

Z, may be evaluated. Consequences of this kind of normal mathematical functional relationship, introduced in Chapter 2 Section 2.3, are reviewed here. Mathematical shorthand for this relation is

$$Z = Z(X, Y) \tag{4.16}$$

read "Z is a function of X and Y." In this statement of the relationship, X and Y are "independent variables" and Z is the "dependent variable" with a value that depends upon the values that may be independently assigned to X and Y. Figure 4.2 illustrates such a relation.

Assume that this function is well behaved, i.e., is smooth and continuous with continuous derivatives. An infinitesimal step in a process may be represented by independent changes in X and Y. Owing to the functional relation, Equation 4.16, the corresponding change in Z may be written:

$$dZ = M \, dX + N \, dY \tag{4.17}$$

where the operator "d" implies the familiar notion of a differential of a function in elementary calculus. The coefficients in this equation are certain explicit partial derivatives:

$$M = \left(\frac{\partial Z}{\partial X}\right)_Y \quad \text{and} \quad N = \left(\frac{\partial Z}{\partial Y}\right)_X \tag{4.18}$$

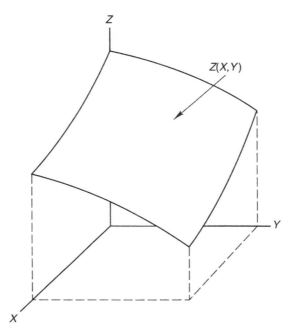

FIGURE 4.2 Geometric representation of the notion "Z is a function of X and Y."

The graphical meaning of these relations is evident in Figure 4.2. This equation is the formal basis for all coefficient relations. Given any thermodynamic relation among differentials of state functions it is possible, by inspection, to write down the coefficient relations. For example, the combined statement of the first and second laws, Equation 4.4, applied to processes for which $\delta W' = 0$, becomes

$$dU = T\,dS - P\,dV \tag{4.19}$$

which has the general form of Equation 4.17. The coefficients in this equation are evidently the partial derivatives:

$$T = \left(\frac{\partial U}{\partial S}\right)_V \qquad \text{and} \qquad -P = \left(\frac{\partial U}{\partial V}\right)_S \tag{4.20}$$

Application of this strategy to the other three combined statements derived from the definitions of the energy functions, Equation 4.6, Equation 4.8, and Equation 4.10, yields

$$T = \left(\frac{\partial H}{\partial S}\right)_P \qquad \text{and} \qquad V = \left(\frac{\partial H}{\partial P}\right)_S \tag{4.21}$$

$$-S = \left(\frac{\partial F}{\partial T}\right)_V \qquad \text{and} \qquad -P = \left(\frac{\partial F}{\partial V}\right)_T \tag{4.22}$$

$$-S = \left(\frac{\partial G}{\partial T}\right)_P \qquad \text{and} \qquad V = \left(\frac{\partial G}{\partial P}\right)_T \tag{4.23}$$

Since the letters in these relations represent thermodynamic quantities, these relations have physical substance.

In thermodynamic applications of this class of relations it is important to specify explicitly not only the variable in the numerator and denominator of each partial derivative but also *the variable that is being held constant*, indicated in the subscript outside the parentheses of the partial derivative. The two derivatives

$$\left(\frac{\partial H}{\partial T}\right)_P \qquad \text{and} \qquad \left(\frac{\partial H}{\partial T}\right)_V$$

are not equal to each other. Indeed, it can be shown that

$$\left(\frac{\partial H}{\partial T}\right)_V = \left(\frac{\partial H}{\partial T}\right)_P + \left(\frac{\partial H}{\partial P}\right)_T \left(\frac{\partial P}{\partial T}\right)_V$$

It is clearly absolutely essential to be explicit about all three variables required in the specification of a partial derivative in thermodynamics.

If the state of the system is a function of more than two variables $[Z = Z(X, Y, U, V, \ldots)]$ the differential generalizes to

$$dZ = M \, dX + N \, dY + P \, dU + R \, dV + \cdots \qquad (4.24)$$

with each coefficient corresponding to a partial derivative of Z. This straightforward generalization will be useful in later chapters where more complex systems are the subject of study.

4.1.4 MAXWELL RELATIONS

This category of thermodynamic relationships also derives from the mathematical properties of state functions. Consider again variables obeying the functional relationship in Equation 4.16 with differential given by Equation 4.17. Consider again the two partial derivatives

$$M = \left(\frac{\partial Z}{\partial X}\right)_Y \qquad \text{and} \qquad N = \left(\frac{\partial Z}{\partial Y}\right)_X$$

Consider the following derivatives:

$$\left(\frac{\partial M}{\partial Y}\right)_X = \left[\frac{\partial}{\partial Y}\left(\frac{\partial Z}{\partial X}\right)_Y\right]_X \qquad \text{and} \qquad \left(\frac{\partial N}{\partial X}\right)_Y = \left[\frac{\partial}{\partial X}\left(\frac{\partial Z}{\partial Y}\right)_X\right]_Y$$

Substitute the coefficient relations for M and N given in Equations 4.18. Note that these two quantities are equal except for the order of differentiation. Thus

$$\left(\frac{\partial M}{\partial Y}\right)_X = \left(\frac{\partial N}{\partial X}\right)_Y \qquad (4.25)$$

forms the mathematical basis for the Maxwell relations in thermodynamics.

In multivariable calculus, relations similar to this are called the Cauchy–Riemann conditions. They are the basis for determining whether a given differential form, e.g., $M \, dX + N \, dY$, is in fact an exact differential, i.e., whether there exists a function $Z(X, Y)$ of which $(M \, dX + N \, dY)$ is the differential. Evidently this will be so if and only if Equation 4.25 is satisfied. In thermodynamics, the establishment that Z, X, and Y are state functions guarantees that the form $(M \, dX + N \, dY)$ is always an exact differential and thus the Maxwell relations apply.

Maxwell relations for any thermodynamic differential form may be written by inspection. As examples, application of Equation 4.25 to the combined statements derived earlier yield:

From $dU = T \, dS - P dV$,

$$-\left(\frac{\partial P}{\partial S}\right)_V = \left(\frac{\partial T}{\partial V}\right)_S$$

From $dH = T\,dS + V\,dP$,

$$\left(\frac{\partial T}{\partial P}\right)_S = \left(\frac{\partial V}{\partial S}\right)_P$$

From $dF = -S\,dT - P\,dV$,

$$-\left(\frac{\partial S}{\partial V}\right)_T = -\left(\frac{\partial P}{\partial T}\right)_V$$

From $dG = -S\,dT + V\,dP$,

$$-\left(\frac{\partial S}{\partial P}\right)_T = \left(\frac{\partial V}{\partial T}\right)_P \qquad (4.26)$$

These relations apply to any system at any condition. These are examples of Maxwell relations, not an exhaustive list. Equations of this form hold for every functional relationship that can be constructed joining any dependent state function to two independent ones.

To illustrate the somewhat unexpected connections between properties, which result from these relations, consider the last in the list, labeled Equation 4.26. From the definition of the coefficient of thermal expansion, Equation 4.11, it follows that:

$$\left(\frac{\partial V}{\partial T}\right)_P = V\alpha$$

Thus, from the last of the Maxwell relations, Equation 4.26,

$$\left(\frac{\partial S}{\partial P}\right)_T = -V\alpha \qquad (4.27)$$

This result demonstrates that the pressure dependence of the entropy of a system is determined by its coefficient of thermal expansion, a result that is by no means intuitively evident.

The Maxwell relations are also readily generalized to systems that require more than two independent variables. Recall Equation 4.24:

$$dZ = M\,dX + N\,dY + P\,dU + R\,dV + \cdots \qquad (4.24)$$

Maxwell relations must hold for each pair of terms in this equation, since each coefficient represents a differential of the state function, Z. Thus, for example,

$$\left(\frac{\partial R}{\partial X}\right)_{Y,U,V,\ldots} = \left(\frac{\partial M}{\partial V}\right)_{X,Y,U,\ldots}; \qquad \left(\frac{\partial R}{\partial Y}\right)_{X,U,V,\ldots} = \left(\frac{\partial N}{\partial V}\right)_{X,Y,U,\ldots} : \text{etc.} \quad (4.28)$$

EXAMPLE 4.1
It will be shown in Section 4.2 that the form of the equation relating entropy to
temperature and pressure is

$$S = S(T, P) \qquad dS = \frac{C_P}{T} dT - V\alpha\, dP$$

Write the coefficient and Maxwell relations for this equation.
The coefficients of dT and dP are, respectively,

$$\left(\frac{\partial S}{\partial T}\right)_P = \frac{C_P}{T} \qquad \left(\frac{\partial S}{\partial P}\right)_T = -V\alpha$$

The Maxwell relation is

$$\left(\frac{\partial\left(\frac{C_P}{T}\right)}{\partial P}\right)_T = \left(\frac{\partial(-V\alpha)}{\partial T}\right)_P$$

Evidently how the heat capacity varies with pressure is related to the coefficient of
thermal expansion for a system.

Two other relationships among partial derivatives that may be defined between
the variables X, Y, and Z may be derived with the aid of results presented in this
section. The *reciprocal relation*:

$$\left(\frac{\partial Z}{\partial X}\right)_Y = \frac{1}{\left(\frac{\partial X}{\partial Z}\right)_Y} \qquad\qquad (4.29)$$

and the *ratio relation* (see Problem 4.2):

$$\left(\frac{\partial Z}{\partial X}\right)_Y\left(\frac{\partial X}{\partial Y}\right)_Z\left(\frac{\partial Y}{\partial Z}\right)_X = -1 \qquad\qquad (4.30)$$

These equations are useful in manipulating partial derivatives in the process of
developing relations between thermodynamic state functions.

As a further step in the development of the structure of thermodynamics, four
classes of relationships between state functions have been presented. The laws of
thermodynamics provide the physical basis for all of these relations. Definitions
introduce energy functions that are convenient in specific applications and a set of

experimental variables that form the core of information needed to solve practical problems. Coefficient and Maxwell relations are consequences of the mathematical properties of differentials of state functions.

The basis for deriving the fifth category of relations, conditions for equilibrium, is presented in Chapter 5.

4.2 GENERAL STRATEGY FOR DERIVING THERMODYNAMIC RELATIONS

The relationships presented and classified in the previous section provide the basis for developing completely general equations between arbitrarily selected thermodynamic state functions. Specifically, for a system which does not do nonmechanical work, i.e., for any system for which $\delta W' = 0$, it is possible to develop an equation relating any function of the state of that system (as a dependent variable) to any pair of other state functions (as independent variables). As was pointed out at the beginning of this chapter, this is the key link in the chain of reasoning needed to translate and solve practical problems that require thermodynamics.

This section presents a general procedure for deriving such relations. The procedure is simple, straightforward, and mathematically rigorous. Since there is no mathematical sleight-of-hand involved, and because the sequence of steps to be followed is clearly defined, it should become possible to solve the most general problem with rigor and confidence.

Extension of this procedure to systems which are more general, for which $\delta W'$ is not zero, is straightforward. Applications to this kind of system, which require the development of more of the apparatus of thermodynamics, are presented in later chapters.

An exhaustive list of the state functions so far defined is given in Table 4.4; others will be introduced as more complex classes of systems are encountered and the thermodynamic apparatus expands.

TABLE 4.4
State Functions Defined in Thermodynamics

Equation of State Variables	
Temperature	T
Pressure	P
Volume	V
Energy Functions	
Internal Energy	U
Enthalpy	H
Helmholtz Free Energy	F
Gibbs Free Energy	G
Entropy	S

The general procedure begins by choosing temperature and pressure as independent variables, since these are most frequently encountered in practical problems, and deriving relations for all of the other state functions in Table 4.4 as functions of T and P. It is shown that once equations are derived for volume $[V = V(T,P)]$ and entropy $[S = S(T,P)]$, expressions for the four energy functions readily follow. Once all of the functions have been expressed in terms of (T,P), conversion to other pairs of independent variables, e.g., express $H = H(V, S)$, requires no mathematics beyond algebra.

4.2.1 ENTROPY AND VOLUME RELATIONS TO T AND P

For every simple substance there exists a relation between its volume, temperature, and pressure, which may be written formally as:

$$V = V(T, P) \tag{4.31}$$

Since these are all state functions, the corresponding differential relation, analogous to the generic Equation 4.17, is

$$dV = M\, dT + N\, dP$$

The coefficient relations for M and N in this equation give:

$$dV = \left(\frac{\partial V}{\partial T}\right)_P dT + \left(\frac{\partial V}{\partial P}\right)_T dP$$

The partial derivatives in this equation may be expressed in terms of the experimental variables α and β defined in Equation 4.11 and Equation 4.12.

$$\left(\frac{\partial V}{\partial T}\right)_P = V\alpha \quad\quad and \quad\quad \left(\frac{\partial V}{\partial P}\right)_T = -V\beta$$

Thus, without any restrictions on the nature of the substance that makes up the system,

$$V = V(T, P) \quad\quad dV = V\alpha\, dT - V\beta\, dP \tag{4.32}$$

This equation may be regarded as a differential form of an equation of state. If experimental information is available for the substance in the system for α and β as functions of temperature and pressure, this equation may be integrated (at the very least numerically) to obtain a complete description of the variation of the volume of the system with its temperature and pressure.

Consider now the analogous problem for the entropy function:

$$S = S(T, P) \tag{4.33}$$

The corresponding differential relation is

$$dS = M \, dT + N \, dP$$

The coefficients in this equation may be written as partial derivatives so that

$$dS = \left(\frac{\partial S}{\partial T}\right)_P dT + \left(\frac{\partial S}{\partial P}\right)_T dP \tag{4.34}$$

The second coefficient, N, has already been evaluated in the consideration of Maxwell relations. Recall Equation 4.26:

$$N = \left(\frac{\partial S}{\partial P}\right)_T = -\left(\frac{\partial V}{\partial T}\right)_P = -V\alpha \tag{4.35}$$

Evaluation of M requires a thermodynamic argument based on the definition of heat capacity.

For any reversible process, the second law connects the heat absorbed with the change in entropy:

$$\delta Q_{\text{rev}} = T \, dS \tag{3.10}$$

With Equation 4.34

$$\delta Q_{\text{rev}} = T\left[\left(\frac{\partial S}{\partial T}\right)_P dT + \left(\frac{\partial S}{\partial P}\right)_T dP\right]$$

The heat absorbed for a constant pressure process, for which $dP = 0$, may be written

$$\delta Q_{\text{rev},P} = T\left(\frac{\partial S}{\partial T}\right)_P dT_P \tag{4.36}$$

Measurements of the reversible heat absorbed for a constant pressure process are the basis for the experimental evaluation of C_P, the heat capacity at constant pressure,

$$\delta Q_{\text{rev},P} \equiv C_P dT_P \tag{4.13}$$

Comparison of these two expressions for heat absorbed at constant pressure demonstrates that the coefficients of dT_P must be the same quantity in both equations. It is concluded that

$$C_P = T\left(\frac{\partial S}{\partial T}\right)_P$$

or,

$$\left(\frac{\partial S}{\partial T}\right)_P = \frac{C_P}{T} \tag{4.37}$$

Thus, a general differential relation for the variation of entropy with temperature and pressure for any simple system is

$$S = S(T, P) \qquad dS = \frac{C_P}{T} dT - V\alpha\, dP \tag{4.38}$$

If C_P and α are known as functions of temperature and pressure for the system under study, the entropy of the system may be computed by integration of this equation as a function of temperature and pressure. There are no restrictions on the validity of this equation other than the limitations imposed by the simplification that $\delta W' = 0$ for all processes considered in this chapter.

Although Equation 4.32 and Equation 4.38 are sufficient to implement the general procedure shortly to be developed, it is also useful to introduce relations between state functions and the experimental variable, C_V. This brief development is introduced here because it parallels that just presented for C_P. Recall the definition of the heat capacity at constant volume, Equation 4.15:

$$\delta Q_{\text{rev},V} = C_V dT_V \tag{4.15}$$

Variables about which information is given in the process used to determine C_V are the temperature (which is measured) and the volume (which is held constant). The heat absorbed is related to the entropy change of the system. This suggests that the state function $S = S(T, V)$ should be examined. The differential form of this function is

$$dS = \left(\frac{\partial S}{\partial T}\right)_V dT + \left(\frac{\partial S}{\partial V}\right)_T dV$$

For a constant volume process, such as is used in determining C_V, $dV = 0$, and the change in entropy is

$$dS_V = \left(\frac{\partial S}{\partial T}\right)_V dT_V$$

Apply again the second law as embodied in Equation 3.10:

$$\delta Q_{\text{rev},V} = T\, dS_V = T\left(\frac{\partial S}{\partial T}\right)_V dS_V \tag{4.39}$$

Comparison of coefficients of Equation 4.39 and Equation 4.15 gives

$$\left(\frac{\partial S}{\partial T}\right)_V = \frac{C_V}{T} \tag{4.40}$$

Thus, C_V may also be expressed in terms of the state functions S and T. Comparison of Equation 4.40 and Equation 4.37 again demonstrates the necessity for specifying the variables that are held constant in taking each derivative. Clearly, since C_P and C_V are not equal to one another, the two entropy derivatives to which they are related must also be different.

4.2.2 ENERGY FUNCTIONS EXPRESSED IN TERMS OF T AND P

The four combined statements of the first and second laws derived in Section 4.1.2 follow directly from the definitions of each of the energy functions and the original version of this statement expressed in terms of dU. The results are summarized here:

$$U = U(S, V); \qquad dU = T\,dS - P\,dV \tag{4.4}$$

$$H = (S, P); \qquad dH = T\,dS + V\,dP \tag{4.6}$$

$$F = F(T, V); \qquad dF = -S\,dT - P\,dV \tag{4.8}$$

$$G = G(T, P); \qquad dG = -S\,dT + V\,dP \tag{4.10}$$

Equation 4.32 and Equation 4.38 provide general differential expressions for dV and dS for any system in terms of dT and dP. Conversion of these equations for the energy functions into expressions having T and P as independent variables merely requires substitution for dV or dS where they appear, followed by algebraic simplification. To illustrate, substitute for dS and dV in Equation 4.4:

$$dU = T\left[\frac{C_P}{T}dT - V\alpha\,dP\right] - P[V\alpha\,dT - V\beta\,dP]$$

Collect like terms:

$$U = (T, P); \qquad dU = (C_P - PV\alpha)dT + V(P\beta - T\alpha)dP \tag{4.41}$$

To convert Equation 4.6 for the enthalpy to a function of T and P, substitute for dS,

$$dH = T\left[\frac{C_P}{T}dT - V\alpha\,dP\right] + V\,dP$$

and collect terms:

$$H = H(T, P); \qquad dH = C_P dT + V(1 - T\alpha)dP \tag{4.42}$$

Equation 4.8 is similarly easily converted to a function of T and P by substituting for dV:

$$dF = -S\,dT - P(V\alpha\,dT - V\beta\,dP)$$

Collect terms:

$$F = F(T,P); \qquad dF = -(S + PV\alpha)dT + PV\beta\,dP \qquad (4.43)$$

The Gibbs free energy function is already expressed in terms of the independent variables T and P in the version originally derived, Equation 4.10.

This collection of all of the state functions expressed in terms of T and P just derived is compiled for quick reference in Table 4.5. Coefficient relations apply to each term in these equations; e.g., in Equation 4.42,

$$\left(\frac{\partial H}{\partial T}\right)_P = C_P \qquad \text{and} \qquad \left(\frac{\partial H}{\partial P}\right)_T = V(1 - T\alpha) \qquad (4.44)$$

as do Maxwell relations, e.g.,

$$\left(\frac{\partial C_P}{\partial P}\right)_T = \left(\frac{\partial[V(1 - T\alpha)]}{\partial T}\right)_P \qquad (4.45)$$

The coefficients in these equations contain the following factors:

T, P (which are the independent variables specified in any application).
Experimental variables, α, β, C_P, assumed to be available in tables or the literature.
S and V, which can be evaluated as functions of T and P, given α, β and C_P.

Thus, if the initial and final states for a process traversed by a system are specified by their temperature and pressure, knowledge of the values of α, β and C_P permits calculation of all of the changes in state functions accompanying that

---

TABLE 4.5
Thermodynamic State Functions Expressed in Terms of the Independent Variables Temperature and Pressure

$V = V(T,P)$	$dV = V\alpha\,dT - V\beta\,dP$	Equation 4.32
$S = S(T,P)$	$dS = \dfrac{C_P}{T}dT - V\alpha\,dP$	Equation 4.38
$U = U(T,P)$	$dU = (C_P - PV\alpha)dT + V(p\beta - T\alpha)dP$	Equation 4.41
$H = H(T,P)$	$dH = C_P dT + V(1 - T\alpha)dP$	Equation 4.42
$F = F(T,P)$	$dF = -(S + PV\alpha)dT + PV\beta\,dP$	Equation 4.43
$G = G(T,P)$	$dG = -S\,dT + V\,dP$	Equation 4.10

process. The power of these equations is demonstrated by the fact that they apply to any system taken through any process.

4.2.3 The General Procedure

Translation of a statement of a practical problem into the language of thermodynamics requires the identification of the independent variables (about which information is given), the dependent variable (about which information is sought), and the development of the relationship that connects these variables. The derivation of such relationships is at the heart of the problem. The treatment presented in the previous section yields an exhaustive list of all such relations for any problem in which the independent variables are temperature and pressure (Table 4.5). The procedure for solving the more general problem, in which any pair of state functions (X,Y) may be the independent variables and any third state function Z the variable sought, is laid out in this section. This part of the problem is considered solved (the relation derived) when the coefficients, M and N, have been evaluated in terms of the experimental variables, α, β, C_P (and/or C_V) and the state functions T, P, S, and V, which can always be computed therefrom.

The step-by-step procedure for deriving such relationships is straightforward once the equations in Table 4.5 have been obtained. It requires no higher mathematics than algebra and, indeed, no thermodynamic thinking is involved, since the manipulations are purely mathematical. There are seven steps in the procedure:

1. Identify the variables: $Z = Z(X,Y)$
2. Write the differential form: $dZ = M\,dX + N\,dY$
3. Use Table 4.5 to express dX and dY in terms of the variables dT and dP, (this is always possible, since Table 4.5 provides an exhaustive list for all of the state functions we have defined):

$$dZ = M[X_T dT + X_P dP] + N[Y_T dT + Y_P dP]$$

   where the coefficients, X_T, X_P, Y_T, Y_P are the coefficients of dT and dP in the expressions for dX and dY in Table 4.5

4. Collect terms:

$$dZ = [M{\cdot}X_T + N{\cdot}Y_T]dT + [M{\cdot}X_P + N{\cdot}Y_P]dP$$

5. Obtain $Z = Z(T,P)$ from Table 4.5:

$$dZ = Z_T dT + Z_P dP$$

6. The equations listed under steps 4 and 5 are alternate expressions for $Z = Z(T,P)$. Therefore, the coefficients of dT and dP in both equations

are, respectively,

$$\left(\frac{\partial Z}{\partial T_P}\right) \quad \text{and} \quad \left(\frac{\partial Z}{\partial P}\right)_T$$

Equate like terms:

$$M \cdot X_T + N \cdot Y_T = Z_T \tag{4.46a}$$

$$M \cdot X_P + N \cdot Y_P = Z_P \tag{4.46b}$$

7. This is a pair of linear simultaneous equations in M and N and may always be solved for these variables either by elimination of one variable then solution for the other, or, in the worst case, by determinants. The resulting expressions for M and N will certainly be expressed in terms of the experimental variables since all of the terms in Table 4.5 (from which X_T, Y_T, Z_T, X_P, Y_P and Z_P are obtained) are expressed in terms of these variables.

This approach to developing relations among state variables has several advantages over alternative methods. It is mathematically rigorous so that it may be applied with confidence. It requires conscious identification of the independent and dependent variables so that the initial statement of the problem must be clear. The equation that results has general applicability and may be used in any situation, avoiding confusion about which relations are special (e.g., are limited in validity to systems composed of ideal gases or to specific processes). Finally, the path from statement of the problem to its solution is straightforward, eliminating potential confusion about the manipulation of the quantities involved, further fostering confidence in the final result.

Since the mathematics involved in this methodology is only algebra, it is easy to make simple mistakes in computation in following the procedure. Once M and N are evaluated for a given problem, it is strongly recommended that a check of their units be carried out. Most algebraic errors will be clearly exposed with this tactic since the units obtained for M and N in step 7 must agree with the units they are required to have in writing step 2. In this comparison, it is only necessary to represent each thermodynamic quantity in terms of generic units involving only combinations of P, V and T. It will be found to be convenient to represent all measures of energy as having units (PV), then entropy has units (PV/T), heat capacities also have units (PV/T), α has units $(1/T)$, β has units $(1/P)$.

In the application of the result to the solution of a numerical problem for a specified system, the actual units that are used will be determined primarily by the units that are used in the database from which the experimental information was obtained. Thus, energy functions may be expressed in joules, calories, ergs, liter-atmospheres, Newton-meters, BTU and the like. Such alternative representations of energy may be interconverted at will using the conversion factors reported in

Appendix A. Similarly, heat capacities have units of (energy/mol K) and may be found carrying any of the energy units displayed above. The coefficients of expansion and compressibility may also appear in a variety of units (α as $1/°F$, e.g., and β as $1\ psi^{-1}$ or $1\ MPa^{-1}$). All correspond to the generic units listed in the last paragraph.

The general procedure has been stated in the abstract; its application will now be made more concrete with a number of examples. The first set of examples (4.2 to 4.6) illustrate the step-by-step procedure to be followed in order to derive the differential expressions that relate a specific dependent variable to a pair of independent variables. Later examples will apply this procedure as a module in the overall problem that begins with a verbal statement of the situation and ends with a numerical result.

EXAMPLE 4.2

Relate the entropy of a system to its temperature and volume.

1. Identify the variables: $S = S(T, V)$
2. Write the differential form: $dS = M\ dT + N\ dV$
3. Convert dV, using Table 4.5:

$$dS = M\ dT + N(V\alpha\ dT - V\beta\ dP)$$

4. Collect terms: $dS = (M + NV\alpha)dT - NV\beta\ dP$
5. Obtain $S = S(T, P)$ from Table 4.5:

$$dS = [C_P/T]dT - V\alpha\ dP$$

6. Compare coefficients: $M + NV\alpha = C_P/T$

$$-NV\beta = -V\alpha$$

7. Solve this pair of equations for M and N:

$$N = \left(\frac{\partial S}{\partial V}\right)_T = \frac{\alpha}{\beta}$$

$$M = \left(\frac{\partial S}{\partial T}\right)_V = \frac{1}{T}\left(C_P - \frac{TV\alpha^2}{\beta}\right)$$

Check the units. From step 2, $M\ dT$ must have the same units as dS, i.e., (PV/T). Thus M has units (PV/T^2). This checks with the units for the value of M derived in step 7. In particular, for the second term in the brackets the units are $(TV(1/T)^2/(1/P) = PV/T)$; multiplication by the factor $(1/T)$ outside the brackets yields the correct units. $N\ dP$ also has units of (PV/T). Thus, N has units (P/T). Substituting the units for α and β gives for $N\ [(1/T)/(1/P)] = P/T$.

Thus, entropy as a function of temperature and volume is given by

$$dS = \frac{1}{T}\left(C_P - \frac{TV\alpha^2}{\beta}\right)dT + \frac{\alpha}{\beta}dV. \qquad (4.47)$$

Incidentally, this result may be used to derive the general relation that exists between C_P and C_V mentioned in Section 4.1.2. By inspection, the coefficient of dT in this equation is

$$\left(\frac{\partial S}{\partial T}\right)_V = \frac{1}{T}\left(C_P - \frac{TV\alpha^2}{\beta}\right)$$

It was shown [Equation 4.40] that this coefficient is related to C_V

$$\left(\frac{\partial S}{\partial T}\right)_V = \frac{C_V}{T}$$

Thus,

$$C_V = C_P - \frac{TV\alpha^2}{\beta} \tag{4.48}$$

Accordingly, if either heat capacity has been measured for a system and α and β are known, the other may be calculated. Since all factors in the second term on the right-hand side are positive C_V is always smaller than C_P.

EXAMPLE 4.3
Relate the entropy of a system to its pressure and volume.

1. Identify the variables: $S = S(P, V)$
2. Write the differential form: d$S = M$ d$P + N$ dV
3. Convert dV, using Table 4.5:

$$dS = M\ dP + N(V\alpha\ dT - V\beta\ dP)$$

4. Collect terms:

$$dS = NV\alpha\ dT + (M - NV\beta)dP$$

5. Obtain $S = S(T, P)$ from Table 4.5:

$$dS = [C_P/T]dT - V\alpha\ dP$$

6. Compare coefficients: $NV\alpha = C_P/T$

$$M - NV\beta = -V\alpha$$

7. Solve this pair of equations for M and N:

$$N = \left(\frac{\partial S}{\partial V}\right)_P = \frac{C_P}{TV\alpha}$$

$$M = \left(\frac{\partial S}{\partial P}\right)_V = \left(\frac{C_P\beta}{T\alpha} - V\alpha\right)$$

Check the units. For M : $[V/T] = [(PV/T)(1/P)/(T/T) - V(1/T)]$
For N : $[P/T] = [(PV/T)/(TV/T)] = P/T$
The units check. Thus, the required relation is

$$dS = \left(\frac{C_P \beta}{T\alpha} - V\alpha\right)dP + \frac{C_P}{TV\alpha}dV \qquad (4.49)$$

EXAMPLE 4.4

Find the relationship that would be needed to compute the change in Helmholtz free energy when the initial and final states are specified by their pressure and volume.

1. Identify the variables: $F = F(P, V)$
2. Write the differential form: $dF = M\,dP + N\,dV$
3. Convert dV, using Table 4.5: $dF = M\,dP + N(V\alpha\,dT - V\beta\,dP)$
4. Collect terms: $dF = NV\alpha\,dT + (M - NV\beta)dP$
5. Obtain $F = F(T, P)$ from Table 4.5:

$$dF = -(S + PV\alpha)\,dT + PV\beta\,dP$$

6. Compare coefficients: $NV\alpha = -(S + PV\alpha)$.

$$M - NV\beta = PV\beta$$

7. Solve this pair of equations for M and N:

$$N = \left(\frac{\partial F}{\partial V}\right)_P = -\left(\frac{S}{V\alpha} + P\right)$$

$$M = \left(\frac{\partial F}{\partial P}\right)_V = -\frac{S\beta}{\alpha}$$

Check the units. For M : $V = (PV/T)(1/P)/(1/T) = V$
For N : $P = [(PV/T)/(V/T) + P] = P$.
The units check. Thus, the required relation is

$$dF = -\frac{S\beta}{\alpha}dP - \left(\frac{S}{V\alpha} + P\right)dV \qquad (4.50)$$

EXAMPLE 4.5

Express the change in enthalpy as a function of volume and entropy.

1. Identify the variables: $H = H(S, V)$
2. Write the differential form: $dH = M\,dS + N\,dV$

3. Convert both dS and dV, using Table 4.5:

$$dH = M[(C_P/T)dT - V\alpha \, dP) + N(V\alpha \, dT - V\beta \, dP)$$

4. Collect terms:

$$dH = [M(C_P/T) + NV\alpha]dT - [MV\alpha + NV\beta]dP$$

5. Obtain $H = H(T, P)$ from Table 4.5:

$$dH = C_P dT + V(1 - T\alpha)dP$$

6. Compare coefficients: $MC_P/T + NV\alpha = C_P$

$$-MV\alpha - NV\beta = V(1 - T\alpha)$$

7. Solve this pair of equations for M and N:
   Solution of this pair of equations is a nontrivial exercise in algebra but it is only algebra.

Determinants might be used or substitution. The simplest approach in this case follows: Multiply the top equation by β and the bottom by α:

$$MC_P\beta/T + NV\alpha\beta = C_P\beta - MV\alpha^2$$
$$- NV\alpha\beta = V\alpha(1 - T\alpha)$$

Add the equations: the term involving N drops out.

$$M[C_P\beta/T - V\alpha^2] = C_P\beta + V\alpha(1 - T\alpha)$$

This result may be simplified by applying Equation 4.48. (It is usually true that if one of the independent variables is volume, expression of the coefficients in terms of C_V rather than C_P will give a result that is algebraically simpler.)

$$M\frac{\beta}{T}\left(C_P - \frac{TV\alpha^2}{\beta}\right) = \beta\left(C_P - \frac{TV\alpha^2}{\beta} + \frac{V\alpha}{\beta}\right)$$

or

$$M\frac{\beta}{T}C_V = \beta\left(C_V + \frac{V\alpha}{\beta}\right)$$

Thus,

$$M = \left(\frac{\partial H}{\partial S}\right)_V = T\left(1 + \frac{V\alpha}{\beta C_V}\right)$$

Substitute this result for M into the original version of the first equation

$$T\left(1 + \frac{V\alpha}{\beta C_V}\right)\frac{C_P}{T} + NV\alpha = C_P$$

Solve for N:

$$N = \left(\frac{\partial H}{\partial V}\right)_S = -\frac{C_P}{\beta C_V}$$

Check the units. For $M: T = T\{1 + V(1/T)/[(PV/T)(1/P)]\}$
For $N: P = (PV/T)/[(PV/T)(1/P)]$
Thus, the differential expression for $H = H(S,V)$ is

$$dH = T\left(1 + \frac{V\alpha}{C_V\beta}\right)dS - \frac{C_P}{C_V\beta}dV \tag{4.51}$$

EXAMPLE 4.6

Derive an expression for the increase in temperature for a process in which the volume of the system is changed at constant entropy. The variable about which information is sought is T; information is given about S and V.

1. Identify the variables: $T = T(S,V)$.

   This case differs from those presented above because the dependent variable, T, is one of the variables that we have chosen as an independent variable in developing the arsenal of relations contained in Table 4.5. In such cases rearrange the governing equation so that T is one of the *dependent* variables:

$$S = S(T,V)$$

2. Find the differential relation:

$$dS = M\,dT + N\,dV$$

3. Evaluate M and N. Then solve the resulting equation for dT. The relation $S = S(T,V)$ was derived in Example 4.2 Combine Equation 4.47 and Equation 4.48:

$$dS = \frac{C_V}{T}dT + \frac{\alpha}{\beta}dV \tag{4.52}$$

4. Rearrange to solve for dT:

$$dT = \frac{T}{C_V}dS - \frac{T\alpha}{C_V\beta}dV \tag{4.53}$$

From the statement of the problem for the process considered $dS = 0$. Thus, for this *isentropic* process,

$$dT_S = -\frac{T\alpha}{C_V\beta}dV_S$$

If the experimental variables are known for the system integration of this equation from the initial to the final volume will yield the change in temperature.

4.2.4 APPLICATION TO AN IDEAL GAS

The gaseous state of matter is frequently encountered in systems of practical importance. Interaction with gases may lead to degradation through oxidation or hot corrosion. The vapor state may provide a medium for controlled addition of material to a system, as in nitriding or carburization in heat treatment or thin film deposition in processing microelectronic devices. Some chemical reactions may be more closely controlled in the vapor state. The apparatus for the description of mixtures of gases, equilibrium between solids or liquids and their vapors and reactions in gases, is developed in later chapters. Applications in this section are limited to *unary gases*.

The *ideal gas model* provides a description of these gaseous systems that is adequate for many practical purposes. The *equation of state* for an ideal gas, i.e., an algebraic equation that relates volume to temperature and pressure, was established through experimental observation of the behavior of gases by the early 19th century. This most important equation in the description of the behavior of gases is

$$PV = nRT \tag{4.54}$$

where P, V and T are as defined previously. R is the gas constant, determined experimentally to be:

$$R = 8.314 \text{ (J/mol K)} = 1.987 \text{ (Cal/mol K)} = 0.08206 \text{ (l-atm/mol K)}$$

$$= 82.06 \text{ (cc-atm/mol K)} \tag{4.55}$$

and n is the number of moles of the gas in the system. If this equation is used to describe a system containing 1 mol of substance, then $n = 1$ and

$$PV = RT \tag{4.56}$$

where V may be interpreted as the *molar volume* of the gas (volume occupied by 1 mol).

The experimental variables may be computed for an ideal gas as follows. Apply the definition of the volume coefficient of thermal expansion, Equation 4.11,

$$\alpha \text{ (ideal gas)} = \frac{1}{V}\left(\frac{\partial V}{\partial T}\right)_P = \frac{1}{R\frac{T}{P}}\left[\frac{\partial}{\partial T}\left(R\frac{T}{P}\right)\right]_P = \left(\frac{P}{RT}\right)\left(\frac{R}{P}\right) = \frac{1}{T} \quad (4.57)$$

The coefficient of compressibility, Equation 4.12,

$$\beta(\text{ideal gas}) = -\frac{1}{V}\left(\frac{\partial V}{\partial P}\right)_T = -\left(\frac{1}{\frac{RT}{P}}\right)\left[\frac{\partial}{\partial P}\left(\frac{RT}{P}\right)\right]_T = -\frac{P}{RT}\left(-\frac{RT}{P^2}\right) = \frac{1}{P} \quad (4.58)$$

The kinetic theory of gases demonstrates that the heat capacity of an ideal gas is independent of temperature and pressure, but depends upon the number of atoms and their configuration in each gas molecule (see Chapter 6). The relation between C_P and C_V expressed for a general system in Equation 4.48 is particularly simple for an ideal gas since the factor

$$\frac{TV\alpha^2}{\beta} = T\left(R\frac{T}{P}\right)\left(\frac{1}{T}\right)^2\left(\frac{1}{\frac{1}{P}}\right) = R$$

Thus, for an ideal gas,

$$C_V = C_P - R \quad (4.59)$$

For a *monatomic* gas (each gas molecule is just a single atom) it is found that

$$C_V(\text{monatomic ideal gas}) = \frac{3}{2}R \quad (4.60)$$

Note that C_V has the same units as R. Insert this value into Equation 4.59 to obtain the corresponding value for C_P:

$$C_P(\text{monatomic ideal gas}) = \frac{3}{2}R + R = \frac{5}{2}R \quad (4.61)$$

An important simplification in the thermodynamic behavior of an ideal gas may be obtained by considering the relation between internal energy and temperature and pressure, $U = U(T,P)$. From Table 4.5,

$$U = U(T,P); \quad dU = (C_P - PV\alpha)dT + V(P\beta - T\alpha)dP \quad (4.41)$$

Apply Equation 4.57 and Equation 4.58 to evaluate the coefficient of dP for an ideal gas:

$$V(P\beta - T\alpha) = V\left[P\left(\frac{1}{P}\right) - T\left(\frac{1}{T}\right)\right] = 0$$

The coefficient of dT may be simplified by evaluating α:

$$C_P - PV\alpha = C_P - PV\left(\frac{1}{T}\right) = C_P - R = C_V$$

Thus, for an ideal gas, Equation 4.41 becomes

$$dU(\text{ideal gas}) = C_V dT + 0dP = C_V dT$$

For any finite change in state from an initial condition given by (T_1, P_1) to a final state (T_2, P_2), the change in internal energy may be obtained by integrating this equation. This integration is simplified because C_V is a constant for an ideal gas. Thus, for any change in state of a monatomic ideal gas,

$$\Delta U(\text{ideal gas}) = C_V(T_2 - T_1) = \frac{3}{2}R(T_2 - T_1) \tag{4.62}$$

It is concluded that *the internal energy of an ideal gas is a function only of its temperature*. This results from the observation that the coefficient of dP in Equation 4.41 is 0; this is true only for ideal gases and for no other substances. The change in internal energy of an ideal gas for any process can, without exception, be computed from Equation 4.62.

A similar result may be obtained for the enthalpy function. Recall Equation 4.42 from Table 4.5.

$$H = H(T, P); \qquad dH = C_P dT + V(1 - T\alpha)dP \tag{4.42}$$

For an ideal gas, the coefficient of dP is found to be

$$V(1 - T\alpha) = V\left[1 - T\left(\frac{1}{T}\right)\right] = 0$$

and Equation 4.42 becomes

$$dH(\text{ideal gas}) = C_P dT + 0dP = C_P dT$$

Integration gives, for a finite change in state:

$$\Delta H(\text{ideal gas}) = C_P(T_2 - T_1) = \frac{5}{2}R(T_2 - T_1) \tag{4.63}$$

from which it is seen that the change in enthalpy of an ideal gas for any process depends only upon the initial and final temperature of the system and no other state function. This unusual result is valid only for an ideal gas because the value of the coefficient of dP is zero.

These results demonstrate that if an ideal gas is subject to any arbitrary process that takes it from state A to state B, the change in internal energy and enthalpy may be computed if the initial and final temperatures are either given or may be computed from the information given. Since U and H are state functions, these results hold for any process connecting the initial and final states, whether reversible or irreversible.

The next series of examples (Example 4.7 to Example 4.11) applies the general procedure and the experimental variables just evaluated to make thermodynamic calculations for processes described for a *monatomic ideal gas*.

EXAMPLE 4.7

Compute the change in entropy when 1 mol of an ideal gas initially at 298 K and one atmosphere pressure is isothermally compressed to 1000 atm.

Identify the variables: $S = S(T, P)$

The differential form for this relationship for a general substance is contained in Table 4.5 since the independent variables are T and P.

$$dS = \frac{C_P}{T} dT - V\alpha \, dP \tag{4.38}$$

For the problem at hand, d$T = 0$. The coefficient of dP may be evaluated for an idea gas by applying Equation 4.57 and Equation 4.58

$$dS_T = -V\alpha \, dP_T = -\left(R\frac{T}{P}\right)\left(\frac{1}{T}\right)dP_T = -\frac{R}{P}dP_T$$

This is the change in entropy for each infinitesimal step in the process. The change in entropy for the whole process is the *integral* of dS_T: from the initial to the final pressure:

$$\Delta S_T = \int_{P_1}^{P_2} \left(-\frac{R}{P}\right)dP = -R \ln P \Big|_{P_1}^{P_2} = -R[\ln P_2 - \ln P_1]$$

$$\Delta S_T(\text{ideal gas}) = -R \ln \frac{P_2}{P_1} \tag{4.64}$$

This expression gives the change in entropy when a monatomic ideal gas experiences a change in pressure at constant temperature. For the current example, $P_1 = 1$ atm and $P_2 = 1000$ atm; thus

$$\Delta S_T = -(8.314 \text{ J/mol K})\ln(1000/1) = -57.4 \text{ J/mol K}$$

The next example introduces the notion of an *adiabatic* process. Imagine a system enclosed in a thermally insulating jacket. Then whatever processes occur in the system, no heat flows across its boundary. Processes that occur in such a system for which $\delta Q = 0$, whether reversible or irreversible, are called *adiabatic*. If, in addition, the process considered is carried out reversibly then, since $\delta Q_{rev} = T \, dS$ and $\delta Q = 0$, it follows the $dS = 0$ for each infinitesimal step. Thus, adiabatic, reversible processes are isentropic, i.e., occur at constant entropy. (This conclusion is not valid for an irreversible adiabatic process since entropy is produced within the system.)

EXAMPLE 4.8

One mole of an ideal gas initially at temperature T_1 and occupying volume V_1 is compressed reversibly and adiabatically to a final volume V_2. Compute the final temperature, T_2 of the system.

Identify the variables: $T = T(S, V)$

This problem was previously treated for the general case in Example 4.6, yielding Equation 4.52:

$$dT = \frac{T}{C_V} dS - \frac{T\alpha}{C_V \beta} dV \qquad (4.52)$$

Since the process under study is *isentropic*, $dS = 0$ and

$$dT_S = -\frac{T\alpha}{C_V \beta} dV_S \qquad (4.65)$$

For an ideal gas $\alpha = 1/T$ and $\beta = 1/P$:

$$dT_S = -\frac{T\left(\dfrac{1}{T}\right)}{C_V\left(\dfrac{1}{P}\right)} dV_S = \frac{P}{C_V} dV_S$$

Eliminate P through the equation of state:

$$dT_S = -\frac{\left(\dfrac{RT}{V}\right)}{C_V} dV_S$$

Separate the variables and integrate from the initial to the final state:

$$\int_{T_1}^{T_2} \frac{dT}{T} = -\int_{V_1}^{V_2} \frac{R}{C_V} \frac{dV}{V}$$

$$\ln\left(\frac{T_2}{T_1}\right) = -\frac{R}{C_V} \ln\left(\frac{V_2}{V_1}\right) = \ln\left(\frac{V_2}{V_1}\right)^{\left(-\frac{R}{C_V}\right)}$$

Two logarithms are equal if their arguments are equal:

$$\frac{T_2}{T_1} = \left(\frac{V_2}{V_1}\right)^{\left(-\frac{R}{C_V}\right)} = \left(\frac{V_1}{V_2}\right)^{\left(\frac{R}{C_V}\right)}$$

Thus, the final temperature at the end of the reversibly adiabatic compression is

$$T_2 = T_1\left(\frac{V_1}{V_2}\right)^{\left(\frac{R}{C_V}\right)} \tag{4.66}$$

Note that if V_2 is larger than V_1, T_2 is smaller than T_1; if a gas expands adiabatically, its temperature drops. This phenomenon is the basis for some refrigeration cycles. Similarly, an adiabatic compression of the gas raises its temperature; thus a bicycle pump gets warm as the tire is inflated. Equation 4.65 shows that this phenomenon is general, for all substances. The minus sign on the right-hand side of this equation implies that the temperature change and volume change are in the opposite direction for any reversible adiabatic process.

Equation 4.66, the adiabatic equation for an ideal gas, may also be expressed in terms of relations of T to P or P to V simply be applying the equation of state.

$$T_2 = T_1\left[\frac{\left(\frac{RT_1}{P_1}\right)}{\left(\frac{RT_2}{P_2}\right)}\right]^{\left(\frac{R}{C_V}\right)}$$

Algebraic manipulation of this equation simplifies to:

$$T_2 = T_1\left(\frac{P_2}{P_1}\right)^{\left(\frac{R}{C_P}\right)} \tag{4.67}$$

which makes use of the previous result that $C_V = C_P - R$ for an ideal gas, Equation 4.59. A similar substitution and manipulation yields the adiabatic relationship expressed in terms of P and V:

$$P_2 = P_1\left(\frac{V_1}{V_2}\right)^{\left(\frac{C_P}{C_V}\right)} \tag{4.68}$$

These three relationships all describe the same process; the reversible adiabatic compression of an ideal gas; they simply use different variables to describe the path.

The "free expansion" of an ideal gas is a favorite homework problem in every thermodynamic text. Picture a system, Figure 4.3a, with an internal partition separating its volume into two equal parts. The partition contains a valve that may be opened as desired. The whole system is a rigid, thermally insulated box so that processes that occur within it are *adiabatic* and no work can be exchanged with the surroundings. Initially, the valve is closed and one side of the system is evacuated; the other contains 1 mol of an ideal gas at some temperature, T_1. The valve is then opened and the gas expands freely and irreversibly into the evacuated side of the chamber, Figure 4.3b.

Since the walls of the container do not move during this process, no work is done on the surroundings by the expanding gas: $\delta W = 0$. Since the system is adiabatic, no heat is exchanged with the surroundings: $\delta Q = 0$. During this process the system is *isolated* from its surroundings: no exchanges of any kind occur. Since no transfers occur between system and surroundings, the contribution to the entropy change due to exchanges with the surroundings is zero. Thus, whatever change in entropy occurs in the gas as it expands must be *entropy production* arising from the irreversible nature of the process which evidently occurs far from equilibrium. According to the second law, this entropy production and thus the change in entropy of the system must be *positive*. The problem set out requires the computation of the entropy production for an irreversible process.

This irreversible process is complex: during the expansion, pressure and temperature gradients exist within the gas, heat and mass flows occur, and turbulent conditions may even obtain. No information is supplied that would permit estimation of these complex details in the evolution of the system toward its final state. How then can the entropy produced be computed?

Recognize that entropy is a state function. Thus, the change in entropy for this complex irreversible process is exactly the same as that for the simplest reversible process that can be imagined that connects the initial and final states. The initial state is known: its volume and temperature (V_1, T_1) are specified. The volume of the final state is also known since the gas fills the whole container and the two compartments have the same volume: thus $V_2 = 2V_1$. To complete the specification of the final state, its temperature must be determined.

Recall that the system is isolated during this process: $\delta Q = \delta W = 0$. Thus, according to the first law of thermodynamics, the internal energy of the system, U, cannot change during the process, $\Delta U = 0$. Since the gas is ideal, the internal energy is determined by its temperature, Equation 4.58. Since the internal energy of the initial and final states is the same, the final temperature must be the same as the initial temperature. Thus, while the temperature distribution may be broad and time dependent during the process, the final temperature of the system when it comes to rest is uniform and must also be T_1. The simplest process that can be visualized that connects the initial state (T_1, V_1) and final state $(T_1, V_2 = 2V_1)$ in this system is thus a reversible, isothermal expansion from V_1 to $2V_1$.

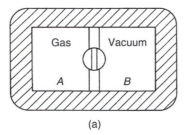   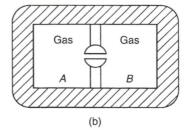

(a)   (b)

FIGURE 4.3 System illustrating the free expansion of an ideal gas. Initially the gas is contained in chamber A; chamber B is evacuated (a). The valve is opened and the gas expands irreversible ("freely") to fill both chambers (b).

EXAMPLE 4.9

Compute the change in entropy when 1 mol of an ideal gas expands freely to twice its volume.

Identify the variables: $S = S(T, V)$

Refer to Example 4.6, Equation 4.52:

$$dS = \frac{C_V}{T} dT + \frac{\alpha}{\beta} dV \qquad (4.52)$$

Set $dT = 0$ for the isothermal reversible process considered and evaluate α and β for an ideal gas:

$$dS_T = \frac{\alpha}{\beta} dV_T = \frac{1}{T} \frac{1}{\left(\frac{1}{P}\right)} dV_T = \frac{P}{T} dV_T = \frac{R}{V} dV_T$$

Integrate from the initial to the final state:

$$\Delta S_T = \int_{V_1}^{V_2} \frac{R}{V} dV = R \ln\left(\frac{V_2}{V_1}\right) = R \ln\left(\frac{2V_1}{V_1}\right)$$

$$\Delta S_T = R \ln 2 = 5.76 \ \frac{J}{mol \ K}$$

In this reversible isothermal process the entropy production is zero and the entropy change computed is completely due to heat transfer across the boundary of the system. In the irreversible process illustrated in Figure 4.3, and connecting the same two end states, the entropy change is completely due to entropy production in the system; there is no entropy transferred across the boundary. The change in entropy for the reversible isothermal process is identical to that for the irreversible free expansion in an isolated system because the initial and final states are the same and entropy is a state function. Thus, the entropy produced in the free expansion of an ideal gas to twice its initial volume is positive, as required by the second law and is equal to $+5.76$ J/mol K.

The heat absorbed and work done on a system may be computed for any reversible process by integrating $T \, dS$ or $-P \, dV$ along the path specified by the sequence of states that defines the process. In the general case this integration is a line integral along a curve that mathematically describes the process. This integration is simplified if one of the variables describing the state of the system is held constant. Example 4.10 illustrates this application of the general procedure.

EXAMPLE 4.10

Derive expressions for the heat absorbed by the system for each of the following classes of reversible processes for 1 mol of an ideal gas:

a. Isothermal change in pressure.
b. Isobaric change in volume.
c. Isochoric (constant volume) change in temperature.

Since the variables sought are process variables, the paths must be specified. Statements 1 to 3 specify paths for three different processes. The heat absorbed in each case is evaluated through $\delta Q_{rev} = T\, dS$.

a. Relationship required: $S = S(T, P)$, Equation 4.38

$$dS = \frac{C_P}{T} dT - V\alpha\, dP$$

Process is isothermal: set $dT = 0$: $dS_T = -V\alpha\, dP_T$

$$dS_T = -\left(\frac{RT}{P}\right)\left(\frac{1}{T}\right)dP_T = -\frac{R}{P}dP_T$$

The heat absorbed:

$$\delta Q_{rev,T} = T\, dS_T = -\frac{RT}{P}dP_T$$

Integrate:

$$Q_{rev,T} = -RT \ln\left(\frac{P_2}{P_1}\right) \tag{4.69}$$

b. Relationship required: $S = S(P, V)$, Equation 4.49

$$dS = \left(\frac{C_P \beta}{T\alpha} - V\alpha\right)dP + \left(\frac{C_P}{TV\alpha}\right)dV \tag{4.49}$$

Isobaric change; set $dP = 0$:

$$dS_P = \frac{C_P}{TV\alpha}dV_P$$

Ideal Gas:

$$dS_P = \frac{C_P}{TV\frac{1}{T}}dV_P = \frac{C_P}{V}dV$$

Heat absorbed:

$$\delta Q_{rev,P} = T\, dS_P = T\frac{C_P}{V}dV_P$$

Substitute for $T = PV/R$:

$$\delta Q_{rev,P} = \left(\frac{PV}{R}\right)\frac{C_P}{V}dV_P = P\frac{C_P}{R}dV_P$$

Integrate; (P is constant):

$$Q_{rev,P} = P\frac{C_P}{R}(V_2 - V_1) \qquad (4.70)$$

Note that this result could also be expressed:

$$Q_{rev,P} = C_P(T_2 - T_1)$$

which is consistent with the definition of heat capacity at constant pressure, where C_P is not a function of temperature.

c. Relationship required: $S = S(T, V)$, Equation 4.52.

$$dS = \frac{C_V}{T}dT + \frac{\alpha}{\beta}dV \qquad (4.52)$$

Isochoric: set $dV = 0$:

$$dS_V = \frac{C_V}{T}dT_V$$

The heat absorbed:

$$\delta Q_{rev,V} = T\,dS_V = T\frac{C_V}{T}dT_V = C_V dT_V$$

Integration gives:

$$Q_{rev,V} = C_V(T_2 - T_1)$$

which is consistent with the definition of heat capacity at constant volume, where C_V is not a function of temperature. This result could also be derived by recognizing that, since for an isochoric process $dV = 0$, $\delta W = 0$, so that $\delta Q = dU_V$. Integration gives the result: $\Delta U_V = Q$. According to Equation 4.62, since the internal energy of an ideal gas depends only upon its temperature, for any process, $\Delta U = C_V(T_2 - T_1)$.

EXAMPLE 4.11

One mole of an ideal gas, initially at 273 K and one atmosphere is contained in a chamber that permits programmed control of its state. Controlled quantities of heat and work are supplied to the system so that its pressure and volume change along a line given by the equation

$$V = 22.4\left(\frac{1}{\text{atm}}\right)P$$

Assume the process is carried out reversibly. Compute the heat required to be supplied to the system to take it to a final pressure of 0.5 atmospheres. What is the final temperature of the gas?

Independent variables are P and V; information is needed about S since the heat absorbed is TdS. Thus, the relation required is $S = S(P, V)$. From Example 4.2, Equation 4.49,

$$dS = \left(\frac{C_P\beta}{T\alpha} - V\alpha\right)dP + \frac{C_P}{TV\alpha}dV \qquad (4.49)$$

For an ideal gas, evaluation of α and β yields

$$dS = \frac{C_P}{V}dV + \frac{C_V}{P}dP$$

In order to compute the heat absorbed for an incremental step along the specified path, it is necessary to use the equation describing the path to express one of the independent variables in terms of the other. From the specified path,

$$V = 22.4P \qquad \text{and} \qquad dV = 22.4\,dP$$

Use these expressions for V and dV in the entropy relation just obtained to obtain an expression for an incremental change in entropy along the controlled path,

$$dS(\text{this path}) = \frac{C_V}{P}dP + \frac{C_P}{(22.4P)}(22.4\,dP) = (C_V + C_P)\frac{1}{P}dP$$

The heat absorbed for this step is

$$\delta Q_{\text{rev}} = T\,dS = T(C_V + C_P)\frac{1}{P}dP$$

In order to integrate this equation to obtain the total heat absorbed, it is necessary to relate T to P along this path. Evidently this requires combination of the equation of state with the equation for the path:

$$T(P) = \frac{PV}{R} = \frac{P}{R}(22.4P) = 22.4\frac{P^2}{R}$$

A check of the units in this result demonstrates that the gas constant, R must be expressed as $R = 0.08206$ (1 atm/mol K) for consistent units. Thus, for the given path the temperature of the system varies with the square of the pressure. Substitute:

$$\delta Q_{\text{rev}} = \left(22.4\frac{P^2}{R}\right)(C_V + C_P)\frac{1}{P}dP = 22.4(C_V + C_P)\frac{P}{R}dP$$

Integration gives the heat absorbed

$$Q_{rev} = \int_{P_1}^{P_2} 22.4\left(\frac{C_V + C_P}{R}\right)P\,dP = 22.4\left(\frac{C_V + C_P}{R}\right)\frac{P^2}{2}\Big]_{P_1}^{P_2}$$

Insert values for C_V and C_P for a monatomic gas, together with the initial and final pressures for the process,

$$Q_{rev} = 22.4\left(\frac{\frac{3}{2}R + \frac{5}{2}R}{R}\right)\frac{[(0.5)^2 - (1.0)^2]}{2} = 22.4 \times 4(-0.375)$$

$$Q_{rev} = -33.6\,l\,atm = -3404\,J$$

The final temperature of the system is the value on the path that corresponds to 0.5 atmospheres of pressure:

$$T_2 = 22.4\left(\frac{1}{atm}\right)\frac{(0.5)^2(atm)^2}{0.08206\left(\frac{l\,atm}{mol\,K}\right)} = 68.2\,K$$

The final volume is $V = 22.4 \times 5 = 11.2\,l$. Evidently in order to lower *both* the pressure and the volume simultaneously, it is necessary to extract a significant quantity of heat from the system.

4.2.5 Applications to Solids and Liquids

Tables 4.1 to 4.3 surveyed the ranges of values of the experimental variables that are typical for solid and liquids. It is observed that, while α, β and C_P are, in general, functions of temperature, pressure, and composition, for a given material these quantities are not strong functions of the state of the system. For many applications that do not involve the gas phase, particularly when only an estimate of the thermodynamic properties is sought, the assumption that α, β and C_P are constants is a useful approximation. Where precise computations are required, the dependence of these quantities upon T and P must be obtained from the literature or experimentally determined. The examples presented in this section illustrate both precise and approximate calculations for selected thermodynamic state functions so that the magnitude of the errors introduced by these approximations can be assessed.

Example 4.12

One mole of nickel initially at 300 K and one atmosphere pressure is taken through two separate processes:

a. An isobaric change in temperature to 1000 K.
b. An isothermal compression to 1000 atm.

Compare the change in enthalpy of the nickel for these two processes.
The experimental variables for nickel (Appendix B) are:

$$V(300 \text{ K}, 1 \text{ atm}) = 6.60\left(\frac{\text{cc}}{\text{mol}}\right); \qquad \alpha = 40 \times 10^{-6}\left(\frac{1}{T}\right); \qquad \beta = 26 \times 10^{-7}\left(\frac{1}{\text{atm}}\right)$$

$$C_P(T) = 11.17 + 37.78 \times 10^{-3}T + \frac{3.18 \times 10^5}{T^2}\left(\frac{\text{J}}{\text{mol K}}\right)$$

Identify the variables: $H = H(T, P)$
Obtain the relationship [Equation 4.42 from Table 4.5]

$$dH = C_P dT + V(1 - T\alpha)dP \qquad (4.52)$$

For process (a) $dP = 0$ and the change in enthalpy is

$$\Delta H = \int_{300 \text{ K}}^{1000 \text{ K}} \left(11.17 + 37.78 \times 10^{-3}T + \frac{3.18 \times 10^5}{T^2}\right)dT$$

$$= \left[11.17T + 37.78 \times 10^{-3}\frac{T^2}{2} + \frac{3.18 \times 10^5}{-T^1}\right]_{300 \text{ K}}^{1000 \text{ K}}$$

The second and third terms on the right-hand side of this equation arise from the
temperature-dependent contribution to the heat capacity. The first term represents the
estimate of ΔH for the process if this temperature-dependent terms were ignored.
Evidently, a significant error would be introduced if the temperature-dependent terms
were ignored.
For process (b) $dT = 0$ and the change in enthalpy is

$$\Delta H = \int_{1}^{1000} V(1 - T\alpha)dP$$

If V and α are approximated as constants, they may be factored out of the integral
so that

$$\Delta H = V(1 - T\alpha)(P_2 - P_1)$$

$$\Delta H = 6.60\frac{\text{cc}}{\text{mol}}\left[1 - 40 \times 10^{-6}\frac{1}{T}300 \text{ K}\right][1000 - 1]\text{atm}\frac{8.314 \text{ J}}{82.06 \text{ cc atm}}$$

Since α typically has values in the range of $10^{-5}(1 \text{ K}^{-1})$, the product $T\alpha$ is in
general small in comparison to 1. Thus, any second-order contributions to α, e.g.,
due to its variation with pressure, will certainly be negligible and the factor
$[1 - T\alpha]$ may be treated as a constant (i.e., not dependent on pressure) without
significant error. The pressure dependence of the molar volume may be expressed to

a first approximation as

$$V(P) = V(P_1)[1 + \beta(P - P_1)]$$

Substitute this result for V in the equation for dH_T

$$dH_T = V(P_1)[1 + \beta(P - P_1)][1 - T\alpha]dP_T$$

Integrate over the pressure range:

$$\Delta H = \int_{P_1}^{P_2} V(P_1)[1 + \beta(P - P_1)](1 - T\alpha)dP$$

$$\Delta H = V(P_1)(1 - T\alpha)\left[P + \beta\frac{(P - P_1)^2}{2} \right]_{P_1}^{P_2}$$

$$\Delta H = V(P_1)[1 - T\alpha]\left[(P_2 - P_1) + \beta\frac{(P_2 - P_1)^2}{2} \right]$$

Insert the numerical values for the problem at hand:

$$\Delta H = (6514 + 8.5) = 6522\frac{1\ \text{atm}}{\text{mol K}} = 660.0 + 0.857 = 660.8\ \frac{\text{J}}{\text{mol}}$$

where the second term traces to the assumed pressure variation of the molar volume and the first term is identical to that computed approximately earlier in this example. It will be noted that the second term is significant only if the pressure change in the system is extremely large, say 100,000 atm. Thus, for most processes, the assumption that V does not change with pressure yields an approximate result that is within the experimental error of measurement of the precise result.

Comparison of the enthalpy changes for processes (a) and (b) in this example reveals another point that has broad application in the behavior of solids and liquids. Process (a) visualizes a change of a factor of 3 in the absolute temperature; in process (b) the pressure is changed by a factor of 1000. The energy change, here reported as the enthalpy, for process (a) is nonetheless about 25 times larger than for process (b). *Thus, for solids and liquids, energy changes associated with thermal influences tend to be much larger than those arising from mechanical influences.*

EXAMPLE 4.13
Compute the change in Gibbs free energy of magnesia (MgO) when 1 mol is heated from 298 (room temperature) to 1300 K at one atmosphere pressure. The properties of MgO obtained from Appendix E are:

$$C_P = 48.99 + 3.43 \times 10^{-3}T - \frac{11.34 \times 10^{-5}}{T^2}\ \left(\frac{\text{J}}{\text{mol K}} \right)$$

$$S_{298}^{\circ} = 26.9\left(\frac{J}{mol\ K}\right)$$

where $S_{(298)}^{\circ}$ is the absolute entropy of MgO at 298 K, obtained by integrating heat capacity data from 0 to 298 K. The variables in the problem:

$$G = G(T,P)$$

From Table 4.5:

$$dG = -S\ dT + V\ dP \qquad (4.10)$$

In this problem,

$$dP = 0; \qquad dG_P = -S\ dT_P$$

The temperature dependence of S cannot be neglected in this case. Thus, in order to integrate this expression, it is first necessary to develop an expression for the variation of entropy with temperature at constant pressure. Recall again Equation 4.38:

$$dS = \frac{C_P}{T}dT - V\alpha\ dP \qquad (4.38)$$

The condition $dP = 0$ gives:

$$dP = 0; \qquad dS_P = \frac{C_P(T)}{T}dT_P$$

Insert the expression for the heat capacity and integrate from 298 K to a variable temperature, T;

$$\int_{298\ K}^{T} dS = S(T) - S_{298}^{\circ} = \int_{298\ K}^{T} \frac{C_P(T)}{T}dT$$

The absolute value of the entropy at any temperature is thus

$$S(T) = S_{298}^{\circ} + \int_{298\ K}^{T} \frac{C_P(T)}{T}dT$$

Insert this expression into

$$\int_{298\ K}^{T} dG_P = G(T) - G(298) = \int_{298\ K}^{T} -S(T)dT_P$$

to give

$$G(T) - G(298) = \int_{298}^{T} -\left[S_{298} + \int_{298}^{T} \frac{C_P(T)}{T}dT\right]dT$$

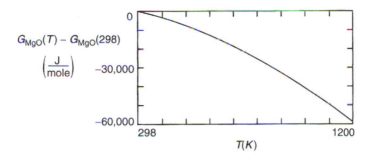

FIGURE 4.4 Variation of Gibbs free energy of MgO with temperature at one atmosphere pressure.

Insert the heat capacity function for MgO and the absolute entropy of MgO at 298 into this equation and perform the indicated double integration. This procedure may be very conveniently carried out in MathCad or a similar mathematical software package. This functional relationship is presented in Figure 4.4.

EXAMPLE 4.14

One mole of copper initially at 700 K and one atmosphere is contained in a thermally insulated jacket. The system is compressed reversibly to a pressure of 10,000 atmospheres. Compute the change in temperature for this reversible adiabatic process.

Identify the variables: $T = T(S, P)$

Rearrange (because the independent variable is T)

$$S = S(T, P)$$

Recall this relation from Table 4.5:

$$dS = \frac{C_P}{T} dT - V \alpha \, dP \tag{4.38}$$

Solve for dT:

$$dT = \frac{T}{C_P} dS + \frac{TV\alpha}{C_P} dP$$

Set $dS = 0$ for this reversible adiabatic process:

$$dT_S = \frac{TV\alpha}{C_P} dP_S$$

Assuming V, α and C_P are constants in the problem, separate the variables:

$$\frac{dT}{T} = \frac{V\alpha}{C_P}dP$$

Integrate:

$$\ln\left(\frac{T_2}{T_1}\right) = \frac{V\alpha}{C_P}(P_2 - P_1)$$

Substitute the properties of copper from Appendix B and compute the result:

$$T_2 = 700\ \text{K}\left[\exp\left(\frac{7.09\dfrac{\text{cc}}{\text{mol}}51 \times 10^{-6}\dfrac{1}{T}}{27.3\dfrac{\text{J}}{\text{mol K}}}(10,000 - 1)\left(\frac{8.314\ \text{J}}{82.06\ \text{cc atm}}\right)\right)\right]$$

$$T_2 = 709\ \text{K}.$$

EXAMPLE 4.15

This example presents an analog to the illustration of the free expansion of an ideal gas presented in Example 4.9. One mole of pure iron is initially at 300 K and 100,000 atm. The system is contained in a thermally insulating jacket. The pressure is suddenly released; the iron expands irreversibly to a final pressure of one atmosphere. Compute the entropy produced in this irreversible process.

Since for an elastic solid, the volume change is very small, even under extreme hydrostatic pressures such as those considered here, it may be assumed that the work done on the system during this expansion is negligibly different from zero. Since the process is also adiabatic, by the first law the internal energy of the system is constant. Further, since there is essentially no exchange with the surroundings, the entropy flow across the boundary of the system is also negligible. Thus, the change in entropy for the process is virtually all entropy produced inside the system; by the second law, this entropy production must be positive.

The entropy production in this irreversible process may be computed by applying again the familiar strategy. Visualize the simplest reversible process that will connect the same initial and final states. Since entropy is a state function, the change for that process will be the same as that for the irreversible process in the problem. Thus, it is necessary to compute the change in entropy associated with a reversible decompression of 1 mol of iron at constant internal energy.

1. Identify the variables: $S = S(U, P)$.
2. Write the differential form: $dS = M\ dU + N\ dP$.
3. Use Equation 4.41 in Table 4.5 to convert dU to a function of T and P:

$$dS = M[(C_P - PV\alpha)dT + V(P\beta - T\alpha)dP] + N\ dP.$$

4. Collect terms:

$$dS = M(C_P - PV\alpha)dT + [MV(P\beta - T\alpha) + N]dP$$

5. Recall Equation 4.38 from Table 4.5:

$$dS = \frac{C_P}{T} dT - V\alpha \, dP \qquad (4.38)$$

6. Compare coefficients

$$M(C_P - PV\alpha) = \frac{C_P}{T}$$

$$MV(P\beta - T\alpha) + N = -V\alpha$$

7. Solve for M and N:

$$M = \left(\frac{\partial S}{\partial U}\right)_P = \frac{C_P}{T(C_P - PV\alpha)}$$

$$N = \left(\frac{\partial S}{\partial P}\right)_U = -\frac{PV(C_P\beta - VT\alpha^2)}{T(C_P - PV\alpha)}$$

The required relation is:

$$dS = \frac{C_P}{T(C_P - PV\alpha)} dU - \frac{PV(C_P\beta - TV\alpha^2)}{T(C_P - PV\alpha)} dP$$

In the problem at hand $dU = 0$ and the change in entropy is

$$dS_U = -\frac{PV(C_P\beta - VT\alpha^2)}{T(C_P - PV\alpha)} dP_U$$

Integration of this equation requires a knowledge of the dependence upon pressure of all of the experimental variables contained in the coefficient of dP_U. However, examination of the numerical values of these quantities shows that the term $PV\alpha$ is small in comparison with C_P, even at the start of the process with $P = 100,000$ atm. The factor in the numerator, $TV\alpha^2$, is even smaller. Thus, to a very good approximation,

$$dS_U = -\frac{PV\beta}{T} dP_U$$

Computation of the change in temperature [based upon $T = T(U,P)$] reveals that the temperature may decrease by about 10 K in this process. This change is small compared with the value of the temperature for the system. If it is also neglected, the entropy produced may be calculated approximately to be:

$$\Delta S = -\left[\frac{\left(7.09\frac{cc}{mol}\,6.6 \times 10^{-7}\frac{1}{atm}\right)}{300\ K}\right]\frac{1}{2}(1^2 - 100,000^2)(atm)^2$$

$$\Delta S = +2.2 \times 10^{-5}\left(\frac{cc\ atm}{mol\ K}\right) = +2.2 \times 10^{-6}\left(\frac{J}{mol\ K}\right).$$

4.3 SUMMARY

Thermodynamic relationships may be usefully classified as:

1. The laws of thermodynamics.
2. Definitions.
3. Coefficient relations.
4. Maxwell relations.
5. Conditions for equilibrium.

The laws of thermodynamics form the physical basis for solving practical problems.

Additional energy functions are defined:

$$H \equiv U + PV; \qquad F \equiv U - TS; \qquad G \equiv U + PV - TS$$

because they provide measures of energy that are convenient for specific classes of processes. Alternate forms of the combined statements of the first and second laws may be derived for these energy functions:

$$dU = T\,dS - P\,dV + \delta W' \tag{4.4}$$

$$dH = T\,dS + V\,dP + \delta W' \tag{4.6}$$

$$dF = -S\,dT - P\,dV + \delta W' \tag{4.8}$$

$$dG = -S\,dT + V\,dP + \delta W' \tag{4.10}$$

A collection of experimental variables is defined:

$$\alpha \equiv \frac{1}{V}\left(\frac{\partial V}{\partial T}\right)_P \tag{4.11}$$

$$\beta \equiv -\frac{1}{V}\left(\frac{\partial V}{\partial P}\right)_T \tag{4.12}$$

$$\delta Q_{\text{rev},P} \equiv C_P dT_P \tag{4.13}$$

$$\delta Q_{\text{rev},V} \equiv C_V dT_V \tag{4.15}$$

These variables provide complete information about changes in the thermodynamic properties of systems for which $\delta W' = 0$; no other empirical information is required.

Coefficient and Maxwell relations derive from the mathematical properties of differentials of functions of several variables. For any thermodynamic relation between differentials of state functions, these relations may be read by inspection of the equation:

$$Z = Z(X, Y) \tag{4.16}$$

$$dZ = M \, dX + N \, dY \tag{4.17}$$

$$M = \left(\frac{\partial Z}{\partial X}\right)_Y; \qquad N = \left(\frac{\partial Z}{\partial Y}\right)_X \tag{4.18}$$

$$\left(\frac{\partial M}{\partial Y}\right)_X = \left(\frac{\partial N}{\partial X}\right)_Y \tag{4.25}$$

A general, rigorous procedure provides an algorithm for deriving the thermodynamic relationship between any dependent variable, Z and any pair of independent variables, (X,Y). This procedure begins with derivation of functional relationships for every other state function in terms of independent variables, T and P. Expressions for the energy functions are easily derived once relationships for $V = V(T, P)$ and $S = S(T, P)$ are obtained:

$$dV = V\alpha \, dT - V\beta \, dP \tag{4.32}$$

$$dS = \frac{C_P}{T} dT - V\alpha \, dP \tag{4.38}$$

The dependence of the energy functions upon T and P may be derived by insertion of these expressions for dS and dV into the combined statements of the first and second laws reviewed at the beginning of this summary:

$$U = U(T, P); \qquad dU = (C_P - PV\alpha)dT + V(P\beta - T\alpha)dP \tag{4.41}$$

$$H = H(T, P); \qquad dH = C_P dT + V(1 - T\alpha)dP \tag{4.42}$$

$$F = F(T, P); \qquad dF = -(S + PV\alpha)dT + PV\beta \, dP \tag{4.43}$$

$$G = G(T, P); \qquad dG = -S \, dT + V \, dP \tag{4.11}$$

These key equations are summarized in Table 4.5. For any given problem with arbitrary variables $Z = Z(X, Y)$ with differential form $dZ = M \, dX + N \, dY$ the general procedure uses these equations to convert dX and dY to differential forms involving dT and dP. Then, dZ is written as a function of dT and dP. The resulting pair of equations for dZ are alternate expressions for $Z = Z(T, P)$. Corresponding coefficients must therefore be equal. Equating corresponding coefficients yields two linear simultaneous equations in M and N. Evaluation of M and N in terms of experimental variables completes the derivation of the required relationship.

To obtain a numerical result for a given practical problem which seeks the change in Z for initial and final states specified by values of X and Y, it is necessary to obtain numerical values for the experimental variables in M and N and integrate, first over dX, then over dY. If the problem at hand requires the evaluation of a process variable involving dZ (either $\delta Q = T\, dS$ or $\delta W = -P\, dV$), it is necessary to carry out the integration along the path which must be described explicitly as a relation between X and Y.

A variety of examples illustrate the procedure. It is demonstrated that its application is particularly simple for an ideal gas, defined by the equation of state:

$$PV = RT \tag{4.56}$$

and a fixed value of heat capacity:

$$C_V = \frac{3}{2}R; \qquad C_P = C_V + R = \frac{5}{2}R$$

for a monatomic gas. These examples demonstrate that the internal energy and the enthalpy of an ideal gas are each a function only of temperature:

$$\Delta U = C_V(T_2 - T_1) \tag{4.62}$$

$$\Delta H = C_P(T_2 - T_1) \tag{4.63}$$

A final set of examples applies to condensed phases demonstrating the generality of the procedure.

HOMEWORK PROBLEMS

Problem 4.1.   Write out the combined statements of the first and second laws for the energy functions, $U = U(S, V)$, $H = H(S, P)$, $F = F(T, V)$ and $G = G(T, P)$. Assume $\delta W'$ is zero:
    a. Write out all eight coefficient relations.
    b. Derive all four Maxwell relations.
for these equations.

Problem 4.2.   Derive the ratio relation Equation 4.30:

$$\left(\frac{\partial Z}{\partial X}\right)_Y \left(\frac{\partial X}{\partial Y}\right)_Z \left(\frac{\partial Y}{\partial Z}\right)_X = -1$$

[*Hint*: Begin with the differential form of $Z = Z(X, Y)$; solve for dX; write the differential form of $X = X(Y, Z)$; compare coefficients.]

Problem 4.3.   The molar volume of Al_2O_3 at 25°C and 1 atm is 25.715 cc/mol. Its coefficient of thermal expansion is $26 \times 10^{-6}\ \text{K}^{-1}$ and the coefficient of compressibility is 8.0×10^{-7} atm^{-1}. Estimate the molar volume of Al_2O_3 at 400°C and 10 kbars pressure (10×10^3 atm).

Problem 4.4. Compare the entropy changes for the following processes:
  a. One gram atom of nickel is heated at one atmosphere from 300 to 1300 K.
  b. One gram atom of nickel at 300 K is isothermally compressed from 1 atm to 100 kbars.
  c. One mole of zirconia is heated at one atmosphere from 300 to 1300 K.
  d. One mole of zirconia at 300 K is isothermally compressed from 1 atm to 100 kbars. (Use $V = 22.0$ (cc mol^{-1}) and $\alpha = 10 \times 10^{-6}$ K^{-1}.)
  e. One mole of oxygen is heated at one atmosphere from 300 to 1300 K. (Assume O_2 is an ideal gas with $C_P = 7/2R$).
  f. One mole of oxygen at 300 K is isothermally compressed from 1 atm to 100 kbars.

What general qualitative conclusions do you draw from these calculations?

Problem 4.5. Express the results obtained for parts (c) and (d) of Problem 4.4 in values *per gram atom* of ZrO_2. (Each mole of zirconia contains three gram atoms of its elements: one gram atom of zirconium and two of oxygen.) How does this result influence the conclusions you made in the comparisons in Problem 4.4?

Use the *general procedure for deriving relationships in thermodynamics* to solve the remaining problems in this chapter. Begin by identifying explicitly the dependent and independent variables in each case.

Problem 4.6. Compute the change in internal energy when 12 l of argon gas at 273 K and 1 atm is compressed to 6 l with the final pressure equal to 10 atm. Solve this problem in two different ways:
  a. Apply the general procedure to evaluate $U = U(P, V)$ for an ideal gas and integrate from initial to final (P,V).
  b. Use the information given to compute the final temperature of the gas and apply the general relation $\Delta U = C_V \Delta T$.

Assume $C_V = (3/2)R$ for this monatomic gas.

Problem 4.7. Compute and plot the surface that represents the relationship for the entropy of nitrogen gas as a function of temperature and pressure in the (P,T) range from (1 atm, 300 K) to (10 atm, 1000 K). Since nitrogen is diatomic, $C_P = (7/2)R$. Assume it obeys the ideal gas law in this domain.

Problem 4.8. Compute and plot the surfaces that represent the variation with pressure and volume of:
  a. The internal energy.
  b. The enthalpy

of 1 mol of nitrogen gas. Cover the range in (P,V) space from (1 atm, 22.4 l) to (10 atm, 8.2 l). Plot constant energy lines on the surfaces. Assume it behaves ideally.

Problem 4.9. Evaluate the partial derivative

$$\left(\frac{\partial H}{\partial G}\right)_S$$

in terms of experimental variables.

Problem 4.10. Derive the relationship that describes the dependence of Helmholtz free energy upon entropy and temperature. Design an experiment which would require this relationship in analyzing the results.

Problem 4.11. Demonstrate that the change in a state function for a process is independent of the path by calculating the change in Gibbs free energy for two processes that change the state of 1 mol of a monatomic ideal gas from (298 K, 1 atm) to (600 K, 1000 atm):
   a. *Process A*: heat the gas at 1 atm from 298 K to 600 K, then compress it at 600 K from 1 atm to 1000 atm to arrive at the final state; (600 K, 1000 atm).
   b. *Process B*: compress the gas from 1 atm to 1000 atm at 298 K, then heat the gas at 1000 atm from 298 K to 600 K to arrive at the final state; (600 K, 1000).

Show that the same result is obtained by applying the definitional relation: $\Delta G = \Delta H - \Delta(TS)$. (Note that $\Delta(TS) \neq T\Delta S$.) (*Hint*: the absolute entropy in the second step of part (b) contains the pressure term.)

Problem 4.12. A system is designed that permits continuous programmed control of the pressure and volume of the gas that it contains. The system is filled with 1 g atom of helium and brought to an initial condition of one atmosphere and 18 l. It is then reversibly compressed to 12 l along a programmed path given by the relationship

$$V = -2P^2 + 20$$

where P is in atmospheres and V is in liters. Compute:
   a. The initial and final temperature of the system.
   b. The heat absorbed by the system.
   c. The work done by the system.
   d. The changes in U, H, F, G, and S.

Problem 4.13. Estimate the pressure increase required to impart 1 J of mechanical work in reversibly compressing 1 mol of silver at room temperature. What pressure rise would be required to impart 1 J of work to 1 mol of alumina at room temperature? For Al_2O_3 take the molar volume to be 25.715 (cc mol^{-1}) and $\beta = 8.0 \times 10^{-7}$ (atm)$^{-1}$.

Problem 4.14. Compute and plot the surface representing the Gibbs free energy of hydrogen gas as a function of temperature and pressure in the range from (298 K, 10^{-10} atm) to (1000 K, 100 atm). Use (298 K, 1 atm) as the zero point for the calculation. The absolute entropy of H_2 at 298 K and 1 atm is 130.57 (J mol K^{-1}); assume that $C_P = 7/2$ R (J mol K^{-1}) is independent of P and T.

Problem 4.15. Use a mathematics applications program, spreadsheet or computer language to program and plot the generic equations for computing the temperature dependence of enthalpy, entropy, and Gibbs free energy as a function of temperature at one atmosphere pressure. Assume as input the absolute entropy of the substance at 298 K and values of a, b and c in the empirical heat capacity expression

$$C_P = a + bT + \frac{c}{T^2}$$

Use the program to compute H, S and G (relative to their values at 298 K) as functions of temperature at one atmosphere for
    a. Argon.
    b. Titanium.
    c. TiO_2.

Problem 4.16. Using the bounding values in Tables 4.1 to 4.3, estimate the range of the magnitude of the difference between C_P and C_V for condensed phases.

Problem 4.17. According to Appendix E, the heat of formation of the compound CoO is $-237,700$ J. Compute the enthalpy change for this reaction as a function of temperature in the range from 300 to 1000 K. What is the maximum error incurring in this range if it is assumed that ΔH is independent of temperature. (*Hint*: $d(\Delta H) = \Delta C_P(T)\, dT$.)

REFERENCES

1. Lide, D.R. (Ed.), *Handbook of Physics and Chemistry*, CRC Press, Boca Raton, FL 1991, OH (Revised annually).
2. Stull, D.R., *JANAF (Joint Army-Navy-Air Force) Thermochemical Tables*, Clearinghouse for Technical and Scientific Information, Springfield, VA, 1965, Addenda, supplements and updates published periodically.
3. Hultgren, R., Orr, R.L., Anderson, P.D., and Kelley, K.K., *Selected Values of Thermodynamic Properties of Metals and Alloys*, Wiley, New York, 1963.
4. *Thermophysical Properties Research Center Data Book*, Thermophysical Properties Research Center, Purdue University, West Layfayette, IN. Data Base.

5. *TRCTHERMO*, Thermodynamic Research Center, Texas Engineering Experiment Station, The Texas A & M University System, College Station, TX. Data Base.

6. Bales, C.W., Pelton, A.D., and Thompson, W.T., *FACT (Facility for the Analysis of Chemical Thermodynamics)*, developers, McGill University, Montreal, Que., Canada. Data Base.

7. Kelley, K.K., *Contributions to the Data on Theoretical Metallurgy, Vol. I–XV*, U.S. Bureau of Mines, (1932–1962). Valuable compilations are long out of print, but may be available in libraries.

5 Equilibrium in Thermodynamic Systems

CONTENTS

This chapter introduces a general principle that is the basis for determining the internal condition of the most complex kind of thermodynamic system when it reaches its equilibrium state. This general criterion for equilibrium has the same level of importance in the development of thermodynamics as do the laws themselves, as implied in Figure 1.4. It is the foundation that provides the strategy for the calculation equilibrium maps: phase diagrams, chemical equilibria in predominance diagrams without and with electrical effects, the role played by capillarity effects, the chemistry of defects in crystals, as well as the primary results of statistical thermodynamics. These applications of this criterion occupy most of the rest of this text.

Some intuitive notions of equilibrium are first presented; these ideas are then formalized to provide a thermodynamic statement of the criterion that determines when a system that is isolated from its surrounding during its approach to equilibrium has attained its equilibrium state:

In an isolated system, the entropy is a maximum at equilibrium.

This thermodynamic extremum principle (the entropy has an extreme value, a maximum) is then formulated mathematically. A set of equations, called the conditions for equilibrium, that describe the relationships that the internal properties of the isolated system must have when it achieves its equilibrium state are derived from this extremum principle. It is then demonstrated that although these conditions for equilibrium are derived for an isolated system, they are valid for any system at equilibrium whether or not it was isolated during its approach to that final condition.

The strategy for obtaining these conditions for equilibrium is illustrated in this chapter by applying it to the simplest case: equilibrium in a unary, two-phase system.

This strategy is applied repeatedly in subsequent chapters to derive the conditions for equilibrium in systems of progressively increasing sophistication. These equations are the basis for making practical calculations about the final resting condition of complex systems.

The chapter ends with a collection of alternate statements of the criterion for equilibrium, each of which is useful in the description of systems with specific external constraints.

5.1 INTUITIVE NOTIONS OF EQUILIBRIUM

The idea that a system changes toward some final ultimate condition and, once there, remains in that condition unless acted upon by some external agent is a familiar one. Such a system is said to come to equilibrium or approach its equilibrium state. The specific nature of that state depends upon the chemical and energy content of the system and that of its surroundings. A system comes to equilibrium in given surroundings. It may only be disturbed from its equilibrium state by changing the state of its surroundings.

A simple application of this idea is illustrated in Figure 5.1. The system acted upon by a set of mechanical forces shown comes to equilibrium when it arrays itself with respect to the forces so that they are in balance. The mathematical statement that describes this condition is particularly simple for this mechanical system.

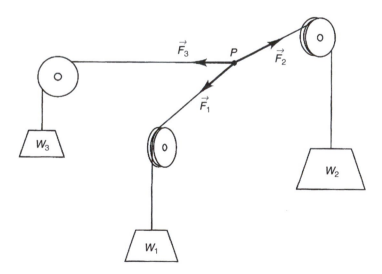

FIGURE 5.1 A simple mechanical system in equilibrium. The vector sum of the forces acting at P is zero.

The vector sum of the forces acting on the point P is zero:

$$\sum_i F_i = 0 \tag{5.1}$$

which implies:

$$\sum_i F_{ix} = 0; \qquad \sum_i F_{iy} = 0; \qquad \sum_i F_{iz} = 0 \tag{5.2}$$

where the subscripts x, y, and z represent components of the force vectors resolved in some orthogonal coordinate system, and the summation over the index i is taken over all of the forces acting on the body. Once the system has found the position in space O which achieves this balance, it will remain fixed in that position until one of the weights is changed. Furthermore, if the system is displaced from its equilibrium position, e.g., if the body is moved to any position in the neighborhood of O, it will return to O.

In thermodynamics, the influences that operate to modify the condition of a system are more general than mechanical forces. Nonetheless, the intuitive notion of an equilibrium condition also applies to such systems. This idea of an equilibrium state has two components:

1. It is a state of rest.
2. It is a state of balance.

The first component means that the condition of the system, no matter what it might be, is time independent. The system has achieved a stationary state. No changes can occur in a system that has come to equilibrium except by the action of influences that originate outside the system/surroundings complex. The second component to the concept means that if the system is perturbed from its equilibrium condition by some outside influence, it will return again to the same condition when it again comes to rest.

5.2 THERMODYNAMIC FORMULATION OF A GENERAL CRITERION FOR EQUILIBRIUM

Systems can experience two distinct classes of conditions that are time invariant, i.e., two classes of stationary states. Either the system is in an equilibrium state or it has achieved a steady state.

A simple example of a steady state is shown in Figure 5.2. A copper rod is surrounded by an insulating jacket except at its ends. One end is placed in contact with a furnace maintained at temperature T_1; the other end contacts a water-cooled plate maintained at temperature T_2. Heat will flow through the rod from left to right and the temperature profile will evolve with time. Eventually, because the external conditions are fixed, the temperature profile will achieve a distribution that no longer

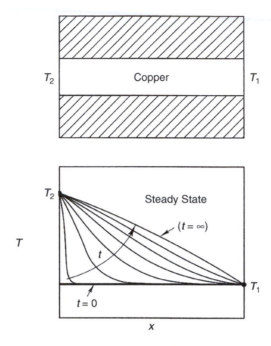

FIGURE 5.2 The thermally insulated copper rod achieves a time invariant (steady state) temperature distribution if and only if the temperatures at the ends are maintained constant with time.

changes with time. To maintain this time invariant condition, it is necessary to supply heat continuously to the left end of the rod and extract heat from the right. The resulting condition is time independent, but it is not an equilibrium condition; it is a steady state.

In order to distinguish a steady state from an equilibrium state, apply a simple but general test. Isolate the system from the surroundings, i.e., surround the system (at least in a thought experiment) with a rigid, impermeable, thermally insulating boundary. No heat, no work, no matter crosses such a boundary. If a system, which is not changing with time, is isolated from its surroundings and changes then begin to occur within the system, then the initial stationary condition was a steady state. This steady state was maintained by flows across the boundary that were cut-off upon isolation. If the ends of the bar in Figure 5.2 are isolated from the heat source and sink, the internal steady state temperature profile begins to change.

In contrast, if upon isolation no changes occur in the internal condition of the system, then the initial time invariant state was an equilibrium state. A system in equilibrium with itself and its surroundings is not exchanging matter or energy with the surroundings. Thus, isolating such a system from its surroundings has no effect upon its internal equilibrium state.

Examine the result of this test for the equilibrium state closely, for it contains an important principle, first stated succinctly by J. Willard Gibbs.[1] If a system is in equilibrium both internally and with its surroundings, then isolating it from its surroundings will produce no change in the internal state of the system. Thus, the internal condition of any system that has achieved equilibrium with its surroundings will be identical with the condition of some other system that has come to the same final equilibrium state but was isolated from its surroundings during its approach to equilibrium. If a criterion can be devised that determines the final resting state for isolated systems and a strategy can be devised for deriving the relationships that characterize the internal condition of that isolated system at equilibrium, then the resulting set of relationships will be general. These relationships will describe any system that comes to equilibrium, whether or not it was isolated from its surroundings during its approach to equilibrium.

This insight is potentially very useful because isolated systems are relatively easy to treat thermodynamically. Changes inside the system are not complicated by exchanges of heat, work, or matter with the surroundings. Focus upon the change in entropy of an isolated system. Since there are no exchanges of any kind with the surroundings, whatever processes occur within the system, no entropy is transferred across the boundary: $\Delta S_{\text{trans}} = 0$. Thus, the total entropy change experienced by an isolated system, no matter what processes produce that change, results entirely from entropy production within the system. The second law of thermodynamics guarantees that entropy production is always positive. It is concluded that, for any real process that occurs within an isolated system, the total entropy of the system can only increase.

This conclusion identifies a thermodynamic property, S, which monitors the direction of possible change in an isolated system. No matter how complicated the system or how irreversible the process because the second law is pervasive, this conclusion applies.

Thus, as an isolated system evolves through its process of spontaneous change, its entropy continually increases. When the system is first isolated from its surroundings by enclosing it in a thermally insulating, rigid, impermeable jacket, its internal energy, volume, and chemistry have a certain set of values. As the isolated system evolves, these values cannot change. When the system finally attains its state of rest, its equilibrium state, the entropy of the system must be the largest value that system may exhibit. If this were not true, i.e., if there existed a state of that isolated system that had a higher entropy than the equilibrium state, then a change from that state toward the equilibrium state would be accompanied by a decrease in entropy. In an isolated system this violates the second law. It is thus concluded that:

> In an isolated system the equilibrium state is the state that has the maximum value of entropy that system can exhibit.

This extremum principle will be referred to as the criterion for equilibrium in an isolated system.

Note that this statement does not imply that the equilibrium state for any thermodynamic system is the state that has the maximum entropy the system can exhibit. This statement only applies if the system was isolated from its surroundings during its approach to equilibrium. If, during its evolution, the system were not isolated from its surroundings, then entropy may be transferred across its boundary.

The change in the total entropy of a system that is not isolated includes a contribution due to transfers, which may be positive or negative. The entropy of such a system may thus increase or decrease and the total entropy of the system is not a valid monitor of the direction of spontaneous change. Such a system will eventually come to equilibrium with its surroundings but it cannot be stated in general that the entropy of its final state is higher than all its previous states as it evolved. The entropy change is guaranteed by the second law to be positive only for systems that are isolated from their surroundings so that the total entropy change derives solely from entropy production.

Return now to the test proposed to determine whether a given system is in equilibrium both internally and with its surroundings. Imagine an arbitrary system exchanging heat, work, and matter with its surroundings as it evolves. No general statement can be made about the expected direction of change of any of its thermodynamic properties as it proceeds spontaneously toward its final state of equilibrium. Eventually the system attains its time invariant state of balance, called its equilibrium state. The internal condition of the system at this point is characterized by relationships (more explicitly, a set of equations) that must exist between its properties, called conditions for equilibrium. Now isolate the system. If the system is in equilibrium with itself and its surroundings, no changes will occur in its internal condition. Thus, the relationships that define the equilibrium state of a general system that approached equilibrium with an arbitrary history will be identical to the set of relations that characterize the conditions for equilibrium in a system that was isolated from its surroundings as it approached that equilibrium state.

This argument, first advanced by Gibbs,[1] demonstrates that although the criterion for equilibrium (the extremum principle) that describes the equilibrium state for the isolated system will not be valid for a general system, the equations between the internal properties of the system derived from that criterion (the conditions for equilibrium) will be the same. It is concluded that:

> The conditions for equilibrium that are derived for an isolated system will be the same as those that hold for any system that achieves an equilibrium state, no matter what the details of the history of the system may be.

This set of equations, the conditions for equilibrium, is derived from the extremum principle, which provides the criterion for equilibrium in an isolated system, i.e., the entropy is a maximum. The strategy for deriving the conditions for equilibrium in an isolated system from the criterion for equilibrium is developed in the next section.

5.3   MATHEMATICAL FORMULATION OF THE GENERAL CONDITIONS FOR EQUILIBRIUM

At first glance, the mathematical problem posed by the principles spelled out in the preceding section is familiar and straightforward. Given a function (in this case the entropy, S) of several variables, find the values of the set of variables that correspond

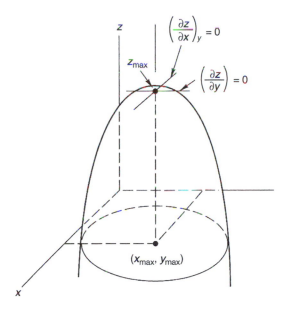

FIGURE 5.3 Illustration of the function z of two variables, x and y, in which z shows an extremum that is a maximum. The maximum value of z, z_{max}, occurs at the combination of dependent variables, (x_{max}, y_{max}).

to an extreme value (in this case a maximum) of the function. However, there are two complicating factors:

1. The entropy may be a function of a very large number of variables (see Chapter 6).
2. The variables that describe the change in entropy are not independent of each other in an isolated system.

More explicitly, the fact that the system is isolated implies that certain relationships exist among some of the variables. The extremum sought in this case is, therefore, a constrained maximum. Although this class of problems is treated in most standard calculus texts in those sections that deal with derivatives of functions of several variables, it is not widely emphasized in calculus courses.

The mathematical nature of the problem can be clarified by first considering a function, z, of two independent variables, x and y.

$$z = z(x, y) \tag{5.3}$$

Figure 5.3 is a geometric illustration of this functional relationship. To find an extreme value of z, write its differential:

$$dz = \left(\frac{\partial z}{\partial x}\right)_y dx + \left(\frac{\partial z}{\partial y}\right)_x dy \tag{5.4}$$

The extreme value occurs at those points on the $z(x, y)$ surface for which both derivatives are simultaneously zero:

$$\left(\frac{\partial z}{\partial x}\right)_y = 0 \qquad \text{and} \qquad \left(\frac{\partial z}{\partial y}\right)_x = 0 \qquad\qquad (5.5)$$

This pair of equations in x and y characterizes the extremum; their simultaneous solution yields a pair of values for the variables, (x_m, y_m), which are the coordinates in the xy plane that locate the extremum. In order to determine whether the extremum is a maximum, minimum, or saddle point, it is necessary to examine the second derivatives of z with respect to x and y. If both second derivatives are negative, then the extremum is a maximum.

These tactics apply to a function, z, of an arbitrarily large number, n, of independent variables:

$$z = z(u, v, x, y, \ldots) \qquad\qquad (5.6)$$

The differential of z has n terms in it:

$$dz = \left(\frac{\partial z}{\partial u}\right)du + \left(\frac{\partial z}{\partial v}\right)dv + \left(\frac{\partial z}{\partial x}\right)dx + \left(\frac{\partial z}{\partial y}\right)dy + \cdots \qquad\qquad (5.7)$$

An extreme value of z evidently occurs when all of the n coefficients in this equation simultaneously vanish, i.e., when

$$\left(\frac{\partial z}{\partial u}\right) = 0; \qquad \left(\frac{\partial z}{\partial v}\right) = 0; \qquad \left(\frac{\partial z}{\partial x}\right) = 0; \qquad \left(\frac{\partial z}{\partial y}\right) = 0; \cdots \qquad (5.8)$$

This set of n simultaneous equations in n unknowns may be solved (at least in principle) for the set of values of the n variables $(u_m, v_m, x_m, y_m, \ldots)$ that are the coordinates that locate the extremum in the function z.

The extrema described by Equation 5.3 to Equation 5.5 and Equation 5.6 to Equation 5.8 are unconstrained extrema because the set of variables that describe z in these two examples is explicitly assumed to be independent variables. In the thermodynamic application at hand, this assumption is not valid; the variables that most conveniently describe the change in entropy in an isolated system are not independent of each other.

If x and y are not independent variables in the first illustration, then there exists an equation relating these two variables; that is, in addition to the relation

$$z = z(x, y)$$

it will be true that:

$$y = y(x) \qquad\qquad (5.9)$$

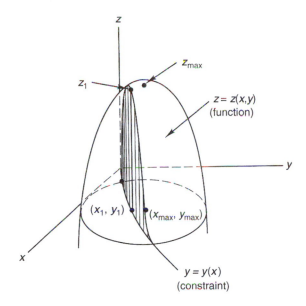

FIGURE 5.4 If the problem of interest involving the function in Figure 5.3 requires that the values of the variables x and y be constrained by a relationship $y = y(x)$, then, for this subset of (x, y) values, the maximum values of z, z_1, is different from the absolute maximum, z_{max}.

This situation is shown in Figure 5.4. Only those values of z that map above the curve describing the relation between the variables x and y are of interest in the problem. The relation $y = y(x)$ may be thought of as a constraint operating on the system that restricts the possible values of x and y. It will be observed in Figure 5.4 that the set of values that z takes on over this allowed domain of values of x and y has a maximum z_1 at the point (x_1, y_1) that is different from z_{max} at (x_{max}, y_{max}). This is a constrained maximum in z subject to the condition $y = y(x)$.

Figure 5.5 illustrates the proposition that a function $z(x, y)$ which is monotonic in x and y, i.e., exhibits no unconstrained extrema, may nonetheless exhibit a constrained maximum z_{max} if the constraining relation $y = y(x)$ is appropriately complicated.

These notions generalize to the case in which z is a function of n variables. The system is represented by a set of equations, the first representing the variation of z with the n variables,

$$z = z(u, v, x, y, \ ...)$$

and an additional m equations $(m < n)$ among the variables

$$u = u(x, y, \ ...)$$
$$v = v(x, y, \ ...)$$

(5.10)

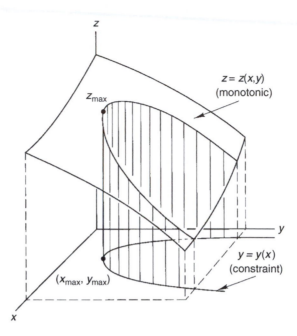

FIGURE 5.5 Even if the function $z(x, y)$ is monotonic, that is, exhibits no extrema, a constrained maximum in z can exist if the constraining relation between x and y has an extremum.

Here, z is the function whose extreme value is sought and the remaining equations are constraints placed upon the variables in z. An extreme value found for z with these auxiliary relations is a constrained extremum.

The general approach to the solution of this class of problems makes use of the constraining Equation 5.10 to eliminate dependent variables in the expression for z. The result expresses z in terms of $(n - m)$ variables which are now independent of each other and the extreme value may be found by applying the strategy in Equation 5.6 to Equation 5.8 for the case of independent variables, i.e., by writing the differential of z in terms of set of independent variables and setting all of the coefficients equal to zero.

EXAMPLE 5.1

Finding a constrained maximum by eliminating dependent variables by direct substitution.

Suppose z is expressed as a function of four variables,

$$z = xu + yv$$

but, in the problem at hand, u and v are related to x and y:

$$u = x + y + 12$$
$$v = x - y - 8$$

To find the extreme value of z subject to these constraints, eliminate the dependent variables u and v in the expression for z by substituting their relationships to the independent variables, x and y:

$$z = x(x + y + 12) + y(x - y - 8)$$

Simplify:

$$z = x^2 + 12x + 2xy - 8y - y^2$$

The variable, z, is now expressed in terms of independent variables, x and y. To find an extremum, form the differential:

$$dz = (2x + 12 + 2y)dx + (2x - 8 - 2y)dy$$

where the coefficients are the partials of z with respect to x and y. Set the coefficients equal to zero:

$$2x + 2y + 12 = 0$$
$$2x - 2y - 8 = 0$$

Solve: $x = -1$, $y = -5$. (Test the result by putting these values into the two linear equations and showing that they are obeyed.) These are the coordinates of the constrained extremum. Substitute these values into the expression for z to obtain the value of z at the extremum: $z = 14$.

EXAMPLE 5.2
Finding a constrained extremum by eliminating the differentials of the dependent variables.

Consider again the equation for z given in Example 5.1 with its attendant constraints. An alternate strategy for finding the extremum in z subject to the constraining relations $u = u(x, y)$, $v = v(x, y)$ begins by taking the differentials of the function z and the constraining equations:

$$dz = u \, dx + x \, du + v \, dy + y \, dv$$
$$du = dx + dy$$
$$dv = dx - dy$$

Eliminate the dependent differentials by substituting for du and dv in the expression for dz:

$$dz = u \, dx + x(dx + dy) + v \, dy + y(dx - dy)$$

Collect terms:

$$dz = (u + x + y)dx + (v + x - y)dy$$

Set the coefficients equal to zero:

$$u + x + y = 0$$
$$v + x - y = 0$$

To solve for x and y, use the constraining equations to evaluate u and v:

$$(x + y + 12) + x + y = 0 \rightarrow 2x + 2y + 12 = 0$$
$$(x - y - 8) + x - y = 0 \rightarrow 2x - 2y - 8 = 0$$

Note that these conditions for the extremum are identical with those derived in Example 5.1 and will thus yield the same solution.

The strategy outlined in Example 5.1 cannot be applied if the constraining equations are not explicit, i.e., if they cannot be solved explicitly for u and v. For example, suppose the equation relating u to x and y has the form:

$$x^2 yu \ln u = 12$$

This equation cannot be solved explicitly to obtain an expression of the form $u = u(x, y)$, to be substituted wherever u appears in the expression for z. The strategy presented in Example 5.2 can be applied, even where the constraining relations are implicit in the dependent variables. For this reason, the strategy employed for finding constrained maxima in the entropy function to obtain conditions for equilibrium will be that presented in Example 5.2.

Now that the mathematical nature of the problem has been presented in detail, it is possible to lay out a general strategy for finding the conditions for equilibrium in a thermodynamic system of arbitrary complexity:

1. Write a differential expression for the change in entropy that the system may experience when taken through an arbitrary process, explicitly including terms for all possible variations in the state of the system.
2. Write differential expressions that describe the constraints that apply to the system because it is considered to be isolated from its surroundings.
3. Use the isolation constraints to eliminate dependent variables in the description of the change in entropy for the system. The resulting differential expression for the change in entropy applies explicitly to the possible changes that may occur in an isolated system with all variables independent.
4. Set the coefficients of each of the differentials in the new expression for dS equal to zero. These equations, which specify an extreme value for S, are the conditions for equilibrium for the system.

This strategy for finding conditions for equilibrium is illustrated for a unary two-phase system in the next section.

5.4 APPLICATION OF THE GENERAL STRATEGY FOR FINDING CONDITIONS FOR THERMODYNAMIC EQUILIBRIUM: THE UNARY TWO-PHASE SYSTEM

Focus on a unary (one component) two-phase nonreacting otherwise simple system. The strategy for dealing with a two-phase system is straightforward. Each of the separate phases can be treated as a one-phase system with its own set of extensive and intensive properties. The apparatus for describing the behavior of such single-phase systems was developed in Chapter 4. The description of a two-phase system simply makes use of the fact that, for all of the extensive properties defined, the value for a system is the sum of the values for each of its parts. Thus, for a two-phase system each of the properties S', V', U', H', F', and G' is the sum of the values of the corresponding quantities for the separate phases.

Because the phase boundary separating two phases is a natural boundary, it is not possible to restrict the flow of matter between the phases. Accordingly, it is necessary to treat each phase as an open system in describing its behavior; that is, to allow for the possibility that the number of moles of each phase may change during an arbitrary process.

The first step in applying the general strategy requires an expression for the change in entropy experienced by the system when it is taken through an arbitrary process. The entropy for the system is the sum of the entropies of its parts; therefore, the change in entropy of the system is

$$dS'_{sys} = d(S'^\alpha + S'^\beta) = dS'^\alpha + dS'^\beta \qquad (5.11)$$

where the superscript α denotes the properties of one of the phases (designated the α phase) and superscript β designates properties of the β phase. It is thus necessary to compute the changes in entropy experienced separately by the α and β parts of the system during an arbitrary process.

Focus first upon the behavior of the α phase. Its internal energy, U'^α, is in general a function of its entropy, S'^α, volume, V'^α, and number of moles, n^α:

$$U'^\alpha = U'^\alpha(S'^\alpha, V'^\alpha, n^\alpha) \qquad (5.12)$$

From the combined statement of the first and second laws for the α phase

$$dU'^\alpha = T^\alpha \, dS'^\alpha - P^\alpha \, dV'^\alpha + \mu^\alpha \, dn^\alpha \qquad (5.13)$$

This equation introduces the coefficient,

$$\mu^\alpha = \left(\frac{\partial U'^\alpha}{\partial n^\alpha} \right)_{S'^\alpha, V'^\alpha} \qquad (5.14)$$

which is called the chemical potential of the component that makes up the α phase. This quantity plays a central role in the description of the chemical aspects of the behavior of matter; the concept will be developed in detail in Chapter 8 and beyond. For the present purpose, which is to illustrate the application of the general strategy for obtaining conditions for equilibrium, a definition, Equation 5.14, is all that is required. Rearranging Equation 5.13 gives an expression for the change in entropy of the α phase:

$$dS'^\alpha = \frac{1}{T^\alpha} dU'^\alpha + \frac{P^\alpha}{T^\alpha} dV'^\alpha - \frac{\mu^\alpha}{T^\alpha} dn^\alpha \qquad (5.15)$$

Identical reasoning for the β phase produces the analogous expression for the change in entropy of the β phase:

$$dS'^\beta = \frac{1}{T^\beta} dU'^\beta + \frac{P^\beta}{T^\beta} dV'^\beta - \frac{\mu^\beta}{T^\beta} dn^\beta \qquad (5.16)$$

Apply Equation 5.11: the expression for the change in entropy of the system has six terms:

$$dS'_{sys} = dS'^\alpha + dS'^\beta$$

$$dS'_{sys} = \frac{1}{T^\alpha} dU'^\alpha + \frac{P^\alpha}{T^\alpha} dV'^\alpha - \frac{\mu^\alpha}{T^\alpha} dn^\alpha \qquad (5.17)$$

$$+ \frac{1}{T^\beta} dU'^\beta + \frac{P^\beta}{T^\beta} dV'^\beta - \frac{\mu^\beta}{T^\beta} dn^\beta$$

The second step in the strategy requires evaluation of the constraints imposed by isolating the system. The boundary of an isolated system is rigid (so no work can be done on or by the system since the force due to the external pressure cannot move), thermally insulating (so no heat can be exchanged with the system) and impermeable (so no matter crosses the boundary of the system). These characteristics place the following constraints upon the state functions that describe the two-phase system:

Because there are no exchanges with the surroundings then, by the first law, the internal energy of the system must remain constant. Thus, whatever processes occur within the system,

$$dU'_{sys} = d(U'^\alpha + U'^\beta) = dU'^\alpha + dU'^\beta = 0 \qquad (5.18)$$

Because the boundary is rigid,

$$dV'_{sys} = d(V'^\alpha + V'^\beta) = dV'^\alpha + dV'^\beta = 0 \qquad (5.19)$$

Because the boundary is impermeable,

$$dn_{sys} = d(n^\alpha + n^\beta) = dn^\alpha + dn^\beta = 0 \qquad (5.20)$$

Thus, the differentials in the six terms that are required in Equations 5.17 to describe the change in entropy of a unary two-phase system when it is taken through an arbitrary process are not independent when the process is carried out in an isolated system. Three equations relate these six variables: Equation 5.18 to Equation 5.20. Use these relations to eliminate three of the six variables in Equation 5.17:

$$dU'^\alpha = -dU'^\beta \tag{5.21}$$

$$dV'^\alpha = -dV'^\beta \tag{5.22}$$

$$dn^\alpha = -dn^\beta \tag{5.23}$$

Internal energy, volume, and moles of material may be exchanged between the phases but are conserved for the isolated system. Substitute these expressions for dU'^β, dV'^β and dn^β in Equation 5.17:

$$
\begin{aligned}
dS'_{sys,iso} &= dS'^\alpha + dS'^\beta \\
&= \frac{1}{T^\alpha}dU'^\alpha + \frac{P^\alpha}{T^\alpha}dV'^\alpha - \frac{\mu^\alpha}{T^\alpha}dn^\alpha + \frac{1}{T^\beta}(-dU'^\alpha) \\
&\quad + \frac{P^\beta}{T^\beta}(-dV'^\alpha) - \frac{\mu^\beta}{T^\beta}(-dn^\alpha)
\end{aligned} \tag{5.24}
$$

Collect terms:

$$dS'_{sys,iso} = \left(\frac{1}{T^\alpha} - \frac{1}{T^\beta}\right)dU'^\alpha + \left(\frac{P^\alpha}{T^\alpha} - \frac{P^\beta}{T^\beta}\right)dV'^\alpha - \left(\frac{\mu^\alpha}{T^\alpha} - \frac{\mu^\beta}{T^\beta}\right)dn^\alpha \tag{5.25}$$

This equation describes the change in entropy accompanying an arbitrary change in state of a unary two-phase system when it is isolated from its surroundings. The function, S', is now expressed in terms of three properties that are independently variable, U'^α, V'^α, and n^α. The conditions that describe the maximum in entropy for such an isolated system are obtained by setting the coefficients of each of the three differentials in Equation 5.25 equal to zero:

$$\frac{1}{T^\alpha} - \frac{1}{T^\beta} = 0 \rightarrow T^\alpha = T^\beta \qquad \text{(thermal equilibrium)} \tag{5.26}$$

$$\frac{P^\alpha}{T^\alpha} - \frac{P^\beta}{T^\beta} = 0 \rightarrow P^\alpha = P^\beta \qquad \text{(mechanical equilibrium)} \tag{5.27}$$

$$\frac{\mu^\alpha}{T^\alpha} - \frac{\mu^\beta}{T^\beta} = 0 \rightarrow \mu^\alpha = \mu^\beta \qquad \text{(chemical equilibrium)} \tag{5.28}$$

When these three conditions for equilibrium are met within this two-phase isolated system, its entropy will be a maximum and the criterion for equilibrium satisfied. (Strictly speaking, it is also necessary to examine the second derivatives in the function to determine that the extremum is a maximum; however, exploration of this aspect of the problem is beyond the scope of this presentation.) Furthermore, it has been shown that the conditions for equilibrium derived for an isolated system are valid equations in an equilibrated system whether it is isolated or not. Thus, Equation 5.26 to Equation 5.28 are general conditions for thermal, mechanical, and chemical equilibrium in a unary two-phase system.

Stated in words, a unary two-phase system is in equilibrium when the temperature, pressure, and chemical potential of both phases are equal.

In Chapter 7, it will be shown how these equations form the basis for computing phase diagrams for unary systems.

5.5 ALTERNATE FORMULATIONS OF THE CRITERION FOR EQUILIBRIUM

The criterion for equilibrium developed in this chapter is chosen as the basis for presentation in the remainder of this text because it follows very directly from the second law of thermodynamics. Other criteria may be formulated, expressed in terms of extreme values other state functions, combined with constraining relations that are appropriate to their application.

For example, consider a system that is constrained to processes in which the entropy, volume, and number of moles of components cannot change. For a system with these constraints it can be demonstrated that the internal energy, U', can only decrease. Thus, the internal energy function is a monitor of the direction of spontaneous change in such a system. As such, a system proceeds toward equilibrium, its internal energy decreases until, at the equilibrium state its internal energy is a minimum. Thus, an alternate criterion for equilibrium may be stated:

> For a system constrained to constant entropy, volume and quantity of each of its components, the internal energy is a minimum at equilibrium.

There are two significant problems in formulating this alternate criterion:

1. Its derivation from the second law is less direct than the maximum entropy criterion.
2. While a system with the desired constraints, i.e., S', V' and n are constant, may be visualized mathematically, it may not be realized practically.

Since the criterion is general, it must apply to systems that approach equilibrium irreversibly. In such cases entropy is produced; the constraint, "S' is a constant," can only be achieved if the entropy flow out of the system is maintained to be equal to the entropy production. The construction of a real system satisfying this constant entropy constraint would incorporate very sophisticated monitoring and control devices. In contrast, the constraints required for the maximum entropy criterion are those of an isolated system, which may be easily visualized and approached in the laboratory.

It may be easily demonstrated that the minimum internal energy criterion (at constant S', V' and n) presented above, yields the same set of conditions for equilibrium as does the maximum entropy criterion (at constant U', V' and n). To find the constrained minimum in internal energy follow the same mathematical strategy. For each phase, the change in internal energy may be written:

$$dU'^\alpha = T^\alpha \, dS'^\alpha - P^\alpha dV'^\alpha + \mu^\alpha \, dn^\alpha \qquad (5.29)$$

and

$$dU'^\beta = T^\beta \, dS'^\beta - P^\beta \, dV'^\beta + \mu^\beta \, dn^\beta \qquad (5.30)$$

For the system,

$$dU'_{sys} = dU'^\alpha + dU'^\beta$$
$$dU'_{sys} = T^\alpha \, dS'^\alpha - P^\alpha \, dV'^\alpha + \mu^\alpha \, dn^\alpha + T^\beta \, dS'^\beta - P^\beta \, dV'^\beta + \mu^\beta \, dn^\beta \qquad (5.31)$$

The constraints for this criterion are:

$$dS'_{sys} = dS'^\alpha + dS'^\beta = 0 \rightarrow dS'^\alpha = -dS'^\beta \qquad (5.32)$$

$$dV'_{sys} = dV'^\alpha + dV'^\beta = 0 \rightarrow dV'^\alpha = -dV'^\beta \qquad (5.33)$$

$$dn_{sys} = dn^\alpha + dn^\beta = 0 \rightarrow dn^\alpha = -dn^\beta \qquad (5.34)$$

Use these constraints to eliminate dependent variables in Equation 5.31

$$dU'_{sys} = T^\alpha \, dS'^\alpha - P^\alpha \, dV'^\alpha + \mu^\alpha \, dn^\alpha + T^\beta(-dS'^\alpha) - P^\beta(-dV'^\alpha) + \mu^\beta(-dn^\alpha)$$

Collect terms

$$dU'_{sys} = (T^\alpha - T^\beta)dS'^\alpha - (P^\alpha - P^\beta)dV'^\alpha + (\mu^\alpha - \mu^\beta)dn^\alpha \qquad (5.35)$$

This expresses the internal energy of the system in terms of independent variables, S'^α, V'^α and n^α. To find the extremum (minimum) in the internal energy, set all the coefficients equal to zero:

$$T^\alpha - T^\beta = 0 \rightarrow T^\alpha = T^\beta \qquad \text{(thermal equilibrium)} \qquad (5.36)$$

$$P^\alpha - P^\beta = 0 \rightarrow P^\alpha = P^\beta \qquad \text{(mechanical equilibrium)} \qquad (5.37)$$

$$\mu^\alpha - \mu^\beta = 0 \rightarrow \mu^\alpha = \mu^\beta \qquad \text{(chemical equilibrium)} \qquad (5.38)$$

These conditions for equilibrium are identical with those obtained from the maximum entropy criterion. In his classic work, Gibbs[2] applied this minimum internal energy criterion for equilibrium; the resulting working equations are the same and the attendant mathematics is slightly simpler. However, the system constraints associated with this criterion are experimentally unrealistic. Accordingly, in this text, the criterion that will be used consistently in arriving at conditions for equilibrium is the maximum entropy-isolated system criterion. The strategy based upon this criterion provides the most direct connection between the laws of thermodynamics and the conditions for equilibrium.

Each of the other energy functions, H, F and G, may be associated with the condition for spontaneous change leading ultimately to a criterion for equilibrium, provided the system under consideration is subject to constraints that are appropriate for the given energy function:

1. In a system constrained to constant entropy and pressure, the direction of spontaneous change is monitored by a decrease in the enthalpy function; H is a minimum at equilibrium.
2. In a system constrained to constant temperature and volume, the Helmholtz free energy function decreases during every spontaneous change; F is a minimum at equilibrium.
3. In a system constrained to constant temperature and pressure, the Gibbs free energy function decreases for every spontaneous change; G is a minimum at equilibrium.

To illustrate the origin of these criteria, consider the last case for which it is assumed the system under consideration is experimentally controlled so that, whatever processes occur, its temperature and pressure are maintained to be constant. Consider two states, I and II, along the isothermal, isobaric path traversed by the system. From the definition of Gibbs free energy, for each of these states,

$$G^{\prime I} = U^{\prime I} + P^{I} V^{\prime I} - T^{I} S^{\prime I} \tag{5.39}$$

$$G^{\prime II} = U^{\prime II} + P^{II} V^{\prime II} - T^{II} S^{\prime II} \tag{5.40}$$

For the constraints imposed, $P^{I} = P^{II} = P$ and $T^{I} = T^{II} = T$. The difference in Gibbs free energy between the states may be written,

$$G^{\prime II} - G^{\prime I} = (U^{\prime II} - U^{\prime I}) + P(V^{\prime II} - V^{\prime I}) - T(S^{\prime II} - S^{\prime I})$$

or

$$\Delta G^{\prime} = \Delta U^{\prime} + P \Delta V^{\prime} - T \Delta S^{\prime} \tag{5.41}$$

The first law of thermodynamics in its most general form evaluates the change in internal energy for any real, irreversible process as

$$\Delta U^{\prime} = Q + W + W^{\prime} \tag{5.42}$$

The other two terms in Equation 5.41 can be expressed in terms of the process variables associate with an isothermal, isobaric reversible process connecting states I and II:

$$W_{\mathrm{rev},T} = -\int_{V_{1'}}^{V_{2'}} P\mathrm{d}V' = -P\int_{V_{1'}}^{V_{2'}} \mathrm{d}'V = -P(V_2 - V_1) = -P\Delta V' \qquad (5.43)$$

The heat absorbed for such a reversible process is given by

$$Q_{\mathrm{rev},P} = \int_{S_{1'}}^{S_{2'}} T\mathrm{d}S' = T\int_{S_{1'}}^{S_{2'}} \mathrm{d}S' = T(S_2 - S_1) = T\Delta S' \qquad (5.44)$$

Insert Equation 5.42 to Equation 5.44 into Equation 5.41:

$$\Delta G' = [Q + W + W'] - Q_{\mathrm{rev}} - W_{\mathrm{rev}}$$

or

$$\Delta G' = [Q - Q_{\mathrm{rev}}] + [W - W_{\mathrm{rev}}] + W' \qquad (5.45)$$

According to the second law, the heat absorbed during a reversible process connecting two end states is larger (algebraically) than for any irreversible process connecting those states: $Q_{\mathrm{rev}} > Q$. Further, the reversible mechanical work is also larger than that for any irreversible process: $W_{\mathrm{rev}} > W$. Finally, it may be argued that any spontaneous process may be arranged so that the system does nonmechanical work on the surroundings: $W' < 0$. When these conclusions are applied term by term to Equation 5.45, it is seen that both terms in brackets are negative and W' is also negative. Thus, for any irreversible spontaneous change in an isothermal, isobaric system, $\Delta G' < 0$. In other words, the Gibbs free energy function decreases for every spontaneous change in a system constrained to constant temperature and pressure. G' may thus be used as a monitor of the direction of spontaneous change in systems so constrained (but only in systems so constrained!). It follows that, when a system controlled at constant temperature and pressure comes to rest at its equilibrium state, its Gibbs free energy will be a minimum.

Since temperature and pressure must be held constant in applying this criterion, it is not possible to deduce from this extremum principle the conditions for thermal and mechanical equilibrium for a system. Only the conditions for chemical equilibrium may be derived from the principle of minimization of the Gibbs free energy. This is another point of argument that favors the maximum entropy-isolated system criterion as a basis for arriving at general conditions for equilibrium.

5.6 SUMMARY

Because the entropy of an isolated system can only increase, the final equilibrium state of an isolated system has the maximum value of entropy that the system may exhibit.

This extremum principle is the criterion for equilibrium in an isolated system.

Mathematical formulation of this constrained maximum leads to the conditions for equilibrium for this isolated system; these conditions are equations relating properties of the system that must be satisfied in order for the extremum to be achieved.

If a system that is not isolated from its surroundings comes to equilibrium with itself and its surroundings then, at that point, isolation of the system from its surroundings will not change the internal condition of the system.

Thus, the conditions for equilibrium derived for an isolated system are general conditions for equilibrium.

Application of this strategy to a unary two-phase system shows that, at equilibrium:

1. The temperatures of the two phases must be the same.
2. The pressures of the two phases must be the same.
3. The chemical potentials of the two phases must be the same.

Alternate conditions for spontaneous change and criteria for equilibrium may be derived in terms of other thermodynamic state functions:

1. For systems constrained to constant S' and V', U' decreases, and is a minimum at equilibrium.
2. For systems constrained to constant S' and P, H' decreases and is a minimum at equilibrium.
3. For systems constrained to constant T and V', F' decreases and is a minimum at equilibrium.
4. For systems constrained to constant T and P, G' decreases, and is a minimum at equilibrium.

Application of any of these criteria leads to the same set of conditions for equilibrium.

HOMEWORK PROBLEMS

5.1 Discuss the meaning of the statement: "An equilibrium state is a state of balance."

5.2 Give three illustrative examples each of
   a. An equilibrium state.
   b. A steady state.

5.3 State in outline form the logical steps that lead to our general criterion for equilibrium.

5.4 Contrast the concepts presented in this chapter as
   (a) Criterion for equilibrium
   (b) Conditions for equilibrium.

5.5 State in words and mathematically the set of relationships implied by the statement that a thermodynamic system is isolated from its surroundings during a process.

5.6 The development presented in this chapter describes the behavior of an isolated system. Systems of practical interest are rarely isolated from their surroundings in their approach toward equilibrium. How is it then possible to conclude that the results of this development are general, i.e., are not limited to an isolated system?

5.7 Find the extreme values of the function

$$z = -(x - 2)^2 - (y - 2)^2 + 4$$

Find the constrained maximum of this function corresponding to the condition

$$y = 2x$$

   a. Solve the problem first by eliminating y as a variable and finding the extreme value of the function $z = z[x, y(x)]$.
   b. Then solve the same problem by writing the differential forms of the two equations and substituting for dy in the expression for dz, then set the coefficient of dx equal to zero.

5.8 The steps in the strategy for finding conditions for equilibrium are:
   a. Write an expression for the change in entropy of the system when it is taken through an arbitrary process.
   b. Write the isolation constraints in differential form.
   c. Use the isolation constraints to eliminate dependent variables in the expression for the entropy.
   d. Collect terms.
   e. Set the coefficients of each differential equal to zero.
   f. Solve these equations for the conditions for equilibrium.
   g. Use the example of a unary two-phase system presented in Section 5.4 to write out each of these steps mathematically.

5.9 The combined statement of the first and second laws for the change in enthalpy of a unary single-phase system may be written:

$$dH' = T dS' + V' dP + \mu\, dn$$

Use this result to write an expression for the change in enthalpy of a two-phase $(\alpha + \beta)$ system. If the entropy, pressure, and total number of moles are constrained to be constant, then the criterion for equilibrium is that the enthalpy is a minimum. Paraphrase the strategy used to deduce the conditions for equilibrium in an isolated system to derive them for a system constrained to constant S', P and n. What happens to the condition for mechanical equilibrium?

REFERENCES

1. Willard Gibbs, J., *The Scientific Papers of J. Willard Gibbs*, Volume 1, Thermodynamics, Dover Publications, New York, p. 62, 1961.
2. Willard Gibbs, J., *The Scientific Papers of J. Willard Gibbs*, Volume 1, Thermodynamics, Dover Publications, New York, pp. 62–65, 1961.

6 Statistical Thermodynamics

CONTENTS

All of the concepts and relationships developed so far in this text visualize a thermodynamic system as consisting of some continuous medium or as a collection of separate media, each of which is continuous. A system is endowed with properties such as heat capacity, coefficient of expansion, an equation of state, and so on. Knowledge of the variation of these properties with the state of the system is sufficient to describe the macroscopic phenomena that the system may experience. This level of description of the behavior of matter is called "phenomenological" thermodynamics. No use has been made of the idea that the substance of this continuum is actually composed of atoms or molecules and that the behavior of the system is somehow related to the properties of the particles that compose it.

Development of the connection between the thermodynamic behavior of macroscopic systems outlined in Figure 1.4 and that of atoms on their submicroscopic scale is useful for a variety of reasons. A new level of understanding of how matter behaves emerges from this connection. At the phenomenological level of description, it is sufficient to know that experimental measurements show that the heat capacity of substance A is different from that of substance B, for example. At the atomic level of description, it is possible to predict not only that substances A and B have different heat capacities but how large the differences might be expected to be. Atomistic models provide a level of explanation for the behavior of a substance; phenomenological information merely provides a consistent description of how it behaves. The atomistic point of view provides a perspective that helps make sense out of the output of the continuum description of matter. In addition, the atomistic description may provide insight into the physical meaning of some of the phenomenological concepts that are unfamiliar. Indeed, much of the atomistic approach developed in this chapter centers on the concept of entropy, raising the level of sophistication of an understanding of the meaning of this elusive phenomenological property.

An atomistic description of the behavior of matter begins with the idea that each atom in the system may be assigned values of properties that describe its condition. For example, each atom in a gas has a position vector, $\mathbf{x}$, and velocity vector, $\mathbf{v}$. The state of an atom in a condensed phase is usually described by its energy, e. Such a description has an obvious problem: one cubic centimeter of a typical condensed phase contains about 10^{22} atoms (about one tenth of a mole). Specification of the properties (e.g., energy level) of 10^{22} atoms is a hopeless task. Even a modern computer listing an entry every picosecond (10^{12} entries per second) would take more than three centuries to make such a list. Specification of the thermodynamic state of a system with such an enormous table listing separately the energy of each atom is called the microstate of the system.

The mathematical discipline that provides the tools necessary to analyze very large collections of numbers is statistics. The primary tool supplied for the purpose is the concept of the distribution function. Atoms that have similar values of properties are lumped together into classes (in this case energy levels); the distribution function simply reports the number of particles in each class (energy level). In this way the quantity of information required to specify the condition of the system is greatly reduced. Description of the behavior of matter in terms of the distribution of particles over their allowable states is at the base of statistical thermodynamics. Specification of the thermodynamic state of a system in terms of such a distribution function is called the macrostate for the system.

This chapter begins with a more detailed development of the concepts of microstate and macrostate just introduced. The number of distinct microstates corresponding to a given macrostate is computed by applying some combinatorial analysis adapted from the field of statistics. A fundamental hypothesis of statistical thermodynamics, due to Boltzmann, is introduced; this hypothesis relates the entropy of the system to the number of microstates that correspond to a given macrostate. Then the general strategy for finding conditions for equilibrium developed in Chapter 5 is applied to this statistical description of the behavior of matter. In this case, the equilibrium condition is described by a specific distribution of atoms over their allowable states, called the Boltzmann distribution function. The important distinguishing physical quantity contained in the description of this equilibrium distribution is found to be the partition function for the system. The partition function for a given system may be evaluated from a list of energy levels that the particles in the system may exhibit. Finally, it is shown that, given the partition function for a system, all of the macroscopic phenomenological properties of the system that we have defined may be computed. This completes the connection between an atomistic model for the system, formulated in terms of the list of allowable energy states, and the experimentally measurable thermodynamic properties. Application of this algorithm is illustrated for the ideal gas model and the Einstein model for a crystalline solid.

6.1   MICROSTATES, MACROSTATES, AND ENTROPY

A unary thermodynamic system is composed of a very large number of particles (atoms or molecules) that are all structurally identical. The condition of such a collection of particles at any instant in time could, in principle, be described by listing the condition of each particle in the array. The phrase, in principle, is included because, in practice, such a specification is not possible as noted in the introduction to this chapter. Notwithstanding that such a description is impractical, the concept of such a specification of the state of the system may be contemplated. Such a description, specifying the condition or state of each particle that composes the system, is defined to be a microstate that the system may exhibit. If any entry in this list is altered, i.e., if any particle changes its condition, the system is considered to be in a different microstate. Evidently, because the number of particles and conditions is large, the number of different microstates that a given system might exhibit in its evolution is enormous. This is illustrated in Table 6.1 for a system consisting of only four particles each of which may exhibit only two states.

TABLE 6.1
Microstates and Macrostates for a Simple System

List of Microstates					
State	ε_1	ε_2	State	ε_1	ε_2
A	abcd	—	I	bc	ad
B	abc	d	J	bd	ac
C	abd	c	K	cd	ab
D	acd	b	L	a	bcd
E	bcd	a	M	b	acd
F	ab	cd	N	c	abd
G	ac	bd	O	d	abc
H	ad	bc	P	—	abcd

List of Macrostates				
	Number of Particles			
State	ε_1	ε_1	Corresponding Microstates	Number	Probability
I	4	0	A	1	1/16
II	3	1	B,C,D,E	4	4/16
III	2	2	F,G,H,I,J,K	6	6/16
IV	1	3	L,M,N,O	4	4/16
V	0	4	P	1	1/16

Particles: a,b,c,d; States: ε_1, ε_2; Number of microstates $= 2^4 = 16$.

Since the particles are assumed to be physically identical, the macroscopically observable behavior of the system is not dependent upon which particles exist in a given state, but merely upon how many particles are in that state. The macroscopic properties will be the same whether particles a and b are in state ε_2, or particles c and d. Thus the microstates listed as B, C, D, and E in Table 6.1 will give the same values for the macroscopic properties of the system. Each of these microstates corresponds to the condition "two particles are in state ε_1 and two particles are in state ε_2." This observation gives rise to a much more efficient and useful way of describing the state of a system at an atomic level, called the macrostate for the system.

To specify the macrostate of a system at a given instant in time, focus not on the particles but on the list of possible conditions or states that the individual atoms may exhibit. In Table 6.1, the number of states is two. More generally, suppose that there are r states that each atom may exhibit. In Table 6.1, specification of a given macrostate requires two numbers: the number in state ε_1 and the number in state ε_2. Specification of the macrostate labeled II requires two numbers: 3, 1. This pair of numbers constitutes a rudimentary distribution function: "3 particles in state ε_1, 1 particle in state ε_2." In the general case, a macrostate is specified by assigning a number of particles to each of the r available atomic states. Let the list of possible states be ε_1, ε_2, ε_3,...,ε_i,...,ε_r. Then a particular macrostate is given by

$$\begin{array}{ccccccc} \varepsilon_1 & \varepsilon_2 & \varepsilon_3 & \cdots & \varepsilon_i & \cdots & \varepsilon_r \\ n_1 & n_2 & n_3 & \cdots & n_i & \cdots & n_r \end{array}$$

where n_i is the number of particles that have energy ε_i. The set of numbers $(n_1,...,n_r)$ is a distribution function specifying how the atoms are distributed over the energy levels. This distribution describes the macrostate for the system.

For the example of four particles that each may exhibit two energy states, Table 6.1 shows that the system may exhibit 16 microstates. The number of possible macrostates is significantly smaller: there are five macrostates possible for the system. Evidently a given macrostate may correspond to a number of different microstates. These relationships are spelled out explicitly for the simple example in Table 6.1. As the number of particles and energy levels is increased, the number of microstates that may correspond to a given macrostate may become very large. For example, for a system consisting of just ten particles distributed over three energy levels, the total number of microstates is $3^{10} = 59,049$. The number of distinguishable macrostates for this system is 60. Thus for this system, on the average, a macrostate corresponds to about 1000 microstates. Real thermodynamic systems may contain 10^{22} particles and 10^{15} energy levels. It may be appreciated that the number of microstates corresponding to the typical macrostate in these cases is almost incredibly large. The number of microstates that correspond to a given macrostate is a central quantity in the development of statistical thermodynamics.

From the atomistic point of view a thermodynamic process, i.e., a change in the macroscopic state of the system, corresponds to a redistribution of the atoms over their allowable states, i.e., to a collection of changes in the number of particles in each atomic energy state. As atoms experience changes in their condition, altering from one state to a neighboring one, the system as a whole evolves through a sequence of microstates. The fraction of its lifetime that each particle spends in a given energy state may in the long run be expected to be the same for all particles, since they are identical. Accordingly, it may be argued that the time that the system spends in any given microstate is the same for all microstates. The time that the system spends in a particular macrostate is the sum of the times that it spends in the various microstates that correspond to that macrostate. The fraction of time it spends in any given macrostate is thus the ratio of the number of microstates that correspond to that macrostate to the total number of microstates that the system is capable of exhibiting. This fraction may be plausibly interpreted as the probability that the system exhibits the given macrostate in any randomly selected instant in time.

The simple system analyzed in Table 6.1 serves to illustrate this principle. There is a total of 16 microstates for this system of four particles and two energy levels (2^4). If all are equally likely, the system spends 1/16th of its time in each microstate. There are five macrostates labeled with Roman numerals. As an example, focus on the macrostate labeled II. It occurs when any one of the four microstates, B, C, D, or E, exists. Since each of these microstates exists 1/16th of the time, macrostate II will occur 4/16 or 1/4 of the time. The probability that macrostate II will be observed at any instant in time is 1/4.

In order to apply these ideas to a thermodynamic system, it is necessary to generalize the computation of the total number of microstates a system may exhibit and the number of microstates that correspond to any given macrostate. Consider a system that contains a large number, N_0, of particles. (For example, for one mole of system N_0 is 6.023×10^{23}.) Suppose that each of these particles may exist in any of a large number of conditions or states: let the number of such states be r. In Table 6.1, N_0 is four and r is two. The total number of microstates that a system may exhibit may be computed through the following exercise in induction. Particle **a** may be placed in any of the r states. Particle **b** may also be independently placed in any of the r states. If these were the only two particles in the system, the number of possible arrangements in the r states would be $r \cdot r$. If a third particle is added to the system, it may also be placed independently in any of the r states. The number of arrangements increases to $r \cdot r \cdot r$. Each particle that is added to the system may be added in any of the r states. Thus, the number of different ways the N_0 particles may be arranged in the r states is r^{N_0}. Even if r is not large, raising it to a power of the order of 10^{23} makes the total number of microstates enormous.

The number of microstates corresponding to each of the five macrostates available to the four-particle two-state system is listed in Table 6.1. For a general system composed of N_0 particles distributed over r states, the description of a particular macrostate has the form of a distribution function:

$$(n_1, n_2, n_3, \ldots, n_i, \ldots, n_r)$$

where n_i is the number of particle in state ε_i. For example, for a 20-particle system with seven energy states this list might read:

$$(1, 3, 4, 6, 2, 3, 1)$$

for a particular macrostate. The number of microstates that correspond to a given macrostate is a standard problem in that branch of the foundations of statistics known as combinatorial analysis. The statement of the problem, "How many microstates correspond to the macrostate $(n_1, n_2, \ldots, n_r)$?" may be restated: "How many different ways can N_0 balls be arranged in r boxes such that there are n_1 balls in the first box, n_2 in the second and so on to n_r balls in the rth box?" The answer to this question, familiar in combinatorial analysis is:

$$\Omega = \frac{N_0!}{n_1! \cdot n_2! \ldots n_r!} \tag{6.1}$$

Ω is thus the number of microstates that correspond to the macrostate given by the set of n_i values in the denominator. The notation $n!$, read "n factorial," represents the product of the sequence of numbers:

$$n! \equiv n \cdot (n-1) \cdot (n-2) \ldots 3 \cdot 2 \cdot 1$$

Note that as n increases, $n!$ increases very rapidly.

Application of Equation 6.1 to each of the five macrostates listed for the system considered in Table 6.1 demonstrates the validity of the formula. Recall that $0! = 1$.

$$\text{Macrostate I}: \ \Omega_I = \frac{4!}{4! 0!} = \frac{(4 \cdot 3 \cdot 2 \cdot 1)}{(4 \cdot 3 \cdot 2 \cdot 1) \cdot (1)} = 1$$

$$\text{Macrostate II}: \ \Omega_{II} = \frac{4!}{3! 1!} = \frac{(4 \cdot 3 \cdot 2 \cdot 1)}{(3 \cdot 2 \cdot 1) \cdot (1)} = 4$$

$$\text{Macrostate III}: \ \Omega_{III} = \frac{4!}{2! 2!} = \frac{(4 \cdot 3 \cdot 2 \cdot 1)}{(2 \cdot 1) \cdot (21)} = 6$$

$$\text{Macrostate IV}: \ \Omega_{IV} = \frac{4!}{1! 3!} = \frac{(4 \cdot 3 \cdot 2 \cdot 1)}{(1) \cdot (3 \cdot 2 \cdot 1)} = 4$$

$$\text{Macrostate V}: \ \Omega_V = \frac{4!}{0! 4!} = \frac{(4 \cdot 3 \cdot 2 \cdot 1)}{(1) \cdot (4 \cdot 3 \cdot 2 \cdot 1)} = 1$$

It has been argued that the probability that the system exists in a given macrostate, interpreted as the fraction of the time that it spends in microstates that correspond to that macrostate, is the ratio of the number of microstates, Ω_J, which correspond to the Jth macrostate to the total number of microstates the system may exhibit. This probability is thus[1],

$$P_J = \frac{\Omega_j}{r^{N_0}} = \frac{N_0!}{\prod_{i=1}^{r} n_i!} \cdot \frac{1}{r^{N_0}} \tag{6.2}$$

The notation $\Pi n_i!$ is shorthand for the product $n_1!n_2!...n_r!$. Macrostates for which Ω_J is large will exist in the system for a greater fraction of time than those for which Ω_J is small. For example, in Table 6.1, macrostate III is six times more probable than states I or V.

Of all of the macrostates that may exist for a system one will contain more microstates than any other. This macrostate has the maximum value of Ω and the maximum probability of appearing at any moment in time. Examination of the behavior of the P_J function for a variety of macrostates demonstrates that, for systems with large numbers of particles, this function has an extremely sharp peak at the macrostate containing the maximum number of microstates (Figure 6.1). Macrostates that are only slightly different from the maximum probability state have a very much smaller probability of being observed. Macrostates that differ significantly from the maximum probability state have negligible probability of coming into existence. Thus, the maximum probability state, or those very near to it, will be observed almost all of the time. If this most likely state is interpreted as the macrostate that corresponds to the equilibrium state for the system, then this hypothesis forms the basis for connecting the statistical, atomistic description of the system with phenomenological thermodynamics.

In phenomenological thermodynamics, the equilibrium state is also character-ized by an extremum: the entropy of an isolated system is a maximum at equilibrium. This correspondence suggests a connection between entropy and Ω, the number of microstates corresponding to a given macrostate. If the functional relationship between these quantities is monotonic (i.e., both either increase together or decrease together) then when one function maximizes, so will the other. Furthermore, if one compares the values of these two quantities for the range of conditions a system might exhibit, it is evident that Ω varies over orders of magnitude of orders of magnitude, while values of entropy vary over, at most,

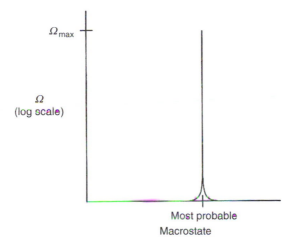

FIGURE 6.1 The probability distribution for macrostates has an extremely sharp peak for systems with large numbers of particles and energy levels.

one or two orders of magnitude. These considerations lead to the basic assumption that connects the atomistic and phenomenological descriptions known as the Boltzmann hypothesis:

$$S = k \ln \Omega \qquad (6.3)$$

where k turns out to be a universal constant, called the Boltzmann constant. It will be demonstrated later that the value of k is simply R/N_0, where R is the ideal gas constant and N_0 is Avagadro's number. Thus, k is simply the value of the gas constant per atom (or molecule) rather than per mole of gas. The consequences of this hypothesis will now be explored.

6.2 CONDITIONS FOR EQUILIBRIUM IN STATISTICAL THERMODYNAMICS

In the atomistic description of the thermodynamic behavior of a system the equilibrium state is that particular macrostate which maximizes the entropy of the system when it is isolated. Thus, the equilibrium state is a particular set of values of the numbers that define macrostates: $(n_1, n_2,...,n_r)_{eq}$. To find this set of numbers and relate them to the original specification of the behavior of the system contained in the list of states the individual atoms may exhibit $(\varepsilon_1, \varepsilon_2,...,\varepsilon_r)$, it is only necessary to apply the general strategy for finding conditions for equilibrium developed in Chapter 5. This strategy involves the following steps:

1. Write an expression for the change in entropy of the system in terms of the variables that define its state. In statistical thermodynamics, these variables are $(n_1, n_2,...,n_r)$.
2. Write expressions for the constraints on the variation of these variables imposed by the limitation of an isolated system.
3. Derive the set of equations that must be satisfied in order for the entropy function to be a maximum, subject to the isolation constraints.

These steps are carried out in this section.

6.2.1 EVALUATION OF ENTROPY

An explicit relation between the entropy of a macrostate and the set of numbers that specify its distribution is obtained by combining Equation 6.1 and Equation 6.3:

$$S = k \ln \left(\frac{N_0!}{\prod_{i=1}^{r} n_i!} \right) \qquad (6.4)$$

The computation of the factorials required to evaluate S is a very time-consuming process, even for a high speed computer. Computation of 1000! requires the computation of 1000 products of numbers. Since the numerator in Equation 6.4 is of the order of 10^{23}, its computation would be impractical. Fortunately, the values of factorials of large numbers may be estimated at very high precision by a simple formula known as *Stirling's approximation*[2]:

$$\ln x! \cong x \ln x - x \tag{6.5}$$

For example, for $x = 100$, $\ln x! = 364$, while $x \ln x - x$ gives 361, an error of less than 1%. The precision of the estimate increases as x increases and is excellent for very large x values, as are characteristic of the n_i quantities in statistical thermodynamics.

Use the properties of logarithms to simplify Equation 6.4. First, the log of a ratio of numbers is the difference of their logs:

$$S = k \left[\ln N_0! - \ln \left(\prod_{i=1}^{r} n_i! \right) \right]$$

In the second term, the log of a product of numbers is the sum of their logs:

$$S = k \left[\ln N_0! - \sum_{i=1}^{r} \ln(n_i!) \right]$$

apply Stirling's approximation to both $\ln x!$ terms in this equation:

$$S = k \left[(N_0 \ln N_0 - N_0) - \sum_{i=1}^{r} (n_i \ln n_i - n_i) \right] \tag{6.6}$$

The summation may be expressed as the difference between two summations:

$$S = k \left[(N_0 \ln N_0 - N_0) - \sum_{i=1}^{r} n_i \ln n_i + \sum_{i=1}^{r} n_i \right]$$

Recognize that the total number of particles in the system is the sum of the numbers in each level:

$$N_0 = \sum_{i=1}^{r} n_i \tag{6.7}$$

Thus,

$$S = k\left[N_0 \ln N_0 - N_0 - \sum_{i=1}^{r} n_i \ln n_i + N_0 \right]$$

or

$$S = k\left[N_0 \ln N_0 - \sum_{i=1}^{r} n_i \ln n_i \right]$$

Substitute Equation 6.7 for the coefficient of the first term:

$$S = k\left[\left(\sum_{i=1}^{r} n_i \right) \ln N_0 - \sum_{i=1}^{r} n_i \ln n_i \right]$$

Combine as a single summation:

$$S = k\left[\sum_{i=1}^{r} n_i (\ln N_0 - \ln n_i) \right]$$

The difference of two logs is the log of their ratio:

$$S = k \sum_{i=1}^{r} n_i \ln\left(\frac{N_0}{n_i} \right)$$

Finally, use the property that the log of an inverse of a ratio is equal to minus the log of the ratio:

$$S = -k \sum_{i=1}^{r} n_i \ln\left(\frac{n_i}{N_0} \right) \tag{6.8}$$

This development thus converts Equation 6.4 from an intractable computation problem involving factorials to a much simpler problem involving logarithms of large numbers. Equation 6.8 may be used to compute the entropy of any macrostate that may exist for any given atomistic model of the system.

In statistical thermodynamics, a process, i.e., a change in the thermodynamic state, is described by a change in the macrostate for the system. More explicitly, a process involves changes in the number of particles in some or all of the energy states, i.e., a redistribution of the particles over the energy levels in the system. The number of particles in some energy levels increases while those in others decrease. Mathematically, a process may be represented by a collection of changes in the n_i's:

$$(\Delta n_1, \Delta n_2, \ldots, \Delta n_i, \ldots, \Delta n_r)$$

Since the number of particles in each energy level is very large, for an infinitesimal change in the macroscopic state of the system these finite changes, Δn_i, may be replaced with infinitesimal changes in the variables, dn_i. Thus, for an infinitesimal change in the thermodynamic state of the system, the corresponding redistribution of the particles among the energy states may be described by a set of infinitesimal changes in the (large) numbers, n_i:

$$(dn_1, dn_2, ..., dn_i, ..., dn_r)$$

Since Equation 6.8 describes the entropy of any macroscopic state, the change in entropy for an arbitrary change in state may be obtained by differentiating this equation.

$$dS = -kd\left[\sum_{i=1}^{r} n_i \ln\left(\frac{n_i}{N_0}\right)\right]$$

$$dS = -k\sum_{i=1}^{r} d\left[n_i \ln\left(\frac{n_i}{N_0}\right)\right]$$

$$dS = -k\sum_{i=1}^{r} d[n_i \ln n_i - n_i \ln N_0]$$

$$dS = -k\sum_{i=1}^{r}\left[\ln n_i \cdot dn_i + n_i\left(\frac{1}{n_i}\right)dn_i - \ln N_0 \cdot dn_i - n_i\left(\frac{1}{N_0}\right)dN_0\right]$$

$$dS = -k\left[\sum_{i=1}^{r}(\ln n_i - \ln N_0)dn_i + \sum_{i=1}^{r} dn_i - \sum_{i=1}^{r}\left(\frac{n_i}{N_0}\right)dN_0\right]$$

From Equation 6.7,

$$dN_0 = \sum_{i=1}^{r} dn_i \quad \text{and} \quad \sum_{i=1}^{r} \frac{n_i}{N_0} = 1$$

Thus,

$$dS = -k\left[\sum_{i=1}^{r}(\ln n_i - \ln N_0)dn_i + dN_0 - 1 \cdot dN_0\right]$$

Finally,

$$dS = -k\sum_{i=1}^{r} \ln\left(\frac{n_i}{N_0}\right)\cdot dn_i \tag{6.9}$$

This expression describes the change in entropy for any process through which the system may be taken; no restrictions have been imposed upon the nature of the redistribution of the particles over their energy levels.

6.2.2 EVALUATION OF THE ISOLATION CONSTRAINTS

Recall that the application of the criterion for equilibrium requires that the system be isolated from its surroundings. This constraint places some restrictions upon the interchanges that may occur during the redistribution of the particles over their energy levels. Isolation from the surroundings implies that, whatever processes occur inside the system, the total number of particles cannot change and the internal energy of the system cannot change. The total number of particles is related to the number in each energy level in Equation 6.7. Computation of the internal energy of the system is straightforward. If ε_i is the energy of a particle in the ith state, then $n_i \varepsilon_i$ is the energy of all of the particles in that state. The total energy of the whole collection is just the sum of the energies in each level:

$$U = \sum_{i=1}^{r} \varepsilon_i n_i \qquad (6.10)$$

Since N_0 and U cannot change in an isolated system, redistributions of particles over the energy levels are subject to the constraints:

$$dU = \sum_{i=1}^{r} \varepsilon_i dn_i = 0 \qquad (6.11)$$

and

$$dN_0 = \sum_{i=1}^{r} dn_i = 0 \qquad (6.12)$$

A second term in the differentiation of the product $\varepsilon_i n_i$ is not required because the energy levels over which the particles are distributed, the ε_i, do not change during a process; processes are viewed to occur by redistributing the particles on a fixed set of energy levels.

6.2.3 THE CONSTRAINED MAXIMUM IN THE ENTROPY FUNCTION

The final step in applying the strategy for finding conditions for equilibrium involves determining the set of values of the variables that yield a maximum in entropy in a system subject to the constraints imposed by isolation. Equation 6.11 and Equation 6.12 must be used to eliminate the dependent variables in the expression for the change in entropy, Equation 6.9. In Equation 6.9 the change in entropy is expressed as a function of r variables, i.e., a value of dn_i for each of the r energy levels. In an isolated system only $(r - 2)$ of these dn_i's are independent because they are related by Equation 6.11 and Equation 6.12. The simple substitution procedure used to eliminate dependent variables in the previous application of this strategy in Chapter 5 cannot be applied in this case; the large number of variables involved makes the algebra in the problem unwieldy.

Fortunately, a general procedure for solving this class of mathematical problems exists, and is known as the method of Lagrange multipliers. For the present purpose, it is sufficient to outline the procedure and apply it to the problem at hand. The steps involved in finding a constrained extremum in a function of many variables (such as S) subject to constraining equations, such as Equation 6.11 and Equation 6.12, are:

1. Multiply each of the differential forms of the constraining equations by an arbitrary constant.
2. Add these forms to the differential of the function whose extreme value is sought.
3. Collect like terms and set the resulting combination of differential forms equal to zero.
4. Set the coefficients of each of the differentials that appear in this equation equal to zero.
5. Solve the resulting set of equations, evaluating the Lagrange multipliers in the process.

This procedure solves the general mathematical problem of finding a constrained extremum. If the function whose extreme value is sought is the entropy, Equation 6.9, and the constraining equations are the isolation constraints, Equation 6.11 and Equation 6.12, then this set of solutions derived from setting the coefficients equal to zero are the conditions for equilibrium for the system.

To apply this strategy to the problem at hand, first multiply the constraint equations by arbitrary constants:

From Equation 6.11:

$$\alpha \cdot dN_0 = \alpha \cdot \sum_{i=1}^{r} dn_i = 0$$

From Equation 6.2:

$$\beta \cdot dU = \beta \cdot \sum_{i=1}^{r} \varepsilon_i dn_i = 0$$

where α and β are the Lagrange multipliers. Add these to the expression for dS, Equation 6.9, and set the result equal to zero:

$$dS + \alpha \cdot dN_0 + \beta \cdot dU + 0 \qquad (6.13)$$

Substitute for dS, dN_0, and dU:

$$-k \sum_{i=1}^{r} \ln\left(\frac{n_i}{N_0}\right) \cdot dn_i + \alpha \cdot \sum_{i=1}^{r} dn_i + \beta \cdot \sum_{i=1}^{r} \varepsilon_i \, dn_i = 0$$

Collect terms:

$$\sum_{i=1}^{r} \left[-k \ln\left(\frac{n_i}{N_0}\right) + \alpha + \beta \varepsilon_i \right] dn_i = 0 \qquad (6.14)$$

This equation contains r terms, each identical in form but each with its own value of ε_i and n_i. To find the maximum in S subject to the isolation constraints, set each of these coefficients equal to zero. This yields r equations of the form

$$-k \ln\left(\frac{n_i}{N_0}\right) + \alpha + \beta \varepsilon_i = 0 \qquad (i = 1, 2, ..., r)$$

Algebraic manipulation of this result gives:

$$\frac{n_i}{N_0} = e^{\frac{\alpha}{k}} \cdot e^{\frac{\beta \varepsilon_i}{k}} \qquad (i = 1, 2, ..., r) \qquad (6.15)$$

This set of r equations constitutes the conditions for equilibrium in the statistical description of thermodynamic behavior. The equilibrium macrostate is character-ized by this relationship: the fraction of particles in the ith energy level (n_i/N_0) is exponentially related to the value of the energy for that level, ε_i.

It remains to evaluate the Lagrange multipliers, α and β, in terms of the thermodynamic properties of the system. The isolation constraints expressed in Equation 6.7 and Equation 6.10 are used for this purpose. Substitute the result obtained in Equation 6.15 into Equation 6.7:

$$\sum_{i=1}^{r} \frac{n_i}{N_0} = 1 = \sum_{i=1}^{r} e^{\frac{\alpha}{k}} \cdot e^{\frac{\beta \varepsilon_i}{k}} = e^{\frac{\alpha}{k}} \cdot \sum_{i=1}^{r} e^{\frac{\beta \varepsilon_i}{k}}$$

Solve for the factor containing α:

$$e^{\frac{\alpha}{k}} = \frac{1}{\displaystyle\sum_{i=1}^{r} e^{\frac{\beta \varepsilon_i}{k}}} \qquad (6.16)$$

The summation in the denominator on the right-hand side of this equation plays a central role in statistical thermodynamics. Define this quantity to be the partition function, $\wp$, for the system.

$$\wp \equiv \sum_{i=1}^{r} e^{\frac{\beta \varepsilon_i}{k}} \tag{6.17}$$

It will be shown that all of the thermodynamic properties of the system may be computed if the partition function is known.

Equation 6.16 permits elimination of the Lagrange multiplier, α, from the expression for the equilibrium distribution, Equation 6.15:

$$\frac{n_i}{N_0} = \frac{1}{\wp} \cdot e^{\frac{\beta \varepsilon_i}{k}} \tag{6.18}$$

It remains to evaluate β.

The Lagrange multiplier associated with the energy constraint may be determined by comparing the expression for the change in entropy for a reversible process computed in statistical thermodynamics with that computed in phenomenological thermodynamics. The statistical thermodynamical expression for entropy change is given in Equation 6.9. A reversible process is a sequence of equilibrium states. Thus, the ratio (n_i/N_0) that appears in Equation 6.9 may be evaluated by applying Equation 6.18:

$$dS = -k \sum_{i=1}^{r} \ln\left(\frac{n_i}{N_0}\right) \cdot dn_i = -k \sum_{i=1}^{r} \ln\left(\frac{1}{\wp} \cdot e^{\frac{\beta \varepsilon_i}{k}}\right) dn_i \tag{6.19}$$

Use the properties of logarithms, including $\ln(e^x) = x$:

$$dS = -k \sum_{i=1}^{r} \left(\frac{\beta \varepsilon_i}{k} - \ln \wp\right) dn_i = -\beta \sum_{i=1}^{r} \varepsilon_i dn_i + k \ln \wp \sum_{i=1}^{r} dn_i$$

Apply Equation 6.11 and Equation 6.12 to identify the summations in this expression:

$$dS = -\beta dU + k \ln \wp dN_0 \tag{6.20}$$

The combined statement of the first and second laws provides the analogous phenomenological expression. For an open system,

$$dU = T\, dS - P\, dV + \mu\, dN_0$$

(N_0 is the total number of atoms in the system; μ is thus the chemical potential per atom.) Solve for dS:

$$dS = \frac{1}{T} dU + \frac{P}{T} dV - \frac{\mu}{T} dN_0 \tag{6.21}$$

The volume term in Equation 6.21 has no counterpart in the statistical expression for the entropy, Equation 6.20, because in this introductory development of statistical thermodynamics it is assumed that the average volume occupied by an atom is the same for all energy levels. Compare Equation 6.20 and Equation 6.21:

$$\beta = -\frac{1}{T} \tag{6.22}$$

$$\mu = kT \ln \wp \tag{6.23}$$

Thus, β is found to be minus the reciprocal of the absolute temperature.

Substitution of this result into Equation 6.17 and Equation 6.18 gives the complete expression for the equilibrium distribution of particles over the energy levels:

$$\frac{n_i}{N_0} = \frac{1}{\wp} e^{-\frac{\varepsilon_i}{kT}} \tag{6.24}$$

where, $\wp$, the partition function is

$$\wp \equiv \sum_{i=1}^{r} e^{-\frac{\varepsilon_i}{kT}} \tag{6.25}$$

A list of the energy levels available to the particles in the system, the set of ε_i's, constitutes a model for the behavior of the atoms in the system. According to Equation 6.25, the partition function for the model may be computed from such a list. Equation 6.24 then yields the number of particles in each energy level at equilibrium. This description of the behavior of the system is complete and detailed: all of the macroscopic thermodynamic properties of the system may be derived from these results.

6.2.4 CALCULATION OF THE MACROSCOPIC PROPERTIES FROM THE PARTITION FUNCTION

Equation 6.8 gives the entropy of a system for any distribution of particles over their energy levels. The value of the entropy for the equilibrium distribution is obtained by substituting Equation 6.24 into Equation 6.8:

$$S = -k \sum_{i=1}^{r} n_i \ln\left(\frac{n_i}{N_0}\right) = -k \sum_{i=1}^{r} n_i \ln\left[\frac{1}{\wp} e^{\frac{\varepsilon_i}{kT}}\right] = -k \sum_{i=1}^{r} n_i \left[-\frac{\varepsilon_i}{kT} - \ln \wp\right]$$

$$= +\frac{k}{kT} \sum_{i=1}^{r} \varepsilon_i n_i + k \ln \wp \sum_{i=1}^{r} n_i$$

The summation in the second term is the total number of particles in the system, N_0; the summation in the first term is the internal energy of the system, U.

$$S = \frac{1}{T}U + kN_0 \ln \wp \tag{6.26}$$

Recall the definition of the Helmholtz free energy function, F, Equation 4.7. Equation 6.26 may be written:

$$F \equiv U - TS = U - T\left[\frac{1}{T}U + kN_0 \ln \wp\right]$$

$$F = -N_0 kT \ln \wp \tag{6.27}$$

Thus, the Helmholtz free energy function may be computed if the partition function is known: no other information is required.

Next, recall the combined statement of the first and second laws for the Helmholtz free energy function, Equation 4.8:

$$dF = -SdT - PdV + \delta W' \tag{4.8}$$

The coefficient relationship corresponding to dT is

$$S = -\left(\frac{\partial F}{\partial T}\right)_V$$

Apply this relationship to Equation 6.27 to compute the entropy of the system in terms of the partition function:

$$S = -\left[\frac{\partial}{\partial T}(-N_0 kT \ln \wp)\right]_V$$

$$S = N_0 k \ln \wp + N_0 kT\left(\frac{\partial \ln \wp}{\partial T}\right)_V \tag{6.28}$$

The internal energy of the system may now be computed by rearranging the definitional relationship for F

$$U = F + TS = -N_0 kT \ln \wp + T\left[N_0 k \ln \wp + N_0 kT\left(\frac{\partial \ln \wp}{\partial T}\right)_V\right]$$

$$U = N_0 kT^2\left(\frac{\partial \ln \wp}{\partial T}\right)_V \tag{6.29}$$

In order to assess the validity of an atomistic model it is necessary to compare its predictions with experimental observations. Since the experimental information that is determined directly in databases is the heat capacity, it will be useful to obtain an expression for C_V in terms of the partition function. Equation 4.40 gives the relationship between heat capacity and the temperature dependence of the internal energy:

$$C_V = \left(\frac{\partial U}{\partial T}\right)_V = 2N_0kT\left(\frac{\partial \ln \wp}{\partial T}\right)_V + N_0kT^2\left(\frac{\partial^2 \ln \wp}{\partial T^2}\right)_V \qquad (6.30)$$

Computation of the remaining thermodynamic state functions, V, H, G, C_p, require a formulation that includes the pressure dependence of the partition function, which is neglected in this introduction to statistical thermodynamics.

Equation 6.27 to Equation 6.30 complete the algorithm. An atomistic model for the thermodynamic behavior of a system begins with a complete list of the energy levels that the particles in the system may exhibit. Given this list and no other information, the partition function for the model may be computed. Given the partition function, all of the macroscopic thermodynamic properties of the system may be computed. Of particular interest is the heat capacity since it provides the most direct test of the predictions of such an atomistic model. Thus, statistical thermodynamics provides an algorithm which converts an atomistic model for the behavior of a system as input to values of the macroscopic thermodynamic properties as output. Examples of the application of this algorithm are presented in the next section.

6.3 APPLICATIONS OF THE ALGORITHM

Three models for the available energy levels of a system are explored. The first visualizes a system that exhibits only two energy levels: this simple model illustrates the procedures involved in pursuing the algorithm. The second, the Einstein model for a crystal, is more realistic and more sophisticated. The third example, the ideal gas model, is mathematically advanced, though physically straightforward: it yields results that are plausible and familiar.

6.3.1 A MODEL WITH TWO ENERGY LEVELS

Consider a system that has N_0 particles that may exist in either of only two energy states, ε_1 and ε_2. Suppose further that ε_2 has a value that is twice that of ε_1. For simplicity, let ε be the energy of a particle in level 1; then $\varepsilon_2 = 2\varepsilon$ in this model. This list,

$$(\varepsilon, 2\varepsilon)$$

completes the description of the model. The equilibrium distribution of the particles between these states may now be computed, as well as its macroscopic thermodynamic properties.

The partition function for this model is given by

$$\wp = \sum_{i=1}^{r} e^{-\frac{\varepsilon_i}{kT}} = e^{-\frac{\varepsilon_1}{kT}} + e^{-\frac{\varepsilon_2}{kT}} = e^{-\frac{\varepsilon}{kT}} + e^{-\frac{2\varepsilon}{kT}} = e^{-\frac{\varepsilon}{kT}}\left[1 + e^{-\frac{\varepsilon}{kT}}\right] \quad (6.31)$$

Apply Equation 6.24 to find the equilibrium distribution of particles.

$$\frac{n_1}{N_0} = \frac{e^{-\frac{\varepsilon_1}{kT}}}{\wp} = \frac{e^{-\frac{\varepsilon}{kT}}}{e^{-\frac{\varepsilon}{kT}}\left[1 + e^{-\frac{\varepsilon}{kT}}\right]} = \frac{1}{\left[1 + e^{-\frac{\varepsilon}{kT}}\right]}$$

$$\frac{n_2}{N_0} = \frac{e^{-\frac{\varepsilon_2}{kT}}}{\wp} = \frac{e^{-\frac{2\varepsilon}{kT}}}{e^{-\frac{\varepsilon}{kT}}\left[1 + e^{-\frac{\varepsilon}{kT}}\right]} = \frac{e^{-\frac{\varepsilon}{kT}}}{\left[1 + e^{-\frac{\varepsilon}{kT}}\right]}$$

The ratio of occupancy of the states is

$$\frac{n_2}{n_1} = e^{-\frac{\varepsilon}{kT}} \quad (6.32)$$

The relative occupancy of the energy levels is determined by the magnitude of ε in comparison with thermal energy represented by kT. At very low temperatures, $\varepsilon/kT \gg 1$, the ratio is small and most of the particles lie in level 1. At sufficiently high temperatures in this system $\varepsilon/kT \ll 1$, approaching zero and the particles tend to become evenly distributed between the two states.

Computation of the thermodynamic properties, applying equations 6.27 to Equation 6.30, requires evaluation of the temperature derivative of the natural logarithm of the partition function. From Equation 6.31, after some manipulation,

$$\ln \wp = -\frac{\varepsilon}{kT} + \ln\left(1 + e^{-\frac{\varepsilon}{kT}}\right) \quad (6.33)$$

$$\left(\frac{\partial \ln \wp}{\partial T}\right)_v = \frac{\varepsilon}{kT^2}\frac{\left[1 + 2e^{-\frac{\varepsilon}{kT}}\right]}{\left[1 + e^{-\frac{\varepsilon}{kT}}\right]} \quad (6.34)$$

Substitution of these results into Equation 6.27 to Equation 6.29 yields, after some algebraic simplification,

$$F = N_0\varepsilon - N_0 kT \ln\left[1 + e^{-\frac{\varepsilon}{kT}}\right] \quad (6.35)$$

$$S = \frac{N_0\varepsilon}{T} \cdot \frac{e^{-\frac{\varepsilon}{kT}}}{\left[1 + e^{-\frac{\varepsilon}{kT}}\right]} + N_0 k \ln\left[1 + e^{-\frac{\varepsilon}{kT}}\right] \quad (6.36)$$

$$U = N_0 \varepsilon \left[\frac{1 + 2e^{-\frac{\varepsilon}{kT}}}{1 + e^{-\frac{\varepsilon}{kT}}} \right] \tag{6.37}$$

The heat capacity may be evaluated from the coefficient relation,

$$C_V = \frac{N_0 \varepsilon^2}{kT^2} \cdot \frac{e^{-\frac{\varepsilon}{kT}}}{\left[1 + e^{-\frac{\varepsilon}{kT}} \right]^2} \tag{6.38}$$

The properties of this atomistic model are determined by a single parameter: ε.

EXAMPLE 6.1

To illustrate the calculation of property changes for a simple process choose a value for this parameter: let $\varepsilon = 2 \times 10^{-20}$ J/atom. Compute the change in entropy for this system when one mole is heated from 300 to 1000 K. From Equation 6.36 the entropy at 300 K is computed to be 0.38 J/mol; at 1000 K, the value is 4.05 J/mol. The change in entropy for the process is thus

$$\Delta S = S_{1000} - S_{300} = 4.05 - 0.38 = 3.66 \left(\frac{J}{mole - K} \right)$$

It should be pointed out that this simple model is not realistic. For example, at low temperatures, C_V approaches infinity, while at high temperatures C_V approaches zero. It is presented here only to illustrate the application of statistical thermodynamics to the computation of macroscopic thermodynamic properties from a proposed atomistic model for the behavior of the system.

6.3.2   EINSTEIN'S MODEL OF A CRYSTAL

Most solids are crystalline. The atoms arrange themselves on a lattice which repeats a simple structural unit in three dimensions. Einstein developed a conceptually simple model as a first attempt to understand the thermodynamic behavior of crystalline solids. Although the model proved inadequate in quantitative comparisons to the behavior of real crystals, it successfully predicts the qualitative form of important aspects of behavior. This theory provided the basis for subsequent descriptions which were more successful and are still being perfected. It is also very useful as an illustration of the application of the statistical thermodynamics algorithm.

To make the model concrete, suppose that the atoms arrange themselves in a simple cubic structure, i.e., the unit cell is a cube with an atom at each corner. Each atom has six nearest neighbors. The energy of the crystal is assumed to reside entirely in the bonds between near neighbor pairs; the total energy of the crystal is viewed as the sum of the energies of the bonds it contains. The bonding between pairs of atoms is modeled as a simple spring with a spring constant that represents the strength of the bond. The atoms vibrate about their equilibrium positions, each connected by springs to its nearest neighbors, Figure 6.2. The energy of the crystal is the kinetic energy of the atoms as they vibrate about their equilibrium positions. It is shown in physics that the kinetic energy of motion of particles connected by a spring is proportional to this vibrational frequency, ν, of the spring.

A cubic crystal consisting of N_0 atoms contains $3N_0$ bonds; each of the six bonds associated with a specific atom is shared between two neighbors. Since all of the springs connecting pairs of atoms are coupled together in the crystal, analysis of coupled oscillators had shown that only certain discrete vibrational frequencies may occur in such a system. Since each frequency is associated with a corresponding energy value, the bond energies in the system may exhibit only discrete values. The bond energies are thus quantized. Einstein showed that the list of allowable energies for bonds could be described by the expression

$$\varepsilon_i = \left(i + \frac{1}{2}\right)h\nu \tag{6.39}$$

where i is an integer, h is Planck's constant (a universal constant equal to 6.624×10^{-27} erg-sec/atom and common in quantum theory), and ν is a

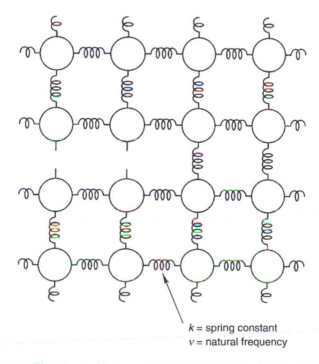

k = spring constant
ν = natural frequency

FIGURE 6.2 The Einstein model for a simple cubic crystal. Atoms vibrate about their equilibrium positions harmonically, as if joined by simple ideal springs.

characteristic vibrational frequency from which all of the frequencies of the bonds may be computed. The characteristic frequency, ν, is related to the spring constant and hence the strength of the binding energy in the crystal. This model has a single adjustable parameter, ν, that may be used in an attempt to explain observed differences between the behavior of the elements. It is not surprising that a one parameter model is not sufficient for this purpose.

The list of energy levels in the crystal given in Equation 6.39 is a complete description of the model. Application of the statistical thermodynamics algorithm permits prediction of the thermodynamic properties of such a system. First, evaluate the partition function:

$$\wp = \sum_{i=0}^{r} e^{-\frac{\varepsilon}{kT}} = \sum_{i=0}^{r} e^{-\frac{(i+\frac{1}{2})h\nu}{kT}} \tag{6.40}$$

Since the contribution from high energy levels (large i) is small this sum is well approximated by the infinite sum:

$$\wp = \sum_{i=0}^{r} e^{-i\frac{h\nu}{kT}} \cdot e^{-\frac{1}{2}\frac{h\nu}{kT}} = e^{-\frac{1}{2}\frac{h\nu}{kT}} \sum_{i=0}^{r} e^{-i\frac{h\nu}{kT}}$$

Let $x = \exp(-(h\nu/kT)$. Then the infinite series in this expression may be recognized as the familiar geometric series with a sum that converges to

$$\sum_{i=0}^{r} x^i = \frac{1}{1-x}$$

Thus, the partition function may be evaluated:

$$\wp = \frac{e^{-\frac{1}{2}\frac{h\nu}{kT}}}{\left[1 - e^{-\frac{h\nu}{kT}}\right]} \tag{6.41}$$

Thus,

$$\ln \wp = -\frac{1}{2}\frac{h\nu}{kT} - \ln\left[1 - e^{-\frac{h\nu}{kT}}\right]$$

Given the partition function, the thermodynamic properties of the system may be computed by applying Equation 6.27 to Equation 6.30. In this calculation the "particles" that exhibit the energy values in Equation 6.29 are the bonds in the system. Thus, for a simple cubic crystal containing N_0 atoms there are $3N_0$

"particles."

$$F = -3N_0 kT \ln \wp - \frac{3}{2} N_0 h\nu + 3N_0 kT \ln \left[1 - e^{-\frac{h\nu}{kT}} \right] \qquad (6.42)$$

$$S = 3 \frac{N_0 h\nu}{T} \left[\frac{e^{-\frac{h\nu}{kT}}}{1 - e^{-\frac{h\nu}{kT}}} \right] - 3N_0 k \ln \left(1 - e^{-\frac{h\nu}{kT}} \right) \qquad (6.43)$$

$$U = \frac{3}{2} N_0 h\nu \left[\frac{1 + e^{-\frac{h\nu}{kT}}}{1 - e^{-\frac{h\nu}{kT}}} \right] \qquad (6.44)$$

$$C_V = 3N_0 k \left(\frac{h\nu}{kT} \right)^2 \frac{e^{-\frac{h\nu}{kT}}}{\left(1 - e^{-\frac{h\nu}{kT}} \right)^2} \qquad (6.45)$$

The validity of this model for the description of the thermodynamic behavior of crystals may be assessed by comparing experimental measurements of the heat capacity as a function of temperature with the curve computed from Equation (6.45). The characteristic frequency, ν, the only adjustable parameter in the theory, may be statistically selected to provide the best fit. Such a comparison is shown in Figure 6.3. The qualitative form of the computed curve agrees well with the experimentally observed heat capacity dependence upon temperature. However, the agreement is not quantitative. More sophisticated theories, incorporating a distribution of values of ν and taking account of contributions to the energy of the crystal other than the kinetic energy of lattice vibrations, have been devised and applied with better success. Nonetheless, this simple, one parameter model provides a plausible basis upon which to build more sophisticated rationalizations of the thermodynamic behavior of crystals.

6.3.3 Monatomic Gas Model

Consider a gas composed of identical particles, each of which is a single atom. Rare gases like argon and neon satisfy this description. The thermodynamic properties of such a gas may be computed from an atomistic model that is based on the assumption that the energy contributed by each particle is simply the kinetic energy associated with its translation through space. The state of a particle is completely determined by its mass, m, velocity, $\mathbf{v}$ and its position $\mathbf{x}$ in space. Since $\mathbf{v}$ and $\mathbf{x}$ are vectors, each with three independent components, it is necessary to specify the values of six variables to specify the state of a gas atom in the system: its position (x, y, z) and its velocity components (v_x, v_y, v_z). The positional variables range over the set of positions that correspond to the volume of the system that contains the gas. The velocity components may vary in principle from $-\infty$ to $+\infty$.

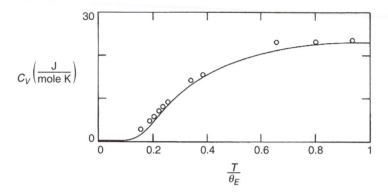

FIGURE 6.3 Comparison of experimentally measured heat capacity for diamond with that computed from the Einstein one parameter model, the Einstein temperature, $\theta_E = (h\nu/kT) = 1320$ K. *Source*: From Kittel, C, *Introduction to Solid State Physics*, 4th ed., Wiley, New York, NY, 1971.

In order to relate macroscopic properties to this atomistic model, it is first necessary to evaluate the partition function for the model. In its definition, the partition function is a sum over all of the states that the particles may exhibit of the values of the quantity exp($-\varepsilon_i/kT$) corresponding to each state, see Equation 6.25. It is assumed that the energy exhibited by a particle in this model is the kinetic energy associated with its motion through space

$$\varepsilon = \frac{1}{2}mv^2 = \frac{1}{2}m(v_x^2 + v_y^2 + v_z^2) \tag{6.46}$$

Imagine that the gas is enclosed in a box the dimensions of which are l_x, l_y, and l_z. In this model the variables are not quantized: the position of a particle may vary continuously within the domain of the enclosing box. The velocity components in principle vary continuously between $-\infty$ and $+\infty$. Because the variables involved are continuous, the summation over all the available states that constitute the partition function must be replaced by an integral over all states. Since the state of a particle is a function of six variables, this is a sextuple integral:

$$\wp = \int_0^{l_z} \int_0^{l_y} \int_0^{l_x} \int_{-\infty}^{\infty} \int_{-\infty}^{\infty} \int_{-\infty}^{\infty} e^{-\frac{\varepsilon}{kT}} dv_x dv_y dv_z dx\, dy\, dz \tag{6.47}$$

Substitute the expression for the energy in this model, Equation 6.46:

$$\wp = \int_0^{l_z} \int_0^{l_y} \int_0^{l_x} \int_{\infty}^{\infty} \int_{\infty}^{\infty} \int_{\infty}^{\infty} e^{-\frac{1}{2kT}m(v_x^2 + v_y^2 + v_z^2)} dv_x\, dv_y\, dv_z\, dx\, dy\, dz \tag{6.48}$$

The exponential in the integrand may be written as the product of three exponentials:

$$\wp = \int_0^{l_z} \int_0^{l_y} \int_0^{l_x} \int_\infty^\infty \int_\infty^\infty \int_\infty^\infty e^{-\frac{mv_x^2}{2kT}} \cdot e^{-\frac{mv_y^2}{2kT}} \cdot e^{-\frac{mv_z^2}{2kT}}\, dv_x\, dv_y\, dv_z\, dx\, dy\, dz \quad (6.49)$$

The position and velocity coordinates are independent of each other. As a consequence, this sextuple integral may be written as the product of six independent integrals, each of which is a function of only one variable.

$$\wp = \int_0^{l_z} \int_0^{l_y} \int_0^{l_x} dx\, dy\, dz \int_\infty^\infty e^{-\frac{mv_x^2}{2kT}}\, dv_x \int_\infty^\infty e^{-\frac{mv_y^2}{2kT}}\, dv_y \int_\infty^\infty e^{\frac{mv_z^2}{2kT}}\, dv_z \quad (6.50)$$

Evaluation of these integrals is amazingly straightforward. The first three integrate to the volume V of the enclosing container, no matter what shape it might have:

$$V = \int_0^{l_z} \int_0^{l_y} \int_0^{l_x} dx\, dy\, dz \quad (6.51)$$

The velocity integrals are all identical and have a form which is familiar in mathematics:

$$\int_{-\infty}^\infty e^{-a^2 x^2}\, dx = \sqrt{\frac{\pi}{a}}$$

where $a^2 = m/2kT$. Thus,

$$\int_{-\infty}^\infty e^{-\frac{m}{2kT}v^2}\, dv = \sqrt{\frac{2\pi kT}{m}}$$

and the partition function for a monatomic ideal gas is

$$\wp = V\left[\frac{2\pi kT}{m}\right]^{\frac{3}{2}} \quad (6.52)$$

In order to evaluate the thermodynamic properties of the gas, the natural logarithm of the partition function:

$$\ln \wp = \ln V + \frac{3}{2}\ln\left(\frac{2\pi k}{m}\right) + \frac{3}{2}\ln T \quad (6.53)$$

and its temperature derivative

$$\left(\frac{\partial \ln \wp}{\partial T}\right)_V = \frac{3}{2} \cdot \frac{1}{T} \tag{6.54}$$

are required. Apply Equation 6.27 to Equation 6.30 to compute the properties:

$$F = -N_0 kT \ln \wp = -N_0 kT \left[\ln V + \frac{3}{2}\ln\left(\frac{2\pi k}{m}\right) + \frac{3}{2}\ln T\right] \tag{6.55}$$

$$S = -\left(\frac{\partial F}{\partial T}\right)_V = N_0 k \ln\left[V\left(\frac{2\pi kT}{m}\right)^{\frac{3}{2}}\right] + \frac{3}{2}N_0 k \tag{6.56}$$

$$U = N_0 kT^2\left(\frac{\partial \ln \wp}{\partial T}\right)_V = N_0 kT^2\left(\frac{3}{2}\cdot\frac{1}{T}\right) = \frac{3}{2}N_0 kT \tag{6.57}$$

$$C_V = \left(\frac{\partial U}{\partial T}\right)_V = \frac{3}{2}N_0 k \tag{6.58}$$

Note that it is deduced from the model that the internal energy of an ideal gas is a function only of its temperature; the heat capacity is found to be independent of temperature. Furthermore, recognizing that, for one mole of the gas $N_0 k = R$, the gas constant, the value of the heat capacity agrees well with experimental values for the monatomic gases in Table 6.2.

Finally, recall again the combined statements of the first and second laws for the Helmholtz free energy:

$$dF = -SdT - PdV + \delta W' \tag{4.8a}$$

TABLE 6.2
Comparison of Heat Capacities of Monatomic and Diatomic Gases with Values Predicted from the Principle of Equipartition of Energy

Monatomic Gas	C_P (J/mol K)	Diatomic Gas	C_P (J/mol K)
Ideal gas	5/2 R = 20.79	Ideal gas	7/2 R = 29.10
Argon	20.72	Chlorine (Cl_2)	33.84
Krypton	20.69	Fluorine (F_2)	31.52
Neon	20.76	Hydrogen (F_2)	28.82
Radon	20.76	Oxygen (O_2)	29.34
Xenon	20.76	Nitrogen (N_2)	29.15

The coefficient relation for the second term is

$$\left(\frac{\partial F}{\partial V}\right)_T = -P \tag{6.59}$$

Apply this relation to the ideal gas model expression for the Helmholtz free energy, Equation 6.55:

$$P = -\left(\frac{\partial F}{\partial V}\right)_T = -\left(\frac{\partial(-N_0 kT \ln \wp)}{\partial V}\right)_T$$

$$= N_0 kT\left[\frac{\partial}{\partial V}\left(\ln V + \frac{3}{2}\ln\left(\frac{2\pi k}{m}\right) + \frac{3}{2}\ln T\right)\right]_T$$

Carry out the operation (the second and third terms are constant for this derivative):

$$P = N_0 kT \cdot \frac{1}{V}$$

Rearrange:

$$PV = N_0 kT = RT \tag{6.60}$$

which is the familiar equation of state developed experimentally for the behavior of simple gases. The coefficients of expansion and compressibility for the gas may be derived directly by applying this result to their definitional relations. Thus, all of the properties of monatomic gases may be explained on the basis of a simple atomistic model in which it is assumed that the energy of each atom is contained in its kinetic energy of translation through space.

If the gas under consideration is composed of molecules containing two or more atoms (e.g., H_2O, CO_2, CH_4, etc.), then there will be contributions to the kinetic energy of the molecule due to motions in addition to the translation of its center of mass. Specifically, a kinetic energy is associated with the rotation of the molecule and with the vibrational displacements of its atoms relative to its center of mass. A contribution may also arise from the motion of the electrons within the molecule. In order to treat polyatomic molecules it is necessary to develop models for the energy states associated with these added kinds of motion and to compute the corresponding contributions for the partition function for the system. The partition function could then be used to compute the heat capacity of a system for a variety of assumed configurations for the molecules that compose it. Comparison with experimental measurements of heat capacities then provides insight into the molecular structure and the behavior of matter at the atomic level.

An interesting simplification arises in the treatment of molecular gases. Kinetic energies of rotation and vibration have the same mathematical form as does the translation energy, though the physical factors involved are different. For example, in describing the kinetic energy of rotation, mass is replaced by the moment of inertia of the molecule and translational velocity is replaced by angular velocity. The total kinetic energy of the molecule has the form:

$$\varepsilon = \sum_{j=1}^{n} b_j v_j^2 \tag{6.61}$$

There is a term for each independent component of motion that the molecule may exhibit. Thus, the upper limit on the sum, n, is the number of independent motion components for the molecule. Clearly, the value of n depends upon the details of the structure of the molecule. There will be one rotational term if the molecule has an axis of symmetry; two if it does not. There will be a vibrational term for each bond in the molecular structure.

The evaluation of the partition function for this case appears complex because it involves integration over three positional variables and n velocity variables:

$$\wp = \int_0^{l_z} \int_0^{l_y} \int_0^{l_x} dx\, dy\, dz \int_\infty^\infty \int_\infty^\infty \cdots \int_\infty^\infty e^{-\frac{1}{kT} \sum_{j=1}^{n} b_j v_j^2} dv_1\, dv_2 ... dv_n \tag{6.62}$$

The integrand may be rewritten as a product of exponentials:

$$\wp = V \cdot \int_{-\infty}^\infty \int_{-\infty}^\infty \cdots \int_{-\infty}^\infty \prod_{j=1}^{n} \left[e^{-\frac{b_j}{kT} v_j^2} \right] dv_1\, dv_2 ... dv_n$$

Since all of these factors are independent (each contains only one variable, v_j) this multiple integral may be written as a product of n integrals, all mathematically identical in form:

$$\wp = V \cdot \prod_{j=1}^{n} \left[\int_{-\infty}^\infty e^{-\frac{b_j}{kT} v_j^2} dv_j \right]$$

This standard form of integral was evaluated for the monatomic case:

$$\int_\infty^\infty e^{-a^2 x^2} dx = \sqrt{\frac{\pi}{a}}$$

where the parameter $a^2 = b_j/kT$. Thus, for each factor in the product of integrals,

$$\int_\infty^\infty e^{-\frac{b_j}{kT} v_j^2} dv_j = \left[\frac{\pi kT}{b_j} \right]^{\frac{1}{2}}$$

The partition function may be written as the product,

$$\wp = V \cdot \prod_{j=1}^{n} \left[\frac{\pi k T}{b_j} \right]^{\frac{1}{2}} \qquad (6.63)$$

The natural logarithm of this function is

$$\ln \wp = \ln V + \sum_{j=1}^{n} \ln \left(\frac{\pi k T}{b_j} \right)^{1/2}$$

which may be written

$$\ln \wp = \ln V + \sum_{j=1}^{n} \frac{1}{2} \ln \left(\frac{\pi k}{b_j} \right) + \sum_{j=1}^{n} \frac{1}{2} \ln T$$

The derivative with respect to temperature at constant volume, required in the evaluation of the internal energy and heat capacity for the gas, is particularly simple:

$$\left(\frac{\partial \ln \wp}{\partial T} \right)_V = 0 + 0 + \sum_{j=1}^{n} \frac{1}{2} \cdot \frac{1}{T} = \frac{n}{2T}$$

The only parameter in the model that remains in this derivative is n, the number of independent components of motion that the system may exhibit. The internal energy of a molecular gas is obtained by inserting this value into Equation 6.29:

$$U = N_0 k T^2 \left(\frac{\partial \ln \wp}{\partial T} \right)_V = N_0 k T^2 \cdot \left(\frac{n}{2T} \right)$$

$$U = n \cdot \frac{1}{2} N_0 k T \qquad (6.64)$$

The heat capacity is the temperature derivative of the internal energy:

$$C_V = n \cdot \frac{1}{2} N_0 k \qquad (6.65)$$

Thus, the heat capacity of a molecular gas is predicted to depend only upon the number of independent components of motion that the molecule can display. Table 6.2 presents some examples that test this prediction.

The principle derived from this application of the statistical thermodynamical algorithm, namely that each independent component of motion of the molecules in the gas contributes the same quantity, $1/2 \, kT$, to the internal energy of the gas, is called the principle of equipartition of energy. Modern analytical tools provide many avenues for the direct exploration of the structure of molecules. Application of this principle provided this kind of insight before these tools developed.

Statistical thermodynamics provides the link between phenomenological information supplied by experiments in classical thermodynamics and the structure and behavior of matter at an atomic level. An explanation for an observation that the heat capacity for a system composed of molecules of gas A is larger than that for a system composed of molecules of gas B is put forward and tested. Patterns of behavior emerge and the intricacies of the behavior of matter are brought to a new level of understanding.

6.4 ALTERNATE STATISTICAL FORMULATIONS

Over the years, experience has accumulated in attempting to apply the algorithm of statistical thermodynamics to an increasing range of classes of systems. It became evident that the specific formulation that computes the number of microstates corresponding to a given macrostate, Equation 6.1, is not adequate to describe all aspects of the behavior of matter. For example, in the formulation presented here no restrictions are placed upon the number of particles that may exist in any energy state in the system. In the treatment of the behavior of electrons, it has been found necessary to invoke the Pauli exclusion principle, which states that no more than two electrons may exhibit the same energy at any instant in time. In distributing electrons in their allowable energy levels this restriction must be strictly obeyed. With this restriction it is clear that the number of possible microstates corresponding to a given macrostate is not given by Equation 6.1; a modified combinatorial analysis must be devised. Since this relationship is at the foundation of the development of statistical thermodynamics, it will lead to different relationships for the equilibrium distribution of electrons over energy levels and to different forms in the predictions of macroscopic thermodynamic properties.

Three different forms of the relation between microstates and macrostates have been prominent in the development of statistical thermodynamics. Each is labeled with the names of the original authors of the formulation. The development presented in this chapter is called Maxwell–Boltzmann statistics. The other developments are referred to as Bose–Einstein and Fermi–Dirac statistics. All are giants in the development of our understanding of how the world works.

The relationship between entropy and the number of microstates per macrostate remains the same in all three developments, the Boltzmann hypothesis, Equation 6.3. Following the general strategy for finding the conditions for equilibrium, i.e., determining the distribution of particles on energy levels that yields the maximum entropy in an isolated system, yields the equilibrium distribution in each case:

Maxwell–Boltzmann distribution:

$$\frac{n_i}{N} = \frac{1}{\wp_{MB} e^{\frac{\varepsilon_i}{kT}}} \qquad (6.66)$$

Bose–Einstein distribution:

$$\frac{n_i}{N} = \frac{1}{\wp_{BE}\left(e^{\frac{\varepsilon_i}{kT}} - 1\right)}$$ (6.67)

Fermi–Dirac distribution:

$$\frac{n_i}{N} = \frac{1}{\wp_{FD}\left(e^{\frac{\varepsilon_i}{kT}} + 1\right)}$$ (6.68)

For precise definitions of the quantities in these equations, consult Ref. 2. The remainder of the algorithm, which computes the macroscopic thermodynamic properties from these distribution functions, is formally identical to that presented in this chapter. Application of these formulations, each in the circumstances for which the underlying statistical assumptions are appropriate, has greatly broadened the scope of application of statistical thermodynamics.

6.5 SUMMARY

A microstate for a system is a specification of the state or condition of each particle in the system. The number of different microstates a given system may exhibit is enormous.

A macrostate for a system is a list of the number of particles existing in each condition that they may exhibit. The number of microstates that correspond to a given macrostate may be computed from combinatorial analysis:

$$\Omega = \frac{N_0!}{n_1! \cdot n_2! \cdots n_r!}$$ (6.1)

Boltzmann hypothesized that the entropy of a macrostate is logarithmically related to the number of microstates it contains:

$$S = k \ln \Omega$$ (6.3)

Application of the general strategy for finding conditions for equilibrium yields, in this case, the Boltzmann distribution function

$$\frac{n_i}{N_0} = \frac{1}{\wp} \cdot e^{-\frac{\varepsilon_i}{kT}}$$ (6.24)

where $\wp$, is the partition function for the system, given by:

$$\wp = \sum_{i=0}^{r} e^{-\frac{\varepsilon_i}{kT}} \tag{6.25}$$

All of the macroscopic thermodynamic properties of the system may be computed from the partition function.

Application of the algorithm that converts a list of energy levels into predictions of thermodynamic properties provided a rudimentary model for crystals and a satisfactory model for the behavior of gases at ordinary conditions.

Maxwell–Boltzmann statistics are inadequate to explain some aspects of the behavior of matter. Alternate formulations, namely Bose–Einstein and Fermi–Dirac statistics, have significantly broadened the range of application of the statistical thermodynamical algorithm.

PROBLEMS

Problem 6.1.   Explain in words your understanding of the difference between phenomenological thermodynamics and statistical thermodynamics.

Problem 6.2.   Show, by listing all of the macrostates, that the number of macrostates possible for a system composed of ten particles that may occupy three energy levels is 60.

Problem 6.3.   Consider a system with two particles a and b that may each exhibit any of four energy levels, ε_1, ε_2, ε_3, and ε_4.
   a. How many microstates might this system exhibit?
   b. Enumerate its microstates.
   c. Use the list of microstates to generate a list of macrostates for this system.
   d. Identify the microstates corresponding to each macrostate.

Problem 6.4.   Calculate the number of microstates corresponding to each of the following combinations:
   a. A system with three particles and four energy levels.
   b. A system with 15 particles and four energy levels.
   c. A system with four particles and 15 energy levels.
   d. A cluster of 50 particles each of which may reside in any of 30 energy levels.
   e. 1000 particles that may reside in 100 energy levels.

Problem 6.5.   Consider a model in which the available energy levels are linearly spaced along the energy axis:

$$\varepsilon_n = \left(n + \frac{1}{2}\right)\varepsilon_0 \quad (n = 0, 1, 2, ..., 9)$$

The system contains ten particles. Consider two macrostates:

$$\{0, 0, 1, 2, 4, 2, 1, 0, 0, 0\}...\text{State I}$$

$$\{0, 1, 1, 2, 2, 2, 1, 1, 0, 0\}...\text{State II}$$

a. Which macrostate has the higher energy?
b. Which macrostate has the higher entropy?
c. Which macrostate is more likely to be observed?

Problem 6.6. Compute the factorials of the following numbers: 10; 30; 60,
a. Directly.
b. Using Stirling's approximation.
c. In each case, compute the error in ln $x!$ that results from the approximation.

Problem 6.7. A system containing 500 particles and 15 energy levels is in the following macrostate:

$$\{14, 18, 27, 38, 51, 78, 67, 54, 32, 27, 23, 20, 19, 17, 15\}$$

This system experiences a process in which the number of particles in each energy level changes by the following amounts:

$$\{0, 0, -1, -1, -2, 0, +1, +1, +2, +2, +1, 0, -1, -1, -1\}$$

Estimate the change in entropy for this process.

Problem 6.8. Using a convenient computer applications package, calculate and plot the partition function for the Einstein model as a function of (T/θ_E). Calculate and plot the heat capacity at constant volume of a simple cubic Einstein crystal as a function of temperature in the range from 0, T, 1000 K. Repeat the calculation for each of the following values of the Einstein temperature, θ_E :

100 K;   200 K;   300 K;   500 K.

Problem 6.9. Use the Einstein model to compute the change in internal energy of crystal when it is heated reversibly at one atmosphere pressure from 90 to 210 K. Assume $\theta_E = 250$ K.

Problem 6.10. Compute the change in entropy when one mole of a monatomic ideal gas is compressed from an initial condition at 273 K and 1 atm to a final condition of 500 K at 3.5 atm.
a Make the calculation using familiar phenomenological thermodynamics.
b Repeat the calculation using the results of statistical thermodynamics. (Hint: first calculate the initial and final volumes.)

Problem 6.11.   At ordinary temperatures and pressures the heat capacity of ammonia (NH_3) is 37 J/mol. Apply the principle of equipartition of energy to speculate on the spatial arrangement of the atoms in the ammonia molecule.

REFERENCES

1. Boas, M.L., *Mathematical Methods in the Physical Sciences*, Wiley, New York, NY, pp. 699–703, 1983.
2. Lee, J.F., Sears, J.W., and Turcotte, P.L., *Statistical Thermodynamics*, Addison Wesley, Reading, MA, pp. 148–165, 1963.

7 Unary Heterogeneous Systems

CONTENTS

In this chapter, the structure of thermodynamics illustrated in Figure 1.4 is implemented to develop equilibrium maps for unary systems. A system may be considered to be unary if, for the range of states under study, it may be considered to consist of a single chemical component. Each of the elements forms a unary system over its full range of existence. Molecular compounds such as CO_2 or H_2O may be treated as unary systems over much of the range of temperatures and pressures normally encountered in the laboratory; under conditions in which they may decompose to form significant quantities of other molecules, they may not be treated as unary systems.

From a thermodynamic point of view a system is homogeneous if it consists of a single phase. More specifically, a system consists of a single phase if its intensive properties are uniform or, at most, vary continuously throughout the system. A heterogeneous system consists of more than one phase; some of the intensive properties will exhibit discontinuities at the boundaries between the phases in a heterogeneous system.

All of the elements may exist in at least three distinct states of matter or phase forms: solid, liquid, gas. In addition, many of the elements exhibit more than one phase form in the solid state. For example, at 1 atm pressure when pure iron is equilibrated below 910°C it exists in a body centered cubic (BCC) crystal structure. Between 910 and 1455°C the equilibrium state for iron is the face centered cubic (FCC) crystal structure. Between 1455°C and its melting point at 1537°C the stable form of iron is again body centered cubic. Different solid phase forms of the same element are called allotropes; the BCC and FCC structures are allotropic forms of iron.

When a piece of pure iron is heated reversibly from room temperature it remains in the BCC structure until the temperature of 910°C is attained. Continued input of heat into the system does not raise the temperature above 910°C. Instead, small crystals of FCC iron nucleate within the BCC structure and begin to grow, consuming the BCC structure. The system requires the input of heat to make this transformation occur. Eventually, all of the BCC structure is replaced by the FCC phase; only then does the continued input of heat begin to raise the temperature of the remade structure. This phase change from BCC to FCC iron at 910°C and 1 atm pressure is called an allotropic transformation. Below 910°C, BCC is the stable phase form; above 910°C, FCC iron is stable. Thus, at 1 atm, 910°C is the limit of stability of both phase forms.

If during the transformation from BCC to FCC iron at 910°C the flow of heat is brought to zero, then the mixture of BCC and FCC iron may exist at equilibrium for an indefinite period. This two-phase system is a heterogeneous system in the thermodynamic usage of the word. The part of the system that is BCC will have the intensive properties of that crystal structure, e.g., molar volume, entropy, internal energy and the like. The part of the structure that is FCC will have the properties of FCC iron. At boundaries where these two crystal structures meet these intensive properties will exhibit a discontinuity.

The allotropic transformation is a special case of a more general concept: the phase transformation, which applies to any change in the phase form of a system. Melting or fusion, which is the change from a solid to a liquid phase and boiling or vaporization, which is the change from a liquid to the vapor state, are familiar examples of phase transformations. Thermodynamic properties associated with these phase changes are compiled for a number of elements in Appendix C. The study of all classes of phase transformations in unary as well as multicomponent systems is extremely important in materials science because these processes control the microstructure of materials. Since many properties that materials exhibit are very sensitive to the microstructure of the material, control of microstructure is tantamount to control of properties.

The set of thermodynamic conditions under which a given phase form is stable for a particular element may be succinctly summarized in an equilibrium map on temperature–pressure coordinates called a phase diagram. Figure 7.1 and Figure 7.2 represent examples of phase diagrams for elements and other systems that may be treated as unary. The lines on these diagrams are called phase boundaries and represent the limits of stability of each of the phase forms that the system exhibits. These lines thus represent conditions, i.e., combinations of temperature and pressure, at which transformations between the impinging phase forms occur. A system that exists at a combination of P and T lying on a phase boundary will, at equilibrium, consist of a mixture of the two phases that the line bounds on the

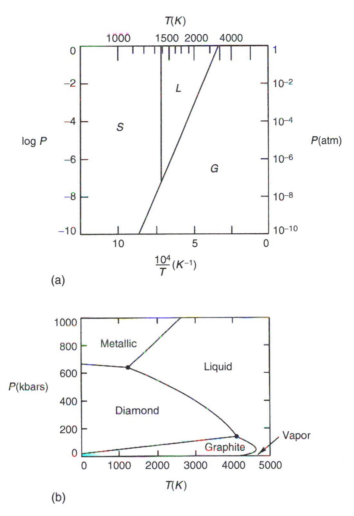

FIGURE 7.1 Typical phase diagrams for elements: (a) copper; (b) carbon.[1] With permission.

diagram. If the system is taken through a reversible process represented by any path on the (P, T) diagram, when that path intersects a phase boundary the change in temperature and pressure will be arrested while the phase transformation occurs. When the change in structure is complete, the change along the (P, T) path will resume.

The limits of stability of the single-phase regions in Figure 7.1 and Figure 7.2 are defined by the conditions under which pairs of phases may coexist at equilibrium. The equations that define the conditions for equilibrium in a unary two-phase system were derived in Chapter 5 as an illustration of the application of the general strategy for finding conditions for equilibrium. In this chapter it will be shown how application of these conditions for equilibrium equations leads to the rules for constructing unary phase diagrams in (P, T) space. Combination of the three

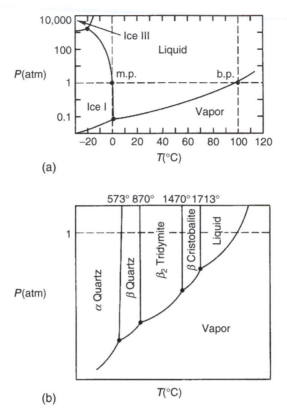

FIGURE 7.2 Pressure–temperature phase diagrams for compounds treated as unary systems: (a) water; (b) SiO_2.[1] With permission.

equations that describe the conditions for equilibrium into a single relationship yields the working equation, called the Clausius–Clapeyron equation, for calculating such unary phase diagrams. Application of this result to the computation of phase boundaries for real systems and the estimation of phase boundaries where information is incomplete is presented for sublimation, vaporization, melting and solid–solid transformations. Inverting the strategy permits the estimation of some thermodynamic properties from phase diagram information.

The simplest representation of regions of stability of the phases in a unary system is obtained when it is plotted in (P, T) coordinates. Alternate representations may be obtained by choosing other combinations of properties as the axes for the plot; e.g., for some purposes plotting the phase diagram in (P, V) or (S, V) coordinates may be useful. All such plots share the same underlying thermodynamic basis. However, each representation provides unique information regarding the relationships of the phases. Also, the rules for constructing the phase diagram are different for different choices of variables. Examples of these alternate constructions are presented in Section 7.6.

7.1 STRUCTURE OF UNARY PHASE DIAGRAMS IN (P,T) SPACE

As shown in Figure 7.1 and Figure 7.2, the representations of the limits of stability of phases on a unary phase diagram in the pressure–temperature plane have the following characteristics:

a. The domain of stability of each single phase is represented by an area
b. The domain of stability for two phases coexisting in equilibrium is a line
c. The domain of stability for three phases existing simultaneously in equilibrium is a triple point, at which three single-phase areas and three two-phase lines meet
d. There are no regions where more then three phases may coexist at equilibrium

These characteristics of unary phase diagrams are direct consequences of the conditions for equilibrium in two- and three-phase structures and the choice of variables (P, T) to represent the state of the system. It will be shown in Chapter 9 that these characteristics may be deduced from a general principle, called the Gibbs Phase Rule, which gives rules of construction for phase diagrams with any number of components and phases. In this illustration of that principle for a unary system it is possible to demonstrate the origins of these construction rules graphically.

7.1.1 CHEMICAL POTENTIAL AND THE GIBBS FREE ENERGY

In order to develop this graphical illustration, it is first necessary to establish the connection between the chemical potential, introduced in Chapter 5, Equation 5.13 and Equation 5.14 and the molar Gibbs free energy. Fortunately, this connection is straightforward. The term involving the chemical potential was added to the apparatus of thermodynamics in order to describe systems that are capable of exchanging matter with their surroundings during a process, i.e., in order to treat open systems. Recall that, in the treatment of open systems, it is necessary to distinguish explicitly the extensive properties of the system, designated by a prime ($'$), from their intensive counterparts, which are left unprimed in this notation. Thus, U' is the internal energy of the system, no matter how many moles it may contain; U is the corresponding molar internal energy. If it is recognized that the number of moles of the component in a unary system may be varied during a process, then the combined statements of the first and second laws must have an additional term describing this possible change in state for the system, Equation 5.13:

$$dU' = TdS' - PdV' + \mu dn \tag{5.13}$$

where

$$\mu = \left(\frac{\partial U'}{\partial n} \right)_{S',V'} \tag{5.14}$$

is the chemical potential of the component in the system. Recall the definition of the Gibbs free energy function. For a system composed of an arbitrary number of moles of substance,

$$G' = U' + PV' - TS' \tag{4.9}$$

Differentiate:

$$dG' = dU' + PdV' + V'dP - TdS' - S'dT$$

For an open, unary system, substitute for dU' with Equation 5.13

$$dG' = (TdS' - PdV' + \mu dn) + PdV' + V'dP - TdS' - S'dT$$

Simplify:

$$dG' = -S'dT + V'dP + \mu dn \tag{7.1}$$

Apply the coefficient relation to the third term:

$$\mu = \left(\frac{\partial G'}{\partial n} \right)_{T,P} \tag{7.2}$$

In this equation, G' is the Gibbs free energy of the system. Let G be the corresponding value of the Gibbs free energy per mole of system, i.e., the molar Gibbs free energy. Then, $G' = nG$ and

$$\mu = \left(\frac{\partial G'}{\partial n} \right)_{T,P} = \left[\frac{\partial (nG)}{\partial n} \right]_{T,P} = G \left(\frac{\partial n}{\partial n} \right)_{T,P} = G \tag{7.3}$$

Thus, in a unary system, the chemical potential of the component in any state is identical with the molar Gibbs free energy for that state. This simple relationship will be invoked frequently in this and subsequent chapters.

7.1.2 CHEMICAL POTENTIAL SURFACES AND THE STRUCTURE OF UNARY PHASE DIAGRAMS

Because $\mu = G$ in a unary system, the dependence of the chemical potential upon temperature and pressure is identical to the variation of molar Gibbs free energy.

$$d\mu = dG = -SdT + VdP \tag{7.4}$$

where the coefficients are the molar entropy and molar volume for the phase form of the substance under study. If the α phase is taken through an arbitrary change in state, for example,

$$d\mu^{\alpha} = -S^{\alpha}dT^{\alpha} + V^{\alpha}dP^{\alpha} \tag{7.5}$$

where the superscript α indicates a property that is evaluated for the α phase. The molar entropy and molar volume of the α phase may be computed as functions of temperature and pressure from heat capacity, expansion and compressibility data as spelled out in Chapter 4. Given database information for S^α and V^α, Equation 7.5 may be integrated, at least in principle, to yield a function

$$\mu^\alpha = \mu^\alpha(T^\alpha, P^\alpha) \tag{7.6}$$

This functional relationship may be visualized graphically as the surface shown in Figure 7.3a. A similar argument may be applied to each phase that may exist in the system. Thus, for the liquid phase a chemical potential surface exists which may be obtained from heat capacity, expansion and compressibility information about the liquid phase:

$$\mu^L = \mu^L(T^L, P^L) \tag{7.7}$$

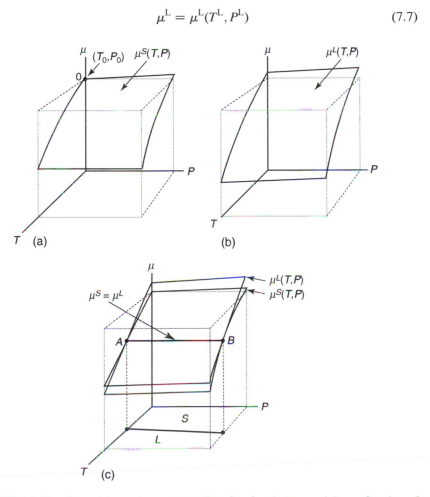

FIGURE 7.3 Sketches of the surface representing the chemical potential as a function of temperature and pressure for the solid phase (a) and the liquid phase (b) are superimposed in (c).

Figure 7.3b is a schematic representation of the chemical potential of the liquid phase as a function of T and P.

The chemical potentials of these two phases may be compared at any given (T, P) combination if and only if the reference state used in their computation is the same. Thus, some condition of the system, defined by a choice for temperature (T_0), pressure (P_0) and phase form (α or L) must be chosen for the system and the chemical potentials of the α and L phase computed relative to that reference state. With this constraint, it is possible to construct surfaces for both of the phases on one graph, Figure 7.3c.

These two surfaces will intersect along a space curve AB. At any point on that space curve the temperatures, pressures and chemical potentials of the two phases are identical. These three conditions,

$$T^\alpha = T^L; \qquad P^\alpha = P^L; \qquad \mu^\alpha = \mu^L$$

are precisely the conditions that must be met in order for the α and liquid phases to coexist in equilibrium, Equation 5.26 to Equation 5.28. Thus, the curve of intersection of the two chemical potential surfaces is the locus of points for which the α and liquid phases are in heterogeneous equilibrium. The projection of this line onto the (P, T) plane is the phase boundary delineating the $(\alpha + L)$ two-phase equilibrium and the limits of stability of the α and liquid phases.

Figure 7.4 illustrates the construction when the chemical potential surface for the gas phase is added to these considerations. The curve of intersection of the α and gas surfaces, COD, represents the $(\alpha + G)$ two-phase equilibrium (the sublimation curve $C'O'D'$ on the $[P, T]$ plane). The curve produced by the intersection of the liquid and gas surfaces, EOF, describes the $(L + G)$ equilibrium (the vaporization curve $E'O'F'$ on the $[P, T]$ plane).

All three surfaces intersect at a single point, O, which is also common to the three curves that represent the two-phase equilibria. At this unique point, temperatures, pressures and chemical potentials of all three phases are the same and the three phases coexist in equilibrium. The projection of this point onto the (P, T) plane, O', is the triple point for the three phases $(\alpha + L + G)$.

7.1.3 CALCULATION OF CHEMICAL POTENTIAL SURFACES

The surface represented by the function $\mu^\alpha(T, P)$ may be computed by integrating Equation 7.5. This integration may be carried out by fixing the value of P and integrating the $-S^\alpha dT$ term over temperature, producing an isobaric section through the surface. Repetition of this process at a sequence of values of P generates the surface. The information required to produce such an isobaric section for the α phase is the temperature dependence of the entropy of the α phase. The absolute value of the entropy of the α phase may be computed from

$$dS_P^\alpha = \frac{C_P^\alpha}{T} dT$$

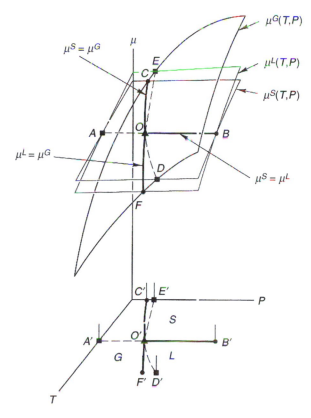

FIGURE 7.4 Superposition of chemical potential surfaces for solid, liquid and gas phases demonstrates the relationship between the structure of the phase diagram and the underlying thermodynamics.

or

$$S^\alpha(T) = S^\alpha_{298} + \int_{298}^T \frac{C_P^\alpha(T)}{T}\,dT \qquad (7.8)$$

For a particular phase, the database information necessary to carry out this integration is S^α_{298} and the heat capacity function, $C_P^\alpha(T)$. Substitute this entropy expression into

$$d\mu^\alpha = dG^\alpha = -\left[S^\alpha_{298} + \int_{298}^T \frac{C_P^\alpha(T)}{T}\,dT \right] dT$$

Integrate again from the reference temperature, 298 K, to a variable temperature, T:

$$\int_{298}^T d\mu^\alpha = \int_{298}^T - \left[S^\alpha_{298} + \int_{298}^T \frac{C_P^\alpha(T)}{T}\,dT \right] dT$$

which gives

$$\mu^{\alpha}(T) - \mu^{\alpha}(298) = G^{\alpha}(T) - G^{\alpha}(298) = \int_{298}^{T} -\left[S_{298}^{\alpha} + \int_{298}^{T} \frac{C_P^{\alpha}(T)}{T} dT \right] dT \quad (7.9)$$

The curve labeled α in Figure 7.5 plots the result of such a calculation.

A similar strategy yields the free energy curve for the liquid phase. Some appropriate bookkeeping must be observed in constructing this curve. The point has been made that comparison of these chemical potential values is only possible if they are all calculated from the same reference state. The reference state chosen for Equation 7.9 is the α phase form at 298 K. If values for the two phases are to be compared, this must also be the reference state for the calculation of the free energy of the liquid phase. Thus, the curve marked L plotted in Figure 7.5 is the function $G^{L}(T) - G^{\alpha}(298)$. The connection between the liquid and solid phases may be made at the melting point, T_m, where the equality of the chemical potentials implies that

$$G^{L}(T_m) = G^{\alpha}(T_m)$$

The quantity plotted in Figure 7.5 may be written as follows:

$$G^{L}(T) - G^{\alpha}(298) = [G^{L}(T_m) - G^{L}(T_m)] + [G^{\alpha}(T_m) - G^{\alpha}(298)] \quad (7.10)$$

The right side merely adds and subtracts equal quantities to the expression on the left side. The quantity in the first bracket on the right side may be evaluated by a straightforward integration from the melting point, T_m, of $dG^{L} = -S^{L}dT$ for the

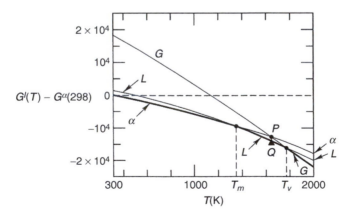

FIGURE 7.5 The temperature dependences of the molar free energy (equivalent to the chemical potential in a unary system) for the solid, liquid and gas phases are computed at 1 atm pressure. Intersections of the curves represent conditions for two-phase equilibrium. All values are computed relative to that for the solid at 298 K.

liquid phase, as was done for the α phase in deriving Equation 7.9:

$$G^{L}(T) - G^{L}(T_m) = \int_{T_m}^{T} - \left[S^{L}(T_m) + \int_{T_m}^{T} \frac{C_P^{L}(T)}{T} dT \right] dT \qquad (7.11)$$

where $S^{L}(T_m)$ is the absolute entropy of the liquid phase at the melting point. This quantity may be computed from the absolute entropy of the α phase at the melting point, i.e., by evaluating $S^{\alpha}(T)$ in Equation 7.8 at the melting point, $T = T_m$, and adding the entropy of fusion at the melting point, $\Delta S^{S \to L}(T_m)$:

$$S^{L}(T_m) = S_{298}^{\alpha} + \int_{298}^{T_m} \frac{C_P^{\alpha}(T)}{T} dT + \Delta S^{\alpha \to L}(T_m) \qquad (7.12)$$

The quantity in the second brackets in Equation 7.10 may be obtained from the curve $G(T)$ for the α phase, Equation 7.9 evaluated at T_m. Equation 7.10 may then be plotted as a function of temperature as the liquid curve in Figure 7.5.

A similar procedure, based upon the connection between the liquid and gas phases at the vaporization temperature, T_v, computes the free energy curve for the gas phase, relative to the pure α phase at 298 K. This curve is labeled G in Figure 7.5. If the solid state exhibited allotropic forms other than α, curves for those phases would be computed in an analogous manner.

7.1.4 COMPETING EQUILIBRIA: METASTABILITY

The general criterion for equilibrium derived in Chapter 5 is based upon the principle that in an isolated system entropy can only increase; thus, at equilibrium in an isolated system the entropy is a maximum. A corollary derived in Section 5.5 demonstrated that in a system constrained to constant temperature and pressure, the Gibbs free energy can only decrease; at equilibrium, the Gibbs free energy is a minimum. The conditions for equilibrium derived from this extremum principle are identical to those derived from the principle that states that the entropy is a maximum at equilibrium in an isolated system, namely the chemical potential, temperature and pressure are uniform at equilibrium. The constructions in Figure 7.4 and on the isobaric section in Figure 7.5 shows how these conditions for equilibrium determine the structure of one-, two- and three-phase fields in unary systems.

The free energy curves on the isobaric section, Figure 7.5, illustrate these relationships. At each temperature in the range from 298 K to T_m the α phase has the lowest free energy and is thus the stable phase form. Between T_m and T_v the liquid curve has the lowest free energy and liquid is stable. Above T_v the gas phase is stable. At any temperature, T_1, the vertical distance between the curves, which reports the difference in free energy between the phase forms at that temperature, gives a quantitative measure of the relative stabilities of the phases.

It will also be noted that the α and gas curves intersect at the point marked P; at that point the molar free energy (or chemical potential) of the solid and gas phases are equal and the conditions for equilibrium between those two phases are satisfied.

However, there exists a separate condition of the system, labeled Q on the liquid curve, which has a lower free energy and is therefore the state of stable equilibrium. The two-phase equilibrium represented by the point P is said to be metastable; the stable equilibrium structure at that temperature is the liquid phase.

A broader perspective may be obtained by examining Figure 7.4. At each point along the curve COD, the chemical potentials, temperatures and pressures of the α and gas phases are the same and the conditions for equilibrium are satisfied with respect to these two phases. However, points on this curve in the segment between O and D lie above the liquid surface at the same temperature and pressure. According to the extremum principle, in a comparison between states that have the same temperature and pressure, the equilibrium state is that for which the Gibbs free energy is lower. The liquid phase has a lower free energy than does an equilibrated mixture of solid and gas in this domain. Thus, for pressure–temperature combinations lying along the segment OD, the liquid phase form is the equilibrium state for the system. Mixtures of $(\alpha + G)$ that satisfy the conditions for equilibrium but do not possess the minimum free energy possible for the system are said to be in a state of metastable equilibrium.

Metastable equilibrium states play a key role in many phase transformations that are important in materials science. The sequence of thermodynamic states through which a system passes as it approaches the ultimate equilibrium condition frequently encounters states of metastable equilibrium. Indeed, if the kinetics of the transformation is slow, a system may encounter a metastable equilibrium condition and remain in that state indefinitely so that the material may be used in its metastable state.

7.1.5 PHASE STABILITY CALCULATIONS

Procedures for computing and optimizing phase diagrams, not only in unary systems, but in systems consisting of two, three or more components, are based upon sophisticated computer software that seek to compute the equilibrium phases present at any temperature and composition in the range of practical interest. (The pressure variable has been largely ignored in these multicomponent systems.) These calculations come under the umbrella of CALPHAD, an international organization that oversees the gathering of data, its optimization and its application to generate the variety of equilibrium maps described in Figure 1.4. The results, which are carefully assessed and evaluated equilibrium maps, are disseminated through journals, such as CALPHAD, the Journal of Phase Diagrams, Metallurgical Transactions, the Journal of the American Ceramic Society, through materials related professional societies and also commercially. These procedures will be discussed in detail in Chapters 8, 9, 10 and 11, which deal with multicomponent systems.

The criterion for equilibrium that is the basis for all such calculations is the one applied in this chapter which states that, in any comparison of systems at the same temperature and pressure, the system with the lowest value of the Gibbs free energy is the stable system. This extremum principle, the minimization of the Gibbs free energy to identify the stable phases, is the basis for the computer calculation of all equilibrium maps. For a given system, which may exhibit many phase forms, these calculations apply database information in two broad categories: phase stabilities of the elements, and mixing behavior of solutions. Thermodynamic models and database information required to describe solution behavior are developed in Chapter 8. In this context the term "phase stability" refers to the Gibbs free energy of the pure element in its various phase forms as a function of temperature, as illustrated in Figure 7.5 and derived from Equations such as Equation 7.9 and Equation 7.11.

The database required contains transformation temperatures, transformation entropies and heat capacity data as a function of temperature. To compute a binary (or higher order phase diagram) it is necessary to be able to evaluate or estimate this information for each of the elements for every phase form that appears in the diagram. This produces a challenging problem at the foundation of these methods.

Consider for example the iron–chromium binary system. According to Appendix C, pure iron has a BCC structure between room temperature and 1184 K, and again from 1665 to 1809 K, its melting point. Between 1184 and 1665 K, the stable form of iron is FCC. Thus, in the phase diagram emanating from the iron side of the diagram there is a range of compositions and temperatures where the FCC phase is stable. At the other end of the diagram, chromium has a BCC structure from room temperature to the melting point at 2130 K. Yet the calculation of the FCC region in the phase diagram requires the relative stability of the FCC structure of pure chromium as part of its input, i.e., the temperature dependence of the Gibbs free energy of FCC chromium. This information cannot be determined directly experimentally because the FCC form of pure chromium cannot be made; FCC chromium is metastable with respect to BCC chromium at all temperatures.

There have been two approaches to handling this problem. The first, called first principle calculations, is based upon intensive calculations that begin with the geometric placement of the atoms in any crystal structure and models the interactions between pairs of atoms that incorporate the electronic structure. So far, this approach has had some success in predicting enthalpies of competing phases, but the calculation of entropy contributions, essential to the evaluation of the Gibbs free energy and the temperature dependence of properties, remains in its infancy. This approach has the potential for developing a deep understanding of the phenomena at the atomic and electronic level that produce phase diagrams, but it is a long way from becoming practical.

The second approach makes use of the thermodynamic relations that permit calculation of phase boundary compositions from phase stability and mixing data (developed in Chapters 9 and 10) and applies them in reverse. By combining phase boundary composition data determined directly experimentally with solution information, e.g., heat of mixing data, one can use these conditions-for-equilibrium equations to extract elemental phase stabilities. Thus, one can estimate the free energy difference between FCC and BCC chromium from the phase boundaries of the FCC–BCC field in the binary Fe–Cr phase diagram, supported by experimental mixing data for this solution. Further, this procedure may be applied to any binary phase diagram in which chromium is one of the elements and the FCC structure has a domain of stability. By including a number of such binary phase diagrams in the database, optimization procedures can be applied to produce best estimate of the Gibbs free energy of FCC chromium as a function of temperature. Once established for pure chromium, this information can be used in calculating and assessing any phase diagram, binary, ternary or higher order, which contains chromium and a domain of stability of an FCC phase.

Phase diagrams of binary, ternary or higher order systems cannot by calculated without this crucial information about the phase stabilities of the pure components.

7.2 THE CLAUSIUS–CLAPEYRON EQUATION

The curves that are the domains of two phase equilibria in (P, T) space in unary systems may each be described mathematically by some function $P = P(T)$. The Clausius–Clapeyron equation, derived in this section, is a differential form of this equation. For any pair of phases that may coexist in the unary system, integration of the Clausius–Clapeyron equation yields a mathematical expression for the corresponding phase boundary on the phase diagram. Repeated application to all of the pairs of phases that may exist in the system yields all possible two-phase domains. These two-phase boundaries define the limits of stability of the single-phase regions. Intersections of the two-phase curves produce triple points where three phases coexist. Thus, the Clausius–Clapeyron equation is the only relation required for calculating and constructing unary phase diagrams.

Consider any pair of phases known to exist in the system, α and β. If the α phase is taken through any arbitrary change in state, its change in chemical potential is given by Equation 7.5:

$$d\mu^{\alpha} = -S^{\alpha}dT^{\alpha} + V^{\alpha}dP^{\alpha} \tag{7.5}$$

If the β phase is taken through an arbitrary change in state, its chemical potential obeys the analogous relation:

$$d\mu^{\beta} = -S^{\beta}dT^{\beta} + V^{\beta}dP^{\beta} \tag{7.13}$$

S^{α} and V^{α} are molar properties of the α phase and S^{β}, V^{β} are those for the β phase. If during this infinitesimal process, α and β are maintained in two-phase equilibrium, then the changes in P, T, and μ of each phase are constrained by the conditions for equilibrium derived in Chapter 5:

$$T^{\alpha} = T^{\beta} \Rightarrow dT^{\alpha} = dT^{\beta} = dT \tag{7.14}$$

$$P^{\alpha} = P^{\beta} \Rightarrow dP^{\alpha} = dP^{\beta} = dP \tag{7.15}$$

$$\mu^{\alpha} = \mu^{\beta} \Rightarrow d\mu^{\alpha} = d\mu^{\beta} = d\mu \tag{7.16}$$

That is, if during the process α and β are maintained in equilibrium, then the relationship that $T^{\alpha} = T^{\beta}$ requires that both phases experience the same temperature change. Similar statements apply to the pressures and chemical potentials.

Equation 7.14 to Equation 7.16 imply that Equation 7.5 and Equation 7.13 may be set equal to each other and that the superscripts may be dropped on the dT and dP factors.

$$d\mu^\alpha = -S^\alpha dT + V^\alpha dP = d\mu^\beta = -S^\beta dT + V^\beta dP$$

Combine like terms:

$$(S^\beta - S^\alpha)dT = (V^\beta - V^\alpha)dP \tag{7.17}$$

The coefficient of dT in this equation is the entropy per mole of the β phase form minus the entropy per mole of the α form. This quantity may be replaced by

$$\Delta S^{\alpha \to \beta} \equiv S^\beta - S^\alpha \tag{7.18}$$

which is the change in entropy accompanying the transformation of one mole of the α phase form to the β form at the temperature and pressure being considered. Similarly,

$$\Delta V^{\alpha \to \beta} \equiv V^\beta - V^\alpha \tag{7.19}$$

is the change in molar volume accompanying the transformation from α to β. Equation 7.17 becomes

$$\Delta S^{\alpha \to \beta} dT = \Delta V^{\alpha \to \beta} dP \tag{7.20}$$

Rearrangement of this equation gives

$$\frac{dP}{dT} = \frac{\Delta S^{\alpha \to \beta}}{\Delta V^{\alpha \to \beta}} \tag{7.21}$$

which is one form of the Clausius–Clapeyron equation. This result implies that at any point on a phase boundary that represents a two-phase equilibrium in a unary $P-T$ diagram, the slope is equal to the ratio of the entropy change to the volume change for the transformation. Since the nature of the two phases, α and β, was not specified, this result is general, applying to all two-phase boundaries. Integration of this equation, which requires information about how ΔS and ΔV vary with temperature and pressure, yields the phase boundary $P = P(T)$ for the $(\alpha + \beta)$ equilibrium.

The entropy change associated with a phase transformation is not measured directly in experiments. Calorimetric measurements yield the heat of transformation, e.g., the heat of fusion or heat of vaporization. Since the transformation occurs isobarically under reversible conditions, the heat of transformation is equal to the enthalpy change for the process:

$$Q^{\alpha \to \beta} = \Delta H^{\alpha \to \beta} = H^\beta - H^\alpha \tag{7.22}$$

This measurement can be connected with the entropy of the transformation through the following development. Recall the definitional relationship for the Gibbs free energy and apply it to each phase:

$$G^\alpha = H^\alpha - T^\alpha S^\alpha$$

$$G^\beta = H^\beta - T^\beta S^\beta$$

Recall that, if α and β are in equilibrium, then $\mu^\alpha = \mu^\beta$. Since the chemical potential is equal to the molar Gibbs free energy for a unary phase, Equation 7.3, this condition implies that $G^\alpha = G^\beta$. Thus

$$G^\alpha = H^\alpha - TS^\alpha = G^\beta = H^\beta - TS^\beta$$

Thermal equilibrium requires that $T^\alpha = T^\beta = T$. Rearrange:

$$H^\beta - H^\alpha = T(S^\beta - S^\alpha)$$

or for any phase transformation in a unary system at any point (P, T) along the curve,

$$\Delta S^{\alpha \to \beta} = \frac{\Delta H^{\alpha \to \beta}}{T} \tag{7.23}$$

By Equation 7.22, the enthalpy change is identical to the heat of transformation. The entropy change for a phase transformation in a unary system is thus the heat of transformation divided by the transformation temperature.

Insert Equation 7.23 into the Clausius–Clapeyron Equation 7.21:

$$\frac{dP}{dT} = \frac{\Delta H^{\alpha \to \beta}}{T \Delta V^{\alpha \to \beta}} \tag{7.24}$$

This is the form of the Clausius–Clapeyron equation that is used most frequently in computing unary phase diagrams.

7.3   INTEGRATION OF THE CLAUSIUS–CLAPEYRON EQUATION

In order to integrate Equation 7.24 to compute a phase boundary, it is necessary to develop expressions for ΔH and ΔV for the phase transformation as functions of temperature and pressure. In Chapter 4, general expressions were derived for the change in all of the state functions accompanying changes in temperature and pressure, Table 4.3. The temperature and pressure dependence of ΔH and ΔV may be related to these results by applying the simple principle that the differential of a sum is the sum of the differentials. Thus, to obtain

$$\Delta H^{\alpha \to \beta} = \Delta H^{\alpha \to \beta}(T, P)$$

use the relation

$$d(\Delta H^{\alpha \to \beta}) = d(H^\beta - H^\alpha) = dH^\beta - dH^\alpha \qquad (7.25)$$

Recall Equation 4.42 in Table 4.3:

$$dH = C_P dT + V(1 - T\alpha)dP \qquad (4.42)$$

In this equation, α is the coefficient of thermal expansion for the system, not to be confused with the superscript α designating a particular phase form in the present application. Evaluate Equation 4.42 separately for the α and β phases, substitute the results into Equation 7.25 and collect like terms.

$$d\Delta H^{\alpha \to \beta} = \Delta C_P dT + \Delta[V(1 - T\alpha)]dP \qquad (7.26)$$

where

$$\Delta C_P \equiv C_P^\beta - C_P^\alpha \qquad (7.27)$$

and

$$\Delta[V(1 - T\alpha)] = [V^\beta(1 - T\alpha^\beta)] - [V^\alpha(1 - T\alpha^\alpha)] \qquad (7.28)$$

Assessment of the coefficient of dP in Equation 7.26 for typical values of the properties involved shows that this term is negligible for pressure changes up to 100,000 atm. Thus, for all practical purposes, the enthalpy of transformation may be considered to be a function of temperature only, given by

$$d\Delta H^{\alpha \to \beta} = \Delta C_P dT \qquad (7.29)$$

Heat capacity data for many substances is generally tabulated with the help of the empirical equation,

$$C_P = a + bT + \frac{c}{T^2} + dT^2 \qquad (7.30)$$

(see Appendix B and E). Each phase has its own values of the constants $a, b, c,$ and d in this equation. The difference in heat capacities for two phases defined in

Equation 7.27 may be written

$$\Delta C_P = \Delta a + \Delta bT + \frac{\Delta c}{T^2} + \Delta dT^2 \tag{7.31}$$

where $\Delta a = a^\beta - a^\alpha$, etc. This result may be used to integrate Equation 7.29 from some reference temperature T_0 to a variable upper limit, T

$$\int_{T_0}^{T} d\Delta H = \int_{T_0}^{T} \Delta C_P(T)dT = \int_{T_0}^{T} [\Delta a + \Delta bT + \frac{\Delta c}{T^2} + \Delta dT^2]dT$$

$$\Delta H(T) - \Delta H(T_0) = \left[\Delta aT + \Delta b\frac{T^2}{2} - \frac{\Delta c}{T} + \Delta d\frac{T^3}{3} \right]_{T_0}^{T}$$

or

$$\Delta H(T) = \Delta aT + \frac{\Delta b}{2}T^2 - \Delta c\frac{1}{T} + \frac{\Delta d}{3}T^3 - \Delta D \tag{7.32}$$

where

$$\Delta D = \Delta H(T_0) - \left[\Delta aT_0 + \frac{\Delta b}{2}T_0^2 - \Delta c\frac{1}{T_0} + \frac{\Delta d}{3}T_0^3 \right] \tag{7.33}$$

This result may be used to evaluate the numerator in Equation 7.24.

Evaluation of the denominator requires computation of the volume change for the transformation as a function of temperature and pressure. To obtain

$$\Delta V^{\alpha \rightarrow \beta} = \Delta V^{\alpha \rightarrow \beta}(T, P)$$

it is convenient to use an approach that is different from that applied to the enthalpy. By definition,

$$\Delta V^{\alpha \rightarrow \beta}(T, P) \equiv V^\beta(T, P) - V^\alpha(T, P) \tag{7.34}$$

If β is the vapor phase and α is either a solid or liquid phase, then $V^\beta \gg V^\alpha$. (At standard temperature and pressure the molar volume of an gas occupies 22,400 cc; the molar volume of a condensed phase is typically approximately 10 cc.) Further, at temperatures and pressures at which the vapor phase is ordinarily encountered for most substances, deviations from ideal gas behavior are negligible. Thus, for the sublimation or vaporization curves on the phase diagram

$$\Delta V^{\alpha \rightarrow G} = V^G - V^\alpha \cong V^G = \frac{RT}{P} \tag{7.35}$$

Insertion of this result into the Clausius–Clapeyron equation, together with the expression for the enthalpy of the transformation, permits calculation of sublimation and vaporization curves.

If α and β are both condensed phases (solids or liquids) then information about the coefficients of expansion and compressibility for the two phases is required to compute ΔV as a function of temperature and pressure. If the pressure range under consideration is of the order of a few tens of atmospheres, the assumption that ΔV is a constant is adequate for most calculations in this case. This assumption is not valid if the phase diagram is to be calculated over a range of many thousands of atmospheres.

Adopt the convention that the β phase in this development is always that which is stable at the higher temperature. Then the numerator in the Clausius–Clapeyron equation (ΔS or ΔH) is positive. The sign of the slope of the $\alpha\beta$ phase boundary is thus determined by the sign of ΔV in the denominator. For sublimation and vaporization, ΔV is a large positive number; slopes are correspondingly small and positive. When both phases are condensed $\Delta V^{\alpha-\beta}$ is usually positive but may be negative; the slope of the corresponding phase boundary has the same sign. The most familiar case in which the volume shrinks in passing from the solid to liquid phase form is water. Ice cubes float in water because the density of ice is less than that of water; the molar volume of ice is larger than that of water and $\Delta V = V(\text{water}) - V(\text{ice})$ is negative. The slope of the (S + L) boundary for water is negative, Figure 7.2.

7.3.1 VAPORIZATION AND SUBLIMATION CURVES

If one of the phases is the vapor or gas phase integration of the Clausius–Clapeyron Equation 7.24 is accomplished by substituting Equation 7.32 for ΔH and Equation 7.35 for ΔV:

$$\frac{dP}{dT} = \frac{\Delta H}{T\Delta V} = \frac{\left[\Delta aT + \dfrac{\Delta b}{2}T^2 - \Delta c\dfrac{1}{T} + \dfrac{\Delta d}{3}T^3 - \Delta D\right]}{T\left(\dfrac{RT}{P}\right)}$$

Separate the variables:

$$\frac{dP}{P} = \frac{1}{R}\left[\frac{\Delta a}{T} + \frac{1}{2}\Delta b - \frac{\Delta c}{T^3} + \frac{1}{3}\Delta dT - \frac{\Delta D}{T^2}\right]$$

Integrate:

$$\ln\frac{P}{P_0} = \frac{1}{R}\left[\Delta a \ln T + \frac{1}{2}\Delta bT + \frac{1}{2}\Delta c\frac{1}{T^2} + \frac{1}{6}\Delta dT^2 + \frac{\Delta D}{T} - C_0\right] \qquad (7.36)$$

where Δa, Δb, Δc, and Δd are obtained from empirical heat capacity information, as defined in Equation 7.31 and

$$C_0 = \Delta a \ln T_0 + \frac{1}{2}\Delta b T_0 + \frac{1}{2}\Delta c \frac{1}{T_0^2} + \frac{1}{6}\Delta d T_0^2 + \frac{\Delta D}{T_0} \qquad (7.37)$$

P_0 and T_0 are coordinates of a known point on the phase boundary. This equation describes all such phase boundaries with precision over the range for which the empirical description of heat capacity and the ideal gas approximation are valid.

The vaporization curve, Equation 7.36, takes on a simpler form if the range of temperatures of interest in a given application is restricted to a few hundred degrees Kelvin. Consider again Equation 7.29 for the variation in the heat of transformation as a function of temperature. The integrated form of this equation may be written:

$$\Delta H(T) = \Delta H(T_0) + \int_{T_0}^{T} \Delta C_P(T)\mathrm{d}T \qquad (7.38)$$

Heats of vaporization and sublimation are typically in the range of 100 KJ. The energy change associated with heating the condensed and vapor phases are typically a few KJ; the difference in energy associated with their heating is even smaller. Thus, unless the temperature range is very large, in practical calculations the second term in Equation 7.38 may be neglected and, at any temperature, $\Delta H(T)$ may be taken to be $\Delta H(T_0)$, that is enthalpy change at the transformation temperature. If ΔH is thus assumed to be independent of temperature, integration of the Clausius–Clapeyron equation is simplified:

$$\frac{\mathrm{d}P}{P} = \frac{\Delta H}{RT^2}\mathrm{d}T$$

and

$$\ln\left(\frac{P}{P_0}\right) = -\frac{\Delta H}{R}\left(\frac{1}{T} - \frac{1}{T_0}\right) \qquad (7.39)$$

This expression predicts that, if the vapor pressure in equilibrium with a condensed phase is measured at a series of temperatures, a plot of the logarithm of the vapor pressure vs. the reciprocal of the temperature should be a straight line with a slope equal to $-\Delta H/R$. Figure 7.6 shows such a plot for many of the elements. All of these curves obey this relationship at pressures below 1 atm. The filled circle on most of these curves represents the triple point; the slight change in slope at that point accompanies the transition from the vaporization curve to the sublimation curve.

Figure 7.6 provides a first example of the use of a thermodynamic understanding of phase equilibria to compute a thermodynamic property of the system from the form of a phase boundary. At pressures below 1 atm it is possible to compute the heat of vaporization (or sublimation) from the slope of the vapor pressure curve. To emphasize the significance of this result, realize that no thermal (calorimetric)

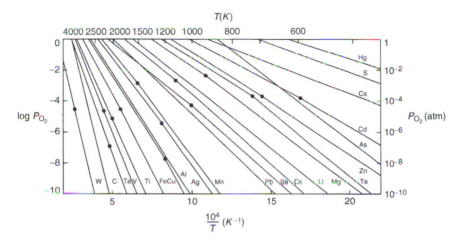

FIGURE 7.6 Compilation of vapor pressure curves for common elements, plotting logarithm of the vapor pressure vs. $1/T$. A change in slope corresponds to the melting point.

measurements are required to evaluate the heat of vaporization. A set of measurements of vapor pressure at a series of temperatures yields this result.

Note also that the slopes of these vaporization curves increase with increasing boiling point of the element. (Recall that the boiling point is defined as the temperature at which liquid and vapor are in equilibrium when the vapor pressure is equal to 1 atm.) This trend may be traced to Trouton's Rule, an empirical observation that the entropy of vaporization of most elements has about the same value: approximately 92 J/g atom. This rule implies that the enthalpy of vaporization, which is related to the entropy by Equation 7.23, increases in proportion to the boiling point. Since the slopes in Figure 7.6 are proportional to the enthalpy of vaporization, they increase with the boiling point of the element.

7.3.2 PHASE BOUNDARIES BETWEEN CONDENSED PHASES

The variation of ΔV, ΔS and ΔH with temperature and pressure is related to differences in heat capacities and coefficients of expansion and compressibility of the two phases involved. An approximate calculation of the phase boundary between condensed phases may be obtained by neglecting the temperature and pressure dependence of these quantities. Integration of Equation 7.21 is then straightforward:

$$P - P_0 = \frac{\Delta S}{\Delta V}(T - T_0) \qquad (7.40)$$

where (P_0, T_0) is a known point on the phase boundary, e.g., the equilibrium temperature at $P = 1$ atm. With these assumptions the phase boundary is a straight line through (P_0, T_0) with slope $\Delta S / \Delta V$.

Alternatively, integration of Equation 7.24 yields the result

$$dP = \frac{\Delta H}{\Delta V} \cdot \frac{dT}{T}$$

$$P - P_0 = \frac{\Delta H}{\Delta V} \cdot \ln\left(\frac{T}{T_0}\right) \tag{7.41}$$

This result may be shown to be equivalent to Equation 7.40 within the bounds of the approximations applied. Expansion of the logarithm gives

$$\ln\left(\frac{T}{T_0}\right) = \left(\frac{T}{T_0} - 1\right) - \frac{1}{2}\left(\frac{T}{T_0} - 1\right)^2 + \frac{1}{3}\left(\frac{T}{T_0} - 1\right)^3 - \cdots$$

Since the range of temperature considered is small compared to T_0, higher order terms may be neglected. Then

$$P - P_0 = \frac{\Delta H}{\Delta V}\left(\frac{T - T_0}{T_0}\right)$$

which is identical to Equation 7.40 since $\Delta S = \Delta H/T_0$.

For precise computations of phase boundaries between condensed phases, experimental values for heat capacities and coefficients of expansion and compressibility are required as functions of temperature and pressure for both phases. Then, the temperature dependence of ΔS or ΔH and ΔV may be computed by applying Equations such as Equation 7.32 and the Clausius–Clapeyron equation subsequently integrated.

7.4 TRIPLE POINTS

On a P–T unary phase diagram, three phases can coexist in equilibrium only at isolated points represented by the single point of intersection of the three chemical potential surfaces, as shown for the solid (α), liquid and gas phases in Figure 7.4. The triple point is also a point of intersection of the three two-phase curves ($\alpha + L$), ($\alpha + G$) and ($L + G$) that form between these three phases. If the system has a fourth phase form with a chemical potential surface lying below this triple point, then the ($\alpha + L + G$) triple point would be metastable and will not appear on the stable phase diagram. If this is not true, then the triple point is stable and will appear. In that case, the configuration around the triple point, deduced from Figure 7.4, is as shown in Figure 7.7. Each intersecting two-phase field has one leg that is stable; the two-phase equilibrium becomes metastable as it extends beyond the triple point. The metastable extensions always lie between the stable legs of the other pair of two-phase equilibria. Thus, in a circuit around a triple point, segments of two-phase equilibrium lines encountered must always alternate between stable and metastable.

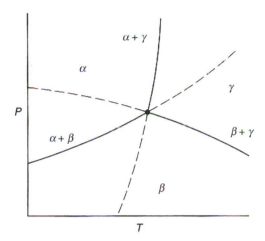

FIGURE 7.7 Sequence of stable and metastable equilibria characteristic of a triple point on a unary $P-T$ phase diagram.

This structure, which may be seen graphically in Figure 7.7, is characteristic of all stable triple points that may exist in the system.

Since a triple point (P_t, T_t) is the intersection of three two-phase equilibrium curves, it is a point that lies on all three lines. Thus, the point (P_t, T_t) simultaneously satisfies the Clausius–Clapeyron for all three of the two-phase equilibria existing among the triplet of phases. The triple point may be computed algebraically by the simultaneous solution of any pair of these equations.

It is also characteristic of a triple point that the properties of the three pairs of phase changes, ΔS, ΔH, and ΔV, are necessarily related since, for example

$$\Delta V^{\alpha \to G} = V^G - V^\alpha = V^G - V^L + V^L - V^\alpha = (V^G - V^L) + (V^L - V^\alpha)$$

$$= \Delta V^{L \to G} + \Delta V^{\alpha \to L} \tag{7.42}$$

Thus, the property change for the solid to gas transformation is the sum of the change from solid to liquid and liquid to gas. This result is simply an application of the principle that requires changes in state functions to be independent of the path. The change in volume (or entropy or enthalpy) for the formation of the vapor directly from the solid is identical to that obtained for a process in which the solid first melts, then the resulting liquid vaporizes.

To illustrate the computation of a triple point, consider again the $(\alpha + L + G)$ three-phase equilibrium. This point is the intersection of the sublimation, melting and vaporization curves. Equation 7.39 gives the mathematical form for the equilibria involving the vapor phase. Let the superscript (^{s}) denote properties of the sublimation curve and (^{v}) those of the vaporization curve. For the solid–vapor equilibrium:

$$P^s = A^s e^{-\Delta H^s / RT} \tag{7.43}$$

For the liquid–vapor equilibrium,

$$P^{v} = A^{v}e^{-\Delta H^{v}/RT} \tag{7.44}$$

The triple point, (P_{t}, T_{t}), lies on both these curves. Thus,

$$P_{t} = A^{s}e^{-\Delta H^{s}/RT_{t}} \qquad \text{and} \qquad P_{t} = A^{v}e^{-\Delta H^{v}/RT_{t}}$$

Solve these two equations for P_{t} and T_{t}, the coordinates of the triple point:

$$T_{t} = \frac{\Delta H^{s} - \Delta H^{v}}{R \ln\left(\dfrac{A^{s}}{A^{v}}\right)} \tag{7.45}$$

$$P_{t} = (A^{s})^{\Delta H^{v}/(\Delta H^{v} - \Delta H^{s})} \cdot (A^{v})^{\Delta H^{s}/(\Delta H^{s} - \Delta H^{v})} \tag{7.46}$$

If values for the constants for the sublimation and vaporization curves are known, these equations permit calculation of the solid–liquid–gas triple point.

A more practical basis for calculating the solid–liquid–gas triple point makes use of the observation that the vaporization and sublimation curves are linear on a plot of the logarithm of the pressure vs. $(1/T)$ in the low pressure range where this triple point appears. Construct the appropriate axes, as shown in Figure 7.8. Note that the $(1/T)$ axis is reversed so that temperature increases from left to right.

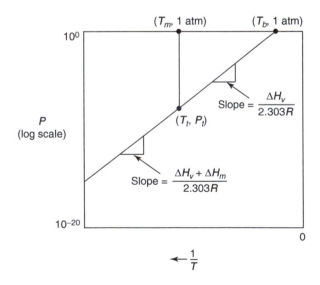

FIGURE 7.8 Illustration of the construction of a P–T phase diagram at pressures below 1 atm. The temperature of the triple point is negligibly different from the melting point at 1 atm.

Caution: this change of direction on the $(1/T)$ axis changes the sign of slopes computed from the equations describing curves on this graph.

Plot the melting point and boiling point (defined to be the temperature at which the equilibrium vapor pressure is 1 atm) on the 1 atm line. Since the total pressure difference between the triple point and 1 atm is a fraction of an atmosphere, the melting temperature will change only by a small fraction of 1°. Thus, on the scale of the diagram, the melting curve is a vertical line; the temperature of the triple point, T_t, is negligibly different from the melting point at 1 atm. Use Equation 7.39 to compute the vapor pressure at this temperature; this is P_t. Draw a straight line from the boiling point to (P_t, T_t); this is the (L + G) two phase line.

The sublimation curve is also a straight line on this plot. Compute the slope of the sublimation line (ΔH^s) from the heat of melting and vaporization by analogy with Equation 7.36:

$$\Delta H^s = \Delta H^m + \Delta H^v \tag{7.47}$$

and construct the line with this slope through the triple point. If there are no allotropic forms for the solid phase, the unary phase diagram in the range below 1 atm is complete.

EXAMPLE 7.1
Phase diagram for silicon. The pertinent properties of pure silicon are given in Appendix C. The melting and boiling points are labeled M and B in Figure 7.9. Substitution of $P = 1$ at $T_b = 2873$ K into Equation 7.44 together with $\Delta H^v = 386$ KJ permits computation of the coefficient, A^v:

$$A^v = P^v e^{\Delta H^v / RT} = (1 \text{ atm}) \cdot e^{386,000/(8.314 \times 2873)} = 1.04 \times 10^7 \text{ atm}$$

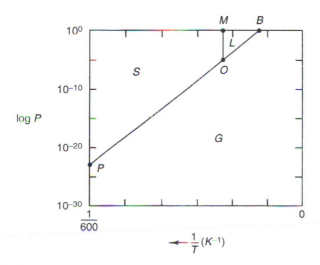

FIGURE 7.9 Illustration of the calculation of the phase diagram at low pressures for silicon.

The vapor pressure curve is given by:

$$P^v = 1.04 \times 10^7 \ (atm) \cdot e^{-386,000/8.314T}$$

Set the triple point temperature equal to the melting point: $T_t = T_m = 1685$ K. Substitute this value into the vapor pressure equation to obtain the pressure at the triple point:

$$P_t = 1.04 \times 10^7 \cdot e^{-386,000/(8.314 \times 1685)} = 1.13 \times 10^{-5} \ atm$$

Plot the triple point (P_t, T_t) as the point O; construct the lines MO and BO. Apply Equation 7.47 to compute the heat of sublimation.

$$\Delta H_s = \Delta H_m + \Delta H_v = 50.2 + 386 = 436 \left(\frac{KJ}{gm \ atom} \right)$$

The triple point (P_t, T_t) also lies on the sublimation curve. Substitute this information, together with the heat of sublimation, into Equation 7.43 to evaluate the coefficient, A^s:

$$A^s = P_t e^{436,000/(8.314 \times 1685)} = 3.75 \times 10^8 \ atm$$

The sublimation curve is given by:

$$P^s = 3.75 \times 10^8 e^{-436,000/8.314T}$$

To plot the sublimation curve, choose a temperature value of 300 K and use the above equation to compute the vapor pressure over the solid at that temperature. This point is labeled P in Figure 7.9. A straight line from O to P is the sublimation curve. Label the solid, liquid and vapor fields. The phase diagram for silicon below 1 atm, computed from melting and boiling points and the heats of fusion and vaporization, is complete. The calculated triple point occurs at $(1.13 \times 10^{-5}$ atm, 1685 K$)$.

7.5　COMPUTER CALCULATIONS OF (P,T) UNARY PHASE DIAGRAMS

Figure 7.10 shows the experimentally determined $P-T$ phase diagram for pure bismuth. On the pressure scale plotted, the vapor phase region is flattened against the temperature axis. A systematic computation of this complicated diagram requires:

a. A list of all of the possible phase forms that may appear: I, II, III,..., L, G
b. Heat capacity, coefficients of expansion and compressibility for each of the phases so that its chemical potential surface, $\mu^l(T,P)$, may be computed
c. Transition temperatures, heat and volume changes for transitions at 1 atm or at other specific conditions so that the relative positions of the chemical potential surfaces may be fixed

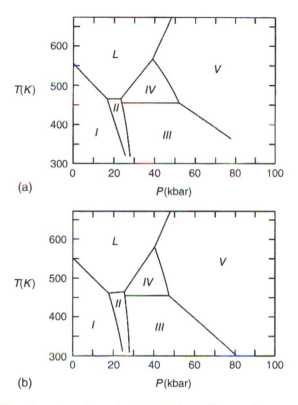

(a)

(b)

FIGURE 7.10 Experimentally determined $P-T$ phase diagram for pure bismuth (a) is compared with the computed diagram in which the temperature and pressure dependence on the parameters is neglected (b).[2]

Each phase has its own chemical potential surface as shown in Figure 7.3a. If there are p phases in the system [$(p - 2)$ solid phase allotropic forms, liquid and vapor], then there are p chemical potential surfaces.

Pairwise intersections of these surfaces describe two-phase equilibria. The number of possible two-phase equilibria is equal to the number of ways in which p things can be arranged two at a time, namely $p!/[2!(p - 2)!]$. Every set of three phases produces a configuration like that shown in Figure 7.4, yielding a triple point qualitatively like that shown in Figure 7.7; there are $p!/[3!(p - 3)!]$ possible triple points in the system. Many, if not most, of these calculated triple points will prove to be metastable, that is, will lie above one or more chemical potential surfaces of phases other than the three defining the triple point. Alternatively, such calculated points may be outside the range of physically meaningful thermodynamic states in the range where either T or P is negative.

For example, in Figure 7.10, bismuth exhibits five solid phase forms, liquid and vapor: $p = 7$. There are $7!/(3!4!) = 35$ triple points possible in this system. On the diagram shown (including the [I–L–G] triple point and an evident extrapolated intersection of the I–II and II–III lines to form a I–II–III triple point below room temperature) there are only six stable triple points. The remaining 29 do not appear. Stable triple points may be identified by comparing the value of the chemical potential characteristic of the three phases involved at that (P_t, T_t) with the chemical potentials of the remaining phases in the system at the same (P_t, T_t). If, and only if, the triple point chemical potential lies below that of all of the remaining phases, then the calculated triple point is stable and will appear in the final diagram.

Once the stable triple points are identified and plotted, the two phase curves that connect them may be constructed, completing the diagram.

7.6 ALTERNATE REPRESENTATIONS OF UNARY PHASE DIAGRAMS

The topology of a unary phase diagram in (P, T) space may be described as a "simple cell structure". The areas of the cells are one-phase fields; cell boundaries are two-phase domains, and cell vertices are triple points representing three-phase equilibria. This simple construction arises because the variables used to describe the state of the phases in the system, P and T, are both "thermodynamic potentials". These properties are precisely those that appear in the conditions for equilibrium. A two phase field $(\alpha + \beta)$ is a line (i.e., a region of zero width) precisely because the pressure and temperature of the α phase in the two phase system is required to be the same as that of the β phase. The corresponding states of the two equilibrated phases are represented by the same point in (P, T) space. Thus the phase boundaries for the α phase, and that for the β phase, coincide.

This will not be true of some property other than P or T is used in the description of the state of the two phases. Suppose the properties of the participating phases are described in terms of their pressure, P, and molar volume, V, i.e., suppose the diagram is plotted in (P, V) space. While the conditions for equilibrium require the pressure to be the same in both phases, the molar volumes, V^α and V^β will not, in general, be the same. The resulting plot of the phase relationships is very different in appearance from the simple (P, T) plot, Figure 7.11.

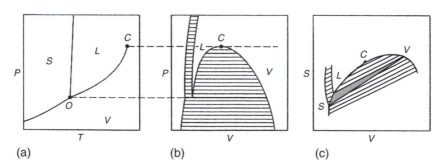

FIGURE 7.11 Alternate constructions for unary phase diagrams compare the plot in (a) (T, P) space with the diagram for the same substance in (V, P) space (b) and (V, S) space (c).

In this representation, a two phase equilibrium is portrayed by a pair of points, one giving the pressure and molar volume of the α phase and the other that of the β phase. The conditions for equilibrium require that the pressure be the same in both phases but the volumes will be different. Thus, a line connecting the pair of points, called a tie line, is horizontal (P is constant) on such a plot. The collection of states that represents all of the possible conditions for equilibrium between the α and β phases consists of two lines, one describing how V^{α} varies with P and the other the variation of V^{β} with P; the space between the lines is filled with horizontal tie lines connecting pairs of equilibrated states.

The liquid + vapor two-phase field terminates at a point C on the (P, T) diagram. The molar volume of the gas phase decreases with increasing pressure and temperature, while that of the liquid increases until, at the critical point C, the molar volumes coincide and the properties of the two phases become indistinguishable. This behavior is clearly represented on a (P, V) phase diagram, Figure 7.11b, but is not particularly evident on the (P, T) diagram where it appears as a single point terminating the vapor pressure curve.

A three-phase field, represented in (P, T) space by a triple point at which three two-phase lines meet, becomes a horizontal tie line at which three two-phase areas intersect in a (P, V) diagram. Each of the three phases has its particular value of molar volume. The line is horizontal because P is a potential, required to be the same in the three phases at equilibrium.

Figure 7.11c represents the same phase diagram in (S, V) space. Neither S nor V is a thermodynamic potential. As a consequence, in an equilibrated two phase structure, both the entropy and the volume will be different in the two phases. Thus, two-phase fields consist of pairs of points connected by tie lines but the tie lines are, in general, not horizontal. At a triple point on the (P, T) diagram, where three-phases are in equilibrium, the values of S and V may be expected to be different for all three phases. Thus, this condition is represented by a tie triangle with corners at the points corresponding to the (S, V) values for each of the three phases.

The reader may have experience with phase diagrams in introductory courses constructed for binary (two component) and ternary (three component) systems. The common binary phase diagram is usually presented for a fixed value of pressure, constructed in (T, X_2) space, where X_2 is the mole fraction of component 2. These diagrams are topologically similar in construction to the (P, V) diagram presented for a unary system in Figure 7.11b precisely because in both cases one variable (T) is a thermodynamic potential while the other (X_2) is not.

Similarly, the usual presentation of a ternary phase diagram, plotted on the Gibbs composition triangle (see Section 9.4.4) at constant T and P has the appearance of the diagram in Figure 7.11c. In this case, the variables are the atom fractions of two of the components, neither of which is a thermodynamic potential. Thus, three-phase fields are triangles and two-phase fields are areas filled with tie lines.

Each of the representations of a unary phase diagram presents specific information about the relationships between the phases of matter that is not contained in other representations. The (P, T)diagram is the simplest representation. The (P, V) diagram helps visualize volume changes associated with processes represented by paths on the diagram as they pass through the fields in the structure but does not present information about the corresponding temperatures. The area under such a path is related to the reversible work done by the system and may thus be useful in understanding cycles in heat engines that involve phase changes such as vaporization and condensation. The (S, V) diagram presents information about yet another property but loses the information about both the pressure and temperature of two- and three-phase equilibria.

7.7 SUMMARY

In a unary system the chemical potential of a phase is identical with its molar free energy. Chemical potential surfaces may, in principle, be computed for each phase form that exists in the system by integrating

$$d\mu = dG = -SdT + VdP \tag{7.4}$$

for that phase using heat capacities and coefficients of expansion and compressibilities obtained from an appropriate database.

Intersections of the chemical potential surfaces of the phases produce two-phase curves and three-phase triple points. Projections of these lines and points onto the (P, T) plane produces the unary phase diagram.

Combination of the conditions for equilibrium between two unary phases, α and β:

$$T^\alpha = T^\beta; \qquad P^\alpha = P^\beta; \qquad \mu^\alpha = \mu^\beta$$

yields the Clausius–Clapeyron equation;

$$\frac{dP}{dT} = \frac{\Delta S^{\alpha \to \beta}}{\Delta V^{\alpha \to \beta}} = \frac{\Delta H^{\alpha \to \beta}}{T \Delta V^{\alpha \to \beta}} \tag{7.21}$$

Integration of this equation gives the $(\alpha + \beta)$ phase boundary.

The intersection of a pair of two phase boundaries that share a common phase (e.g., $[\alpha + \beta]$ and $[\alpha + L]$ identifies the triple point at which the three phases (e.g., $[\alpha + \beta + L]$) coexist at equilibrium.

Knowledge of the relationships that exist between thermodynamic properties and phase diagrams permits calculation of some thermodynamic properties from phase boundaries. For example, the heat of vaporization may be computed from the vapor pressure curve.

Unary phase diagrams may be estimated from very limited data by assuming the vapor phase behaves ideally and ΔH is independent of temperature.

More precise calculations of phase diagrams are available through applications software developed for this purpose.

If the phase diagram is plotted in (P, T) space, both of which are thermodynamic potentials, i.e., are explicitly required to be equal in the conditions for equilibrium, then a simple cell structure results. If one of the variables chosen is not a potential, two-phase fields broaden into areas filled with horizontal (constant potential) tie lines. If neither of the variables representing the state of the phases in the system is a potential, then three-phase fields broaden into tie triangles and the tie lines in the two-phase fields are not constrained to any fixed direction.

PROBLEMS

Problem 7.1. In developing the conditions for equilibrium in a unary heterogeneous system, why is it necessary to first devise the apparatus for handling open systems?

Problem 7.2. Consider a system consisting of three phases, S, L and G. Follow the sequence of steps below to derive the conditions for equilibrium in this three-phase system.
  a. Write expressions for the change in entropy for each phase separately
  b. Write out the expression for the change in entropy for the three-phase system without constraints
  c. Write out the isolation constraints for this three-phase system including the conservation of energy, volume and total number of moles
  d. Use the isolation constraints to write dU'^G, dV'^G, and dn^G in terms of the other variables
  e. Use these relations to eliminate dU'^G, dV'^G, and dn^G from the expression for the entropy of the three-phase system
  f. Collect terms
  g. Set the coefficients of the six independent differentials equal to zero
  h. Write the conditions for equilibrium

Problem 7.3. Compute and plot the chemical potential surface $\mu(T, P)$ for a monatomic ideal gas in the range 5 K $< T <$ 1000 K and 10^{-5} atm $< P <$ 10 atm. Assume the gas is helium with $S_{298} = 126.04$ J/mol K.
  a. On the same graph sketch plausible surfaces for the chemical potential of the liquid and solid phases as functions of temperature and pressure
  b. Use these sketches to illustrate the construction of the phase diagram for a system that exhibits these three phases

Problem 7.4. Prior to Equation 7.29, it is asserted that the pressure-dependent term in $\Delta H(T, P)$ is negligible for ordinary pressures. Look up some typical values for the thermodynamic properties involved in this pressure-dependent term and identify the conditions under which it is, in fact, negligible.

Problem 7.5. Sketch curves representing the variation of the molar Gibbs free energy with temperature at the pressure corresponding to a triple point for an element. Repeat this sketch for a pressure slightly above and slightly below the triple point.

Problem 7.6. A relatively useful and simple integrated form of the Clausius–Clapeyron equation is obtained (Equation 7.39) if the temperature dependence of the heat of vaporization is neglected. Estimate the temperature interval over which this assumption is valid for a typical element such as nickel.

Problem 7.7. At 1 atm pressure, pure water ice melts at $0°C$. At 10 atm, the melting point is found to be $-0.08°C$. The density of water at $0°C$ is 1.000 g/cc, while that of ice is 0.917 g/cc. From this information, estimate the entropy of fusion of ice.

Problem 7.8. Estimate the melting point of the HCP (ε) phase form of pure titanium at 1 atm pressure. Note that ε is metastable above 1155 K at 1 atm.

Problem 7.9. At 1 atm pressure pure germanium, melts at 1232 K and boils at 2980 K. The pressure at the triple point (S,L,G) is 8.4×10^{-8} atm. Estimate the heat of vaporization of germanium.

Problem 7.10. Assuming that the pressure dependence of the transformation temperatures may be neglected below 1 atm, calculate the pressures at the triple points (α, γ, G), (γ, δ, G) and (δ, L, G) for pure iron. Assume also that all heats of transformation are temperature independent.

Problem 7.11. Thallium exists in the following forms: vapor (V), liquid (L), face centered cubic (α), body centered cubic (β) and hexagonal (ε). Estimate and plot a phase diagram for thallium from the following information:

Transformation	ΔH (J/mole)	ΔS (J/mole K)	ΔV (cc/mole)
$L \rightarrow V$	152,000	—	—
$\beta \rightarrow L$	4300	7.4	0.39
$\varepsilon \rightarrow \beta$	380	0.75	-0.02
$\alpha \rightarrow \varepsilon$	-320	0.17	0.10
$\alpha \rightarrow \beta$	60	0.92	0.08

At 1 atm the stable phase forms are:
ε (0–500 K); β (500–576 K); L (576–1730 K); above 1730 K (vapor). (The FCC α form is stable only at low temperatures and high pressures.)
1. List all of the possible two phase equilibria that could exist in this system.
2. Compute and plot the unary phase diagram for thallium in the range from 10^{-15} to 1 atm and from 300 to 2000 K. For the part of the diagram above 1 atm plot (T, P) axes. Below 1 atm plot the axes as $(-1/T, \log P)$.

Assume $\Delta c_P = 0$ for all transformations. Consider only the following equilibria: $(\alpha + \beta)$, $(\alpha + \varepsilon)$, $(\beta + \varepsilon)$, $(\beta + L)$, $\beta + V)$, $(\varepsilon + V)$ and $(L + V)$; the others are metastable.

Problem 7.12. Suppose that the formation of the ε phase in thallium is sluggish and, during times that are practical for experimental measurements, it does not nucleate and form. Recalculate the metastable phase diagram for thallium below 1 atm pressure assuming the ε phase is absent.

Problem 7.13. Sketch a plausible phase diagram for bismuth, Figure 7.10, on (P, V) axes. Make a similar sketch in the (S, V) plane.

Problem 7.14. According to Appendix B and C, the element hafnium exists in four-phase forms at ambient pressure: ε (hexagonal) 298–2016 K; β (BCC) 2016–2504 K; L (liquid) 2504–4870 K) and G (gas) above 4870 K. Given the following additional information (*Source*: From Kubaschewski, O., Alcock, C.B., and Spencer, P.J., in *Materials Thermochemistry*, 6th ed., Pergamon Press, 1993, p. 281): $C_P^\beta(T) = 10.29 + 10.77 \times 10^{-3}T$; $C_P^L(T) = 33.47$ (J/mol K) and assuming the vapor behaves ideally, compute and plot the Gibbs free energy of hafnium as a function of temperature, relative to the hexagonal phase at 298 K for all four phases. The result may be viewed as a phase stability calculation for hafnium.

REFERENCES

1. Kingery, W.D., Bowen, H.K., and Uhlman, D.R., *Introduction to Ceramics*, 2nd ed., John Wiley and Sons, New York, 1976.
2. Kaufman, L. and Bernstein, H., *Computer Calculations of Phase Diagrams with Special Reference to Refractory Metals*, Academic Press, New York, 1970.

8 Multicomponent Homogeneous Nonreacting Systems: Solutions

CONTENTS

Application of the thermodynamics displayed in Figure 1.4 to yield equilibrium maps for systems that contain more than one independent chemical component begins with the development of a formalism for describing thermodynamic properties of such multicomponent systems. This chapter lays out that development.

The chemical content of a system is most directly described by specifying the number of moles of each chemical component that it contains; n_k is the number of moles of component k, which is an *extensive* property of the system. Recall that 1 mol is simply a fixed number of units of the component, namely Avagadro's number, $N_0 = 6.023 \times 10^{23}$, of atoms or molecular units. The corresponding intensive property that defines not the content but the *composition* of the system is the *mole fraction* of component k, written X_k and defined to be n_k/n_T, where n_T is the total number of moles of all of the components in the system.

Systems may alter the number of moles of each component that they contain in two ways:

1. Atoms or molecules may be transferred across the boundary of the system.
2. Chemical reactions may occur within the boundary of the system.

Reacting systems are treated in Chapter 11. *Open multicomponent nonreacting systems*, which have a boundary that permits the flow of matter, are the subject of this chapter.

The central thermodynamic concept required for the description of multicomponent systems is the chemical potential, μ_k, introduced for unary systems in Chapter 5. In illustrating the strategy for finding conditions for equilibrium, it was deduced that the condition for chemical equilibrium in a unary two phase system is equality of the chemical potentials of the single component in the two phases. Chemical potential may also be defined and evaluated for each component in a multicomponent mixture or solution. Analogous definitions for properties of components in mixtures may also be devised for other thermodynamic properties, such as volume, entropy, etc.

In order to define and evaluate these properties, it is necessary to devise a way of assigning an appropriate part of the total value of a thermodynamic property of a multicomponent system to each of the components it contains. A strategy for making such a distribution among the components is developed first in this chapter and leads to the general notion of *partial molal properties*. Each extensive property, U', S', V', H', F', G', has its corresponding set of partial molal properties for the components in a solution. Procedures for evaluating these partial molal properties from experimental information are derived, together with relationships that exist among these properties.

The concept of the *activity* of a component in a solution and a closely related concept, the *activity coefficient*, are also developed in this chapter. Because these quantities are defined in terms of the chemical potential, they also occupy a central role in the description of the thermodynamic behavior of solutions. Indeed, it is demonstrated that if experimental values for the chemical potential or the activity or the activity coefficient of any one of the components in the system is known as a function of temperature, pressure, and composition, then *all* of the thermodynamic properties of the solution may be computed as a function of temperature, pressure, and composition.

The behavior of the partial molal properties when the system is nearly pure, i.e., in *dilute solutions*, can be shown to be general and simple. These laws of dilute solutions are presented and discussed.

The remainder of the chapter develops some models for solution behavior that have come to be recognized as standard for reporting database information about solutions, illustrates strategies for computing thermodynamic properties from such models, and reviews applications where such models may be useful.

8.1 PARTIAL MOLAL PROPERTIES

This development of a strategy for appropriately assigning part of the total value of an extensive thermodynamic property to each component in a multicomponent system is general and can be applied to all of the extensive properties that have been defined. It is perhaps easiest to visualize this idea in its application to the volume of the system.

Consider a solution that contains c components or independently variable chemical species. Some of these components may be elements, some molecules. The system has some total volume, V'. The problem at hand is, "What is a useful way of assigning a part of the total volume to each of the components present?" How many cubic centimeters does it make sense to associate with component 1, 2, etc.? There are a number of ways of making such an assignment of volumes. For example, the total volume could simply be multiplied by the fraction of the total number of molecules in the system which are component 1, 2, etc. Thus, if half of the atoms in the system were component A, one quarter B, and one quarter C, then half of the volume would be attributed to A, one fourth to B, and one fourth to C. The problem with this assignment is evident: it assumes that the volumes per atom of A, B, and C are all equal. Differences in the contribution of the atoms of A, B, and C cannot be deduced.

A variety of other schemes could be devised; however, the strategy based upon the definition of *partial molal properties* given below has proved to be most useful for reasons that will become clear as the concept is developed.

8.1.1 DEFINITION OF PARTIAL MOLAL PROPERTIES

The volume of a system is a state function. If the system is multicomponent and open, then in addition to volume changes that may result from changing the temperature and pressure of the system alterations in volume may occur independently (i.e., even if the temperature and pressure are held constant) if material is added to the system or removed from it. Thus the state function, V', is a function not only of T and P but of the number of moles of each of the components in the system: $n_1, n_2, ..., n_c$. Mathematically,

$$V' = V'(T, P, n_1, n_2, ..., n_c) \tag{8.1}$$

If the system is taken through an arbitrary infinitesimal change in state, which explicitly includes the possibility of changing the number of moles of each of the components, the change in volume may be written:

$$dV' = \left(\frac{\partial V'}{\partial T}\right)_{P,n_k} dT + \left(\frac{\partial V'}{\partial P}\right)_{T,n_k} dP + \left(\frac{\partial V'}{\partial n_1}\right)_{T,P,n_2...,n_c} dn_1$$

$$+ \left(\frac{\partial V'}{\partial n_2}\right)_{T,P,n_1,n_3,...,n_c} dn_2 + \cdots + \left(\frac{\partial V'}{\partial n_c}\right)_{T,P,n_1,n_2,...,n_{c-1}} dn_c \quad (8.2)$$

Write the string of similar terms as a sum[1]:

$$dV' = \left(\frac{\partial V'}{\partial T}\right)_{P,n_k} dT + \left(\frac{\partial V'}{\partial P}\right)_{T,n_k} dP + \sum_{k=1}^{c} \left(\frac{\partial V'}{\partial n_k}\right)_{T,P,n_j \neq n_k} \quad (8.3)$$

The coefficients of dT and dP remain simply related to the coefficients of thermal expansion and compressibility defined in Equation 4.11 and Equation 4.12, now applied to expansions and contractions of the solution under consideration at fixed composition. The coefficient of each of the changes in number of moles may be written:

$$\bar{V}_k \equiv \left(\frac{\partial V'}{\partial n_k}\right)_{T,P,n_j \neq n_k} \quad (k = 1, 2, ..., c) \quad (8.4)$$

There is a coefficient for each of the components in the system. These quantities are defined to be the *partial molal volumes* for each of the components in the system. The units of each is (volume/mol).

An analogous definition may be devised for any of the extensive properties of the system. Use the symbol B' for any of the properties U', S', V', H', F', G'. Then, for an arbitrary change in chemical content, temperature and pressure, the change in the extensive property B' is

$$dB' = M \, dT + N \, dP + \sum_{k=1}^{c} \bar{B}_k \, dn_k \quad (8.5)$$

The "partial molal B for component k" is the corresponding coefficient of dn_k:

$$\bar{B}_k \equiv \left(\frac{\partial B'}{\partial n_k}\right)_{T,P,n_j \neq n_k} \quad (k = 1, 2, ...c) \quad (8.6)$$

---

[1] The notation in the subscript of the composition term, $n_j \neq n_k$, is shorthand to describe the condition that, in taking these derivatives, the numbers of moles of all of the components except n_k are held constant.

Thus, the additional thermodynamic apparatus required to treat multicomponent systems consists of a set of terms, one for each component in the system, with appropriately defined coefficients which are the partial molal properties.

8.1.2 CONSEQUENCES OF THE DEFINITION OF PARTIAL MOLAL PROPERTIES

The first consequence of the definition of partial molal properties is contained in Equation 8.5; the change in the total property for the system for an arbitrary change in composition is the sum of the changes contributed by each component. Additional consequences are derived in this section.

Consider a process in which the temperature and pressure are held constant and the system is formed by adding n_1 moles of component 1, n_2 moles of 2, n_3 of 3 and so on, until a final state consisting of a homogeneous mixture of all of the components at the initial temperature and pressure is achieved. At any step during the process, Equation 8.3 applies with $dT = 0$ and $dP = 0$:

$$dV'_{T,P} = \sum_{k=1}^{c} \bar{V}_k dn_k \qquad (8.7)$$

Computation of the change in volume for the whole finite process requires integration of this equation. The integration procedure will depend upon a knowledge of the process sequence, i.e., in what order the components are added to the mixture. In this case, integration of Equation 8.7 will be complicated by the fact that, for example, as component 2 is added to a system initially containing n_1 moles of component 1, the composition, specified by the mole fraction of component X_2, changes continuously. Hence, the integrands for the two terms, $\bar{V}_1$ and $\bar{V}_2$, will change throughout this step in the process. In order to carry out the integration it would seem that complete knowledge of how $\bar{V}_1$ and $\bar{V}_2$ vary with composition is required; no simple, general equation may be expected to result from this integration.

An alternate strategy for integrating Equation 8.7 makes use of two principles:

1. Since $\bar{V}_k$ is an intensive property, it can only depend upon other intensive properties.
2. Changes in state functions may be computed by concocting the simplest reversible path between two end states and computing the change for that path.

For the process under consideration, visualize the addition of all c components simultaneously in the proportions found in the final mixture. Thus, during the process the intensive properties (T, P, and the set of X_k values) remain fixed and each of the $\bar{V}_k$'s is constant. In this case, integration is straightforward:

$$V' = \sum_{k=1}^{c} \int_0^{n_k} \bar{V}_k dn_k = \sum_{k=1}^{c} \bar{V}_k \int_0^{n_k} dn_k$$

$$V' = \sum_{k=1}^{c} \bar{V}_k n_k \qquad (8.8)$$

This conclusion, that the total value of the volume for the system is the weighted sum of the partial molal volumes, may be extended without complication to any extensive property:

$$B' = \sum_{k=1}^{c} \bar{B}_k n_k \qquad (8.9)$$

Accordingly, one consequence of the definition of partial molal properties is the most rudimentary requirement of any strategy for assigning a part of a total property to each of the components: the sum of the contributions must add up to the whole.

A third consequence of the definition of partial molal properties is referred to as the Gibbs–Duhem equation. Beginning with Equation 8.9, compute the differential of B', still holding temperature and pressure constant in order to focus upon the role of change in chemical content.

$$dB' = d\left[\sum_{k=1}^{c} (\bar{B}_k n_k) \right] = \sum_{k=1}^{c} d(\bar{B}_k n_k)$$

since the differential of a sum is the sum of the differentials. Differentiating the product $(\bar{B}_k n_k)$ yields

$$dB' = \sum_{k=1}^{c} [\bar{B}_k dn_k + n_k d\bar{B}_k]$$

which may be written:

$$dB' = \sum_{k=1}^{c} \bar{B}_k dn_k + \sum_{k=1}^{c} n_k d\bar{B}_k$$

Compare this result with Equation 8.5 (remember T and P remain constant in this treatment). The first summation is equal to the left side of the equation. Accordingly, the second summation must be zero:

$$\sum_{k=1}^{c} n_k d\bar{B}_k = 0 \qquad (8.10)$$

This result is called the Gibbs–Duhem equation. It demonstrates that the partial molal properties are not all independent. In particular, in a binary system in which this equation has only two terms, this equation provides the basis for computing values of partial molal properties for one of the components when values for the other have been determined. This procedure, known as a "Gibbs–Duhem integration", is developed in a later section. More generally, it can be shown that, given a partial molal property as a function of composition in a multicomponent system, values of the partial molal properties for all of the other components may be computed.

8.1.3 THE MIXING PROCESS

Temperature, pressure, volume and, according to the third law, entropy, all have absolute values in thermodynamics. In contrast, there is no universally valid state of the system for which any of the energy functions, U', H', F' and G', have a zero value. Energies of a system are always evaluated *with respect to some reference state*. Problems that involve these functions in general deal only with *changes* in their values for processes. In the treatment of multicomponent open systems the most common process considered in defining the energy functions for a solution is called the *mixing process*.

The initial state for the mixing process is a collection of containers, each holding a given quantity of a pure component in some specified phase form (gas, liquid, solid in some crystal form) and at the temperature and pressure of the solution to be formed. This initial state of any particular component is called its *reference* state for the formation of the solution. It is this state to which values for the energy and other thermodynamic functions of the solution will be referred. Reference states are particularly important in comparing energies of solutions of different phase forms, e.g., solid and liquid solutions. Such comparisons are at the base of the construction of phase diagrams (see Chapter 10). Evidently, such comparisons are valid only if the reference states for each component is chosen to be the *same* for both solutions being compared. The mixing process is the change in state experienced by the system when appropriate amounts of the pure components in their reference states are mixed together to form a homogeneous solution, which is brought to the same temperature and pressure as the initial state. Thus the mixing process is the formation of a solution from its pure components at constant temperature and pressure.

Let the superscript $(^0)$ denote values of properties in the reference state. Thus B_k^0 is the value per mole of the property B for pure component k. The total value of the property B' for the solution formed is given by Equation 8.9. The value of the B^0 for the initial unmixed state is the sum of the values for each of the pure components:

$$B'^0 = \sum_{k=1}^{c} B_k^0 n_k \qquad (8.11)$$

The change in B when $(n_1, n_2, ..., n_c)$ moles of pure $1, 2, ..., c$ are mixed at constant temperature and pressure is the value for the final state minus that for the initial state:

$$\Delta B'_{\text{mix}} = B'_{\text{soln}} - B'^0$$

$$\Delta B'_{\text{mix}} = \sum_{k=1}^{c} \bar{B}_k n_k - \sum_{k=1}^{c} \bar{B}_k^0 n_k = \sum_{k=1}^{c} (\bar{B}_k - \bar{B}_k^0) n_k \qquad (8.12)$$

Introduce the notation

$$\Delta \bar{B}_k \equiv \bar{B}_k - \bar{B}_k^0 \qquad (8.13)$$

which measures the change experienced by 1 mol of component k when it is transferred from its reference state to the surroundings it experiences in the solution with the composition under consideration. Equation 8.12 may be written

$$\Delta \bar{B}_{\text{mix}} = \sum_{k=1}^{c} \Delta \bar{B}_k n_k \qquad (8.14)$$

Thus, $\Delta B'_{\text{mix}}$ is the weighted sum of the changes experienced in the mixing process by the individual components.

The variation of $\Delta B'_{\text{mix}}$ with the composition of the solution may be computed by differentiating Equation 8.12:

$$d\Delta \bar{B}_{\text{mix}} = \sum_{k=1}^{c} [\bar{B}_k dn_k + n_k d\bar{B}_k - \bar{B}_k^0 dn_k - n_k d\bar{B}_k^0]$$

The summation of the second term on the right side is zero by the Gibbs–Duhem Equation 8.10. The fourth term is zero because the $\bar{B}_k^0$ values are properties of the reference state and are not altered by changing the composition of the solution, which is the process under consideration. Thus,

$$d\Delta \bar{B}_{\text{mix}} = \sum_{k=1}^{c} (\bar{B}_k - \bar{B}_k^0) dn_k = \sum_{k=1}^{c} \Delta \bar{B}_k dn_k \qquad (8.15)$$

Differentiate Equation 8.14 completely.

$$d\Delta B'_{\text{mix}} = \sum_{k=1}^{c} (\Delta \bar{B}_k dn_k + n_k d\Delta \bar{B}_k)$$

The first term on the right side is identical to Equation 8.15. Accordingly, the sum over the second terms must be equal to zero:

$$\sum_{k=1}^{c} n_k d\Delta \bar{B}_k = 0 \qquad (8.16)$$

This result is a form of the Gibbs–Duhem Equation 8.10 applied to the mixing process. Thus the three consequences of the definition of partial molal properties embodied in Equation 8.5, Equation 8.9 and Equation 8.10 have analogs for the mixing process in Equation 8.14 to Equation 8.16.

8.1.4 MOLAR VALUES OF THE PROPERTIES OF MIXTURES

Because the definition of partial molal properties involves consideration of an open system, i.e., the change in B' is reported per unit change in the number of moles of component k, the sets of equations derived above are formulated for a system containing an arbitrary number of moles of the components. It is frequently useful to normalize the description of properties of mixtures and express them on the basis of 1 mol of solution formed. This is achieved in a straightforward way by dividing the expressions for the equations derived up to now by n_T, the total number of moles of all of the components in the solution. The *value per mole* of each property is designated as before by dropping the $(')$ to give U, S, V, H, F, G. Dividing both sides of Equation 8.5, Equation 8.9 and Equation 8.10 by n_T yields:

$$dB = \sum_{k=1}^{c} \bar{B}_k dX_k \tag{8.17}$$

$$B = \sum_{k=1}^{c} \bar{B}_k X_k \tag{8.18}$$

$$\sum_{k=1}^{c} X_k d\bar{B}_k = 0 \tag{8.19}$$

Similar operations performed on Equation 8.14 to Equation 8.16 yield expressions for the molar values for the mixing process:

$$d\Delta B_{\text{mix}} = \sum_{k=1}^{c} \Delta\bar{B}_k dX_k \tag{8.20}$$

$$\Delta B_{\text{mix}} = \sum_{k=1}^{c} \Delta\bar{B}_k X_k \tag{8.21}$$

$$\sum_{k=1}^{c} X_k d\Delta\bar{B}_k = 0 \tag{8.22}$$

All of these results are summarized in Table 8.1.

8.2 EVALUATION OF PARTIAL MOLAL PROPERTIES

Partial molal properties may be evaluated from experimental data of two broad types:

1. Measurements of the corresponding total property of the solution, B or ΔB_{mix}, as a function of composition.
2. Measurements of the partial molal property for one of the components, $\bar{B}_k$, as a function of composition.

TABLE 8.1
Summary of Consequences of the Definition of
Partial Molal Properties

Arbitrary Quantity of System	Per Mole of System
$$\Delta B'_{mix} = \sum_{k=1}^{c} \Delta \bar{B}_k n_k$$	$$\Delta B_{mix} = \sum_{k=1}^{c} \Delta \bar{B}_k X_k$$
$$d\Delta B'_{mix} = \sum_{k=1}^{c} \Delta \bar{B}_k dn_k$$	$$d\Delta B_{mix} = \sum_{k=1}^{c} \Delta \bar{B}_k dX_k$$
$$\sum_{k=1}^{c} n_k d\Delta \bar{B}_k = 0$$	$$\sum_{k=1}^{c} X_k d\Delta \bar{B}_k = 0$$

In the developments in this section, the analysis is limited to two-component (binary) systems. Methods exist for extending these analyses to multicomponent systems.[1]

8.2.1 PARTIAL MOLAL PROPERTIES FROM TOTAL PROPERTIES

If the total value of a property, B, or ΔB_{mix}, is known as a function of composition for a solution at some temperature and pressure, then it is possible to compute from this information the values of the corresponding partial molal properties, $\bar{B}_k$ or $\Delta \bar{B}_k$, of each of the components in the mixture as a function of composition. For a binary system Equation 8.20 and Equation 8.21 each has only two terms:

$$d\Delta B_{mix} = \Delta \bar{B}_1 dX_1 + \Delta \bar{B}_2 dX_2 \qquad (8.23)$$

$$\Delta B_{mix} = \Delta \bar{B}_1 X_1 + \Delta \bar{B}_2 X_2 \qquad (8.24)$$

The mole fractions sum to 1:

$$X_1 + X_2 = 1 \qquad (8.25)$$

Accordingly,

$$dX_1 + dX_2 = 0$$

and

$$dX_1 = -dX_2 \qquad (8.26)$$

Equation 8.23 may be written

$$d\Delta B_{mix} = \Delta\bar{B}_1(-dX_2) + \Delta\bar{B}_2 dX_2 = (\Delta\bar{B}_2 - \Delta\bar{B}_1)dX_2$$

The coefficient of dX_2 in this equation is the derivative[2]

$$\frac{d\Delta B_{mix}}{dX_2} = \Delta\bar{B}_2 - \Delta\bar{B}_1 \qquad (8.27)$$

Equation 8.24 and Equation 8.27 may be considered to be two simultaneous linear equations in the unknowns $\Delta\bar{B}_1$ and $\Delta\bar{B}_2$. Solve Equation 8.27 for $\Delta\bar{B}_1$:

$$\Delta\bar{B}_1 = \Delta\bar{B}_2 - \frac{d\Delta B_{mix}}{dX_2}$$

Substitute the result into Equation 8.24

$$\Delta B_{mix} = X_1\left[\Delta\bar{B}_2 - \frac{d\Delta B_{mix}}{dX_2}\right] + X_2\Delta\bar{B}_2 = (X_1 + X_2)\Delta\bar{B}_2 - X_1\frac{d\Delta B_{mix}}{dX_2}$$

Solve for $\Delta\bar{B}_2$, noting that $X_1 + X_2 = 1$,

$$\Delta\bar{B}_2 = \Delta B_{mix} + (1 - X_2)\frac{d\Delta B_{mix}}{dX_2} \qquad (8.28)$$

The analogous result holds for component 1:

$$\Delta\bar{B}_1 = \Delta B_{mix} + (1 - X_1)\frac{d\Delta B_{mix}}{dX_1} \qquad (8.29)$$

since interchange of the subscripts in Equation 8.23 and Equation 8.27 does not alter them.

In order to apply this result to a system in which ΔB_{mix} has been determined experimentally as a function of composition it is necessary to perform a statistical analysis to fit the data with a mathematical function and take its derivative. Note that

$$\frac{d\Delta B_{mix}}{dX_2} = \frac{d\Delta B_{mix}}{dX_1}\frac{dX_1}{dX_2} = -\frac{d\Delta B_{mix}}{dX_1} \qquad (8.30)$$

---

[2] Strictly speaking, this derivative is the partial derivative taken at constant temperature and pressure but, since T and P are held constant throughout this discussion, it may be treated in this context as a total derivative.

Substitution of the function and its derivative into Equation 8.28 and Equation 8.29 then yields both partial molal properties as a function of composition.

EXAMPLE 8.1

Compute the partial molal enthalpies of the two components in a binary solution with an enthalpy of mixing given by

$$\Delta H_{mix} = aX_1X_2 \tag{8.31}$$

This is the simplest possible mathematical form for any function ΔB_{mix}, since such a function is required to pass through zero at the pure components ($X_1 = 1$ and $X_2 = 1$) where the change on mixing is zero. Recognize that there is only one independent mole fraction because they are related through Equation 8.25. Thus this equation could be written as a function of only X_2:

$$\Delta H_{mix} = a(1 - X_2)X_2 = a(X_2 - X_2^2)$$

To evaluate the partial molal enthalpies, it is necessary to compute the terms in Equation 8.28 and Equation 8.29. Compute the derivative:

$$\frac{d\Delta H_{mix}}{dX_2} = a(1 - 2X_2) \tag{8.32}$$

Note that this total derivative is *not the same* as the partial derivative,

$$\left(\frac{\partial \Delta H_{mix}}{\partial X_2}\right)_{X_1} = aX_1 = a(1 - X_2)$$

This partial derivative may be computed mathematically but has no thermodynamic meaning since it is impossible to vary X_2 while holding X_1 constant in a binary system: $X_2 = (1 - X_1)$.

To find the partial molal enthalpy for component 2, substitute expressions in Equation 8.31 and Equation 8.32 into Equation 8.28.

$$\Delta \bar{H}_2 = [aX_1X_2] + X_1[a(1 - 2X_2)]$$

The factor aX_1 is common to both terms:

$$\Delta \bar{H}_2 = aX_1[X_2 + 1 - 2X_2] = aX_1(1 - X_2) = aX_1X_1$$

$$\Delta \bar{H}_2 = aX_1^2 \tag{8.33}$$

It happens in this case, because Equation 8.31 is symmetrical in the variables X_1 and X_2, that the analogous expression for component 1 has the same form.

$$\Delta \bar{H}_1 = aX_2^2 \tag{8.34}$$

In this introductory example, $\Delta \bar{H}_1$ and $\Delta \bar{H}_2$ have the same mathematical form because ΔH_{mix} is a symmetrical function of X_1 and X_2, i.e., interchanging the subscripts 1 and 2 in aX_1X_2 yields the same function. This will not generally be true; as a consequence, the functional form of $\Delta \bar{H}_1$ will in general be different from that found for $\Delta \bar{H}_2$.

A check of the procedure may be obtained by substituting the two partial molal enthalpies into the expression for the total change on mixing:

$$\Delta H_{mix} = X_1(aX_2^2) + X_2(aX_1^2) = aX_1X_2(X_2 + X_1)$$

Since $(X_1 + X_2) = 1$,

$$\Delta H_{mix} = aX_1X_2$$

which recovers Equation 8.31.

8.2.2 GRAPHICAL DETERMINATION OF PARTIAL MOLAL PROPERTIES

Figure 8.1 plots ΔB_{mix} as a function of the mole fraction of component 2. As usual, B may be any extensive property. The value of the partial molal B for component 2 at any composition P may be obtained from this information by applying Equation 8.28. Each of the factors in this equation may be represented graphically in Figure 8.1. The slope of the curve at P is the total derivative $(d\Delta B_{mix}/dX_2)$. Graphically, this slope is the ratio of the length BC to PB.

$$\frac{d\Delta B_{mix}}{dX_2} = \frac{BC}{PB}$$

The coefficient of the derivative in Equation 8.28, $(1 - X_2)$, is the line segment PB on the graph:

$$(1 - X_2) = PB$$

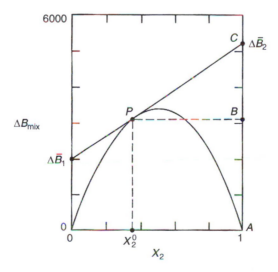

FIGURE 8.1 Graphical representation of the relation between total mixing properties (ΔB_{mix}) and the partial molal properties of the components, ($\Delta \bar{B}_1$ and $\Delta \bar{B}_2$). Noting the intercepts of the tangent line as the point P moves along the curve provides an easy visualization of how the partial molal properties vary with composition.

The first term in the equation, ΔB_{mix} at P, is equal to the length of the segment AB on the graph:

$$\Delta B_{mix} = AB$$

Combining these factors in Equation 8.28 gives the graphical representation of $\Delta \bar{B}_2$.

$$\Delta \bar{B}_2 = \Delta B_{mix} + (1 - X_2)\frac{d\Delta B_{mix}}{dX_2} = AB + PB\frac{BC}{PB} = AB + BC$$

$$\Delta \bar{B}_2 = AC$$

The sum of segments AB and BC is the segment AC, which is the intercept of the tangent line drawn at P on the two-side of the graph. The same argument, changing subscripts and applying Equation 8.29, shows that $\Delta \bar{B}_1$ is the intercept of the same tangent line on the ordinate on the one-side of the graph. Thus to determine both partial molal properties for a solution of a given composition, construct the tangent to the ΔB_{mix} curve at that composition and read the intercepts on the two sides of the graph.

This graphical procedure is not recommended for use in place of the numerical calculation of partial molal properties originally described. However, this graphical evaluation provides an extremely useful tool for visualizing how the partial molal properties vary with composition and for comparing properties of two solutions of the same components that are different phase forms, e.g., solid vs. liquid solutions. Such comparisons are particularly useful in developing the thermodynamic understanding of binary phase diagrams presented in Chapter 10.

8.2.3 EVALUATION OF THE PMPs OF ONE COMPONENT FROM MEASURED VALUES OF PMPs OF THE OTHER

For some solution properties, particularly the Gibbs free energy, it is more convenient to measure the value of one of the partial molal properties than to measure the total change on mixing. In this section it will be demonstrated that, given such information about component 2, the corresponding partial molal property for component 1 may be computed. Once both partial molal properties are known, the total value for the mixture may be computed.

The strategy is based upon integration of the Gibbs–Duhem Equation 8.22. For a binary system,

$$X_1 d\Delta \bar{B}_1 + X_2 d\Delta \bar{B}_2 = 0 \qquad (8.35)$$

Rearrange to solve for $d\Delta \bar{B}_1$:

$$d\Delta \bar{B}_1 = -\frac{X_2}{X_1}d\Delta \bar{B}_2$$

At $X_2 = 0$ (pure component 1) $\Delta\bar{B}_1$ is zero since $\bar{B}_1$ approaches B_1^0 in this limit. Integration from this lower limit to a variable composition, X_2 as an upper limit gives

$$\Delta\bar{B}_1 = \int_{X_2=0}^{X_2} - \frac{X_2}{X_1} d\Delta\bar{B}_2$$

It is assumed that input for this calculation is the functional relationship between $\Delta\bar{B}_2$ and composition, usually obtained by statistical fit of a functional form to a set of experimental data points. Thus,

$$d\Delta\bar{B}_2 = \frac{d\Delta\bar{B}_2}{dX_2} dX_2$$

and the integrated form of the Gibbs–Duhem equation may be written:

$$\Delta\bar{B}_1 = -\int_{X_2=0}^{X_2} \frac{X_2}{X_1} \frac{d\Delta\bar{B}_2}{dX_2} dX_2 \qquad (8.36)$$

It is crucial in this calculation to recognize that there is only one independent compositional variable in this binary system. Thus, the derivative in the integrand is the total derivative with respect to X_2; as argued previously, this is not the same as the partial derivative with respect to X_2 at constant X_1.

EXAMPLE 8.2
Invert Example 8.1 by assuming the input information is the result obtained for the partial molal enthalpy of component 2, Equation 8.33.

$$\Delta\bar{H}_2 = aX_1^2 \qquad (8.33)$$

Compute the partial molal enthalpy for component 1 and the enthalpy of mixing for this solution.
In order to apply Equation 8.36 it is first necessary to evaluate the total derivative from the input function.

$$\frac{d\Delta H_2}{dX_2} = a2X_1 \frac{dX_1}{dX_2} = -2aX_1$$

Substitute this result into Equation 8.36 and integrate.

$$\Delta\bar{H}_1 = -\int_0^{X_2} \frac{X_2}{X_1}(-2aX_1)dX_2$$

$$\Delta\bar{H}_1 = 2a\int_0^{X_2} X_2 dX_2 = 2a\left(\frac{X_2^2}{2}\right)\Bigg|_0^{X_2} = aX_2^2$$

which is the result that was obtained in Equation 8.34. Insert both partial molal properties into Equation 8.21 to compute the total heat of mixing:

$$\Delta H_{\text{mix}} = X_1 \Delta \bar{H}_1 + X_2 \Delta \bar{H}_2 = X_1(aX_2^2) + X_2(aX_1^2)$$

Simplify:

$$\Delta H_{\text{mix}} = aX_1 X_2(X_2 + X_1) = aX_1 X_2$$

These results agree with those that were derived from an alternate input assumption in Example 8.1.

It is thus clear that if any one of the three properties, ΔB_{mix}, $\Delta \bar{B}_1$ or $\Delta \bar{B}_2$, is known as a function of composition, the other two may be computed. If ΔB_{mix} is given as a function of composition, Equation 8.28 and Equation 8.29 are used to compute $\Delta \bar{B}_1$ and $\Delta \bar{B}_2$. Alternatively, if either of the partial molal properties, say $\Delta \bar{B}_1$, is given as a function of composition, the other two properties may be computed by the Gibbs–Duhem integration, Equation 8.36, and the condition that ΔB_{mix} is the weighted sum of the partial molal properties, Equation 8.21.

8.3 RELATIONSHIPS AMONG PARTIAL MOLAL PROPERTIES

Each of the four categories in the classification of relationships developed for the total properties of systems in Chapter 4, the laws, definitions, coefficient relations, and Maxwell relations, has a counterpart in relations among partial molal properties of the components in a system. Most of these relationships may be derived by applying what might be called the "partial molal operator" to the relations among the total properties. This operator, which derives partial molal properties from their corresponding total extensive property, is contained implicitly in the definition of partial molal properties, Equation 8.6; define the operator to be

$$\left(\frac{\partial}{\partial n_k} \right)_{T,P,n_j}$$

The notation n_j is understood to be shorthand for the whole set of compositional variables, except n_k. Application of this operator to any total property, B', yields the corresponding partial molal property.

As an example of the application of this strategy to a *definitional relationship*, recall the definition of enthalpy:

$$H' = U' + PV' \tag{8.37}$$

Apply the partial molal operator to both sides of this equation:

$$\left(\frac{\partial H'}{\partial n_k}\right)_{T,P,n_j} = \left(\frac{\partial U'}{\partial n_k}\right)_{T,P,n_j} + P\left(\frac{\partial V'}{\partial n_k}\right)_{T,P,n_j} + V'\left(\frac{\partial P}{\partial n_k}\right)_{T,P,n_j}$$

Note that the last term is zero since the pressure is defined to be constant in the partial molal operator. Apply the general definition of partial molal properties, Equation 8.6:

$$\bar{H}_k = \bar{U}_k + P\bar{V}_k \tag{8.38}$$

which is the relationship among the partial molal properties of the components in a solution that mimics the definition of enthalpy. Application of the same strategy to the definitions of Helmholtz and Gibbs free energies gives their counterparts:

$$\bar{F}_k = \bar{U}_k - T\bar{S}_k \tag{8.39}$$

$$\bar{G}_k = \bar{H}_k - T\bar{S}_k \tag{8.40}$$

To illustrate the application of this strategy to *coefficient relationships* derived for total properties, recall the combined statements of the first and second laws for the Gibbs free energy function, Equation 4.10:

$$dG' = -S'dT + V'dP + \delta W' \tag{4.10}$$

The coefficient relations for this equation are

$$-S' = \left(\frac{\partial G'}{\partial T}\right)_{P,n_k} \quad \text{and} \quad V' = \left(\frac{\partial G'}{\partial P}\right)_{T,n_k} \tag{8.41}$$

where the compositional variables, contained implicitly in $\delta W'$ in this equation, are explicitly held constant in taking these derivatives. Apply the partial molal operator to both sides of these equations:

$$-\left(\frac{\partial S'}{\partial n_k}\right)_{T,P,n_j} = \left[\frac{\partial}{\partial n_k}\left(\frac{\partial G'}{\partial T}\right)_{P,n_k}\right]_{T,P,n_j} \quad \text{and}$$

$$\left(\frac{\partial V'}{\partial n_k}\right)_{T,P,n_j} = \left[\frac{\partial}{\partial n_k}\left(\frac{\partial G'}{\partial P}\right)_{T,n_k}\right]_{T,P,n_j}$$

Interchange the order of differentiation on the right sides of both equations:

$$-\left(\frac{\partial S'}{\partial n_k}\right)_{T,P,n_j} = \left[\frac{\partial}{\partial T}\left(\frac{\partial G'}{\partial n_k}\right)_{T,P,n_j}\right]_{P,n_k} \quad \text{and}$$

$$\left(\frac{\partial V'}{\partial n_k}\right)_{T,P,n_j} = \left[\frac{\partial}{\partial P}\left(\frac{\partial G'}{\partial n_k}\right)_{T,P,n_j}\right]_{T,P,n_k}$$

Examine the resulting relations and identify the partial molal properties they contain.

$$\bar{S}_k = \left(\frac{\partial \bar{G}_k}{\partial T}\right)_{P,n_k} \quad \text{and} \quad \bar{V}_k = \left(\frac{\partial \bar{G}_k}{\partial P}\right)_{T,n_k} \tag{8.42}$$

Thus the coefficient relations have counterparts in relations among the partial molal properties.

The Maxwell relation corresponding to Equation 4.10 is

$$-\left(\frac{\partial S'}{\partial P}\right)_{T,n_k} = \left(\frac{\partial V'}{\partial T}\right)_{P,n_k} \tag{8.43}$$

Again, apply the partial molal operator to both sides of this equation:

$$-\left[\frac{\partial}{\partial n_k}\left(\frac{\partial S'}{\partial P}\right)_{T,n_k}\right]_{T,P,n_j} = \left[\frac{\partial}{\partial n_k}\left(\frac{\partial V'}{\partial T}\right)_{P,n_k}\right]_{T,P,n_j}$$

Interchange the order of differentiation:

$$-\left[\frac{\partial}{\partial P}\left(\frac{\partial S'}{\partial n_k}\right)_{T,P,n_j}\right]_{T,n_k} = \left[\frac{\partial}{\partial T}\left(\frac{\partial V'}{\partial n_k}\right)_{T,P,n_j}\right]_{P,n_k}$$

Recognize the partial molal properties contained:

$$-\left(\frac{\partial \bar{S}_k}{\partial P}\right)_{T,n_k} = \left(\frac{\partial \bar{V}_k}{\partial T}\right)_{P,n_k} \tag{8.44}$$

which is the analog of the Maxwell relation, Equation 8.43.

In order to obtain a relation among partial molal properties that is the analog of a combined statement of the first and second laws, consider the variation of the partial molal Gibbs free energy of component k with temperature and pressure, keeping the composition of the solution constant:

$$\bar{G}_k = \bar{G}_k(T,P) \tag{8.45}$$

Since $\bar{G}_k$ is a state function, its differential is

$$d\bar{G}_k = \left(\frac{\partial \bar{G}_k}{\partial T}\right)_{P,n_k} dT + \left(\frac{\partial \bar{G}_k}{\partial P}\right)_{T,n_k} dP$$

These coefficients have been evaluated in Equation 8.42; substitute the results:

$$d\bar{G}_k = -\bar{S}_k dT + \bar{V}_k dP \qquad (8.46)$$

Other forms of the combined statement may be derived from this relation. To obtain the analog for the enthalpy function, take the differential of the definitional relation, Equation 8.40, and set it equal to the right side of Equation 8.46:

$$d\bar{G}_k = d\bar{H}_k - T\, d\bar{S}_k - \bar{S}_k T = -\bar{S}_k dT + \bar{V}_k dP$$

Rearrange:

$$d\bar{H}_k = T\, d\bar{S}_k + \bar{V}_k dP \qquad (8.47)$$

which is the partial molal property analog to the combined statement of the first and second laws for the enthalpy function, Equation 4.6.

These examples serve to illustrate the potential for generating a host of relations among the partial molal properties of solutions. Indeed, virtually every relation derived in Chapter 4 has such a counterpart. These relationships are not widely applied, however, primarily because the focus of attention in the description of solutions is upon their variation with temperature and pressure since these are the variables that are usually controlled in studies of multicomponent solutions.

8.4   CHEMICAL POTENTIAL IN MULTICOMPONENT SYSTEMS

The idea of the chemical potential was introduced first in Chapter 5 in applying the strategy for finding conditions for equilibrium to unary systems. In this section, the concept is broadened to include the description of multicomponent systems. In the process it is demonstrated that if the chemical potential is known as a function of temperature, pressure, and composition for one of the components, *then all of the partial molal and total properties of the system may be computed.*

The thermodynamic state of an open homogeneous multicomponent system requires specification of $(c + 2)$ variables because the state of the system may be altered by changing the number of moles of each of the c independent compositional variables for the system. The internal energy may be written

$$U' = U'(S', V', n_1, n_2, \ldots, n_k, \ldots, n_c) \qquad (8.48)$$

The change in U' for an arbitrary change in the state of the system has $(c + 2)$ terms:

$$dU' = T\,dS' - P\,dV' + \mu_1 dn_1 + \mu_2 dn_2 + \cdots + \mu_c dn_c$$

$$dU' = T\,dS' - P\,dV' + \sum_{k=1}^{c} \mu_k dn_k \tag{8.49}$$

Define the coefficient of each compositional variable to be the *chemical potential* for that component. Thus,

$$\mu_k \equiv \left(\frac{\partial U'}{\partial n_k}\right)_{S',V',n_j} \tag{8.50}$$

Compare Equation 8.49 with the original combined statement for the first and second laws, Equation 4.4. For open multicomponent systems the additional apparatus required is evidently contained in

$$\delta W' = \sum_{k=1}^{c} \mu_k dn_k \tag{8.51}$$

Expressions for the other three energy functions follow from their combined statements, Equation 4.6, Equation 4.8 and Equation 4.10:

$$dH' = T\,dS' + V'dP + \sum_{k=1}^{c} \mu_k dn_k \tag{8.52}$$

$$dF' = -S'dT - P\,dV' + \sum_{k=1}^{c} \mu_k dn_k \tag{8.53}$$

$$dG' = -S'dT + V'dP + \sum_{k=1}^{c} \mu_k dn_k \tag{8.54}$$

Application of the coefficient relation to each of these equations shows that the chemical potential may be expressed as any of the following four derivatives:

$$\mu_k = \left(\frac{\partial U'}{\partial n_k}\right)_{S',V',n_j} = \left(\frac{\partial H'}{\partial n_k}\right)_{S',P,n_j} = \left(\frac{\partial F'}{\partial n_k}\right)_{T,V',n_j} = \left(\frac{\partial G'}{\partial n_k}\right)_{T,P,n_j} \tag{8.55}$$

All four of these derivatives appear to be similar to partial molal quantities; however, only one is in fact a partial molal property, namely the Gibbs free energy derivative:

$$\mu_k = \left(\frac{\partial G'}{\partial n_k}\right)_{T,P,n_j} = \bar{G}_k \tag{8.56}$$

None of the other expressions for μ_k are partial molal properties because, in their evaluations, temperature and pressure are not held constant. This constraint, (T, P) constant, is explicitly contained in the definition of partial molal properties, Equation 8.6. Thus,

$$\mu_k = \left(\frac{\partial H'}{\partial n_k} \right)_{S',P,n_j} \neq \bar{H}_k$$

and

$$\bar{H}_k = \left(\frac{\partial H'}{\partial n_k} \right)_{T,P,n_j} \neq \mu_k$$

The equality of chemical potential and the partial molal Gibbs free energy provides the basis for expressing all of the partial molal properties of component k in terms of its chemical potential. From the coefficient relations, Equation 8.42,

$$\bar{S}_k = -\left(\frac{\partial \bar{G}_k}{\partial T} \right)_{P,n_k} = -\left(\frac{\partial \mu_k}{\partial T} \right)_{P,n_k} \tag{8.57}$$

$$\bar{V}_k = \left(\frac{\partial \bar{G}_k}{\partial P} \right)_{T,n_k} = \left(\frac{\partial \mu_k}{\partial P} \right)_{T,n_k} \tag{8.58}$$

The enthalpy may be evaluated by rearranging the definitional relation Equation 8.40:

$$\bar{H}_k = \bar{G}_k + T\bar{S}_k$$

$$\bar{H}_k = \mu_k - T\left(\frac{\partial \mu_k}{\partial T} \right)_{P,n_k} \tag{8.59}$$

The partial molal internal energy is also obtained from the definitional relation:

$$\bar{U}_k = \bar{H}_k - P\bar{V}_k = \mu_k - T\left(\frac{\partial \mu_k}{\partial T} \right)_{P,n_k} - P\left(\frac{\partial \mu_k}{\partial P} \right)_{T,n_k} \tag{8.60}$$

and the Helmholtz free energy function from Equation 8.39:

$$\bar{F}_k = \bar{U}_k - T\bar{S}_k = \mu_k - P\left(\frac{\partial \mu_k}{\partial P} \right)_{T,n_k} \tag{8.61}$$

Thus if the chemical potential of component k is measured for a solution of a given composition as a function of temperature and pressure so that its temperature and pressure derivatives may be computed, all of the partial molal properties of component k may be evaluated. These relationships are summarized in Table 8.2. Relationships analogous to each of these expressions may be written for the set of $\Delta \bar{B}_k$'s simply by

---

TABLE 8.2
Relations of Partial Molal Properties to the Chemical Potential

$$\bar{G}_k = \mu_k$$

$$\bar{S}_k = -\left(\frac{\partial \mu_k}{\partial T}\right)_{P,n_k}$$

$$\bar{V}_k = \left(\frac{\partial \mu_k}{\partial P}\right)_{T,n_k}$$

$$\bar{H}_k = \mu_k - T\left(\frac{\partial \mu_k}{\partial T}\right)_{P,n_k}$$

$$\bar{U}_k = \mu_k - T\left(\frac{\partial \mu_k}{\partial T}\right)_{P,n_k} - P\left(\frac{\partial \mu_k}{\partial P}\right)_{T,n_k}$$

$$\bar{F}_k = \mu_k - P\left(\frac{\partial \mu_k}{\partial P}\right)_{T,n_k}$$

---

substituting $\Delta\mu_k = \mu_k - \mu_k^0$, where μ_k^0 is the chemical potential of k in its reference state, wherever μ_k appears.

Again limit consideration to a binary system. Equality of chemical potential with the partial molal Gibbs free energy is the basis for computing the chemical potential of one component from experimental measurements of the other through a Gibbs–Duhem integration. The Gibbs–Duhem equation for the partial molal Gibbs free energy is

$$X_1 d\Delta\bar{G}_1 + X_2 d\Delta\bar{G}_2 = 0$$

Since $\Delta\bar{G}_k = \Delta\mu_k$, $d\Delta\bar{G}_k = d\Delta\mu_k$:

$$X_1 d\Delta\mu_1 + X_2 d\Delta\mu_2 = 0 \qquad (8.62)$$

The integrated form of this Gibbs–Duhem equation for the chemical potential is adapted from Equation 8.36:

$$\Delta\mu = -\int_{X_2=0}^{X_2} \frac{X_2}{X_1} \frac{d\Delta\mu_2}{dX_2} dX_2 \qquad (8.63)$$

This result gives $\Delta\mu_1$ when $\Delta\mu_2$ is known as a function of composition at any temperature and pressure. Thus, for a binary system, given the chemical potential of one of the components as a function of temperature, pressure, and composition, all of the partial molal properties of both components may be computed. With this information, values for the total properties of the solution may be calculated by applying Equation 8.18.

Extensions of the Gibbs–Duhem integration strategy exist for multicomponent systems. It has been demonstrated that given values of a partial molal property for one component in a multicomponent system as a function of composition at any temperature and pressure, it is possible to compute the value for that property for *all* of the remaining components; no additional information is required. A review of these procedures is presented in Lupis.[2] Thus if $\Delta\mu_k$ is determined for one component, the chemical potential may be computed for all of the others. Combination of this information with the relationships in Table 8.2 demonstrates that given the chemical potential of component k as a function of temperature, pressure, and composition, all of the partial molal properties of all of the components and all of the total properties may be evaluated. The thermodynamic behavior of the system is completely determined.

Since the chemical potential is equal to the partial molal Gibbs free energy, the variation of chemical potential with temperature and pressure is identical to that for $\bar{G}_k$. This relationship was derived in Section 8.3, Equation 8.46:

$$d\bar{G}_k = d\mu_k = -\bar{S}_k d\bar{T} + \bar{V}_k dP \tag{8.64}$$

The corresponding relation holds for $\Delta\bar{G}_k$:

$$d\Delta\bar{G}_k = d\Delta\mu_k = -\Delta\bar{S}_k dT + \Delta\bar{V}_k dP \tag{8.65}$$

8.5   FUGACITIES, ACTIVITIES, AND ACTIVITY COEFFICIENTS

Experimental measurements of the thermodynamic behavior of solutions are not aimed at the direct determination of chemical potentials, although this quantity lies at the core of the description of such systems. Common practice measures another property, called the *activity of component* k, which is defined in terms of the chemical potential by the equation:

$$\mu_k - \mu_k^0 = \Delta\mu_k \equiv RT \ln a_k \tag{8.66}$$

The argument of the logarithm, a_k is the activity of k in a solution at a given temperature, pressure, and composition; R and T have their usual meaning. Activity is a unitless quantity, as is the mole fraction of component k. A closely related quantity, called the *fugacity*, is defined for mixtures of gases.

Another convenient measure of solution behavior, called the *activity coefficient of component* k, written γ_k, is defined by the equation:

$$a_k = \gamma_k X_k \tag{8.67}$$

From Equation 8.66,

$$\mu_k - \mu_k^0 = RT \ln \gamma_k X_k \qquad (8.68)$$

The activity coefficient is also a unitless quantity.[3]

The origin of the term "activity" for a_k may be made clear through Equation 8.67. If $\gamma_k = 1$, the activity of component k is equal to its mole fraction and the behavior of k, from the point of view of its chemical potential, is determined completely by its composition. If $\gamma_k > 1$, $a_k > X_k$ and in the evaluation of its chemical potential, component k "acts as if" the solution contains more of k than the mole fraction suggests. Similarly, if $\gamma_k < 1$ so that $a_k < X_k$, the component acts as if there is less of it present than the composition suggests.

The origin of the logarithmic form of the relation of chemical potential to activity is developed in this section. As the activity concept is applied, first to heterogeneous systems in Chapters 9 and 10, then to reacting systems in Chapter 11 and finally to complex systems in Chapters 12 and 13, the utility and convenience of this choice for the form of the relationship will be made abundantly clear.

8.5.1 PROPERTIES OF IDEAL GAS MIXTURES

It is not surprising that the earliest attempts to understand the thermodynamics of mixtures focused upon the behavior of mixtures of ideal gases. Calculation of the changes in thermodynamic properties for the mixing process in this case is straightforward.

For a collection of ideal gases the mixing process is the change in state accompanying the process in which n_k moles of each of the pure gases, at a pressure P and temperature T, are mixed to form a homogeneous solution at the same temperature and total pressure. Figure 8.2 may be used to visualize this process. The initial condition of the system is represented in Figure 8.2(a); the box that will ultimately contain the mixture of gases is partitioned into segments, each of which contains the number of moles of one of the pure gases that will form the mixture. Since by definition the mixing process occurs at constant temperature and pressure, the gas in each of these compartments is at the temperature, T, and pressure, P, of the final mixture. In Figure 8.2(b) the partitions have been removed, the components mix and the homogeneous gas mixture is ultimately formed. Focus upon the process experienced by one of the components, k. Initially, this component is in its pure state defined by (T, P). In Chapter 6, it was demonstrated that the particles in an ideal gas mixture do not interact; all of their energy is associated with kinetic energy of their motion.

---

[3] Alternate definitions of the activity coefficient are in wide use in the chemical literature depending upon the quantity used to report composition of the solution. For example, if composition is reported in molar concentration, c_k (moles of k/cc of solution) an "activity coefficient," ϕ_k, may be defined by an equation analogous to Equation 8.67:

$$a_k = \phi_k c_k$$

In this case, since a_k is unitless, ϕ_k must have units of (cc/mol). The unitless coefficient defined in Equation 8.67 will be used throughout this text.

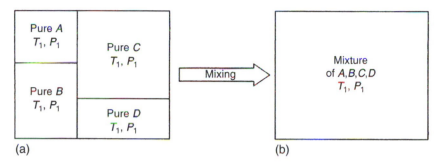

FIGURE 8.2 Illustration of the mixing process. (a) The initial state is a collection of systems each containing an appropriate number of moles of one of the pure components at the temperature T_1 and the pressure P_1. (b) The partitions are removed and the components mix to form a single homogeneous solution at the same temperature T_1 and total pressure P_1.

Accordingly, in the mixed state each component in the mixture behaves as if it occupies the entire volume of the system with no other components present. If the total pressure is P then, according to Dalton's law of partial pressures, each component contributes to that pressure in proportion to the relative number of molecules that it represents. Each component may be viewed as exerting a *partial pressure*, P_k, given by

$$P_k = X_k P \tag{8.69}$$

The total pressure is the sum of these partial pressures:

$$P = \sum_{k=1}^{c} P_k = \sum_{k=1}^{c} X_k P = P \sum_{k=1}^{c} X_k = P$$

since the sum of the mole fractions is one.

Focus upon the change experienced by an individual component k during the mixing process. Initially, as pure k it exists at the temperature T and pressure P. In the mixture component k remains at the temperature T and exerts a pressure equal to $P_k = X_k P$. Thus, for molecules of component k the mixing process is equivalent to an isothermal expansion from an initial pressure P to a final pressure P_k. The change in chemical potential experienced by component k may be obtained by integrating Equation 8.64 at constant temperature,

$$d\mu_k = -\bar{S}_k dT + \bar{V}_k dP = \bar{V}_k dP \tag{8.70}$$

$dT = 0$ because the mixing process is isothermal. Integration requires evaluation of the partial molal volume of component k in an ideal gas mixture. The volume of an ideal gas mixture is

$$V' = n_T \frac{RT}{P} = \frac{RT}{P} \sum_{k=1}^{c} n_k$$

From the definition of partial molal properties,

$$\bar{V}_k = \left(\frac{\partial V'}{\partial n_k} \right)_{T,P,n_j} = \frac{RT}{P} \tag{8.71}$$

Insert this result into Equation 8.70 and integrate:

$$\mu_k - \mu_k^0 = \int_P^{P_k} \bar{V}_k dP = \int_P^{P_k} \frac{RT}{P} dP = RT \ln \frac{P_k}{P}$$

Recall that the partial pressure, $P_k = X_k P$:

$$\Delta \mu_k = RT \ln X_k = \Delta \bar{G}_k \tag{8.72}$$

Compare this result with the definitions of activity and activity coefficient, Equation 8.66 and Equation 8.67. Evidently, for any component in an ideal gas mixture, the activity is equal to the mole fraction and the activity coefficient is equal to one.

The relations summarized in Table 8.2 may be applied to evaluate all of the other partial molal properties of an ideal gas mixture. The derivative with respect to temperature gives the entropy:

$$\left(\frac{\partial \Delta \bar{G}_k}{\partial T} \right)_{P, n_k} = R \ln X_k \tag{8.73}$$

The pressure derivative is zero:

$$\left(\frac{\partial \Delta \bar{G}_k}{\partial P} \right)_{T, n_k} = 0 \tag{8.74}$$

Substitution of these results into the equations summarized in Table 8.2 gives

$$\Delta \bar{S}_k = -\left(\frac{\partial \Delta \mu_k}{\partial T} \right)_{P, n_k} = -R \ln X_k \tag{8.75}$$

$$\Delta \bar{V}_k = \left(\frac{\partial \Delta \mu_k}{\partial P} \right)_{T, n_k} = 0 \tag{8.76}$$

$$\Delta \bar{H}_k = \Delta \mu_k + T \left(\frac{\partial \Delta \mu_k}{\partial T} \right)_{P, n_k} = RT \ln X_k + T(-R \ln X_k) = 0 \tag{8.77}$$

$$\Delta \bar{U}_k = \Delta \bar{H}_k - P \Delta \bar{V}_k = 0 - 0 = 0 \tag{8.78}$$

$$\Delta \bar{F}_k = \Delta \bar{U}_k - T \Delta \bar{S}_k = 0 - T(-R \ln X_k) = RT \ln X_k \tag{8.79}$$

Although these results have been derived for mixtures of gases, they have been adapted to the description of liquid and solid solutions. Mixtures obeying these relations, whether solid, liquid, or gas, are in general called *ideal solutions*. These equations and the corresponding summed relations for the total properties of ideal solutions are summarized in Table 8.3.

This collection of relationships can be use to evaluate all of the properties of an ideal solution if its temperature and composition are given. In the formation of an

TABLE 8.3
Properties of an Ideal Solution

Partial Molal Property	Total Property
$\Delta \bar{G}_k = RT \ln X_k$	$\Delta G_{\text{mix}} = RT \sum_{k=1}^{c} X_k \ln X_k$
$\Delta \bar{S}_k = -R \ln X_k$	$\Delta S_{\text{mix}} = -R \sum_{k=1}^{c} X_k \ln X_k$
$\Delta \bar{V}_k = 0$	$\Delta V_{\text{mix}} = 0$
$\Delta \bar{H}_k = 0$	$\Delta H_{\text{mix}} = 0$
$\Delta \bar{U}_k = 0$	$\Delta U_{\text{mix}} = 0$
$\Delta \bar{F}_k = RT \ln X_k$	$\Delta F_{\text{mix}} = RT \sum_{k=1}^{c} X_k \ln X_k$

ideal solution there is no heat of mixing ($\Delta H_{\text{mix}} = 0$), no volume change ($\Delta V_{\text{mix}} = 0$) and no change in internal energy ($\Delta U_{\text{mix}} = 0$). The effects that differ from zero (ΔG_{mix}, ΔF_{mix}, and ΔS_{mix}) all derive from the change in entropy experienced by the components in moving from the unmixed state to the homogeneous solution. Since no heat is transferred into the system during the mixing process, there is no entropy exchange with the surroundings for an ideal gas. The entropy change computed from Equation 8.75 is the entropy produced by this irreversible mixing process.

This mixing behavior is summarized for a binary system in Figure 8.3 which plots all of the properties of an ideal solution as a function of composition and temperature. Three characteristics of this behavior may be noted:

1. All of the plots are symmetrical with respect to composition; substitution of 1 for 2 and vice versa produces the same equations.
2. Slopes of the plots of ΔS_{mix}, ΔG_{mix}, and ΔF_{mix} vs. composition are vertical at the sides of the diagram because they contain the logarithm function. The derivative of $\ln(x)$ is $1/x$, which approaches ∞ as x approaches zero.
3. The entropy of mixing is independent of temperature; at any given composition ΔG_{mix} and ΔF_{mix} vary linearly with the absolute temperature.

This ideal solution model has proven to be most useful as a basis for comparison of the properties of a real solution with those that an ideal solution would exhibit

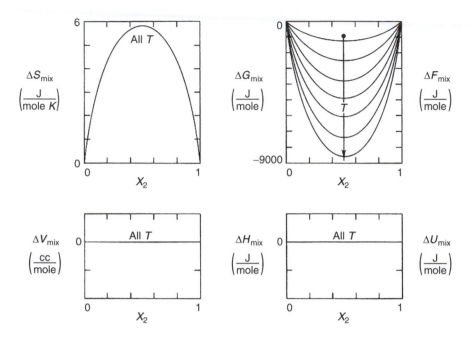

FIGURE 8.3 Properties of an ideal solution including ΔG_{mix}, ΔS_{mix} ΔH_{mix}, ΔV_{mix}, ΔU_{mix}, and ΔF_{mix}. Maximum temperature shown is 1400 K.

at the same temperature, pressure, and composition. Indeed, as developed in Section 8.5.4, it is very useful to characterize the behavior of real solutions in terms of their "departure from ideal solution behavior."

8.5.2 MIXTURES OF REAL GASES: FUGACITY

Attempts to apply the ideal gas model to mixtures of real gases revealed deviations from the predicted behavior. For many practical applications, these deviations can be neglected. For precise analyses or in property ranges for which the ideal gas model is not a viable approximation, it is necessary to devise a more general formalism for describing the mixing behavior of real gases. One strategy for handling this extension to real gas mixtures is based upon the concept of the *fugacity*.

The change in chemical potential for component k upon mixing may be obtained in general by integrating Equation 8.64. For real gases, the partial molal volume is not given by Equation 8.71, but must be determined experimentally. Define a function, α_k, which reports the deviation of the measured partial molal volume at a given temperature and pressure from that which would be computed for an ideal gas of the same temperature, pressure and composition:

$$\alpha_k = \bar{V}_k - \frac{RT}{P} \qquad (8.80)$$

The change in chemical potential during mixing for a component in a mixture of real gases is

$$\Delta\mu_k = \int_P^{P_k} \left(\alpha_k - \frac{RT}{P} \right) dP$$

$$\Delta\mu_k = \int_P^{P_k} \left(\alpha_k dP + RT \ln\frac{P_k}{P} \right) \tag{8.81}$$

Thus, if α_k is determined, the chemical potential may be evaluated.

The fugacity, written f_k, is a property of a component in a gas mixture which has units of pressure and is defined so that the form of the equation for the chemical potential change on mixing, Equation 8.72, is retained:

$$\mu_k - \mu_k^0 = \Delta\mu_k \equiv RT \ln\frac{f_k}{P} \tag{8.82}$$

The fugacity thus occupies the role that the partial pressure plays in ideal gas mixtures. The fugacity of a component in a real gas may be evaluated by setting Equations 8.81 and Equation 8.82 equal to each other:

$$RT \ln\frac{f_k}{P} = \int_P^{P_k} \alpha_k dP + RT \ln\frac{P_k}{P}$$

and solving for f_k. Some algebraic manipulation yields

$$f_k = P_k \exp\left[\int_P^{P_k} \alpha_k dP \right] \tag{8.83}$$

As the deviation from ideal behavior, reported in α_k, approaches zero, the fugacity of component k approaches its partial pressure.

Determination of the fugacity of one of the components in a gas mixture as a function of composition, temperature, and pressure permits calculation of the chemical potential for that component through Equation 8.82. It was demonstrated in Section 8.4 that if the chemical potential is known as a function of temperature, pressure, and composition for one of the components, then all of the properties of the solution may be evaluated. Thus measurement of the fugacity of one component over a range of temperature, pressure, and composition is sufficient to describe the behavior of real gas mixtures completely in that range.

8.5.3 ACTIVITY AND THE BEHAVIOR OF REAL SOLUTIONS

Comparison of the definition of activity, Equation 8.66, with that for fugacity, Equation 8.82, demonstrates that these quantities are simply related;

$$a_k = \frac{f_k}{P} \tag{8.84}$$

where P is the pressure in the reference state for component k. Either quantity provides a viable description of the behavior of mixtures of gases. However, the fugacity concept is limited in its application to gas mixtures because its definition endows it with the character of a pressure. The activity concept is not constrained by such an association and so can be applied to describe the characteristics of gas, liquid, and solid solutions.

Recall the definition of activity, Equation 8.66:

$$\Delta \mu_k = RT \ln a_k = \Delta \bar{G}_k \tag{8.85}$$

Table 8.2 may be used to relate all of the partial molal properties of a component in a mixture to its activity. The temperature derivative is

$$\left(\frac{\partial \Delta \mu_k}{\partial T} \right)_{P,n_k} = R \ln a_k + RT \left(\frac{\partial \ln a_k}{\partial T} \right)_{P,n_k} \tag{8.86}$$

The derivative with respect to pressure,

$$\left(\frac{\partial \Delta \mu_k}{\partial P} \right)_{T,n_k} = RT \left(\frac{\partial \ln a_k}{\partial P} \right)_{T,n_k} \tag{8.87}$$

Substitution of these results into the equations presented in Table 8.2 produces the set of equations summarized in Table 8.4.

Focus again on a binary system. If the activity of component 2 is measured as a function of composition, then the activity of component 1 may be computed through a Gibbs–Duhem integration. Combine the definition of activity, Equation 8.66 with the Gibbs–Duhem integration for the chemical potential, Equation 8.63:

$$\ln a_1 = - \int_{X_2=0}^{X_2} \frac{X_2}{X_1} \frac{d \ln a_2}{dX_2} dX_2 \tag{8.93}$$

Some difficulties may arise in carrying out this integration because as $a_2 \rightarrow 0$, $\ln a_2 \rightarrow -\infty$. Analytical techniques are available that circumvent this problem.[3,4]

Extensions of the Gibbs–Duhem strategy to multicomponent systems permit the calculation of the activities of all of the components from experimental measurements for one component. Thus, like chemical potential and fugacity, complete knowledge of the activity for one component is sufficient to characterize the solution thermodynamics of a multicomponent system.

TABLE 8.4
Relationships between Partial Molal Properties
of Component *k* and Its Activity

$$\Delta \bar{G}_k = RT \ln a_k \qquad (8.85)$$

$$\Delta \bar{S}_k = -R \ln a_k + RT\left(\frac{\partial \ln a_k}{\partial T}\right)_{P,n_k} \qquad (8.88)$$

$$\Delta \bar{V}_k = RT\left(\frac{\partial \ln a_k}{\partial P}\right)_{T,n_k} \qquad (8.89)$$

$$\Delta \bar{H}_k = -RT^2\left(\frac{\partial \ln a_k}{\partial T}\right)_{P,n_k} \qquad (8.90)$$

$$\Delta \bar{U}_k = -RT^2\left(\frac{\partial \ln a_k}{\partial T}\right)_{P,n_k} - PRT\left(\frac{\partial \ln a_k}{\partial P}\right)_{T,n_k} \qquad (8.91)$$

$$\Delta \bar{F}_k = -RT \ln a_k - PRT\left(\frac{\partial \ln a_k}{\partial P}\right)_{T,n_k} \qquad (8.92)$$

8.5.4 USE OF THE ACTIVITY COEFFICIENT TO DESCRIBE THE BEHAVIOR OF REAL SOLUTIONS

For reasons that will become evident in this section, the description of solution thermodynamics based upon the activity coefficient is perhaps the most convenient of those that are reviewed in this chapter. The definition of the activity coefficient γ_k of component k and its relation to chemical potential were presented in Equation 8.67 and Equation 8.68:

$$a_k \equiv \gamma_k X_k \qquad (8.67)$$

$$\Delta \mu_k = RT \ln \gamma_k X_k = \Delta \bar{G}_k \qquad (8.68)$$

The activity coefficient is in general a function of temperature, pressure, and composition and must be determined experimentally for each solution. Using the properties of the logarithm function, Equation 8.68 may be written:

$$\Delta \mu_k = \Delta \bar{G}_k = RT \ln \gamma_k + RT \ln X_k \qquad (8.94)$$

The second term on the right side is the partial molal Gibbs free energy of mixing for an ideal solution. The first term may then be characterized as reporting the "departure from ideal behavior" of component k in the mixture. These two terms are defined to be the "excess" (designated by a superscript (^{xs}) and "ideal" designated (^{id})) contributions to the partial molal Gibbs free energy of k. The total Gibbs free energy of mixing may be written:

$$\Delta \bar{G}_k = \Delta \bar{G}_k^{xs} + \Delta \bar{G}_k^{id} \tag{8.95}$$

where

$$\Delta \bar{G}_k^{xs} = RT \ln \gamma_k \tag{8.96}$$

and

$$\Delta \bar{G}_k^{id} = RT \ln X_k \tag{8.97}$$

If the activity coefficient is larger than 1, component k "acts" as if it has more k than the composition suggests. In this case $\ln \gamma_k$ and consequently the excess free energy is positive and the system is said to exhibit a "positive departure from ideal behavior." For $\gamma_k < 1$, $\Delta \bar{G}_k^{xs}$ is negative and a "negative departure" describes the behavior of the component.

In order to express the remaining partial molal properties in terms of the activity coefficient, it is necessary to apply the definition, Equation 8.94, to the set of relations presented in Table 8.2. The temperature derivative of Equation 8.94 gives:

$$\left(\frac{\partial \Delta \mu_k}{\partial T} \right)_{P,n_k} = R \ln \gamma_k + RT \left(\frac{\partial \ln \gamma_k}{\partial T} \right)_{P,n_k} + R \ln X_k \tag{8.98}$$

The derivative with respect to pressure,

$$\left(\frac{\partial \Delta \mu_k}{\partial P} \right)_{T,n_k} = RT \left(\frac{\partial \ln \gamma_k}{\partial P} \right)_{T,n_k} \tag{8.99}$$

Substitution of these results into the equations in Table 8.2 yield the expressions for the partial molal properties summarized in Table 8.5.

The expression for any one of these partial molal properties may be decomposed into a set of terms which involve only the activity coefficient and may be called the "excess" part of the property and an "ideal" part that is identical with the value for an ideal solution. In the case of the volume, enthalpy, and internal energy it will be recalled that the corresponding changes for an ideal solution are zero; thus for these functions the total property is all "excess".

In a binary system, a Gibbs–Duhem equation may be devised for the activity coefficients. Recall again the Gibbs–Duhem equation for chemical potentials:

$$X_1 d\Delta \mu_1 + X_2 d\Delta \mu_2 = 0 \tag{8.62}$$

TABLE 8.5

Relationships between the Partial Molal Properties of Component k and Its Activity Coefficient

$$\Delta \bar{G}_k = RT \ln \gamma_k + RT \ln X_k \qquad (8.94)$$

$$\Delta \bar{S}_k = -R \ln \gamma_k + RT\left(\frac{\partial \ln \gamma_k}{\partial T}\right)_{P,n_k} - R \ln X_k \qquad (8.100)$$

$$\Delta \bar{V}_k = RT\left(\frac{\partial \ln \gamma_k}{\partial P}\right)_{T,n_k} + 0 \qquad (8.101)$$

$$> \Delta \bar{H}_k = -RT^2\left(\frac{\partial \ln \gamma_k}{\partial T}\right)_{P,n_k} + 0 \qquad (8.102)$$

$$\Delta \bar{U}_k = RT^2\left(\frac{\partial \ln \gamma_k}{\partial T}\right)_{P,n_k} - PRT\left(\frac{\partial \ln \gamma_k}{\partial P}\right)_{T,n_k} + 0 \qquad (8.103)$$

$$\Delta \bar{F}_k = RT \ln \gamma_k - PRT\left(\frac{\partial \ln \gamma_k}{\partial P}\right)_{T,n_k} + RT \ln X_k \qquad (8.104)$$

---

and the relation between chemical potential and activity coefficient, Equation 8.94. Write the differential of the latter equation:

$$d\Delta \mu_k = RT(d \ln \gamma_k + d \ln X_k)$$

Substitute the result into Equation 8.62:

$$X_1 RT(d \ln \gamma_1 + d \ln X_1) + X_2 RT(d \ln \gamma_2 + d \ln X_2) = 0$$

Note that

$$X_1 d \ln X_1 + X_2 d \ln X_2 = X_1 \frac{dX_1}{X_1} + X_2 \frac{dX_2}{X_2} = dX_1 + dX_2 = 0$$

Simplify

$$X_1 d \ln \gamma_1 + X_2 d \ln \gamma_2 = 0 \qquad (8.100)$$

which is the Gibbs–Duhem equation for the activity coefficients in a binary system. Its integrated form is similar to that for the activity, Equation 8.93:

$$\ln \gamma_1 = -\int_{X_2=0}^{X_2} \frac{X_2}{X_1} \frac{d \ln \gamma_2}{dX_2} dX_2 \qquad (8.101)$$

Thus to compute the activity coefficient of component 1 from a set of measurements for component 2, devise a statistically fit function to the data for component 2 and substitute this result into Equation 8.101.

Extensions of the Gibbs–Duhem strategy similar to those that apply to the activity function exist for computing activity coefficients of all of the components in a multicomponent system from measurements for one of the components. Thus a model or direct experimental determination of the activity coefficient for one component provides the information necessary to compute all of the solution properties of the system. Of the variety of formal descriptions that satisfy this purpose presented in this chapter, that based upon the activity coefficient provides the best access to an understanding of thermodynamic behavior of solutions.

8.6   THE BEHAVIOR OF DILUTE SOLUTIONS

Begin with pure component 1 and visualize the addition of a few atoms of component 2 to form a dilute solution of the *solute*, component 2, in the *solvent*, component 1. In this range of compositions, the average solvent atom experiences the same surroundings that it had in the pure state; only a very small fraction of solvent atoms neighbor solute atoms. Thus the only significant influence that the addition of solute atoms has upon the properties of solvent atoms is to slightly reduce their number. Accordingly, the solvent atoms act as if they were in an ideal solution. This behavior gives rise to an experimentally observed limiting law that applies to all solutions, called *Raoult's law for the solvent*, here assumed to be component 1:

$$\lim_{X_1 \to 1} a_1 = X_1 \tag{8.102}$$

In the same composition range, every solute atom is completely surrounded by solvent atoms. That is, until a sufficient number of solute atoms is added so that their spheres of influence begin to interact, each solute atom added to the solution will make the same contribution to the properties of the system. In this range, the average properties of the solute atoms will also be proportional to their concentration. However, this relationship will clearly be different for different solutes added to the same solvent; this proportionality is thus characterized by a constant that is specific to the solute–solvent combination in the system. This behavior gives rise to an experimentally observed limiting law that also applies to all solutions, called *Henry's law for the solute*, here assumed to be component 2:

$$\lim_{X_2 \to 0} a_2 = \gamma_2^0 X_2 \tag{8.103}$$

The coefficient, γ_2^0, called the *Henry's law constant*, is the activity coefficient for the solute and is seen to be independent of composition in the dilute range. The value of this constant depends upon both the solute and the solvent in the system and, for a given solute/solvent combination, varies with temperature and pressure. If the value of the Henry's law constant is determined as a function of temperature and pressure, then all of the thermodynamic properties of a dilute solution may be computed.

These limiting laws are illustrated graphically in Figure 8.4, which show plots of the activities of both components in a binary system as a function of composition at a fixed temperature and pressure. The dilute solution limits correspond to behavior of these plots at the sides of these diagrams. The activity plot for the solvent approaches 1 in the limit of the pure solvent along the line that has a slope of 1 in accordance with Equation 8.102. For the solute, the activity approaches zero in the dilute limit along a line that has a slope equal to the Henry's law constant for that solution, in accordance with Equation 8.103. The range of composition over which the solution obeys these limiting laws, i.e., is a "dilute" solution, varies widely among systems that have been studied, ranging from a few parts per million to several percent.

The two limiting laws are not independent of each other. It can be shown that if either is assumed, then the other may be derived through a Gibbs–Duhem integration of the activity coefficient.

These laws have practical utility because they are general, i.e., they are not based upon a model but are valid for all systems in the dilute range. Many practical applications involve dilute solutions. These laws permit predictions of some aspects of the behavior of the system because the form of the dependence of properties upon composition can be derived without experimental evaluation of the Henry's law coefficient for the system. Examples illustrating their application to the calculations of phase diagrams and to chemical equilibria in dilute reacting systems are presented in Chapters 10 and 11.

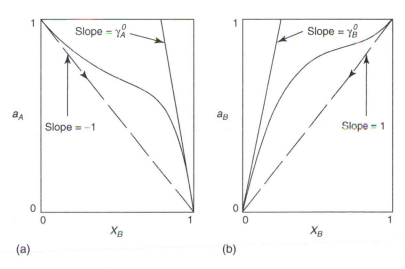

FIGURE 8.4 Variation of activity with composition illustrates the limiting laws for dilute solution behavior.

8.7 SOLUTION MODELS

The simplest model for solution behavior is the ideal solution described in Section 8.5.1. This model contains no adjustable parameters; given the composition and temperature of the solution, all of its properties may be computed. As a consequence, the ideal solution model is incapable of describing differences that may exist between two solutions of the same composition at the same temperature. This section presents some solution models that expand upon this simple model, introducing an increasing number of parameters as the level of sophistication and flexibility is increased.

8.7.1 REGULAR SOLUTION MODELS

One class of solution models is based upon the concept of the regular solution. In its simplest form, the definition of a regular solution contains two components:

1. The entropy of mixing is the same as that for an ideal solution:

$$\Delta S_{mix}^{rs} = \Delta S_{mix}^{id} = -R \sum_{k=1}^{c} X_k \ln X_k \qquad (8.104)$$

   for all components at all temperatures and pressures.
2. The enthalpy of solution is not zero, as in an ideal solution, but is some function of composition.

$$\Delta H_{mix}^{rs} = \Delta H_{mix}^{rs}(X_2, X_3, ..., X_c) \qquad (8.105)$$

A statement equivalent to (1) above is: the excess entropy of mixing is zero for a regular solution.[4] As a consequence of this definition, since the excess partial molal Gibbs free energy is

$$\Delta G_{mix}^{xs} = \Delta H_{mix}^{xs} - T\Delta S_{mix}^{xs}$$

then

$$(\Delta G_{mix}^{xs})^{rs} = (\Delta H_{mix}^{xs})^{rs} - T(0) = \Delta H_{mix}(X_2, X_3, ...) \qquad (8.106)$$

Thus, in a regular solution, because the excess entropy of mixing is defined to be zero, the excess Gibbs free energy is equal to the enthalpy of mixing, which is a function of composition, but not temperature. (Since the temperature derivative of the excess free energy is equal to minus the excess entropy of mixing, Equation 8.42, the temperature derivative of ΔG_{mix}^{xs}, and hence ΔH_{mix} must be zero.)

---

[4] In more sophisticated versions of the regular solution formalism this assumption is relaxed to allow incorporation of a configurational contribution to the excess entropy.[5]

Accordingly, the sign of the departure from ideal behavior (positive or negative) that characterizes the behavior of this class of solution models is determined by the sign of the heat of mixing.

It follows from the definition of the activity coefficient and Equation 8.95 that

$$\Delta \bar{G}_k^{xs} = RT \ln \gamma_k = \Delta \bar{H}_k \qquad (8.107)$$

so that the activity coefficient may be evaluated from the model for the heat of mixing:

$$\gamma_k = e^{\Delta \bar{H}_k / RT} \qquad (8.108)$$

It was shown in Section 8.5.4 that if the activity coefficient is known, then all of the properties of a solution may be calculated. Thus, application of the regular solution model focuses upon the evaluation of the heat of mixing as a function of composition.

The simplest regular solution model that can be devised contains a single adjustable parameter in its description of the heat of mixing:

$$\Delta H_{mix} = a_0 X_1 X_2 = \Delta G_{mix}^{xs} \qquad (8.109)$$

where a_0 is a constant (not to be confused with a_k, the activity of component k). Mathematical forms simpler than this expression are inadmissible because mixing properties must pass through zero at $X_1 = 0$ and $X_2 = 0$. The Gibbs free energy of mixing obtained from this model is

$$\Delta G_{mix} = a_0 X_1 X_2 + RT(X_1 \ln X_1 + X_2 \ln X_2) \qquad (8.110)$$

The sign of the single adjustable parameter, a_0, determines the sign of the excess free energy of mixing and thus the nature of the departure from ideal behavior.

Figure 8.5 illustrates the variation of S, ΔH, and ΔG vs. composition and temperature for this model for a positive departure from ideal behavior. This figure demonstrates the essential symmetry of the properties computed from this model with respect to composition. Note that ΔS and ΔH are not functions of temperature in this model; the temperature dependence of ΔG_{mix} is completely contained in the coefficient of the $T\Delta S_{mix}^{id}$ term. At a given composition, the free energy of mixing varies linearly with temperature. For positive departures from ideal behavior at low temperatures it is observed that the free energy of mixing curve develops an additional maximum and minimum. In Chapter 10, it will be demonstrated that this behavior leads to the development of a *miscibility gap* in the solution which appears as a particular construction of a two-phase field on a phase diagram containing such a solution.

The expression for ΔH_{mix} contained in Equation 8.109 was used in Example 8.1 as an application of the procedure for computing partial molal properties from the total properties of a solution. It was demonstrated that:

$$\Delta \bar{H}_1 = \Delta \bar{G}_1^{xs} = a_0 X_2^2 \qquad \text{and} \qquad \Delta \bar{H}_2 = \Delta \bar{G}_2^{xs} = a_0 X_1^2 \qquad (8.111)$$

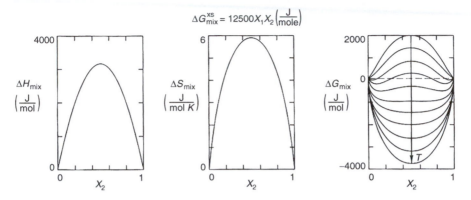

FIGURE 8.5 Variation of thermodynamics mixing properties with composition and temperature for the simplest regular solution model for a positive departure from ideal behavior with $a_0 = 12{,}500$ J/mol. Maximum temperature shown is 1200 K.

Thus, according to Equation 8.108, the activity coefficients for such a solution are:

$$\gamma_1 = e^{a_0 X_2^2/RT} \qquad \text{and} \qquad \gamma_2 = e^{a_0 X_1^2/RT} \tag{8.112}$$

Again it may be seen that the sign of the parameter a_0 determines whether the activity coefficients are greater or less than 1 and hence determines the sign of the departure from ideal behavior that characterizes the model.

Consider the values of the activity coefficients in the limit of dilute solutions. For example, for a dilute solution of solute component 2 in the solvent component 1 this limit is characterized by $X_2 \to 0$ or $X_1 \to 1$. From Equation 8.112, in this limit the activity coefficient for the solute component 2 becomes

$$\gamma_2^0 = e^{\alpha_0/RT} \tag{8.113}$$

A similar evaluation for the dilute solution at the other end of the composition range yields the Henry's law coefficient for component 1 when it is dilute in component 2:

$$\gamma_1^0 = e^{\alpha_0/RT} \tag{8.114}$$

Thus, this simple solution model gives the same value of the Henry's law coefficient at both ends of the composition range. This result exposes a clear limitation in the application of this one parameter solution model that derives from the symmetrical form that it assumes for the heat of mixing, Equation 8.109. It implies that the properties associated with an atom of component 2 surrounded by atoms of 1 in a dilute solution rich in component 1 are identical with the properties of an atom of component 1 when it is surrounded by 2 atoms in a dilute solution rich in component 2. Since components 1 and 2 are different atoms, this situation is in general unrealistic, although it may provide a useful approximation when these components are similar.

In spite of its evident limitations, this simplest of the regular solution models has proven useful as a teaching tool for visualizing the behavior of solutions. For example, all of the variety of classes of phase diagrams that can be constructed may be obtained by applying this model to each phase form that may exist in the system, although quantitative fits to real phase diagrams require more sophisticated models. Initial attempts at computer calculations of phase diagrams were formulated on the basis of this model.[6] These applications are demonstrated in detail in Chapter 10.

It is shown in Section 8.7.3 that the mathematical form of the simplest regular solution model can be derived from an atomic model (the quasichemical model) in which the energy of the system resides in the bonds between neighboring atoms, and these bond energies are independent of the composition of the solution. A more flexible model can be developed if it is postulated that the bond energies vary linearly with composition. This subregular solution model has an excess Gibbs energy given by[6]:

$$\Delta H_{mix} = \Delta G_{mix}^{xs} = X_1 X_2 (\alpha_0 X_1 + \alpha_1 X_2) \tag{8.115}$$

The flexibility of the regular solution model may be further increased by adding terms to the expression for the heat of mixing and introducing an additional model parameter with each term.

8.7.2 Modeling Real Solutions

Nonregular solution models build on the formulations in the last section by assigning a temperature dependence to the coefficients in the equations:

$$\Delta H_{mix} = \Delta G_{mix}^{xs} = X_1 X_2 [\alpha_0(T) X_1 + \alpha_1(T) X_2] \tag{8.116}$$

Regular and subregular solution models are applied in the calculation of phase diagrams for those phases that are described as random substitutional phases in which atoms of any of the components can occupy any lattice site. For multicomponent solutions, the most widely applied form of such models in the phase diagram literature are based upon the Redlich–Kister equation[7–10]:

$$\Delta G_{mix}^{xs} = \sum_{i=1}^{c} \sum_{j>i}^{c} X_i X_j \sum_{v=0}^{n} \Omega_{ij}^v (X_i - X_j)^v \tag{8.117}$$

where Ω_{ij}^v is a temperature-dependent binary interaction parameter dependent on the value of the exponent v. This solution model becomes regular for $v = 0$ and subregular when $v = 1$.

Many intermediate phases on phase diagrams exhibit a tendency toward ordering, a tendency for specific sites (e.g., the body center site in a BCC crystal) to be occupied by a specific component. In ordered structures, the crystal lattice may be visualized as consisting of two (or more in complex cases) sublattices (Figure 8.6). This subject is discussed in detail in Chapter 13. It is not uncommon for such

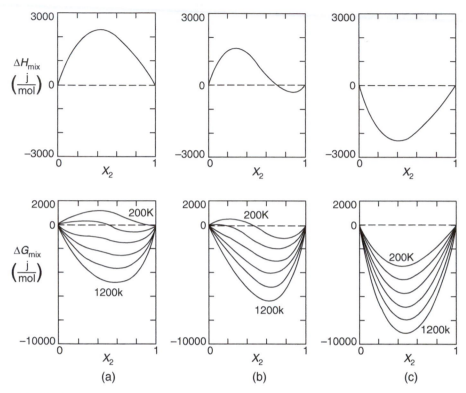

FIGURE 8.6 Mixing properties of a two parameter regular solution model with $\Delta H_{mix} = X_1 X_2 (a_1 X_1 + a_2 X_2)$. (a) $a_1 = 12{,}500$, $a_2 = 5500$; (b) $a_1 = 12{,}500$, $a_2 = -5500$; (c) $a_1 = -12{,}500$, $a_2 = -5500$.

intermediate phases to be stoichiometric, or nearly so, in which one sublattice is occupied almost exclusively by A atoms and the other sublattice by B atoms, giving rise to a formula composition $A_u B_v$ based on the ratio of sublattice sites in the crystal structure, v/u. Solutions which exhibit a tendency toward ordering are described with a sublattice solution model in which the compositional variables are the site fractions (fraction of sites on each sublattice) occupied by each component. Mixing behavior of the components is described on each sublattice and then combined to produce a model for the ordered solution. The mathematics involved is intricate and requires a significantly larger number of parameters to be evaluated in the database for such a solution model.[9]

8.7.3 ATOMISTIC MODELS FOR SOLUTION BEHAVIOR

The parameters contained in these models may be endowed with physical significance on the basis of atomistic models for the behavior of solutions. The most direct of these atomistic models has been called the *quasichemical theory of solutions*, which is developed in some detail in this section. Other approaches, based upon the statistical thermodynamics of the distribution of atoms over the sites in a crystal are more sophisticated.[4-9]

The quasichemical theory of solutions bears its name because it views a solution as a large molecule, with each pair of adjacent atoms treated is if connected by a chemical bond. In a binary system consisting of two kinds of atoms A and B, there are possibly only three classes of nearest-neighbor pairs of atoms that form such bonds:

$$A–A; \qquad B–B; \qquad A–B$$

Each type of bond is assumed to be endowed with its characteristic value of energy:

$$e_{AA}; \qquad e_{BB}; \qquad e_{AB}$$

In each case, the value of the energy corresponds to the formation of the bond from atoms that are originally so far apart that they are not interacting, i.e., when they are in the vapor state. Since this process is the opposite of the vaporization of the bond, which requires a large input of heat, it is expected that bond energies will be large negative numbers.

Figure 8.7 is a plot of the variation of the energy of a system comprised of two atoms as a function of their separation distance, x. The slope of this plot is the negative of the force acting between the pair. Large values of x correspond to two atoms in the vapor state; since the vapor is the reference state for the plot, the energy of the pair is defined to be zero for large x. As the atoms are brought together, their electron clouds begin to interact and they are attracted toward each other. At very small x the ion cores begin to interact and the atoms are repelled. The distance d, which corresponds to the minimum energy of the pair, is the equilibrium separation distance between the atoms in the condensed state. The corresponding energy, e_{ij}, a large negative number on this plot, is the energy associated with each bond of type ij in the solution.

In the quasichemical viewpoint, all of the internal energy of the solution is assumed to be contained in these interactions between neighboring pairs. This is a

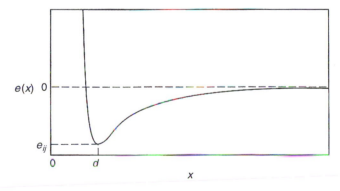

FIGURE 8.7 Variation of the energy of a pair of atoms as a function of their distance of separation. The equilibrium distance, d, between the pair of atoms is characterized by the minimum in their energy.

relatively naive view in comparison with more rigorous treatments of the formation of crystals from the vapor in which the electron cloud develops a distribution of energy values associated with a density of allowable states. One consequence of this simple view is that the energy of a particular bond is independent of all of the surrounding atoms except the pair that form the bond. Thus, the energy of each bond type is independent of composition. Let the numbers of each type of bond in the system be respectively

$$P_{AA}; \quad P_{BB}; \quad P_{AB}$$

Then the internal energy of the solution, i.e., the energy change associated with condensing the solution from a vapor of the same composition, is

$$U_{\text{soln}} = P_{AA}e_{AA} + P_{BB}e_{BB} + P_{AB}e_{AB} \tag{8.118}$$

The number of bonds of each type is related to the composition and the coordination number, z, of the solution. The coordination number is the number of nearest neighbors that surround an atom in a crystal. (In a liquid solution, there is a distribution of coordination numbers; z may be taken to be the average number of nearest neighbors in this case.) In a simple cubic crystal lattice $z = 6$; for body-centered cubic crystals, $z = 8$; in face-centered cubic and hexagonal close-packed structures, $z = 12$. In a system containing N_0 atoms the total number of bonds, P_T is

$$P_T = \frac{1}{2}N_0 z \tag{8.119}$$

The factor of 1/2 is necessary because each bond is shared by two atoms; if one simply multiplied the number of atoms by the average number of near neighbors, each bond would be counted twice.

For a system of a given composition, the numbers of bonds of each of the three types are not independently variable. Each atom of A is at one end of a near–neighbor pair or bond. Each AA bond contains two ends incident upon A atoms and each AB bond contains one A-end. The total number of bond ends that terminate upon A atoms is equal to the total number of A atoms multiplied by z, the number of bond ends per atom. Thus,

$$2P_{AA} + P_{AB} = zN_A = zN_0 X_A \tag{8.120}$$

The analogous argument focused on the B atoms in the system gives

$$2P_{BB} + P_{AB} = zN_B = zN_0 X_B \tag{8.121}$$

Solve these two equations for P_{AA} and P_{BB}:

$$P_{AA} = \frac{1}{2}[X_A N_0 z - P_{AB}] \tag{8.122}$$

$$P_{BB} = \frac{1}{2}[X_B N_0 z - P_{AB}] \tag{8.123}$$

Thus, in the characterization of the number of bonds of each type in a binary system it is only necessary to evaluate the number of unlike bonds, P_{AB}. There are no simplifying assumptions about the nature of the arrangement of atoms in the system in this description. Equation 8.122 and Equation 8.123 may be used to eliminate P_{AA} and P_{BB} in the expression for the energy of the solution, Equation 8.118.

$$U_{soln} = P_{AB}\left[e_{AB} - \frac{1}{2}(e_{AA} + e_{BB})\right] + \frac{1}{2}N_0 z[X_A e_{AA} + X_B e_{BB}] \tag{8.124}$$

The change in internal energy for the mixing process for the solution is defined to be

$$\Delta U_{mix} = U_{soln} - [X_A U_A^0 + X_B U_B^0] \tag{8.125}$$

where U_A^0 and U_B^0 are the internal energies per mole of pure A and pure B, each assumed to have the same crystal structure as the solution. Consider N_0 atoms of pure A. The system contains only AA bonds. The number of bonds is $(1/2)N_0 z$; the energy to form each from a vapor of pure A is e_{AA}. Thus,

$$U_A^0 = \frac{1}{2}N_0 z e_{AA} \tag{8.126}$$

For pure B,

$$U_B^0 = \frac{1}{2}N_0 z e_{BB} \tag{8.127}$$

Substitute these results into Equation 8.125,

$$\Delta U_{mix} = P_{AB}\left[e_{AB} - \frac{1}{2}(e_{AA} + e_{BB})\right] + \frac{1}{2}N_0 z[X_A e_{AA} + X_B e_{BB}]$$
$$- \left[X_A \frac{1}{2}N_0 z e_{AA} + X_B \frac{1}{2}N_0 z e_{BB}\right]$$

and simplify:

$$\Delta U_{mix} = P_{AB}\left[e_{AB} - \frac{1}{2}(e_{AA} + e_{BB})\right] \tag{8.128}$$

For condensed phases, the internal energy of mixing is negligibly different from the enthalpy of mixing since

$$\Delta H_{mix} - \Delta U_{mix} + P\Delta V_{mix} \cong \Delta U_{mix} \tag{8.129}$$

so that Equation 8.128 may be taken as the heat of mixing for the solution. Thus the quasichemical theory predicts that the heat of mixing of a solution is proportional to the number of unlike bonds it contains and to a parameter that reports the difference in energy between unlike (AB) bonds and the average energy of like bonds in the structure. This result is general in the sense that it assumes no restrictions on how the atoms are arranged in the solution. The primary limitation in the theory is the assignment of all energy effects to near–neighbor pairs, so that e_{AA}, e_{BB}, and e_{AB} are independent of composition.

The number of unlike bonds in the system, P_{AB}, takes on central significance in this approach to modeling thermodynamic behavior of solutions. In a mixture of a given composition, the number of unlike near–neighbor pairs reflects the arrangement of atoms in the system. For example, if like atoms tend to cluster together, then P_{AB} will be small, Figure 8.8(a). If, on the other hand, atoms of the two types find themselves in an ordered structure, then most near–neighbor pairs will be unlike atoms, Figure 8.8(c). In a "random" mixture, Figure 8.8(b), P_{AB} will have some intermediate value. It is useful to use this concept of a random mixture as a point of reference in describing atomic arrangements in mixtures since ideal solutions, which are also used as a reference in describing behavior, can be shown to be random mixtures. Furthermore, the computation of P_{AB} for a random mixture is straightforward. Systems with P_{AB} values that are less than the random value for their composition are said to be exhibiting a tendency toward *clustering*. Those for which P_{AB} is larger than the random value evince a tendency toward *ordering*.

The development that follows is limited to the description of random solutions. In the end, it will be shown that this leads to an inconsistent result, except for an ideal solution. More specifically, it is shown that if the heat of mixing is not zero, then the arrangement of atoms cannot be random. This implies that the regular solution model, as described in Section 8.7.1, contains a conceptual flaw; more sophisticated versions of the regular solution model are required to describe the behavior of real solutions. Nonetheless, the regular solution model remains useful for two reasons:

1. It provides a valid approximate description for some real solutions, particularly at high temperatures where the entropy term $(T\Delta S_{mix})$ dominates in the Gibbs free energy of mixing.
2. It provides a mathematically tractable model for introducing concepts that are fundamental to the understanding of its more sophisticated versions.

The arrangement of atoms in a solution is defined to be random if the probability that any site chosen at random is occupied by an A atom is simply equal to the fraction of all of the sites in the structure that are occupied by A atoms, X_A. There is no preference for A atoms to occupy a particular site in the system. Similarly, the probability that a site is occupied by a B atom is X_B in a random mixture. Consider a pair of neighboring sites, labeled I and II. The atoms on these sites will produce an AA bond if two simultaneous independent events are satisfied: site I is occupied by an A atom and site II is occupied by an A atom. Thus, the probability that the I–II

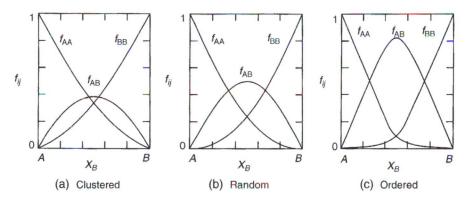

FIGURE 8.8 Variation of the number of each of the bond types in a binary solution for (a) a clustered structure, (b) a random structure, and (c) and ordered structure.

site combination is an AA bond is the product of the probability that site I is occupied by an A atom and the site II is similarly occupied.

$$f_{AA} = X_A X_A = X_A^2 \tag{8.130}$$

A BB bond results if both sites are occupied by B atoms:

$$f_{BB} = X_B X_B = X_B^2 \tag{8.131}$$

An AB bond may result from either of two distinguishable arrangements: site I has an A atom and site II has a B, or site I has a B atom and site II has an A. Thus,

$$f_{AB} = X_A X_B + X_B X_A = 2X_A X_B \tag{8.132}$$

These bond probabilities may be interpreted as the fraction of bonds in the system that belong to each type. As a check of these formulations, the sum of these fractions should equal 1:

$$f_{AA} + f_{BB} + f_{AB} = X_A^2 X_B^2 + 2X_A X_B = (X_A + X_B)^2 = 1$$

The number of bonds of each type in a random mixture is simply the fraction of bonds of that type multiplied by the total number of bonds, $(1/2)N_0 z$. Of particular interest is the number of unlike bonds:

$$P_{AB} = \frac{1}{2} N_0 z f_{AB} = N_0 z X_A X_B \tag{8.133}$$

Insertion of this computation of the number of unlike bonds into Equation 8.128 gives the heat of mixing for a random mixture:

$$\Delta H_{mix} = N_0 z X_A X_B \left[e_{AB} - \frac{1}{2}(e_{AA} + e_{BB}) \right] \tag{8.134}$$

In order to focus upon the composition dependence in this expression, it may be written

$$\Delta H_{mix} = \alpha_0 X_A X_B \tag{8.135}$$

in which

$$\alpha_0 = N_0 z \left[e_{AB} - \frac{1}{2}(e_{AA} + e_{BB}) \right]$$
(8.136)

The entropy of mixing of a random solution can be shown to be identical with the ideal entropy of mixing. In the solution, N_A atoms of A and N_B atoms of B are distributed independently on the N_0 available sites. The number of ways in which this arrangement may be made is

$$\Omega = \frac{N_0!}{N_A! N_B!}$$
(8.137)

Apply the Boltzmann hypothesis, Equation 6.3 to compute the entropy of this distribution:

$$S = k \ln \Omega = k \ln \frac{N_0!}{N_A! N_B!} = k[\ln N_0! - (\ln N_A! + \ln N_B!)]$$

Use Stirling's approximation

$$S = \lfloor k(N_0 \ln N_0 - N_0) - (N_A \ln N_A - N_A) - (N_B \ln N_B - N_B) \rfloor$$

Simplify the result, recognizing that $N_0 = N_A + N_B$:

$$S = k \lfloor (N_A + N_B) \ln N_0 - N_A \ln N_A - N_B \ln N_B \rfloor$$

$$= k \lfloor - N_A (\ln N_A - \ln N_0) - N_B (\ln N_B - \ln N_0) \rfloor$$

$$= -k \left[N_A \ln \frac{N_A}{N_0} + N_B \ln \frac{N_B}{N_0} \right]$$

$$= -k N_0 (X_A \ln X_A + X_B \ln X_B)$$

$$S = -R(X_A \ln X_A + X_B \ln X_B)$$
(8.138)

Since the value of Ω for the unmixed components is 1, this result may be interpreted as the entropy of mixing of a random solution, ΔS_{mix}. This expression is identical with the entropy of mixing of an ideal solution. From the definition of a regular solution it is seen that this is also the entropy of mixing of a regular solution.

The free energy of mixing of a random solution may be computed by inserting Equation 8.135 and Equation 8.138 into the definitional relationship:

$$\Delta G_{mix} = \Delta H_{mix} - T \Delta S_{mix}$$

$$\Delta G_{mix} = a_0 X_A X_B + RT(X_A \ln X_A + X_B \ln X_B)$$
(8.139)

Compare this result with Equation 8.110, the simplest of the regular solution models. It is seen that this model implies that the atoms are arranged randomly in the solution and that the model parameter a_0 is determined by the energies of the three types of bonds in the system.

The first term on the right side of Equation 8.139 is also the excess free energy of mixing and thus determines the sign of the departure of the behavior of the system from an ideal solution. Examination of the form deduced for a_0, Equation 8.136 shows that all factors are positive except the quantity contained in brackets. Figure 8.9 illustrates possible combinations of bond energies that determine the sign of a_0. In mathematical terms,

Positive Departure:

$$a_0 > 0 \Rightarrow e_{AB} > \frac{1}{2}(e_{AA} + e_{BB}) \tag{8.140}$$

Negative Departure:

$$a_0 < 0 \Rightarrow e_{AB} < \frac{1}{2}(e_{AA} + e_{BB}) \tag{8.141}$$

Ideal Solution:

$$a_0 = 0 \Rightarrow e_{AB} = \frac{1}{2}(e_{AA} + e_{BB}) \tag{8.142}$$

These considerations point to a conceptual difficulty in the strict application of the regular solution model with its implicit assumption of random mixing embodied in the definition. Consider, for example, a random solution exhibiting a positive departure from ideal behavior. According to the quasichemical theory that produced the inequality 8.140, in such a system like bonds have a lower energy than unlike

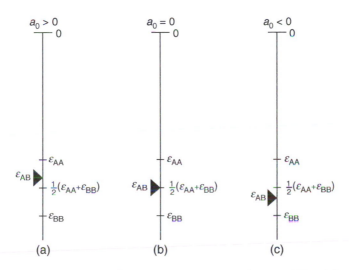

FIGURE 8.9 Illustration of the relative values of bond energies that lead to (a) positive, (b) zero, and (c) negative departures from ideal solution behavior.

bonds. Thus, the system could lower its heat of mixing, and hence the free energy of mixing, by increasing the number of like bonds above the random value given by Equation 8.133. In a system constrained to constant temperature and pressure, a process that lowers the Gibbs free energy is a spontaneous process. Thus, the atoms may be expected to arrange themselves in configurations that have more like bonds than a random solution; i.e., the system tends to exhibit clustering. If this process occurs, then the arrangement of atoms is no longer random and the entropy of mixing is not simply the random value that corresponds to ideal and regular solutions.

The analogous argument may be made for a system that exhibits a negative departure from ideal behavior. Inequality 8.141 requires that like bonds have a lower energy than unlike bonds in such a system. The enthalpy and thus Gibbs free energy, may be lowered by increasing the number of unlike bonds, i.e., by tending to produce an ordered arrangement of atoms.

More sophisticated models have been devised to incorporate deviations from randomness into a computation of the entropy of mixing. Swalin[5] provides an introductory discussion of short- and long-range order parameters and their relation to the entropy of mixing of ordered solutions. Models based upon statistical thermodynamics and the description of clusters of atoms, rather than just near–neighbor pairs, include the central atoms model discussed by Lupis[7] and the cluster variation model due to DeFontaine.[8] Statistically based models that predict the behavior of interstitial solutions in crystals are also discussed by Lupis.[9] Devereux[4] and Lupis[7] also present extensions of these models to ternary and higher-order systems. However, the regular solution model remains a useful tool for introducing the subject of solution models and the strategies involved in their development.

8.8   SUMMARY

Partial molal properties of a component in a mixture are defined by

$$\bar{B}_k \equiv \left(\frac{\partial B^l}{\partial n_k} \right)_{T,P,n_j \neq n_k} \tag{8.6}$$

Consequences of this definition, summarized in Table 8.1, include:

$$d\Delta B_{\text{mix}} = \sum_{k=1}^{c} \Delta \bar{B}_k dX_k \tag{8.20}$$

Partial molal properties of components in a binary system may be computed from the corresponding total property of the solution:

$$\Delta \bar{B}_k = \Delta B_{\text{mix}} + (1 - X_k) \frac{d\Delta B_{\text{mix}}}{dX_k} \tag{8.28}$$

A Gibbs–Duhem integration

$$\Delta \bar{B}_1 = -\int_{X_2=0}^{X_2} \frac{X_2}{X_1} \frac{d\Delta \bar{B}_2}{dX_2} dX_2 \tag{8.36}$$

yields the partial molal property of component 1 when values are known for component 2.

Relationships may be derived among partial molal properties that are analogs to the laws, the definitions, coefficient, and Maxwell relations.

All of the properties of a solution may be computed if either the chemical potential, the activity, defined by

$$\mu_k - \mu_k^0 \equiv RT \ln a_k \qquad (8.66)$$

or the activity coefficient, defined by

$$a_k \equiv \gamma_k X_k \qquad (8.67)$$

is known as a function of composition. These relationships are summarized in Tables 8.2 to 8.5. The definition of the activity coefficient permits decomposition of each partial molal property, as well as each total property of a solution, into an ideal and an excess component. The focus of theories of solutions, and of experimental programs designed to develop databases for solutions, is on the evaluation of the excess free energy of mixing of the solution.

A hierarchy of solution models, beginning with the ideal and regular solutions and progressing through a series of increasingly complex phenomenological solution models, provides a useful basis for the analysis of the behavior of real solutions.

The quasichemical model for solutions yields an expression for the heat of mixing in terms of the bond energies in the system:

$$\Delta H_{\text{mix}} = P_{AB}\left[e_{AB} - \frac{1}{2}(e_{AA} + e_{BB})\right] \qquad (8.128)$$

Application of this model to random solutions demonstrates an essential flaw in the simple regular solution model, but nonetheless yields insight into the connection between atomic arrangements and departures from ideal behavior.

HOMEWORK PROBLEMS

Problem 8.1. Titanium metal is capable of dissolving up to 30 atomic percent oxygen. Consider a solid solution in the system Ti$-$O containing an atom fraction, $X_O = 0.12$. The molar volume of this alloy is 10.68 cc/mol. Calculate:

   a. The weight percent of O in the solution.
   b. The molar concentration (mol/cc) of O in the solution.
   c. The mass concentration (gm/cc) of O in the solution.

Use these calculations to deduce general expressions for weight percent, molar, and mass concentrations of a component in a binary solution in terms of the atom fraction, X_2, the molar volume, V, and the molecular weights, MW$_1$ and MW$_2$, of the elements involved.

Problem 8.2. Review the consequences of the definition of partial molal properties that make it a convenient measure of the contribution of each component to the total value of the thermodynamic properties of a solution.

Problem 8.3. Given that the volume change on mixing of a solution obeys the relation

$$\Delta V_{mix} = 2.7 X_1 X_2^2 \left(\frac{cc}{mol} \right)$$

  a. Derive expressions for the partial molal volumes of each of the components as functions of composition.
  b. Demonstrate that your result is correct by using it to compute ΔV_{mix}, demonstrating that the equation above is recovered.

Problem 8.4. Use the partial molal volumes computed in Problem 8.3 to demonstrate that the Gibbs–Duhem equation holds for these properties in this system.

Problem 8.5. In the system Pandemonium (Pn)–Condominium (Cn), the partial molal heat of mixing of Pandemonium may be fitted by the expression

$$\Delta \bar{H}_{Pn} = 12,500 X_{Pn}^2 X_{Cn} \left(\frac{J}{mol} \right)$$

Calculate and plot the function that describes the variation of the heat of mixing with composition for this system.

Problem 8.6. For an ideal solution it is known that, for component 2,

$$\Delta \bar{G}_2 = RT \ln X_2$$

Use the Gibbs–Duhem integration to derive corresponding relation for component 1.

Problem 8.7. Recall the definitional relationship for the enthalpy function:

$$H' = U' + PV'$$

Use the partial molal operator and the definition of the properties of the pure components to derive the analogous relationship,

$$\Delta \bar{H}_k = \Delta \bar{U}_k + P \Delta \bar{V}_k$$

Problem 8.8. The excess free energy of mixing in face-centered cubic solid solutions of aluminum and zinc is well described by the relation:

$$\Delta G^{\alpha}_{mix} = X_{Al}X_{Zn}(9600X_{Zn} + 13200X_{Al})\left(1 - \frac{T}{4000}\right)$$

Compute and plot curves for ΔG_{mix} as a function of composition for a sequence of temperatures ranging from 300 to 700 K.

Problem 8.9. Using the relation given in Problem 8.8, calculate and plot the activity of Zn in an FCC solid solution of these elements at 500K.

Problem 8.10. The excess Gibbs free energy of mixing of the system A–B is given

$$\Delta G^{xs}_{mix} = a(1 - bT)(1 - cP)X_A X_B$$

where $a = 12,500$ (J/mol), $b = 2 \times 10^{-4}$ K^{-1} and $c = 7 \times 10^{-5}$ (atm)$^{-1}$.
   a. Derive expressions for *all* of the mixing properties for this system.
   b. Use these relations to evaluate all of the properties of the solution at 100 atm pressure, 550 K and a composition $X_B = 0.35$.

Problem 8.11. Criticize the following reported finding: "The system A–B forms a regular solution at high temperatures with the heat of mixing found to obey the relation,

$$\Delta H_{mix} = -14,500X_A X_B\left(1 - \frac{350}{T}\right)$$

Problem 8.12. Given that Henry's law holds for the solute in a dilute real solution, derive Raoult's law for the solvent.

Problem 8.13. The system A–B forms a regular solution with the heat of mixing given by:

$$\Delta H_{mix} = -13,500X_A X_B\left(\frac{J}{mol}\right)$$

   a. Derive expressions for the Henry's law constant for A as a solute in B and B as a solute in A.
   b. Plot both Henry's law constants as a function of temperature.

Problem 8.14. The system A–B may be described by the quasichemical model. The heat of vaporization of pure A is 98,700 (J/mol); that of pure B is 127,000 (J/mol). At a solution composition $X_B = 0.40$ at 750 K the activity of component B is found to be 0.53. Estimate the energy of an AB bond in this system.

Problem 8.15. The system A–B exhibits a tendency toward ordering, with the number of unlike near–neighbor pairs larger than the random value by 30% at all compositions. Compute and plot the number of AA, AB, and BB bonds in the system.

Problem 8.16. The A–B system exhibits a measured heat of mixing given by the relationship:

$$\Delta H_{\text{mix}} = X_A X_B (7,500 X_A + 18,200 X_B)$$

The bond energies (J/bond) for this system are estimated to be:

$$e_{AA} = -6.5 \times 10^{-20}; \qquad e_{BB} = -5.3 \times 10^{-20}; \qquad e_{AB} = -5.4 \times 10^{-20}$$

Compute and plot the fractions of AA, AB, and BB bonds in the system as a function of composition.

REFERENCES

1. Lupis, C.H.P., *Chemical Thermodynamics of Materials*, Elsevier Science Publishing Company Inc., New York, pp. 56–57, 1983.
2. Lupis, C.H.P., *Chemical Thermodynamics of Materials*, Elsevier Science Publishing Company Inc., New York, pp. 282–285, 1983.
3. Darken, L.S. and Gurry, R.W., *Physical Chemistry of Metals*, McGraw-Hill Book Co., New York, 1951.
4. Owen, F.D., *Topics in Metallurgical Thermodynamics*, Wiley, New York, pp. 282–286, 1983.
5. Swalin, R.A., *Thermodynamics of Solids*, Wiley, New York, pp. 148–164, 1972.
6. Kaufman, L. and Bernstein, H., *Computer Calculations of Phase Diagrams*, Academic Press, New York, 1970.
7. Lupis, C.H.P., *Computer Calculations of Phase Diagrams*, Academic Press, New York, pp. 438–475, 1970.
8. DeFontaine, D., Configurational thermodynamics of solid solutions, *Solid State Physics*, Vol. 34, Ehrenreich, H., Seitz, F., and Turnbull, D., Eds., Academic Press, New York, 1979.
9. Lupis, C.H.P., Configurational thermodynamics of solid solutions, *Solid State Physics*, Vol. 34, Ehrenreich, H., Seitz, F., and Turnbull, D., Eds., Academic Press, New York, pp. 477–503, 1979.
10. Saunders, N. and Miodownik, A.P., *CALPHAD, Calculation of Phase Diagrams, A Comprehensive Guide*, Elsevier Science, Oxford, p. 96, 1998.

9 Multicomponent Heterogeneous Systems

CONTENTS

Chapters 9 and 10 illustrate how the structure of thermodynamics given in Figure 1.4 provides the basis for generating the first of the equilibrium maps displayed in that figure: phase diagrams. The general criterion for equilibrium is applied through the strategy for finding conditions for equilibrium to yield the working equations that form the basis for computing and plotting these equilibrium maps for multi-component multiphase systems.

The logical progression through the hierarchy of classes of thermodynamic systems builds upon the treatment of unary heterogeneous systems in Chapter 7 and multicomponent homogeneous systems in Chapter 8 to the class of systems that are the subject of this chapter: multicomponent, heterogeneous systems. This class of systems is important in materials science because most commercial materials contain a number of components in a microstructure that is an aggregate of two or more phases. Control of composition and the arrangement of the phases in the structure is tantamount to controlling properties. Also, many reacting systems involve the interaction of components in more than one phase. For example, oxidation of a metal involves at least three phases: the metal, the gas containing oxygen and the oxide that forms. Microelectronic devices consist of multi-component connectors, sometimes layers of two or three phases, that connect electronically active components that are another phase, all laid down on a substrate that is yet another phase. It is evident that the treatment of multicomponent, multiphase systems has broad application in technology and science.

The apparatus necessary for describing multicomponent, multiphase systems is developed in this chapter. This apparatus is then applied in the general strategy for finding conditions for equilibrium. This set of relations that must exist between the thermodynamic properties when such a system is in equilibrium is then used as a basis for deriving the classic Gibbs phase rule. The Gibbs phase rule is a very general relationship, which is the basis for the construction of phase diagrams, the primary thinking tool used to understand the behavior of multiphase, multi-component systems. The construction of phase diagrams is illustrated for a variety of representations of unary, binary and ternary systems.

The conditions for equilibrium derived in this chapter are the set of equations that relate the limit of stability of phases, graphically presented in a phase diagram, to the thermodynamics of the system. Thus, as was shown for unary two-phase systems in Chapter 7, calculation of phase diagrams from thermodynamic information starts with these equations. The same relationships provide a basis for estimating thermodynamic properties of some of the phases involved from an experimental determination of the phase diagram. Connections of this kind were also demonstrated in Chapter 7 for phase boundaries in a unary system, where it was shown that the heat of vaporization could be determined from the phase boundary between liquid and vapor. Similar applications are illustrated in calculations of phase diagrams for binary and ternary systems in Chapter 10.

9.1 THE DESCRIPTION OF MULTIPHASE MULTICOMPONENT NONREACTING SYSTEMS

Chapter 8 showed that the thermodynamic apparatus necessary to describe a homogeneous multicomponent system consisted of the addition of a string of terms that incorporates variation of the number of moles of each of the components in the system. For a system with c components, the combined statement of the first and second laws becomes:

$$dU' = TdS' - PdV' + \sum_{k=1}^{c} \mu_k dn_k \qquad (8.49)$$

Consider a system that is made up of p distinguishable phase forms, each of which contains the same set of c components. Figure 9.1 illustrates a microstructure with three phases, each containing two components. Each of these phases, viewed as a system, exchanges heat, work and matter with its surroundings, i.e., with the other phases in the system and the exterior of the whole system. Focus upon the α phase, Figure 9.1b. Equation 8.49 describes the change in internal energy experienced by the α phase when the state of the whole multiphase system is

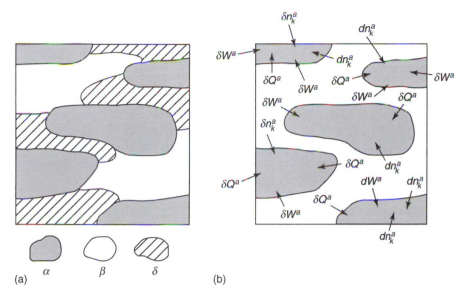

FIGURE 9.1 A microstructure with three phases (a). The α phase exchanges heat, work and components with the other phases and with the surroundings (b).

altered in an arbitrary way. Use the superscript (α) to denote properties of the α phase:

$$dU'^{\alpha} = T^{\alpha}dS'^{\alpha} - P^{\alpha}dV'^{\alpha} + \sum_{k=1}^{c} \mu_k^{\alpha}dn_k^{\alpha} \qquad (9.1)$$

A relationship like this holds in each phase in the system.[1]

The treatment of the behavior of the whole multiphase system focuses upon the extensive properties of the system, as illustrated in Chapter 7. The strategy is simple and makes use of the definition of extensive properties. For the extensive properties, V', S', U', H', F' and G', the value of the property for the system is the sum of the values for the separate parts:

$$B'_{sys} = B'^{I} + B'^{II} + \cdots + B'^{\alpha} + \cdots + B'^{p}$$

$$B'_{sys} = \sum_{\alpha=1}^{p} B'^{\alpha} \qquad (9.2)$$

---

[1] In order to write these equations it is necessary to assume that the intensive properties of the α phase, i.e., T^{α}, P^{α} and μ_k^{α}, are uniform within the phase. This is equivalent to assuming that each phase in the system is in internal equilibrium; exchanges between the phases derive from differences that may exist in their intensive properties. The treatment of phases with nonuniform intensive properties is the subject of Chapter 14, Continuous Systems.

where B' may be any of the extensive properties. If the system is taken through an arbitrary change in state, then the change in B'_{sys} is simply the sum of the changes that each phase experiences because the differential of a sum is the sum of the differentials:

$$dB'_{sys} = d\left(\sum_{\alpha=1}^{p} B^{\alpha}\right) = \sum_{\alpha=1}^{p} dB'^{\alpha} \tag{9.3}$$

This simple principle is the basis for handling the description of multiphase systems.

As an application of this strategy consider the change in internal energy. The change in internal energy of the multiphase system when it is taken through an arbitrary infinitesimal process is

$$dU'_{sys} = \sum_{\alpha=1}^{p} dU'^{\alpha} \tag{9.4}$$

The change in internal energy experienced by each phase is determined by its own properties and is given by Equation 9.1

$$dU'_{sys} = \sum_{\alpha=1}^{p} \left[T^{\alpha} dS'^{\alpha} - P^{\alpha} dV'^{\alpha} + \sum_{k=1}^{c} \mu_k^{\alpha} dn_k^{\alpha} \right] \tag{9.5}$$

The entropy change for the system plays a key role in the strategy for deriving the conditions for equilibrium developed in the next section. For a single multi-component phase, the change in entropy for an arbitrary change in state is obtained by solving Equation 9.1 for dS'^{α} :

$$dS'^{\alpha} = \frac{1}{T^{\alpha}} dU'^{\alpha} + \frac{P^{\alpha}}{T^{\alpha}} dV'^{\alpha} - \frac{1}{T^{\alpha}} \sum_{k=1}^{c} \mu_k^{\alpha} dn_k^{\alpha} \tag{9.6}$$

When the whole multiphase system is taken through an arbitrary process, its change in entropy is given by Equation 9.3,

$$dS'_{sys} = \sum_{\alpha=1}^{p} dS'^{\alpha} \tag{9.7}$$

Substitute Equation 9.6:

$$dS'_{sys} = \sum_{\alpha=1}^{p} \left[\frac{1}{T^{\alpha}} dU'^{\alpha} + \frac{P^{\alpha}}{T^{\alpha}} dV'^{\alpha} - \frac{1}{T^{\alpha}} \sum_{k=1}^{c} \mu_k^{\alpha} dn_k^{\alpha} \right] \tag{9.8}$$

Thus, the thermodynamic apparatus needed to describe multiphase systems is based simply on the definition of extensive properties and the mathematical relationship that states that the differential of a sum is the sum of the differentials.

9.2 CONDITIONS FOR EQUILIBRIUM

Recall again the general criterion for equilibrium and the associated strategy for finding conditions for equilibrium presented in Chapter 5. The criterion states that, in an isolated system, the entropy of the system is a maximum at equilibrium. Implementing this principle requires derivation of the condition of the system that yields a maximum in the entropy function when its variations are constrained by the conditions that it remain isolated from its surroundings.

In order to derive the conditions for equilibrium in a system containing p phases it is only necessary to consider the equilibrium between any two of these phases. The results obtained for a two-phase system may then be extended to a system with an arbitrary number of phases by some simple inductive reasoning. Thus the system under consideration consists of two phases, α and β, each containing c components. The change in entropy of each phase is given by Equation 9.6 with appropriate superscripts. If the two-phase system is taken through an arbitrary change in state, the change in entropy is given by

$$dS'_{\text{sys}} = dS'^{\alpha} + dS'^{\beta} \tag{9.9}$$

or, more explicitly,

$$dS'_{\text{sys}} = \frac{1}{T^{\alpha}} dU'^{\alpha} + \frac{P^{\alpha}}{T^{\alpha}} dV'^{\alpha} - \frac{1}{T^{\alpha}} \sum_{k=1}^{c} \mu_k^{\alpha} dn_k^{\alpha} + \frac{1}{T^{\beta}} dU'^{\beta} + \frac{P^{\beta}}{T^{\beta}} dV'^{\beta}$$
$$- \frac{1}{T^{\beta}} \sum_{k=1}^{c} \mu_k^{\beta} dn_k^{\beta} \tag{9.10}$$

If the system is isolated from its surroundings, then not all of the $2(2+c)$ variables in this equation are independent; some are related through the isolation constraints.

If during this process the two-phase system is isolated from its surroundings then, whatever changes occur within the system, its internal energy, U'_{sys}, volume, V'_{sys} and the number moles of each of its components, $n_{k,\text{sys}}$, cannot change. These isolation constraints may be stated in mathematical form:

$$dU'_{\text{sys}} = 0 = dU'^{\alpha} + dU'^{\beta} \Rightarrow dU'^{\beta} = -dU'^{\alpha} \tag{9.11}$$

$$dV'_{\text{sys}} = 0 = dV'^{\alpha} + dV'^{\beta} \Rightarrow dV'^{\beta} = -dV'^{\alpha} \tag{9.12}$$

$$dn_{k,\text{sys}} = 0 = dn_k^{\alpha} + dn_k^{\beta} \Rightarrow dn_k^{\beta} = -dn_k^{\alpha} \tag{9.13}$$

Evidently, whatever changes occur in one of these variables must be compensated by an opposite change in the other since the total values of the properties for the system are constrained not to change because the system is isolated.

Equation 9.11 to Equation 9.13 may be used to eliminate dependent variables in the expression for the change in entropy, Equation 9.10, applied to an isolated system. Substitute for dU'^β, dV'^β and dn_k^β :

$$dS'_{sys,iso} = \frac{1}{T^\alpha}dU'^\alpha + \frac{P^\alpha}{T^\alpha}dV'^\alpha - \frac{1}{T^\alpha}\sum_{k=1}^{c}\mu_k^\alpha dn_k^\alpha + \frac{1}{T^\beta}(-dU'^\alpha) + \frac{P^\beta}{T^\beta}(-dV'^\alpha)$$

$$- \frac{1}{T^\beta}\sum_{k=1}^{c}\mu_k^\beta(-dn_k^\alpha)$$

Collect terms.

$$dS'_{sys,iso} = \left(\frac{1}{T^\alpha} - \frac{1}{T^\beta}\right)dU'^\alpha + \left(\frac{P^\alpha}{T^\alpha} - \frac{P^\beta}{T^\beta}\right)dV'^\alpha - \sum_{k=1}^{c}\left(\frac{\mu_k^\alpha}{T^\alpha} - \frac{\mu_k^\beta}{T^\beta}\right) \quad (9.14)$$

These $(2 + c)$ variables remaining may be changed independently in an isolated system. The condition for an extremum is obtained by setting each of the coefficients in this expression equal to zero:

$$\frac{1}{T^\alpha} - \frac{1}{T^\beta} = 0 \Rightarrow T^\alpha = T^\beta \qquad \text{(Thermal Equilibrium)} \qquad (9.15)$$

$$\frac{P^\alpha}{T^\alpha} - \frac{P^\beta}{T^\beta} = 0 \Rightarrow P^\alpha = P^\beta \qquad \text{(Mechancial Equilibrium)} \qquad (9.16)$$

$$\frac{\mu_k^\alpha}{T^\alpha} - \frac{\mu_k^\beta}{T^\beta} = 0 \Rightarrow \mu_k^\alpha = \mu_k^\beta \qquad \text{(Chemical Equilibrium)} \qquad (9.17)$$

An equation like Equation 9.17 holds for each of the c components in the system. These equations represent the conditions that must be satisfied in order for any two phases to coexist in thermodynamic equilibrium. They are the conditions for equilibrium in a multicomponent two-phase system.

In order to extend this result to a system containing an arbitrary number of phases, it is only necessary to consider the relationship between the phases two at a time. For example, consider a system consisting of three phases, α, β and ε. When this system attains equilibrium, α is equilibrated with β, β with ε and α with ε. Consider a system composed of the two phases β and ε. Equilibrium in this two-phase system will be achieved when equations like Equation 9.15 to Equation 9.17 are satisfied for a $(\beta + \varepsilon)$ system:

$$T^\beta = T^\varepsilon; \quad P^\beta = P^\varepsilon; \quad \mu_k^\beta = \mu_k^\varepsilon \ (k = 1, 2, ..., c) \qquad (9.18)$$

Combining these equations with those obtained for the $(\alpha + \beta)$ equilibrium gives the conditions for equilibrium in a three-phase system:

$$T^\alpha = T^\beta = T^\varepsilon \qquad (9.19)$$

$$P^\alpha = P^\beta = P^\varepsilon \tag{9.20}$$

$$\mu_k^\alpha = \mu_k^\beta = \mu_k^\varepsilon \ (k = 1, 2, ..., c) \tag{9.21}$$

In a system composed of p phases at equilibrium, each pair of phases is in equilibrium and relationships corresponding to Equation 9.15 to Equation 9.17 exist for that pair. Thus the set of conditions that must be satisfied when a system composed of p phases and c components comes to equilibrium is

$$T^I = T^{II} = \cdots = T^\alpha = \cdots = T^p \tag{9.22}$$

$$P^I = P^{II} = \cdots = P^\alpha = \cdots = P^p \tag{9.23}$$

$$\mu_1^I = \mu_1^{II} = \cdots = \mu_1^\alpha = \cdots = \mu_1^p \tag{9.24a}$$

$$\mu_2^I = \mu_2^{II} = \cdots = \mu_2^\alpha = \cdots = \mu_2^p \tag{9.24b}$$

$$\cdots \quad \cdots \quad \cdots \quad \cdots$$

$$\cdots \quad \cdots \quad \cdots \quad \cdots$$

$$\cdots \quad \cdots \quad \cdots \quad \cdots$$

$$\mu_c^I = \mu_c^{II} = \cdots = \mu_c^\alpha = \cdots = \mu_c^p \tag{9.24c}$$

Stated in words, in order for a multicomponent, multiphase system to come to equilibrium, the temperature, the pressure and the chemical potential of each component must be the same in all of the phases. These equations form the basis for the construction and calculation of phase diagrams.

9.3 THE GIBBS PHASE RULE

Perhaps the most rudimentary piece of information about a complicated thermodynamic system is the number of independent variables that are required to describe its state. In a unary, homogeneous, closed, nonreacting, otherwise simple system this number is two: if, for example, the temperature and pressure are specified, the state of the system is determined. In order to describe the state of a solution it was found necessary to specify its composition through a set of mole fractions in addition to its temperature and pressure. However, as the complexity of a system increases, the number of variables that must be assigned values in order to fix its state does not necessarily increase in proportion. For example, in Chapter 7, it was shown that in a unary two-phase system at equilibrium, relations exist between the variables so that if the pressure in one of the phases is specified, the pressure in the other phase must be the same and, furthermore, the temperature of both phases is determined. In a three-phase unary system, there is no freedom of choice in assigning values to state variables; the pressure and temperature of all three phases is determined.

Evidently the answer to the question, "How many independent variables characterize this system?," is not trivial. In his classic paper published in 1883, titled "On the Equilibrium of Heterogeneous Substances," J. Willard Gibbs presents the solution to this basic problem as part of his development of the foundations of chemical thermodynamics. The number of independent variables that a system has is the number of variables to which values must be assigned in order to define its thermodynamic state. Gibbs called this number the number of degrees of freedom, f, for the system.

The term degrees of freedom is borrowed from mathematical usage where it is applied in the consideration of systems of several equations relating many variables. As a simple example, consider the following system of equations:

$$4x + 3y - z = 9$$

$$5x - 9y + 2z = -3$$

$$x + y + z = 1$$

There are three relationships among three variables. A unique set of values for the variables x, y and z may be found so that this system of equations is satisfied. There is no freedom of choice for these variables; their values are determined for this system of equations; its number of degrees of freedom is zero.

Consider next the set of equations,

$$2u - 3v + x - y + 3z = 10$$

$$-3u + 5v - 2x + 4y + z = -8$$

These two equations in five unknowns cannot be solved to yield unique values for the five variables. Indeed, it is possible to choose arbitrary values for, say, u, v and z and then find values of x and y for which the system of equations is satisfied. This system of two equations in five variables thus has three independent variables; mathematically, it is said to exhibit three degrees of freedom.

These considerations are not limited to systems of linear equations. Every system of well-behaved algebraic equations exhibits this behavior. In general, a system of n such equations in n variables has at least one solution. The solution may not be unique; for example, a quadratic equation in a single unknown has two solutions, both of which satisfy the equation. Computer software based upon numerical iteration techniques that can supply such solutions with efficiency is widely available. However, a system consisting of n equations relating m variables, where $m > n$, does **not** have at least one solution; it has $(m - n)$ too many variables. If arbitrary numerical values are independently assigned to these $(m - n)$ variables, then the system is left with n equations in n unknowns and can be solved.

For a system of n equations among m variables one may define the number of degrees of freedom,

$$f = m - n \qquad (9.25)$$

This is the number of variables to which values may be freely assigned without jeopardizing the validity of the equations describing the behavior in the system. Thus f may be thought of as the number of independent variables in a system of equations; the remaining n variables, which may be computed once the f independent values have been assigned, are then dependent variables.

With this mathematical concept in mind, the number of degrees of freedom of a multicomponent, heterogeneous, nonreacting, otherwise simple system may be computed. It is necessary to enumerate the variables that the system has, list the number of equations that relate these variables when the system has attained equilibrium and subtract. Since the conditions for equilibrium are formulated in terms of the intensive properties of the system (T, P, μ_k), it will be necessary to formulate these counts of variables and equations in terms of intensive properties.

Consider first one phase in the p-phase system. The state of this phase may be fixed by assigning values to its temperature, pressure and composition. It is important to recognize that, for the description in terms of intensive properties, there are $(c - 1)$ compositional variables, X_k, because in any given phase the mole fractions of the components sum to one. Thus the variables that describe its state are $(T, P, X_2, X_3, ..., X_c)$; here it is assumed that X_1 has been computed from the remaining mole fractions.[2] The number of variables in this list is $[2 + (c - 1)] = [1 + c]$. Since all of the phases are assumed to contain the same components, each phase is described by this number of variables. For the system of p phases, each with c components, the variables are:

$$T^I, P^I, X_2^I, X_3^I, ..., X_c^I \qquad \text{For Phase } I$$

$$T^{II}, P^{II}, X_2^{II}, X_3^{II}, ..., X_c^{II} \qquad \text{For Phase } II$$

$$\cdots \qquad\qquad \cdots$$

$$\cdots \qquad\qquad \cdots$$

$$T^\alpha, P^\alpha, X_2^\alpha, X_3^\alpha, ..., X_c^\alpha \qquad \text{For Phase } \alpha$$

$$\cdots \qquad\qquad \cdots$$

$$\cdots \qquad\qquad \cdots$$

$$T^p, P^p, X_2^p, X_3^p, ..., X_c^p \qquad \text{For Phase } p$$

---

[2] Some texts list the intensive properties appropriate to the description of a single phases as $(T, P, \mu_2, \mu_3, ..., \mu_c)$, since these are the variables explicitly contained in the conditions for equilibrium. Note that μ_1 is also not independent of the other chemical potentials, since they are related in a given phase by the Gibbs–Duhem equation. In either case, the number of variables required to specify the state of a single phase is the same: $c + 1$.

Each of the rows has $(1 + c)$ variables listed; there are p rows, one for each phase. The total number of variables in the system is thus

$$m = p(1 + c) \qquad (9.26)$$

When the system comes to equilibrium the conditions for equilibrium, Equation 9.22 to Equation 9.24a–c, must be satisfied. The number of independent equations contained in this system of equations may be obtained by simply counting equal signs. Each row contains $(p - 1)$ independent equations; there are $(2 + c)$ rows of equations representing conditions for thermal, mechanical and chemical equilibrium. Thus, the number of equations relating the variables in the system is

$$n = (p - 1)(2 + c) \qquad (9.27)$$

The number of degrees of freedom in this system of equations is obtained by substituting Equation 9.26 and Equation 9.27 into Equation 9.25:

$$f = m - n = [p(1 + c)] - [(p - 1)(2 + c)]$$

$$f = c - p + 2 \qquad (9.28)$$

which is the Gibbs phase rule. Thus the calculation of the number of independent variables in a system consisting of p phases and c components is straightforward. The implications of this result for the construction of phase diagrams are developed in the next section.

9.4   THE STRUCTURE OF PHASE DIAGRAMS

Phase diagrams are graphical representations of the domains of stability of the various classes of structures (one, two, three phase, etc.) that may exist in a system at equilibrium. Phase diagrams are most commonly constructed in temperature–pressure-composition space; other coordinate systems, though not yet as widely used, may find increasing practical application.

For single-phase regions in a unary system the number of degrees of freedom, $f = 1 - 1 + 2 = 2$ (e.g., T and P); for binary systems, $f = 2 - 1 + 2 = 3$. In a c-component system, f for a single-phase region is $f = c - 1 + 2 = c + 1$. Further, for any given system, the phase rule shows that f decreases as the number of phases, p, is increased. It is thus clear from the phase rule that the regions of stability of single phases have the highest number of degrees of freedom in any system with a given number of components. That is, single-phase regions require the largest number of variables for their specification. Accordingly, the graphical space in which the phase diagram is constructed must have $(c + 1)$ independent coordinates so that the full range of behavior of the single-phase regions may be represented. In a unary system the diagram may be plotted in (T, P) space; binary diagrams require a three-dimensional space, most commonly with (T, P, X_2) coordinates; complete representation of a ternary system requires a four-dimensional space (e.g., T, P, X_2, X_3); and so on.

The printed page is two dimensional: thus most quantitative phase diagrams are represented as sections taken through the multidimensional space required for their full representation. (Projected views of three-dimensional diagrams are useful for teaching purposes or to visualize the relationships among phases described by three variables. However useful such views are, they cannot be used to read quantitative information about the domains of the phases.) A section is obtained by choosing to fix a value for one or more of the independent variables in the system. Phase diagrams for unary, binary and ternary systems plotted with the most commonly chosen variables are shown in Figure 9.2. Since $c + 1 = 2$ for a unary system, the complete representation of all possible states may be plotted on the printed page (Figure 9.2a). Most binary diagrams are plotted for a constant value of the pressure on the system, usually chosen to be one atmosphere (Figure 9.2b). Ternary diagrams are represented for a fixed value of both temperature and pressure (Figure 9.2c).

These sections will be well-behaved (easy to interpret) if they are taken at constant values of the thermodynamic potentials, T, P and/or μ_k, i.e., the variables explicitly contained in the conditions for equilibrium, Equation 9.22 to Equation 9.24a–c. In two-phase, three-phase or higher-order regions on the diagram, it is precisely these variables that are required to be constant by the conditions for equilibrium. The states of the phases participating in these multiphase equilibria will all lie in the plane of the diagram, if and only if, the diagram is plotted by holding constant one or more of these potentials since equilibrated states will all have the same values of the variables that have been chosen to be constant. Thus the compositions, temperatures, and pressures of the three phases that coexist in equilibrium in a ternary system (Figure 9.2c), are all represented in the plane of the diagram. Sections taken holding other variables constant (e.g., X_2 or X_3, or S or V) will have multiphase regions in which the points representing the participating states

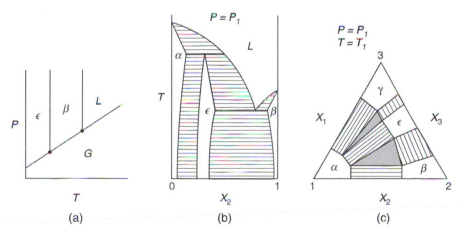

FIGURE 9.2 Common representations of (a) unary ($c = 1$), (b) binary ($c = 2$) and (c) ternary ($c = 3$) phase diagrams.

are out of the plane representing the diagram; values of these variables for phases in equilibrium are in general not the same. The relationships among equilibrated states are therefore not represented on a section obtained by holding, e.g., V or X_2, constant.

9.4.1 PHASE DIAGRAMS PLOTTED IN THERMODYNAMIC POTENTIAL SPACE

The simplest form of phase diagram for any number of components is obtained when it is plotted in a space with thermodynamic potentials $(T, P, \mu_2, \mu_3, ..., \mu_c)$ as coordinates. It is convenient to replace each chemical potential axis with the corresponding value of activity.[3] Thus representation of a phase diagram in $(T, P, a_2, ..., a_c)$ space yields the simplest, though not necessarily most useful, representation. A phase diagram for a c-component system plotted in potential space is a simple cell structure. Further, sections through such diagrams, obtained by assigning a constant value to one of the potentials (e.g., P, or T or μ_k), are also simple cell structures. The notion embodied in the phrase "cell structure" is demonstrated in Figure 9.3 for a binary system. A three-dimensional (P, T, a_2) space is required for full representation of a binary system. The five phases presumed to exist in this system are α, β, and ε solid allotropic forms and gas (G) and liquid (L). Each of the five one-phase domains that exist in this system is represented by the volumes of the cells in this space. Two-phase fields are the surfaces that separate these cells; the cell boundaries. Three-phase fields are triple lines where three cells are incident upon each other, and four-phase fields are the discrete quadruple points at which four cells meet.

Phase diagrams plotted in potential space have this simple configuration precisely because the variables describing the state of each phase are the thermodynamic potentials which, according to the conditions for equilibrium, are equal for two or more phases in equilibrium. Figure 9.3b is a section through the phase diagram shown in Figure 9.3a at constant pressure. The point Q lying on the line separating the α and ε domains in Figure 9.3b represents one combination of the three variables $(T, P$ and $a_2)$ at which the two phases α and ε coexist in equilibrium. The combination of the three variables that describe the state of the α phase $(T^\alpha, P^\alpha$ and $a_2^\alpha)$ in this two-phase system and the combination that represents the state of the ε phase $(T^\varepsilon, P^\varepsilon$ and $a_2^\varepsilon)$ are the same point because at equilibrium $T^\alpha = T^\varepsilon$, $P^\alpha = P^\varepsilon$ and, from $\mu_2^\alpha = \mu_2^\varepsilon$, it follows that $a_2^\alpha = a_2^\varepsilon$ (assuming the reference state for component 2 is the same for both phases). In contrast, if the variables chosen to plot the diagram were $(T, P$ and $X_2)$ (Figure 9.3c), then the same equilibrium state between the α and ε phases is represented by a pair of separated points, Q_1 and Q_2, since X_2^α and X_2^ε, the **compositions**, are not the same for the two equilibrated phases.

---

[3] For dilute solutions the chemical potential μ_k of a solute approaches $-\infty$ as X_k approaches zero; a chemical potential axis thus spans the range from $-\infty$ to 0. In the same situation the activity of a solute, a_k, simply approaches zero as X_k approaches zero, and the activity axis varies between 0 and 1.

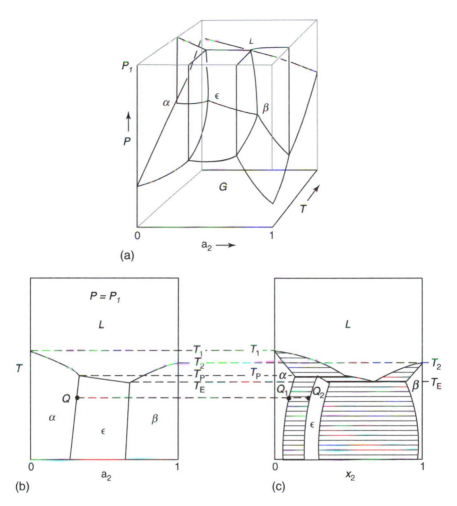

FIGURE 9.3 (a) A binary phase diagram plotted in coordinates that are thermodynamic potentials (P, T and a_2 in this case) is a simple cell structure. (b) A section taken at a constant potential, pressure in this case, is also a cell structure. (c) The constant pressure section shown in part (b) represented in the familiar (X_2, T) space.

9.4.2 UNARY SYSTEMS

The structure of phase diagrams in unary systems has been explored in Section 7.6. Figure 9.4 is analogous to Figure 7.12 and shows three representations of the same system in coordinates for which:

a. Both are thermodynamic potentials (T, P) (Figure 9.4a).
b. One is a potential and the other is not (V, P) (Figure 9.4b).
c. Neither coordinate is a potential (V, S) (Figure 9.4c).

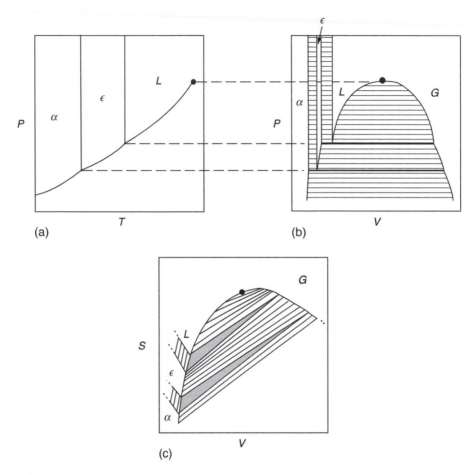

FIGURE 9.4 Three representations of the same unary phase diagram: (a) both variables are potentials (T, P); (b) one variable is a potential, P in (V, P) and (c) neither variable (S nor V) is a potential.

The diagram in (T, P) space is a simple cell structure with cells for single-phase regions, linear cell boundaries representing states of two-phase equilibrium, and triple points depicting the domains of coexistence of three phases. In (V, P) space (Figure 9.4b), two-phase regions broaden into areas with individual coexisting equilibrium states represented as a pair of points on each of the boundaries connected by horizontal (constant potential, P) lines, called tie lines. Three-phase equilibria consist of three separated points in (V, P) space (the molar volumes of the three phases are different), all lying on a tie line that is horizontal because three phases in equilibrium have the same pressure.

In (V, S) space (Figure 9.4c), domains of stability of two phases also consist of areas on the diagram. Any pair of phases in equilibrium is represented by two points on opposite boundaries of a two-phase region [e.g., (V^α, S^α) and $(V^\varepsilon, S^\varepsilon)$]. However, the tie line that connects these states is in general not horizontal because S^α and V^α

are not equal to S^ε and V^ε. Three-phase fields are represented by tie triangles because none of the three phases are expected to have the same values of S and V.

It is now possible to place these constructions in context with the Gibbs phase rule. In a unary system, $c = 1$ and the Gibbs phase rule becomes

$$f_1 = 1 - p + 2 = 3 - p \tag{9.29}$$

An exhaustive list of the possibilities is:

$$p = 1 : f_1 = 3 - 1 = 2$$

$$p = 2 : f_1 = 3 - 2 = 1$$

$$p = 3 : f_1 = 3 - 3 = 0$$

It is not possible for more than three phases to coexist at equilibrium in a unary system.

In (T, P) space the number of degrees of freedom for a given number of phases is equal to the dimensionality of the domain that may contain that number of phases. A single phase, for which $f_1 = 2$, is represented by a two-dimensional domain or an area on the diagram. A two-phase region for which $f_1 = 1$ is represented by a one-dimensional domain; i.e., by a curve on the graph. In Chapter 7 it was demonstrated that this line is given by the Clausius-Clapeyron equation. Three-phase regions, for which $f_1 = 0$, appear as triple points where three two-phase lines intersect. These constructions for two- and three-phase fields will now be analyzed.

Specification of the states of a pair of phases requires four variables: $(T^\alpha, P^\alpha, T^\varepsilon, P^\varepsilon)$. If these states are equilibrated, these variables are related by three equations $(T^\alpha = T^\varepsilon, P^\alpha = P^\varepsilon, \mu^\alpha = \mu^\varepsilon)$. The system has $(4 - 3) = 1$ degree of freedom. If a value is assigned to any of the four variables, say T^α, the other three may be computed from the three equations. By incrementing the value assigned to the independent variable T^α the complete set of states in which α and ε may exist in equilibrium is obtained. The representation of the values of T^α and P^α that correspond to this set of states is a curve $P^\alpha = P^\alpha(T^\alpha)$ in (T, P) space. The set of states corresponding to the ε phase is also represented by a curve, $P^\varepsilon = P^\varepsilon(T^\varepsilon)$. In (T, P) space these two curves coincide because the conditions for two-phase equilibrium require that $T^\alpha = T^\varepsilon$ and $P^\alpha = P^\varepsilon$. Thus the locus of points corresponding to the $(\alpha + \varepsilon)$ equilibrium states appears as a single curve in (T, P) space.

Specification of a state in which three phases coexist requires evaluation of the states of all three phases. This requires the values for six variables, e.g., $(T^\alpha, P^\alpha, T^\varepsilon, P^\varepsilon, T^G, P^G)$ for an $(\alpha + \varepsilon + G)$ field. If these three phases are in equilibrium, there are six relations among these variables: $(T^\alpha = T^\varepsilon = T^G; P^\alpha = P^\varepsilon = P^G; \mu^\alpha = \mu^\varepsilon = \mu^G)$. This system of six variables and six equations has no degrees of freedom, no independent variables. The states of the three phases are uniquely determined. If the states are represented in terms of T and P, then the values of these variables describing each of the three phases coincide; the three different states plot as the same point in (T, P) space. Since the conditions for equilibrium for these three phases include as a subset the conditions for equilibrium between the α and ε phase, those for ε and G, and those for α and G, these three coincident points must also lie on the corresponding two-phase curves. Thus the curves that represent these three two-phase equilibria must all pass through the point that describes the three-phase equilibrium.

A unary phase diagram in (T, P) space is seen to consist of one-phase areas bounded by two-phase lines that intersect three at a time in triple points. The overall construction (Figure 9.4a), has the configuration of a simple cell structure.

The same system has a different appearance and provides a different set of information if plotted in (V, P) space (Figure 9.4b). In each of the cases considered, i.e., one-, two- and three-phase equilibria, the number of variables and the number of relations remain the same. Thus single-phase regions in a unary system have two degrees of freedom; however, in this representation the variables V and P have been chosen to describe the state rather than T and P. Single-phase regions plot as areas.

The specification of the state of two phases requires four variables, now taken to be $(V^\alpha, P^\alpha, V^\varepsilon, P^\varepsilon)$. If the two phases are in equilibrium, three relations hold among these variables: $(T^\alpha = T^\varepsilon; P^\alpha = P^\varepsilon; \mu^\alpha = \mu^\varepsilon)$. There is $4 - 3 = 1$ independent variable, one degree of freedom. If any of the four variables is specified, say V^α, the other three may be computed from database information such as that developed in Chapter 4 for single phases. This requires that relations for T and μ (G for a unary system) as functions of V must be known for each phase; Chapter 4 provides a general procedure for deriving such relations. By incrementing V^α over the full range of interest, the collection of all of the states in which α and ε are in equilibrium is obtained. The representation of the set of values that correspond to states that the α phase may exhibit in equilibrium with ε plots as a curve $P^\alpha = P^\alpha(V^\alpha)$ in (V, P) space. Corresponding states for the ε phase plot as $P^\varepsilon = P^\varepsilon(V^\varepsilon)$. Unlike the plot in (T, P) space, these two curves do not coincide. A given equilibrated $(\alpha + \varepsilon)$ state plots as two discrete points, (V^α, P^α) for the α phase and $(V^\varepsilon, P^\varepsilon)$ for the ε phase, connected by a tie line, which shows that they are related. The conditions for equilibrium do not require that V^α and V^ε be equal. However, these conditions do require that $P^\alpha = P^\varepsilon$. Thus a pair of points that represent states of α and ε that coexist in equilibrium will have the same value of P and, if P is plotted on the y-axis, the tie line will be horizontal. Accordingly, in this representation, and any plot in which one variable is a potential and the other is not, two phase fields plot as areas. The curves bounding the areas are the possible states for the two phases that may exist in equilibrium. Pairs of states of the two phases that are in equilibrium are connected by a tie line which is horizontal in this construction.

Since three-phase equilibria possess zero degrees of freedom in a unary system, the six variables required to specify the states of the three phases $(V^\alpha, P^\alpha, V^\varepsilon, P^\varepsilon V^G, P^G)$, are determined by the six conditions for equilibrium. The state of each phase is represented by a point in (V, P) space. The conditions for mechanical equilibrium, $P^\alpha = P^\varepsilon = P^G$, require that these three points all have the same pressure. If P is the y-axis, the three equilibrium states all lie on the same horizontal line. Subsets of the conditions for three-phase equilibrium correspond to the conditions for two-phase equilibrium; the three two-phase equilibria $(\alpha + \varepsilon)$,

$(\alpha + G)$ and $(\varepsilon + G)$ must intersect at this three-phase tie line. Thus three-phase equilibria in (V, P) space, or any space in which one of the variables is a potential and the other is not, appear as three points connected by a horizontal tie line at which the three corresponding two-phase fields intersect.

A unary system represented in (V, P) space (more generally, a space in which one variable is a potential and the other is not) has one-phase fields that are areas. The two-phase fields consist of areas bounded by curves that represent all of the possible states of the two phases that may exist in equilibrium: specific equilibrated states are connected by horizontal (more generally, constant potential) tie lines that fill the two-phase field. Three-phase equilibria are represented by three points on a horizontal tie line.

The case in which neither of the variables used to describe states of the systems is a potential is illustrated by choosing V and S as variables (Figure 9.4c). The constructions are analogous to the (V, P) representation outlined above with one important exception; the tie lines that connect corresponding states are not constrained to be horizontal. One-phase fields are areas, as in all of the cases examined. Two-phase fields consist of two curves, one of the form $S^\alpha = S^\alpha(V^\alpha)$, the other $S^\varepsilon = S^\varepsilon(V^\varepsilon)$. Specific two-phase systems are represented by a pair of points, one (S^α, V^α) on the α side of the region, the other $(S^\varepsilon, V^\varepsilon)$ on the ε side. These points define a tie line that may be arbitrarily oriented in the (S, V) plane.

A three-phase equilibrium is represented by three points that not only do not lie on a horizontal line but are not collinear at all. These three points define a tie triangle in this representation (Figure 9.4c). The lines that form the sides of the triangle connecting α with ε, α with G and ε with G are each terminal tie lines of the corresponding two-phase field.

Plotted in a space in which neither of the state variables are potentials, a unary phase diagram has areas for one-phase fields, pairs of curves connected by continuously rotating tie lines for two-phase regions, and three-phase regions represented by tie triangles formed by the intersection of three two-phase fields.

The constructions that have been described in detail for a unary system apply to two-dimensional sections through binary, ternary and higher-order systems, provided only that the sections are formed by holding one or more of the thermodynamic potentials constant. The topology of each plot depends upon the nature of the variables used to represent the state of the system. If both variables are potentials, then a cell structure with the rules of construction represented in Figure 9.4a will result. If only one is a potential, the structure will have the appearance of Figure 9.4b. If neither is a potential, then the rules governing the construction of Figure 9.4c will apply. These rules of construction have their roots in the mathematics of graphical representation of systems of equations, and not in the thermodynamics of the systems involved.

9.4.3 BINARY PHASE DIAGRAMS

The complete representation of phase relationships in a binary system requires a space with $c + 1 = 2 + 1 = 3$ dimensions. The simplest structure results when the three variables chosen are potentials: (P, T, a_2). Figure 9.3a shows an example of the kind of cell structure obtained for a binary system that shows three solid phases, α β and ε, in addition to the liquid and gas phases. Figure 9.5a shows another example of such a cell structure; in this simpler case the system only exhibits two solid phase forms, α and β. Figure 9.5b is a sketch of the same system plotted in (T, P, X_2) space, i.e., with the activity of the independent component replaced by its mole fraction. The (T, P, X_2) diagram is clearly more complicated than the cell structure. It is not possible to read quantitative information from either of these diagrams, which are projections of a three-dimensional representation on the two-dimensional printed page.

Quantitative information about phase relationships in binary systems is generally represented by taking a section through the three-dimensional diagram at constant potential, usually at constant pressure. Constructing an isobaric section is equivalent to assigning a value to one of the independent variables, P, for the system. Figure 9.6a,b shows sections through the phase diagrams sketched in Figure 9.5 at the fixed values of P. Figure 9.6c is the same diagram sketched as it would appear in (V, X_2) space. Since these graphs are two dimensional, the information they present can be read directly. Note that the topology of the three diagrams in Figure 9.6 is comparable to the three diagrams sketched in Figure 9.4.

The rules that govern the construction of one-, two- and three-phase fields in both Figures 9.4 and 9.6 have identical origins as presented in detail in Section 9.4.2 for unary systems because the number of degrees of freedom for one-, two- and three-phase regions is the same in both plots. For a binary system $c = 2$ and the Gibbs phase rule, Equation 9.28 becomes

$$f_2 = 2 - p + 2 = 4 - p \qquad (9.30)$$

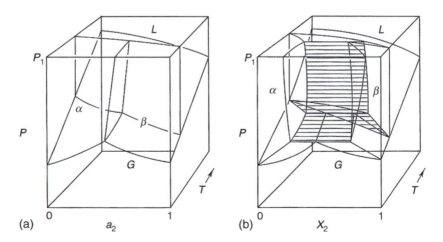

FIGURE 9.5 A complete phase diagram for a binary system that has two solid phase forms, α and β, represented as a simple cell structure in potential space (T, P, a_2) (a) and (b) as a more complicated structure in (T, P, X_2) space.

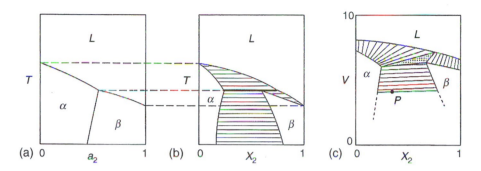

FIGURE 9.6 Constant pressure sections through the diagrams sketched in Figure 9.5 represented in (a) (T, a_2) potential space, (b) (T, X_2) space and (c) (V, X_2) space. The most common form of binary-phase diagrams is given in (b).

Choose a value for one of the variables (P); this reduces the number of degrees of freedom by one:

$$f_2' = 3 - p \tag{9.31}$$

This result is identical to Equation 9.29 for a unary system.

9.4.4 TERNARY PHASE DIAGRAMS

Since for a ternary system $c = 3$, the number of independent variables required to specify the state of a single phase is $c + 1 = 4$. Complete representation of a ternary system requires a four-dimensional space. While such spaces can be manipulated mathematically, they cannot be visualized in a world limited to three dimensions. Fixing one variable, e.g., P, yields a diagram that can be constructed in three dimensions. Textbooks and compilations of diagrams may display these three-dimensional diagrams as projections into the two-dimensional printed page. It has been pointed out that such diagrams may be pedagogically useful; however, it is not possible to read quantitative information on such projections.

In order to represent quantitative information on the printed page it is necessary to reduce the number of degrees of freedom for a one-phase system to two. For a ternary system this is accomplished by assigning values to two of the four variables that describe its state. It has been pointed out that diagrams retain simple rules of interpretation if and only if the variables that are fixed are thermodynamic potentials. Accordingly, in the representation of phase relationships in ternary systems on the two-dimensional printed page, temperature and pressure are usually chosen to be held constant.

Figure 9.7 shows three representations of the same phase diagram for a ternary system plotted on isobaric, isothermal sections. Depiction of the variation of the behavior of the system with temperature requires a series of such diagrams constructed at a sequence of temperatures. The topology and rules of construction of each of these three representations of ternary phase equilibria are identical with

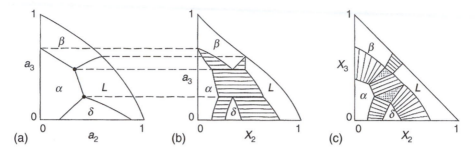

FIGURE 9.7 An isobaric, isothermal section through a four-dimensional ternary-phase diagram yields two-dimensional diagrams which can be plotted (a) in (a_2, a_3) space, (b) in (X_2, a_3) space, or (c) in (X_2, X_3) space.

those presented for unary systems in Figure 9.4. For a ternary system $c = 3$ and the number of degrees of freedom is

$$f_3 = 3 - p + 2 = 5 - p \qquad (9.32)$$

Assigning values to two of the variables in the system reduces the number of degrees of freedom on a section to

$$f'_3 = 5 - p - 2 = 3 - p \qquad (9.33)$$

a result which is identical to Equation 9.29 for a unary system.

The most common format for presentation of ternary phase diagrams fixes T and P and uses X_2 and X_3, the compositions, as variables to describe the state of the system, (Figure 9.7c). Two different versions of the presentation of the composition variables are in common use (Figure 9.8). In Figure 9.8a the two independent composition variables, here taken as X_2 and X_3, are plotted on orthogonal axes in the standard Cartesian system. Since the dependent composition, $X_1 = 1 - (X_2 + X_3)$, lines of constant X_1 plot at a slope of -1. The origin $(0,0)$ corresponds to pure component 1. The boundaries of the diagram, defined by the fact that physically meaningful ranges of the three composition variables are restricted between 0 and 1, are the two Cartesian axes and the line with a slope of -1 passing through $(1,0)$ and $(0,1)$. The composition corresponding to any point in this triangle is easily read from the two axes. This representation of ternary compositions has found application in describing the behavior of dilute solutions or near one corner of the system where one of the components is readily viewed as the dependent compositional variable.

The second representation of composition in ternary systems, illustrated in Figure 9.8b, is more widely used in displaying ternary phase diagrams. This compositional coordinate system is called the Gibbs triangle. It is constructed as an equilateral triangle with sides of unit length. The interior of the triangle is constructed with lines at 0, 60 and 120°, i.e., lines parallel to the sides of the triangle. Each corner represents a pure component; each side represents a binary system in which the concentration of the component at the opposite corner is zero.

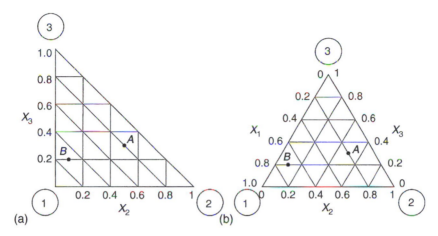

FIGURE 9.8 Alternate representations for the composition plane for isobaric isothermal sections through ternary systems: (a) orthogonal composition axes and (b) the Gibbs triangle. The compositions A and B are the same on both diagrams.

Lines parallel to the side opposite to a corner for any given component are the locus of points within the composition space for which that component has a constant value. In Figure 9.8b the lines at $0°$ represent constant values of component 3, with X_3 labeled on the $2-3$ side of the diagram. The lines at $60°$ represent constant values of the mole fraction of component 2 with numerical values for X_2 labeled on the $1-2$ side; those at $120°$ are constant X_1 values, as labeled on the $1-3$ side of the diagram. The composition of a particular point in the space may be read by constructing lines parallel to these three directions and reading the scales on the sides of the diagram. For the point labeled A in this diagram, the composition is $(X_1 = 0.2, X_2 = 0.5, X_3 = 0.3)$. For the point B, the values are seen to be $(X_1 = 0.7, X_2 = 0.1, X_3 = 0.2)$. These compositions are plotted for comparison on Figure 9.8a.

The Gibbs triangle is widely used to describe ternary compositions because it gives equal weight to all three components. The Cartesian coordinate system tends to assign particular importance to the dependent composition variable, here shown as component 1. Some analyses of behavior require evaluation of slopes of phase boundaries; derivatives taken in the Cartesian system are more easily interpreted than in the Gibbs triangle system.

The most common representation of phase relationships in ternary systems plots the composition axes as mole fractions (or in atomic percent, which is equivalent) on a Gibbs triangle at a fixed temperature and pressure. Since the mole fraction axes are not potentials, these diagrams share the principles of construction and interpretation with those illustrated for a unary system in Figure 9.4c and for a binary system in Figure 9.6c. Single-phase fields are areas with two degrees of freedom. Two-phase regions consist of a pair of curves with corresponding equilibrated states connected by tie lines, consistent with this single degree of freedom. Three-phase regions are invariant with no degrees of freedom; the fixed compositions of the three phases form the corners of a tie triangle.

9.5 THE INTERPRETATION OF PHASE DIAGRAMS

In plotting the phase diagrams in Figure 9.4, Figure 9.6 and Figure 9.7 the maximum number of degrees of freedom has been reduced to two; this choice was made so the diagrams could be presented on the printed page. Any state of the system may be represented by a point in the diagram obtained by assigning a value to each of the two variables on the axes. For every such point the phase diagram tells:

a. How many phases will exist in the system at equilibrium.
b. Which phases they are.
c. Their thermodynamic states expressed in terms of the variables on the axes on the diagram.
d. For diagrams represented in parts (b) and (c) of these figures, the relative amounts of these phases.

Information about the relative amounts of the phases at equilibrium is not available for diagrams plotted in potential coordinates.

In the simple cell structures shown in part (a) of each of these figures a point may lie in one of the areas, on the boundary between areas, or at a triple point. Within any area, the equilibrium state consists of a single phase corresponding to the label for that area with properties corresponding to the coordinates of the point chosen. If the point lies on a boundary, the equilibrium structure consists of the two phases that the boundary separates, each with the properties corresponding to the point. If the point lies on a triple point in the phase diagram, the equilibrium state consists of the three phases that meet at that point, each with the values of the potentials corresponding to the triple point.

Phase diagrams that are constructed in a two-variable space with one axis a potential, and the other not, obey the rules of construction shown in part (b) of Figure 9.4, Figure 9.6 and Figure 9.7. If a point representing the state of the system lies in a single-phase region, then the equilibrium state consists of that single phase with the properties corresponding to the plotted point. If a point representing a state of interest lies within the area of a two-phase field, then it must lie on a particular tie line (Figure 9.9), since the envelope of tie lines fills the two-phase area. A system with the properties designated by the point P in Figure 9.9, will consist of the two phases ε and L when it comes to equilibrium. These phases will have the properties designated by the points A and B. Thus their temperatures will be $T^\varepsilon = T^L$ (the potentials are equal at equilibrium) and their compositions are X_2^ε and X_2^L. The relative amounts of the two phases that will exist at equilibrium may also be determined in this case by applying the lever rule.

9.5.1 THE LEVER RULE FOR TIE LINES

The point P in Figure 9.9 illustrates the lever rule. P represents an equilibrium state of the system specified in terms of its average composition, X_2^0, and its temperature T. This state consists of a mixture of two phases, ε, with properties $(X_2^\varepsilon, T^\varepsilon)$ and the liquid with properties (X_2^L, T^L). In this particular case T is a potential and the

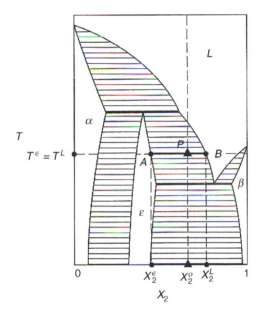

FIGURE 9.9 The line AB is a tie line connecting equilibrium compositions for the ε and liquid phases at the temperature of the line. Relative amounts of the two phases can be determined by the lever rule.

conditions for equilibrium require that $T^\varepsilon = T^L$. The relative amounts of the ε and liquid phases are given by the lever rule, which will now be derived.

Suppose that this system has n_T total moles of the components. When it achieves equilibrium it will be a mixture of the ε and L phases. Let n^ε be the number of moles of the components contained in the ε phase at equilibrium and n^L the number of moles of the L phase. The average composition of the two-phase mixture is given by X_2^0; the total number of moles of component 2 in the system is thus $n_T X_2^0$. The number of moles of component 2 in the ε phase in the system is $n^\varepsilon X_2^\varepsilon$; that of the L phase is $n^L X_2^L$. Conservation of component 2 requires that:

$$n_T X_2^0 = n^\varepsilon X_2^\varepsilon + n^L X_2^L \tag{9.34}$$

Divide both sides of the equation by n_T:

$$X_2^0 = \frac{n^\varepsilon}{n^T} X_2^\varepsilon + \frac{n^L}{n^T} X_2^L = f^\varepsilon X_2^\varepsilon + f^L X_2^L$$

where f^ε and f^L are the fractions of all of the moles of components in the system that are in the ε and L phases when the system comes to equilibrium. Since these are fractions, $f^\varepsilon = 1 - f^L$ and

$$X_2^0 = f^\varepsilon X_2^\varepsilon + (1 - f^\varepsilon) X_2^L = X_2^L + f^\varepsilon (X_2^\varepsilon - X_2^L)$$

Solve this equation for the fraction of the total number of moles in the system that will be in the ε phase at equilibrium:

$$f^\varepsilon = \frac{X_2^L - X_2^0}{X_2^L - X_2^\varepsilon} \tag{9.35}$$

The fraction of moles in the L phases is

$$f^L = 1 - f^\varepsilon = \frac{X_2^0 - X_2^\varepsilon}{X_2^L - X_2^\varepsilon} \tag{9.36}$$

This result may be interpreted graphically in Figure 9.9 as the ratio of two lengths on the tie line:

$$f^\varepsilon = \frac{PB}{AB} \quad \text{and} \quad f^L = \frac{AP}{AB} \tag{9.37}$$

For the point P in Figure 9.9 the relative number of moles contained in the ε phase, f^ε, is given by (PB/AB) and f^L is given by (AP/AB). These arguments yield a simple way of evaluating the quantities of the phases that correspond to the state of a system represented by a point P in a two-phase field. The lever rule may be applied to any tie line in any two-phase field, including those illustrated in part (c) of Figure 9.4, Figure 9.6 and Figure 9.7.

The lever rule gives the relative amounts of the two phases that the tie line connects in moles of phase per mole of system if the properties used to designate the states of the phases are expressed in values per mole. This is the case in all of the examples treated in this chapter. For some purposes it is useful to plot phase diagrams using other units, e.g., weight percent rather than mole percent or atom fraction, or specific volume (volume per gram) rather than molar volume. In these cases the lever rule still applies but the fractions f^α and f^β obtained are weight fractions of the α and β phases.

With this in mind it is possible to illustrate why this construction is called the lever rule. If the quantities are reported in terms of weight or mass, then the lever rule may be visualized through a simple mechanical analog. Imagine the tie line to be a rigid lever and the point P to be its fulcrum, (Figure 9.10). Pans for the α and β phases are hung from the ends of the lever at the points A and B, respectively.

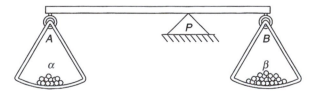

FIGURE 9.10 Mechanical illustration of the lever rule that applies to tie lines in two-phase fields.

For any location of the fulcrum, P, $f^\alpha = (PB/AB)$ and $f^\beta = (AP/AB)$ give the relative weights of the phases that must be placed in the pans to balance the system. If the fulcrum is near the α side of the lever, a small quantity of the β phase will balance a large amount of α. If P is near the β end, a small amount of α balances a large quantity of β. For a sequence of systems for which the state of the two-phase mixture represented by P moves from A to B, the fraction of the number of moles in the system contained in the β phase at equilibrium changes from 0 to 1. This analog does not apply strictly when quantities are reported in terms of the number of moles of the components because in the mechanical analog the balance is achieved in a gravitational field which acts on the mass of the phases rather than their number of moles. The geometrical construction is valid for either mass or molar units. The mechanical analog that aids in visualizing the construction is valid only for mass units.

The lever rule construction applies to every tie line that exists in a two-phase field no matter how many variables are used to specify the states of the two phases involved. For example, in a ternary system the complete specification of the state of a phase requires four variables, e.g., $(T^\alpha, P^\alpha, X_2^\alpha, X_3^\alpha)$, needing a four-dimensional space for its representation. Two-phase fields in this four-dimensional phase diagram are four-dimensional volumes filled with a collection of tie lines connecting pairs of equilibrated states. A state of the system represented by the point P would be described by values of the four variables (T, P, X_2, X_3). The point P must lie on some tie line in the two-phase field connecting states for the α and β phases, each represented by a point A and B in this four-dimensional space. The relative amounts of the two phases existing at equilibrium is still given by the ratios expressed in Equation 9.37.

9.5.2 THE LEVER RULE FOR TIE TRIANGLES

In the representation of phase diagrams in part (a) of Figure 9.4, Figure 9.6 and Figure 9.7 where both axes are potentials, three-phase fields plot as a single point; in the diagrams as plotted in parts (b) of these figures three-phase fields are horizontal lines. The identities of the three phases that coexist and their thermodynamic states may be read from these representations. However, it is not possible to deduce the relative quantities of the three phases that coexist at equilibrium in these representations.

In contrast, if the diagram were plotted on axes in which neither of the variables are potentials, the three-phase regions appear as tie triangles (see part (c) of Figure 9.4, Figure 9.6 and Figure 9.7). A representative three-phase region, a tie triangle, is shown in Figure 9.11. Any state of the system P that lies within the triangle will be composed of the three phases, α, β and ε, at equilibrium. The states of the three phases are given by the corners of the triangle, labeled A, B and C. The relative amounts of the α, β and ε phases corresponding to P are given by a generalization of the lever rule, based upon the same principle of conservation of moles of each of the components. This construction is illustrated without proof in Figure 9.11.

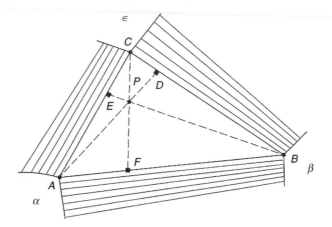

FIGURE 9.11 Use of the tie triangle to determined relative amounts of the three phases in equilibrium for a selected average composition at *P* in a ternary system.

To determine the quantity of a particular phase present in an equilibrated system represented by the point *P*, draw lines from each of the corners through *P* as in Figure 9.11. It can be shown that the fraction of the number of moles in the structure contained in each of the three phases is:

a. f^α, given by the ratio of the segment lengths (PD/AD).
b. f^β, given by the ratio (PE/BE).
c. f^ε, given by the ratio (PF/CF).

This result essentially derives from the requirement that the number of moles in the whole system is the sum of the number in each phase.

A mechanical analog may be visualized for the tie triangle if the quantities involved are reported in terms of the weight of the phases. Imagine the triangle is a rigid frame with a fulcrum located at *P* and pans hanging from each of the points *A*, *B* and *C* (Figure 9.12). The fractions computed from this construction are the relative weights of each of the three phases that would keep the system in balance.

9.6   APPLICATIONS OF PHASE DIAGRAMS IN MATERIALS SCIENCE

Each of the kinds of phase diagrams reviewed in this chapter provides its own view of the information contained in a multicomponent, multiphase system. The information displayed about the phases involved is determined by the choice of axes on which the diagram is plotted. It has been demonstrated that whether both, one, or none of the state variables chosen to represent the system are thermodynamic potentials determines the rules governing the construction and interpretation of the resulting diagram. For any given state of the system, any point on the diagram, the phase diagram tells how many phases are present, which ones they are, and what

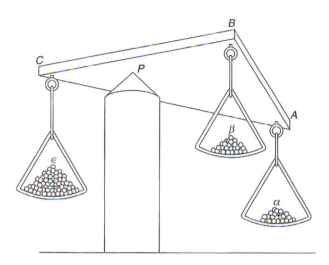

FIGURE 9.12 Mechanical illustration of the generalization of the lever rule applied to a tie triangle in a ternary system.

their properties are expressed in terms of the variables chosen to describe their states when that system comes to equilibrium. For some representations lever rules provide, in addition, information about the relative amounts of the phases in two- or three-phase regions.

Perhaps the most common application of phase diagrams as a tool in materials science is in attempts to understand the changes in internal structure, i.e., microstructure, that may accompany some process to which a material is subject. If the process is carried out slowly so that the system is essentially equilibrated at each step along the way, then each successive state along the path may be represented as a point on the corresponding phase diagram. As the state of the system moves on the diagram through the various one-, two- and three-phase fields, the phase diagram predicts what the state of the system will be at each point. This information may be used to provide insight into the changes that must occur in the system in order to produce each new condition predicted by the phase diagram.

A classic example is provided by the slow solidification of a liquid material system with composition X_2^0 (Figure 9.13). Initially, the material is melted and equilibrated in the liquid state at some temperature, T_0. As the casting cools, its temperature drops. The state of the system crosses into the $(\alpha + L)$ field. At the temperature T_1 the phase diagram reports the equilibrium state as a small quantity of the solid α phase with composition X_2^α in a liquid of a composition slightly richer in component 2 than X_2^0. This requires that particles of the α phase must nucleate and begin to grow. As the temperature reduces toward T_2, the balance shifts on the lever rule, changing the composition of the α phase and increasing its amount at the expense of the liquid. The phase diagram predicts that at T_2 the equilibrated system will consist of about 40% α and 60% liquid. The liquid will have the composition, X_2^c. At T_3, just below the three-phase field, the system consists of a mixture of two solid phases, $(\alpha + \beta)$, with relative amounts about 55% α and 45% β determined

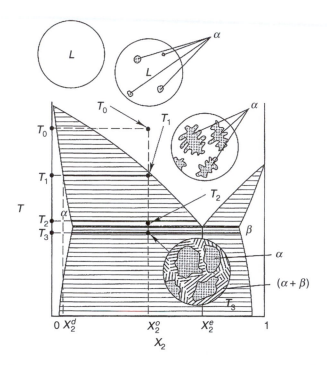

FIGURE 9.13 Sequence of states and microstructural evolution predicted for the solidification of a two-component mixture with a hypoeutectoid composition X_2^0 in a system that exhibits a eutectic phase diagram.

from the lever rule. It is readily inferred that the liquid that existed at T_2 has solidified, isothermally (the temperature change from T_2 to T_3 is arbitrarily small during this process) producing a mixture of α and β phases. This final solidification process usually occurs by nucleation of the two solid phases and their simultaneous cooperative growth into the liquid phase, consuming it. The final structure is composed of the 40% that existed as α at T_2 with the remaining 60% of the system a mixture of α and β phases, sometimes as alternating platelets. Thus the phase diagram provides a framework for understanding the evolution of the microstructure during this solidification process. The pattern of behavior that would result for materials of different compositions (X_2^0 values) may be readily deduced. Materials that exhibit this kind of phase diagram, called a eutectic system, are in fact observed to develop these microstructures when slowly solidified.

The strategy involved in predicting the sequence of events outlined in the last paragraph is straightforward. Plot each successive state that occurs in the process on the phase diagram for the system. Use the diagram to answer the questions, "Which phases are present?," and "How much of each?" Then ask, "What changes in structure must the system experience to proceed from its previous condition to its present one?" Finally, ask, "What processes are required to achieve this change in structure?" This strategy may be applied to systems with phase diagrams of arbitrary complexity.

This simple strategy applies if the system is changing state slowly so that each successive state is not importantly different from equilibrium. Usually the rates of change are such that the material is never in its equilibrium state during processing. Composition, temperature and pressure (more generally, stress) gradients exist in the system. Its state cannot be represented by a point in a state space because its intensive properties are not uniform. In this case, the phase diagram may provide a guide to the state toward which the system is changing, and may still predict, although qualitatively, an overall sequence of changes that the system may be expected to traverse.

Even where the system is far from equilibrium the phase diagram may provide useful quantitative information in the analysis of the microstructural processes the material experiences. As an example, consider the phase diagram shown in Figure 9.14a. The material with composition X_2^0 is first heated and held until it equilibrates at the temperature T_a. It is then quenched to the temperature T_p and held

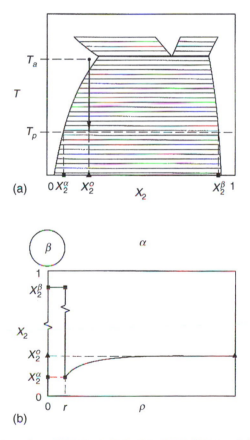

(a)

(b)

FIGURE 9.14 The two-phase field $(\alpha + \beta)$ in this phase diagram, (a) is characteristic of systems that may exhibit precipitation processes. Part (b) sketches the composition profile that may be expected to exist in the matrix α phase near a growing β precipitate particle. The radial distance from the center of the β particle into the α matrix is ρ.

for a length of time. If the quench is rapid so that no transformation occurs during the temperature change, the starting structure at the reaction temperature, T_p, is 100% α phase with composition X_2^0. This structure is supersaturated; i.e., it contains more solute than X_2^α, the maximum amount that the α phase can hold when equilibrated at T_p. The equilibrium state at T_p is a mixture of α and β with the compositions, respectively, X_2^α and X_2^β; the lever rule predicts the final structure will contain about 10% β and 90% α. Thus the state toward which the system is striving is a dispersion of β particles in a matrix of the α phase.

Figure 9.14b is a sketch of the composition profile that would be expected to exist during the growth of a β particle. In the matrix α, far from the particle ($\rho \gg r$), the composition has not yet changed and remains at X_2^0. One of the essential uses of phase diagrams is in the prediction of the composition values at a two-phase $(\alpha + \beta)$ interface during growth. The principle that is applied is the assumption of local equilibrium at the interface. More explicitly, this means that the conditions for thermodynamic equilibrium apply locally in a small volume element that straddles the interface, containing a volume α on one side and β on the other. This principle asserts that there are no discontinuities in the thermodynamic potentials, T, P and μ_k, across the interface, i.e., the conditions for thermodynamic equilibrium are satisfied locally at the interface. Since the phase diagram is a graphical representation of these conditions, at one atmosphere pressure and $T = T_p$, these conditions are satisfied uniquely by the compositions X_2^α and X_2^β obtained from the phase diagram in Figure 9.14a. Thus under the assumption of local equilibrium at the $\alpha\beta$ interface, the composition is X_2^α on the α side and X_2^β on the β side of the interface. These composition values provide the boundary conditions for the solution of the diffusion problem which ultimately predicts the rate at which the particle grows. The resulting composition distribution sketched in Figure 9.14b shows a concentration gradient in the α phase. Solute (component 2 in this case) will flow by diffusion down the gradient toward the growing β particle. Since the β particle is rich in component 2, this flow is precisely what is needed to support the growth of the β particle. While other factors may complicate the growth behavior of particles in real systems (interfaces tend to have a structure that must be maintained during growth; slow interface reaction rates may violate the assumption of local equilibrium; capillarity or stress induced effects may operate) this growth scenario is in fact found in simple precipitation reactions.

One representation of an isothermal, isobaric section through a ternary phase diagram may be constructed by choosing one compositional variable as a chemical potential, e.g., μ_3. Figure 9.15a shows an example of such a diagram for the Ni–Co–O system.[1] The rules of construction of such a diagram correspond to Figure 9.7b. The other compositional axis is most conveniently chosen to be the ratio $X_{Ni}/(X_{Ni} + X_{Co})$. This variable is chosen to represent composition because for a fixed value of oxygen composition it varies from 0 to 1 as the possible range of compositions is traversed. The potential utility of this diagram becomes apparent when it is recognized that tie lines are described by the condition $\mu_O^\alpha = \mu_O^\beta$, for example. Thus on such a plot with μ_O as the ordinate, all of the tie lines are horizontal, i.e., correspond to a fixed value of μ_O. Further, because one of the conditions for three-phase equilibrium is $\mu_O^\alpha = \mu_O^\beta = \mu_O^\varepsilon$, three-phase regions will

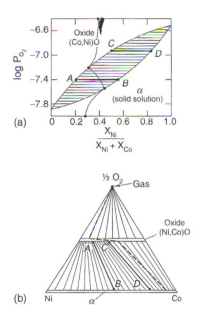

FIGURE 9.15 Isobaric, isothermal ternary diagram for the Ni–Co–O system is plotted in (a) a composition-chemical potential (oxygen potential) space and (b) in a compositional space.[1]

also appear as horizontal lines. The consequence of these observations is that a ternary isotherm plotted with chemical potential as one axis exhibits the same constructions as does an ordinary and familiar binary (T, X_2) diagram. Figure 9.15b contrasts the standard Gibbs triangle construction with mole fraction axes with the same diagram plotted on the axes defined in this paragraph. The version of a ternary phase diagram shown in Figure 9.15a finds particular application when one component is a reactive gas like oxygen, chlorine, or nitrogen. In such a system controlling partial pressures of the gaseous component is equivalent to controlling its chemical potential; indeed, the ordinate on the diagram may be replaced by $\log P_{O_2}$, which for a fixed temperature is proportional to μ_O. The analysis of behavior under controlled atmospheres is thus simplified with this version of ternary diagram.

Figure 9.16 illustrates a similar plot for a slightly more complex ternary system, Fe–Cr–O.[1]

9.7 SUMMARY

The thermodynamic apparatus for describing multiphase systems is based upon the principle that, for extensive properties, the value for the system is the sum of the values contributed by each phase. The change in any extensive property for a multiphase system is then simply the sum of the changes in that property occurring in each phase.

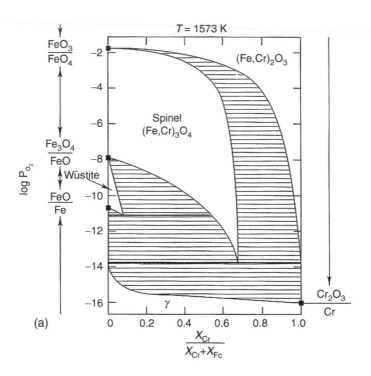

(a)

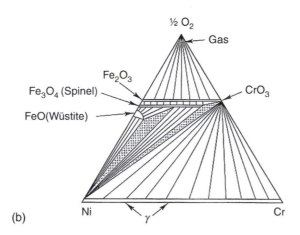

(b)

FIGURE 9.16 (a) Potential-composition diagram and (b) composition space phase diagram for the ternary system Fe–Cr–O.[1]

Application of the general strategy for finding conditions for equilibrium in a multiphase, multicomponent system yields the results:

$$T^I = T^{II} = \cdots = T^\alpha = \cdots = T^p \qquad (9.22)$$

$$P^I = P^{II} = \cdots = P^\alpha = \cdots = P^p \qquad (9.23)$$

$$\mu_1^I = \mu_1^{II} = \cdots = \mu_1^\alpha = \cdots = \mu_1^p \qquad (9.24a)$$

$$\mu_2^I = \mu_2^{II} = \cdots = \mu_2^\alpha = \cdots = \mu_2^p \qquad (9.24b)$$

$$\cdots \quad \cdots \quad \cdots \quad \cdots$$

$$\cdots \quad \cdots \quad \cdots \quad \cdots$$

$$\cdots \quad \cdots \quad \cdots \quad \cdots$$

$$\mu_c^I = \mu_c^{II} = \cdots = \mu_c^\alpha = \cdots = \mu_c^p \qquad (9.24c)$$

The number of degrees of freedom in a multiphase multicomponent system is the number of independent variables it exhibits and is defined to be the difference between the number of variables and the number of relations corresponding to the conditions for equilibrium in the system. Gibbs showed that for a system with c components and p phases,

$$f = c - p + 2 \qquad (9.28)$$

The domains of stability of phases in unary and multicomponent systems may be displayed in a variety of ways. Quantitative representation of such information is usually limited to two-dimensional plots achieved by fixing an appropriate number of the potentials in the system:

  a. If both axes are potentials, then one-phase regions are areas, two-phase domains are lines and three-phase regions are points; the diagram is a simple cell structure.
  b. If only one axis is a potential (e.g., T in binary isobaric diagrams or μ_3 in ternary isobaric isotherms) and the other axis is not a thermodynamic potential, then single-phase regions remain areas but two-phase regions consist of a pair of phase boundaries connected by horizontal tie lines, and three-phase regions appear as a single horizontal tie line.
  c. If neither axis of the diagram is a potential, then two-phase regions have a structure similar to that found in the case with one axis plotted as a potential but the tie lines are not constrained to be horizontal; three-phase regions appear as tie triangles.

In two- and three-phase fields the relative amounts of the phases that exist in equilibrium may be obtained by applying the lever rule to a tie line or a tie triangle.

In general, two-phase regions bound single-phase domains and intersect to produce three-phase equilibria. Thus the calculation of phase diagrams need only focus upon two-phase equilibria.

Phase diagrams are an important tool in interpreting microstructural processes that occur in materials. If processing is slow enough the phase diagram exhibits the sequence of states the system traverses. For processes further from equilibrium, the assumption of local equilibrium applied at interfaces between the phases invokes phase diagram information to visualize boundary conditions for the flows that occur.

HOMEWORK PROBLEMS

Problem 9.1 Use Equation 9.3 to write out a general expression for the change in volume of a three-phase ($\alpha + \beta + \gamma$), two-component (A and B) system. Include all 12 terms.

Problem 9.2 Follow the general strategy for finding conditions for equilibrium for the system described in Problem 9.1.
  a. Write out an explicit expression for the change in entropy.
  b. Write the isolation constraints for this three-phase system.
  c. Use the isolation constraints to eliminate dependent variables in the expression for the entropy.
  d. Collect like terms.
  e. Count the number of independent variables.
  f. Set the coefficients equal to zero.
  g. Write the conditions for equilibrium.

Problem 9.3 Consider a system with components Cu, Ni and Zn that contains four phases: α, β, ε and L.
  a. List the variables required to specify the state of each phase: count them.
  b. List the conditions for equilibrium for this system; count them.
  c. From these counts, compute the number of degrees of freedom.
  d. Compare this result with the number of degrees of freedom computed from the Gibbs phase rule.

Problem 9.4 Sketch the phase diagram for pure water in (P, V) space. Be careful to incorporate the observation that solid water shrinks upon conversion to the liquid state. Discuss complications in the structure of the diagram that derive from this fact.

Problem 9.5 Sketch a (a_2, P) section through the phase diagram for the binary system shown in Figure 9.5a at a constant temperature. Choose a temperature that lies between the triple points of the pure components. Use the resulting cell structure to sketch a plausible (X_2, T) diagram for this system.

Problem 9.6 Consider the phase diagram drawn in Figure 10.20, plotted in (T, X_2) space. Sketch a plausible phase diagram for this system in (T, a_2) coordinates.

Problem 9.7 Sketch an isothermal isobaric phase diagram for the $A-B-C$ system in (a_B, a_C) space; assume this system exhibits four different phases at the temperature of interest. Sketch equivalent phase diagrams for this system:
  a. in (X_B, a_C) space
  b. on the Gibbs triangle in (X_B, X_C) space

Problem 9.8 Sketch or otherwise reproduce the phase diagram shown in Figure 9.7c. In the two and three fields construct contours that represent a constant molar fraction of the phases involved. Use the lever rule to construct contours at $f^{\mathrm{I}} = 0.2$, 0.4, 0.6 and 0.8.

Problem 9.9 Prove the lever rule construction for tie triangles.

Problem 9.10 A Co–Ni alloy with $X_{\mathrm{Ni}} = 0.20$ is heated in air at 1600 K. The phase diagram for this system is shown in Figure 9.15. The oxygen potential must vary from nearly 0 in the gas phase to a large negative number in the alloy. This variation can contain no discontinuities if the oxidation process is diffusion controlled. The curve on Figure 9.15 represents the sequence of states the system has at some point in the oxidation process.
   a. Sketch a microstructure that could correspond to this sequence.
   b. Label the interfaces in your microstructure and evaluate the compositions at the interfaces in the system.

Problem 9.11 Figure 9.14 illustrates a phase diagram and composition profile that may characterize the early stages of precipitation of β from α for an alloy of composition X_2^0 that was solution treated then quenched to the temperature T_{p}.
   a. Sketch a β precipitate particle in its α matrix.
   b. Sketch the composition profile that would result if this sample were taken to equilibrium at T_{p}.
   c. Use the structure in part (b) as the starting structure. Imagine the sample is heated rapidly (upquenched) to T_{a} and held. Use the principle of local equilibrium and the phase diagram to determine the interface compositions at T_{a}. Sketch the composition profile that must develop from the starting structure obtained in part (b).
   d. Argue that this composition is precisely what is needed to dissolve the β particle.

REFERENCES

1. Pelton, A.D. and Schmalzreid, H., *Metall. Trans.*, 4, 1395, 1973.

10 Thermodynamics of Phase Diagrams

CONTENTS

A phase diagram is a map of the equilibrium states for the phase forms and their combinations that may exist in any system. A point in this space, which represents a state of the system that is of interest in a particular application, lies within a specific domain on the map. Reading the map then tells you that, at that state when it comes to equilibrium:

1. What phases are present.
2. The states of those phases.
3. The relative quantities of each phase.

Phase diagrams are a primary thinking tool in materials science because they provide the basis for predicting or interpreting the changes in internal structure of a material that accompany processing or service. A system that is equilibrated in some initial domain on the map and then placed in another domain with a different equilibrium structure, e.g., by simply changing its temperature, will undergo a series of microstructural transformations that take it toward its new equilibrium state, which is given by the phase diagram. Microstructures with optimum properties may be selected by interrupting this process and quenching in the structure, which is then the state in which the material is used. In an inverse application of this strategy, a material that has failed or malfunctioned may be examined in the context of its phase diagram to infer what processes it may have experienced and thus unravel the sequence of events that led to its failure. The phase diagram is often the first place to look in any attempt to understand how a material behaves.

In recent years, there has been a rapid expansion in the development of strategies for calculating phase diagrams. While a significant fraction of binary metallic and ceramic phase diagrams has been explored experimentally, the information is not always complete or consistent. Coupling such studies with thermodynamic measurements and model calculations provides a firm basis for establishing measured diagrams and for extrapolating into domains for which measurements are not available. Since the number of possible ternary systems is much larger than binary systems, only a small fraction of all of the possible ternary systems has been explored. Thus strategies that permit extrapolation from binary information to estimate ternary and higher order phase diagrams are potentially very useful. Also, a knowledge of the underlying thermodynamics provides the basis for understanding the factors that determine the kinetics of processes like nucleation and growth during microstructural transformations. The connections between phase diagrams and the underlying thermodynamics are developed in detail in this chapter.

The most useful thinking tool for visualizing these connections is the free energy–composition or $(G-X)$ diagram. The interplay between the free energy of mixing of the phases in a structure and the form of the phase diagram is developed to illustrate how two- and three-phase fields are generated in binary phase diagrams. The role played by intermediate phases in significantly affecting the structure of phase diagrams is explored. The peculiar $(G-X)$ curve that leads to the formation of a miscibility gap in a phase diagram is presented. Calculations of phase diagrams based upon simple solution models illustrate these developments and lay the foundation for presenting computer calculations of phase diagrams from thermodynamic databases and sophisticated solution models. The connections thus established permit inverting the strategy to estimate thermodynamic properties from experimentally determined phase diagrams. Generalization of these concepts to describe ternary phase diagrams is undertaken at the end of the chapter.

10.1   FREE ENERGY–COMPOSITION $(G-X)$ DIAGRAMS

Visualization of the connection between domains of stability of the phases that may exist in a system and the underlying thermodynamics is most conveniently accomplished with the free energy–composition $(G-X)$ diagram. For a given phase this diagram is a plot of the Gibbs free energy of mixing vs. the mole fraction of component 2 at a fixed pressure and temperature. Figure 8.3 illustrates such a plot for an ideal solution at a series of temperatures. Each of these curves has the familiar mathematical form:

$$\Delta G_{mix} = RT(X_1 \ln X_1 + X_2 \ln X_2) \tag{10.1}$$

(see Table 8.3). It is symmetrical about $X_2 = 0.5$ and has a vertical slope at $X_2 = 0$ and $X_2 = 1$. The value at the minimum (at $X_2 = 0.5$) is $-RT \ln 2$; its magnitude at the minimum thus increases linearly with absolute temperature.

The form of the $(G{-}X)$ curve for real solutions is given by

$$\Delta G_{mix} = \Delta G_{mix}^{xs} + RT(X_1 \ln X_1 + X_2 \ln X_2) \tag{10.2}$$

where the excess free energy of mixing may be positive or negative and in general depends upon composition, temperature and pressure. Figure 10.1 illustrates how the ideal free energy of mixing curve may be altered by the excess free energy contribution. A variety of models for the excess free energy were examined in Chapter 8.

Each phase that may exist in a system has its own free energy–composition diagram. The competition for domains of stability of the phases and interactions that produce two- and three-phase fields that separate them, may be visualized by comparing the $(G{-}X)$ curves for all of the phases in the system. In order for such a comparison of free energies of mixing to be valid it is absolutely essential that the energetics of each component in all of the phases be referred to the same reference state. This point is self evident; however, its implementation is not trivial.

10.1.1 Reference States for $G{-}X$ Curves

The $(G{-}X)$ curves that may be constructed for each phase are derived for the mixing process in which the solution is formed from the pure components in some initial unmixed condition. This initial condition for each component is called the reference

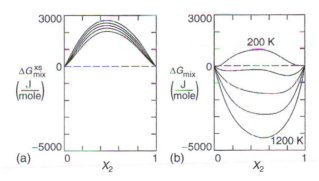

FIGURE 10.1 Free energy–composition diagrams for a flexible three parameter solution model plotted as a function of temperature.

state for that component. Specification of the reference state requires assignment of four attributes:

1. Pressure.
2. Temperature.
3. Composition.
4. Phase form.

A reference state must be defined for each component in the system. The choices made for these four factors in the solution models developed in Chapter 8 were implicit; it is now necessary to be explicit about these assumptions. The expressions derived for the free energy of mixing in those solution models assumed that the condition of each component at the start of the mixing process was:

1. Pressure = pressure in the solution.
2. Temperature = temperature of the solution.
3. Composition = pure component.
4. Phase form = phase form of the solution.

Thus, in describing a liquid solution of components 1 and 2 at one atmosphere and 750 K, the ideal solution model assumes that the reference state for component 1 is pure liquid 1 at 750 K and one atmosphere; that for component 2 is assumed to be pure liquid 2 at 750 K and one atmosphere. If the ideal solution model were used to describe a solid solution of 1 and 2 at 750 K and one atmosphere, the implicitly assumed reference state for 1 is pure solid 1 at 750 K and one atmosphere; that for 2 is pure solid 2 at 750 K and one atmosphere. A direct comparison of the energetics of mixing of 1 and 2 in the solid solution with that in the liquid solution on the basis of these models is not valid.

Such a comparison requires that the choice of reference state for component 1 be the same for both the liquid and solid solutions and that the choice made for 2 must be the same for both solutions. In order to achieve this comparison, it is necessary to change the reference states in the calculation of mixing behavior of one of the solutions. The choice of which reference states are chosen and which are changed is a matter of convenience; it is only essential that they be made the same for component 1 in both phases and the state chosen for 2 must also be the same in both phases. The procedure that must be applied to change a reference state is straightforward; it requires a knowledge of the difference in Gibbs free energies between the two reference states. Expressions for the Gibbs free energy of mixing for a solution model may be written in a way that makes the choice of reference state explicit:

$$\Delta G_{\text{mix}}^{\alpha}\{\alpha; \alpha\} = X_1^{\alpha}(\bar{G}_1^{\alpha} - G_1^{0\alpha}) + X_2^{\alpha}(\bar{G}_2^{\alpha} - G_2^{0\alpha}) \tag{10.3}$$

$$\Delta G_{\text{mix}}^{L}\{L; L\} = X_1^{L}(\bar{G}_1^{L} - G_1^{0L}) + X_2^{L}(\bar{G}_2^{L} - G_2^{0L}) \tag{10.4}$$

The notation in brackets on the left sides of these equations indicates the choice of reference states for components 1 and 2, respectively, and this is further made explicit by the superscripts on the G^0 values on the right sides of these equations. In order to compare the mixing behavior of these two solutions, it is necessary to make the reference states the same. This can be accomplished in any of four ways:

	Solid Solution	Liquid Solution
I	$\{\alpha, \alpha\}$	$\{\alpha, \alpha\}$
II	$\{L, L\}$	$\{L, L\}$
III	$\{\alpha, L\}$	$\{\alpha, L\}$
IV	$\{L, \alpha\}$	$\{L, \alpha\}$

Any of these four choices satisfies the requirements that the reference state for component 1 is the same in both phases and the reference state for component 2 is the same in both phases. Each also requires that two reference states be changed in Equation 10.3 and Equation 10.4.

To illustrate the procedure for changing reference states, choose the reference state labeled III. It is therefore required to evaluate

$$\Delta G^\alpha_{\text{mix}}\{\alpha; L\} = X^\alpha_1(\bar{G}^\alpha_1 - G^{0\alpha}_1) + X^\alpha_2(\bar{G}^\alpha_2 - G^{0L}_2) \qquad (10.5)$$

$$\Delta G^L_{\text{mix}}\{\alpha; L\} = X^L_1(\bar{G}^L_1 - G^{0\alpha}_1) + X^L_2(\bar{G}^L_2 - G^{0L}_2) \qquad (10.6)$$

for this choice of reference states. Careful comparison with Equation 10.3 shows that the $G^{0\alpha}_2$ term has been replaced by G^{0L}_2; in Equation 10.4, the G^{0L}_1 term has been replaced by $G^{0\alpha}_1$, as required by the specified reference states on the left side of these equations. In this pair of equations, the G^0 terms are the same for component 1 (both are phase) and the same for component 2 (both are L). Mixing behavior of these two phases, when formulated in this way, can now be compared.

It remains to connect Equation 10.5 with the corresponding solution model Equation 10.3 with its assumed reference states. In order to achieve this, write Equation 10.5 and simply add and subtract $G^{0\alpha}_2$ inside the brackets in the second term:

$$\Delta G^\alpha_{\text{mix}}\{\alpha; L\} = X^\alpha_1(\bar{G}^\alpha_1 - G^{0\alpha}_1) + X^\alpha_2(\bar{G}^\alpha_2 - G^{0\alpha}_2 + G^{0\alpha}_2 - G^{0L}_2)$$
$$= X^\alpha_1(\bar{G}^\alpha_1 - G^{0\alpha}_1) + X^\alpha_2(\bar{G}^\alpha_2 - G^{0\alpha}_2) + X^\alpha_2(G^{0\alpha}_2 - G^{0L}_2)$$

The first two terms on the right side are identical with the right side of Equation 10.3; that is, the first two terms represent the value for ΔG_{mix} that would be computed from the solution model. The last term in parentheses involves

$$G^{0\alpha}_2 - G^{0L}_2 = \Delta G^{0L \to \alpha}_2$$

which is the change in free energy per mole when pure component 2 is transformed from the liquid to the α phase form at the temperature and pressure of interest. Equation 10.5 may be written:

$$\Delta G_{mix}^{\alpha}\{\alpha; L\} = \Delta G_{mix}^{\alpha}\{\alpha; \alpha\} + X_2^{\alpha}\Delta G_2^{0L\rightarrow\alpha} \tag{10.7}$$

For the liquid solution, a similar strategy yields

$$\Delta G_{mix}^{L}\{\alpha; L\} = \Delta G_{mix}^{L}\{L; L\} + X_1^{L}\Delta G_1^{0\alpha\rightarrow L} \tag{10.8}$$

First terms on the right sides of these equations may be computed from a solution model, e.g., Equation 10.1 or more generally Equation 10.2. The second terms require information about the difference in Gibbs free energy between the liquid and α phases for the pure components.

Figure 10.2a illustrates the effect that changing the reference state for component 2 has upon the shape of the $(G-X)$ curve for the α phase. It is evident

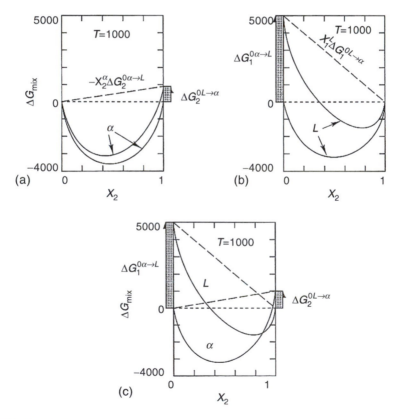

FIGURE 10.2 Effect of change in reference state of pure 1 and pure 2 upon the shape of the $G(X)$ diagram for (a) the solid solution, α, and (b) the liquid solution, L. The comparison of mixing behavior in (c) is now valid because the reference states are consistent.

from Equation 10.7 that a change in the reference state in this case adds a term that varies linearly with mole fraction. In Figure 10.2a, the dashed line is a plot of that linear term. The endpoints of the line occur at

$$X_2 = 0, \qquad \Delta G_{mix} = 0$$

$$X_2 = 1, \qquad \Delta G_{mix} = \Delta G_2^{0L \to \alpha}$$

The resulting ΔG_{mix}^{α} curve hangs from the points $(0,0)$ and $(1, \Delta G_2^{0L \to \alpha})$ on the sides of the diagram. The construction derived from the change in reference state for the liquid phase is shown in Figure 10.2b. Here the hanging points correspond to the evaluation of the second term in Equation 10.8 at $X_2 = 0$ and 1, and are, respectively, $(0, \Delta G_1^{0\alpha \to L})$ and $(1,0)$. Both curves now report the free energy of mixing for their respective phases starting from pure 1 in the α phase form and pure 2 in the liquid phase form. These mixing curves can now be legitimately superimposed on the same diagram, Figure 10.2c and the mixing behavior of the α and liquid solutions may be compared.

These discussions have centered on one of the four possible choices for consistent reference states as an example. Alternate choice will add different linear terms to the expressions for $\Delta G_{mix}\{i,j\}$ and different dashed lines to the $(G-X)$ graphs. As a general principle, the choice of reference states for the components in a particular phase determines the hanging points for the corresponding $(G-X)$ curve at $X_2 = 0$ and $X_2 = 1$. If for a given component the reference state has the same phase form as the solution being considered, then the $(G-X)$ curve will hang from the origin. If the phase form of the reference state is different from that of the solution, then the curve will hang from an ordinate value corresponding to the free energy difference between the phase form of the solution and that of the reference state.

A change in reference state for a solution alters the calculated values of the activities and activity coefficients for the components in that solution. This is implicit in the definition of the activity which contains the reference state value for the chemical potential. Recall the definition:

$$\mu_k - \mu_k^0 = RT \ln a_k \tag{8.66}$$

Apply it to component 2 in the α phase:

$$\mu_2^{\alpha} - \mu_2^{0\alpha} = \overline{G_2^{\alpha}} - G_2^{0\alpha} = RT \ln a_2^{\alpha} \tag{10.9}$$

Now suppose the activity of component 2 in the α phase is required in a context in which it is desirable to define the reference state for component 2 to be some other phase, say the β phase:

$$\overline{G_2^{\alpha}} - G_2^{0\beta} = RT \ln a_2'^{\alpha} \tag{10.10}$$

where d'^{α}_2 is evidently not the same as a^{α}_2. Add and subtract $G^{0\alpha}_2$ to the left side of this equation:

$$\overline{G^{\alpha}_2} - G^{0\alpha}_2 + G^{0\alpha}_2 - G^{0\beta}_2 = RT \ln d'^{\alpha}_2$$

$$RT \ln a^{\alpha}_2 + \Delta G^{0\beta \rightarrow \alpha}_2 = RT \ln d'^{\alpha}_2$$

Solve for the new activity value:

$$d'^{\alpha}_2 = a^{\alpha}_2 e^{\Delta G^{0\beta \rightarrow \alpha}_2 / RT} \tag{10.11}$$

Substitution of the definition of activity coefficient demonstrates that this correction for change in reference state is essentially contained in the activity coefficient.

$$d'^{\alpha}_2 = \gamma'^{\alpha}_2 X^{\alpha}_2 = \gamma^{\alpha}_2 X^{\alpha}_2 e^{\Delta G^{0\beta \rightarrow \alpha}_2 / RT}$$
$$\gamma'^{\alpha}_2 = \gamma^{\alpha}_2 e^{\Delta G^{0\beta \rightarrow \alpha}_2 / RT} \tag{10.12}$$

Changing the reference state for a component thus has the effect of multiplying the activity (or activity coefficient) at any composition by a constant factor for a composition series at a fixed temperature. Graphically, this stretches the ordinate by a constant factor, Figure 10.3. The value of the factor may vary with temperature in a complicated way, as determined by the temperature dependence of the exponential factor in Equation 10.12.

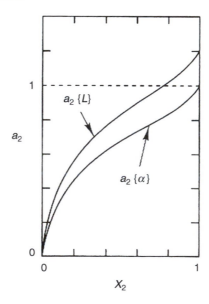

FIGURE 10.3 Variation of the activity of component 2 in the α phase with composition computed from a solution model for the reference choice $\{\alpha\}$ is compared with the same information computed for the reference choice $\{L\}$. In this model $G^{0\alpha \rightarrow L}_2$ is negative.

10.1.2   THE COMMON TANGENT CONSTRUCTION AND TWO PHASE EQUILIBRIUM

In Chapter 9, it was demonstrated that the structure of a two-phase field in a binary system is determined by the conditions for equilibrium:

$$T^\alpha = T^\beta; \qquad P^\alpha = P^\beta; \qquad \mu_1^\alpha = \mu_1^\beta; \qquad \mu_2^\alpha = \mu_2^\beta$$

A graphical representation of this set of equations on a $(G-X)$ diagram is illustrated by the common tangent construction in Figure 10.4. The diagram is drawn at a fixed value of pressure and temperature. Thus the conditions for mechanical and thermal equilibrium between the pair of phases is valid for all points on the graph. The conditions for chemical equilibrium between the α and liquid phases are satisfied uniquely at the pair of compositions corresponding to the points N and P in Figure 10.4.

To demonstrate this assertion, recall the construction in Figure 8.1. Given ΔB_{mix} for any extensive property as a function of composition, the partial molal properties, $\Delta\bar{B}_1$ and $\Delta\bar{B}_2$, corresponding to any composition X_2 may be read as the intercepts on the sides of the diagram of the tangent line drawn to the curve at X_2. On a $(G-X)$ plot, these intercepts are the partial molal Gibbs free energies of the components, which are equal to their chemical potentials, reported relative to the reference states for which the diagram is plotted. Because the $(G-X)$ curves for the two phases cross, it is possible to find a unique pair of compositions, labeled X_2^α and X_2^L in Figure 10.4, through which a common tangent line, i.e., a single line that is tangent to both curves, may be constructed. The intercepts for that line are labeled A and B in Figure 10.4. By the construction presented in Figure 8.1, the point A represents the chemical potential for component 1 in an α solution of composition X_2^α and at

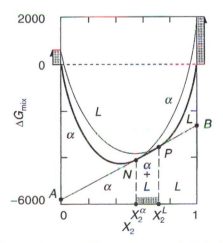

FIGURE 10.4 The $G-X$ curves for two different phases are plotted with consistent reference states. If these curves cross, then a unique line can be drawn that is tangent to both curves. The compositions at these tangent points satisfy the conditions for equilibrium between the two phases.

the same time gives the chemical potential of component 1 in a liquid solution of composition X_2^L. More explicitly,

$$\Delta\mu_1^\alpha = \mu_1^\alpha(X_2^\alpha) - \mu_1^{0\alpha} = \Delta\mu_1^L = \mu_1^L(X_2^L) - \mu_1^{0\alpha}$$

Note that on this graph the reference states for component 1 were carefully chosen to be the same (the α phase form in this case). Thus,

$$\mu_1^\alpha(X_2^\alpha) = \mu_1^L(X_2^L) \qquad\qquad (10.13)$$

for the compositions corresponding to N and P. The common tangent line also gives the same intercept for the α and liquid phases on the two-side of the diagram at the point B. Read the intercepts:

$$\Delta\mu_2^\alpha = \mu_2^\alpha(X_2^\alpha) - \mu_2^{0L} = \Delta\mu_2^L = \mu_2^L(X_2^L) - \mu_2^{0L}$$

Since the reference state for component 2 is chosen to be the liquid phase form for both solutions,

$$\mu_2^\alpha(X_2^\alpha) = \mu_2^L(X_2^L) \qquad\qquad (10.14)$$

Equation 10.13 and Equation 10.14 are precisely the conditions for chemical equilibrium between the two phases. Thus the construction of a tangent line that is common to two $(G-X)$ curves for a pair of phases identifies the compositions of those two phases that coexist in equilibrium at the temperature and pressure for which the $(G-X)$ curves are drawn. These compositions represent the states of the two phases that are at opposite ends of the tie line that marks the two-phase field in the phase diagram for the pressure and temperature chosen.

It is possible to view the relation between the common tangent construction and two-phase equilibria in a different light. It was shown in Chapter 5 that the criterion for equilibrium in a system that is constrained to constant temperature and pressure is a minimum in the Gibbs free energy function. Stated another way, in a comparison of all of the possible states that a system may exhibit that have the same temperature and pressure, the equilibrium state is that which has the lowest value of the Gibbs free energy. $(G-X)$ curves represent the thermodynamic behavior of solutions at constant temperature and pressure; thus this criterion applies to this representation of solution behavior.

As a simple illustration of the application of this criterion, consider the $(G-X)$ curve for a single solid phase, α (Figure 10.5). Focus upon a system with the average composition X_2^0. It is possible to prepare a system with this average composition as a mechanical mixture of two solid solutions with compositions and free energies per mole corresponding to A and B in Figure 10.5. The lever rule may be applied to give the relative amounts of these solutions that would be contained in mechanical mixture of two solid solutions in order to have an average composition X_2^0. By applying the lever rule it is seen that a mixture of (EF/DF) parts of solution A and

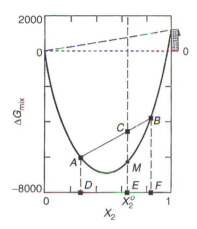

FIGURE 10.5 If the $G-X$ curve for a solution is concave upward at every point then, at any composition X_2^0, a single homogeneous solution has a lower free energy than mechanical mixture of two solutions with the same average composition.

(DE/DF) parts of solution B will have the required average composition. The free energy of this mechanical mixture lies along the line AB at the point C. The point M represents the free energy of a single homogeneous solid solution of the same composition. M lies below C, i.e., has a lower Gibbs free energy than this mechanical mixture of two solutions. Because the $(G-X)$ curve is everywhere concave upward, this conclusion will hold for every pair of solution compositions A and B that straddle the composition X_2^0 so that a mixture of some proportion may yield the average composition X_2^0. Thus the point M, which corresponds to the free energy of a single homogeneous solution of composition X_2^0, has the lowest value of Gibbs free energy of all of the possible configurations that the system could assume.

Now apply the same principle to a system that may exhibit regions of stability for two different phases (Figure 10.6). According to the common tangent construction illustrated in Figure 10.4, the compositions at N and P may exist in two-phase equilibrium. Consider a system that has the composition X_2^0 lying between X_2^α and X_2^β. A system that is a mechanical mixture of α and β of these two compositions will have a free energy of mixing from the reference states that lies along the line joining N and P. For a mixture with proportions (EF/DF) of α and (DE/DF) of β, which has an average composition given by E, the free energy is given by the point M. Consider first a system that is all α phase at the composition X_2^0; its free energy of mixing is designated by the point A in Figure 10.6. Similarly, a system which is all β phase of this composition has a free energy given by point B. Systems having the average composition X_2^0 that are prepared from any combination of mixtures of solutions of the α and β phases, all having free energies of mixing that lie along the lines joining their composition points on the α and β curves, will have Gibbs free energy values lying above the point M. It is concluded that the two phase mixture represented by point M, consisting of (EF/DF) parts of α of composition X_2^α and (DF/EF) parts of β of composition X_2^β, has the lowest value of the Gibbs free energy of all possible configurations that have the average

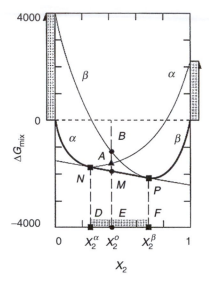

FIGURE 10.6 Consistent $G-X$ curves are plotted for two different phases, α and β. For any composition X_2^0 between the points of common tangency (N and P) a two-phase mixture of α and β with compositions X_2^α and X_2^β has a lower free energy (M) than either homogeneous phase at X_2^0 (α at A and β at B), or any mixture of solution compositions other than X_2^α and X_2^β.

composition X_2^0. Thus this two-phase mixture is the equilibrium state for a system of this composition, a result which is in agreement with the conclusion derived from the equality of the chemical potentials in Figure 10.4.

As a general principle, if a system exhibits an arbitrary number of phases, Figure 10.7, then the sequence of equilibrium configurations for the system is a

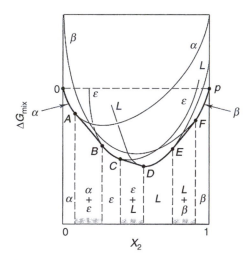

FIGURE 10.7 Taut string construction shows the sequence of equilibrium conditions across the composition range in a system that may exhibit four different phases, α, β, ε, and L.

combination of single and two-phase regions given by the curve segments and common tangent lines that trace the minimum possible value of ΔG_{mix} across the composition range. In Figure 10.7, this trace is given by the segments OA–AB–BC–CD–DE–EF–FP. This representation has been called the taut string construction. The free energy curves of the phases are viewed as rigid and a string tied at O and through P is pulled taut around these curves. The alternating curved and straight segments of the curve formed by the string gives the state at each composition that has the minimum Gibbs free energy.

10.1.3 Two-Phase Fields on Binary Phase Diagrams

The common tangent construction gives the compositions at the phase boundaries of a two-phase field at a fixed temperature. To generate the complete two-phase field on an isobaric binary phase diagram, it is only necessary to repeat this construction at a set of temperatures that span the range for which the diagram is plotted. In developing this construction, it is found that two different aspects of the thermodynamic behavior determine the configuration:

1. The mixing behavior of the two competing phases reported in the excess and ideal free energies of mixing, Equation 10.2.
2. The temperature variation of the relative stabilities of the two phase forms for the pure components as reported in their difference in molar Gibbs free energy.

The relative stabilities of the phase forms for the pure components may be obtained from information about the variation of the Gibbs free energy of the phases with temperature at constant pressure. It was demonstrated in Section 7.1.3 that this information could be computed for any component given heat capacities of the phases, entropies of the transformations, and the absolute entropy of one of the phases at 298 K. Figure 10.8 is a sketch of this information as a function of temperature for the solid and liquid phases for (a) component 1 and (b) component 2. In each case the curves cross at the melting point where the free energies are equal. Below the melting point, α has the lower free energy and is stable; above the melting point the liquid phase is stable. At any temperature, T_1, a quantitative measure of the difference in Gibbs free energy between the phases is given by the vertical distance between the two curves.

Examples of the variation of mixing behavior with composition and temperature for a given phase were presented for some solution models in Figure 8.3 to Figure 8.5. Since $\Delta G_{mix} = \Delta H_{mix} - T\Delta S_{mix}$, as the temperature increases, the contribution from the magnitude of the second term increases and the free energy of mixing becomes more negative. A qualitatively different mixing pattern develops at sufficiently low temperatures for systems with a positive departure from ideal behavior, Figure 8.4. Consequences of this unusual behavior are explored at the end of this section.

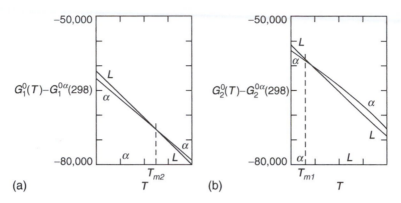

FIGURE 10.8 Molar free energies are plotted as a function of temperature for the liquid and solid phases for (a) pure component 1 and (b) pure component 2.

The pattern of thermodynamic behavior that generates a two-phase $(\alpha + L)$ field may be displayed by computing and plotting $(G\text{--}X)$ curves at a sequence of temperatures that lie between the melting points, Figure 10.9. At each temperature for each component, choose as reference state the phase form that is stable for that component. As a convention for plotting such diagrams, choose component 1 to be the one with the higher melting point. With this convention, in the temperature range between the melting points the reference state of pure 1 is the solid phase, α, and that for component 2 is the liquid phase, L. Thus on each $(G\text{--}X)$ diagram, the mixing curve for α passes through 0 on the one-side of the diagram and that for the liquid passes through 0 on the two-side. At a particular temperature, T, the hanging point for the α curve on the two-side of the diagram is obtained from Figure 10.8b, by reading the free energy difference between liquid and solid at T. Similarly, the hanging point for the liquid curve on the one-side lies a distance above the origin corresponding to the vertical distance between solid and liquid curves plotted for component 1 in Figure 10.8a, evaluated at T. The mixing curves adapted from figures like Figure 8.3 to Figure 8.5, with the straight line additions corresponding to the reference state corrections, hang from these points.

At the melting point of component 1 the α and L curves intersect at the origin because the hanging points for component 1 coincide, $G_1^{0\alpha} = G_1^{0L}$ in Figure 10.8a. As the temperature decreases, the hanging points on the one-side move apart while those on the two-side move toward each other. As a consequence, the point at which the two mixing curves cross moves across the composition scale on the diagram. Since it is a geometric necessity that the common tangent construction straddles the crossing point for the two curves, the composition interval between the tangents,

FIGURE 10.9 (a)–(e) Temperature sequence of $G\text{--}X$ curves demonstrates the pattern of thermodynamic behavior that generates a two-phase field on a binary phase diagram in (X_2, T) space. $a_0^\alpha = 1000$ (J/mol): $a_0^L = -6000$ (J/mol).

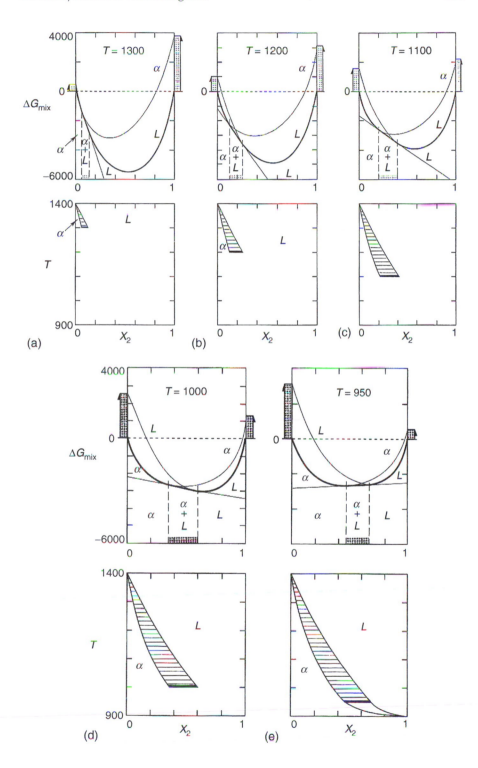

which defines the two-phase field, moves across the diagram from left to right as the temperature decreases. Finally, at the melting point of component 2 the crossing point for the α and L curves coincides with the origin on the two-side of the diagram.

This pattern of behavior is characteristic of every monotonic two-phase field, no matter what the phases may be. The quantitative details depend upon the components and phases involved but the qualitative pattern is pervasive. However, not all two-phase fields are monotonic; some exhibit an extremum (maximum or minimum) in the phase boundaries. Figure 10.10 shows a two-phase field with a minimum together with a set of plausible $(G-X)$ diagrams that would produce this configuration. In this case the behavior of the pure components is similar to that which characterizes monotonic two-phase fields. The minimum derives primarily from a difference in mixing behavior for the α and L phases. The liquid phase has a much more negative departure from ideal behavior than does the solid. As a result, for temperatures below the melting point of component 2, the liquid curve crosses the solid curve twice, Figure 10.10b. Associated with each crossing is a common tangent construction and thus a two-phase field. With continued decreasing temperature, the liquid phase becomes increasingly more unstable with respect to the solid for the pure components, causing the liquid curve to move upward relative to the curve for the solid phase. The crossing points and their associated two phase fields move toward each other. At T_3 the two mixing curves touch each other at a single point; the pair of two-phase fields have merged to a single point at this temperature. At lower temperatures, the solid curve lies below the liquid at all compositions and the liquid phase does not exist at equilibrium. These constructions demonstrate the characteristics of such two-phase fields. Both phase boundaries, the liquidus and the solidus, must have minima that coincide in temperature and composition.

Two-phase fields that exhibit a maximum (or, in fact a maximum and a minimum) are observed and arise from a similar construction.

Another configuration of two-phase field that is frequently encountered in phase diagrams is called the miscibility gap. This class of two-phase field is unique in that it develops in a particular range of compositions and temperatures from a single phase, rather than from the competition between two-phase forms, as is characteristic of the fields in Figures 10.9 and 10.10. Miscibility gaps develop at sufficiently low temperatures within phases that exhibit a positive departure from ideal behavior. Figure 10.11 illustrates this behavior.

Assume that the excess free energy is positive and not very sensitive to temperature. Then, as the temperature is decreased, the contribution from the ideal free energy of mixing, $\Delta G_{mix}^{id} = -T\Delta S_{mix}^{id}$, decreases in magnitude and at sufficiently low temperature the total free energy of mixing curve develops an undulation, Figure 10.11b. The temperature at which this undulation first appears is called the critical temperature for the miscibility gap, usually labeled T_c. A common tangent line may be constructed; in this case the tangents occur at two points on the same curve. The minimum free energy arguments apply and a mixture of two solutions of the same phase form but different compositions, given by the tangent points, has a lower free energy in the composition interval between the tangent

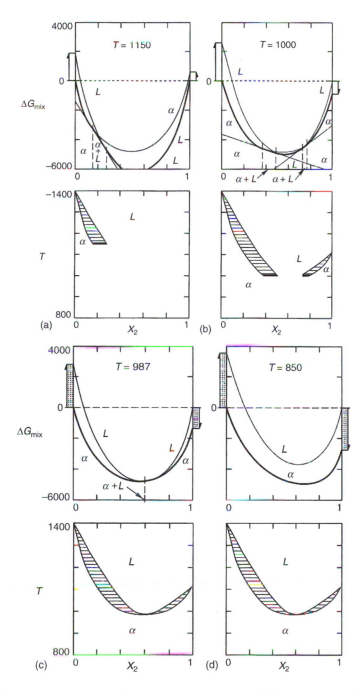

FIGURE 10.10 (a)–(d) Pattern of thermodynamic behavior that generates a two-phase field with a minimum. $a_0^\alpha = 6000$ (J/mol) : $a_0^L = -2000$ (J/mol).

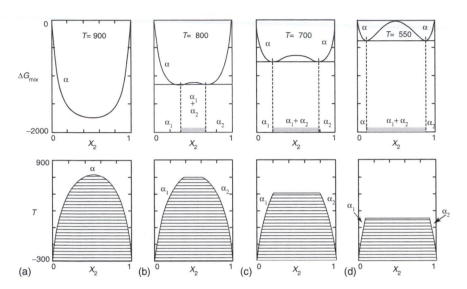

FIGURE 10.11 Pattern of thermodynamic behavior that generates a miscibility gap. $a_0^\alpha = 13,700$ (J/mol).

points than any other mixture of solutions. Thus the equilibrium condition has the unusual characteristic that it is a two-phase mixture (the compositions are discretely different) but both phases have the same structure (e.g., liquid, or FCC, or BCC, etc.). Further decrease in temperature further reduces the ideal mixing contribution, expands the undulation and with it the width of the two-phase field.

An interesting property of miscibility gaps is illustrated in Figure 10.12. The undulation in the $(G-X)$ curve contains a region, bounded by the inflection points M and N, in which the free energy of mixing curve is concave downward. As the temperature is increased, the inflection points move toward each other and merge at the critical point. The domain inside the miscibility gap thus generated is called the spinodal region; it is a characteristic of all miscibility gap structures. The boundary of the spinodal region is identified with the inflection points on the $G-X$ curve; from analytical geometry it will be recalled that the inflection points are given by the condition that the second derivative of the function that the curve represents is zero.

Solutions existing in this range are capable of unusual and useful behavior. Consider, for example, the variation of chemical potential with composition in this region. In normal solutions with $(G-X)$ curves that are concave upward, the chemical potential of component 2 increases with concentration of 2, Figure 10.13a. This may be visualized by recalling the graphical construction that permits determination of partial molal properties from the intercept of a line tangent to the mixing curve, Figure 8.1. If the curve is concave upward, as is usually the case (see, for example, any of the curves in Figure 10.4 to Figure 10.9), the tangent to a point on the curve rotates counterclockwise as the point moves toward the two-side of the

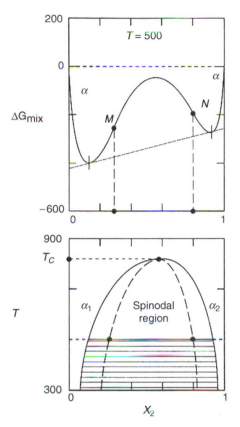

FIGURE 10.12 The thermodynamic basis for the spinodal region inside a miscibility gap.

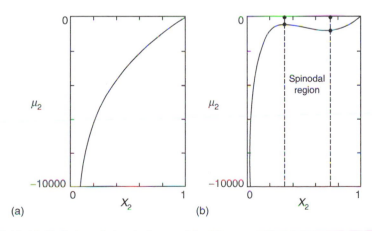

FIGURE 10.13 Variation of chemical potential with composition for component 2 (a) in a normal system and (b) with a miscibility gap and a spinodal region. Both curves are computed from simple regular solution models at 650 K. For (a), $a_0^\alpha = 4000(\text{J/mol})$; for (b), $a_0^\alpha = 12,500(\text{J/mol})$.

diagram. Thus the intercept, which gives $\Delta\mu_2$, moves upward; $\Delta\mu_2$ increases with X_2 as in Figure 10.13a. Inside the spinodal region, where the $(G-X)$ curve is locally concave downward, as the composition moves toward the two-side of the diagram the tangent rotates clockwise, Figure 10.12b. Thus within this region an increase in concentration of component 2 leads to a decrease in its chemical potential, Figure 10.13b.

One consequence of this unusual behavior is uphill diffusion inside the spinodal region. Normally, if a system does not have a uniform composition, atoms move in a direction to eliminate the nonuniformity and flow of a component occurs from regions rich in that component to regions that are less rich. This flow of atoms of one component through the system is called diffusion. In ordinary systems, in which the chemical potential and concentration increase or decrease together, this direction of flow also corresponds to solute atom motion from regions of high chemical potential to regions of low chemical potential, i.e., down the chemical potential gradient. Since the ultimate equilibrium condition of the system is a uniform chemical potential (as will be demonstrated rigorously in Chapter 14), this flow is spontaneous, taking the system toward equilibrium because it reduces differences in chemical potential. Inside the spinodal region, where the chemical potential and composition vary in opposite directions, a flow from a zone with high chemical potential toward one with a lower value is still spontaneous; the system acts to eliminate differences in chemical potential. However, this case corresponds to motion of solute atoms from a region with less concentration of the component toward a region that is more concentrated, i.e., flow is up the concentration gradient. Thus the drift of the system toward equilibrium is accompanied by the development of an increasingly nonuniform composition distribution. The system spontaneously unmixes.

Another way of visualizing this phenomenon is shown in Figure 10.14. Suppose that a system with composition X_2^0 is initially equilibrated at a temperature T_a above the critical temperature of the miscibility gap so that it has a uniform composition. The sample is then quenched to the temperature T_1, which places it inside the spinodal region. A sample that has a uniform composition X_2^0 existing at the temperature T_1 has a free energy of mixing given by the point P in Figure 10.14. Now consider a mixture of two solutions, indicated by the points M and N in this figure. The free energy of this nonuniform mixture lies along the line MN and, if the average composition of the system is X_2^0, is given by the point Q. Note that Q lies below P because the curve is concave downward. Thus the mixture of two solutions (M + N) is more stable than the uniform system at P. Since this will be true even when M and N are arbitrarily close to P, even infinitesimal fluctuations in composition will be stable relative to a uniform system. Fluctuations in composition that form will amplify with time by uphill diffusion, ultimately forming the stable two-phase mixture with compositions given by the common tangent construction at R and S. This process can only occur inside a spinodal region and is called spinodal decomposition. The resulting structure may be technologically useful because the decomposition typically occurs at low temperatures and produces a very fine, nanometer-scale microstructure.

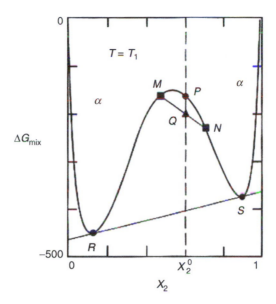

FIGURE 10.14 Inside the spinodal region a mixture (Q) of two solutions (M and N) are more stable than an initially uniform system (P). This spontaneous unmixing of an initially uniform phase is called a spinodal decomposition.

10.1.4 THREE-PHASE EQUILIBRIA

Consider next a binary system that is capable of forming three different phases, e.g., α, L and β, where the crystal structure of β is different from that of α. The description of the thermodynamic behavior of such a system requires information about the mixing behavior and relative stabilities of the pure components for all three phases. It is then possible to plot, on the same $(G–X)$ diagram, ΔG_{mix} curves for each of the three phases and compare their free energies. In this comparison, it is essential that the behavior of component 1 in all three phases be referred to the same reference state; the same statement must also hold for component 2.

Interactions between the three mixing curves two at a time will produce the three two-phase fields that are possible in this system: $(\alpha + L)$, $(\beta + L)$ and $(\alpha + \beta)$. If at a given temperature a pair of the (G/X) curves cross, then the common tangent construction exists for that pair of curves and the phase boundaries may be obtained by the construction developed in the last section. The full two-phase field may be generated by scanning over the temperature range of interest in the phase diagram. As described in Chapter 9, intersections of these two-phase fields produce the three-phase equilibria that may exist in an isobaric binary phase diagram.

The two-phase field interactions that lead to a eutectic diagram are shown in Figure 10.15a. The $(\alpha + L)$ and $(\beta + L)$ fields slope in opposite directions. The point E at which their liquidus curves intersect defines the eutectic composition and temperature. A constant temperature line through E represents the equilibrium $(\alpha + \beta + L)$; it intersects the solidus curve for the $(\alpha + L)$ field at A and the solidus

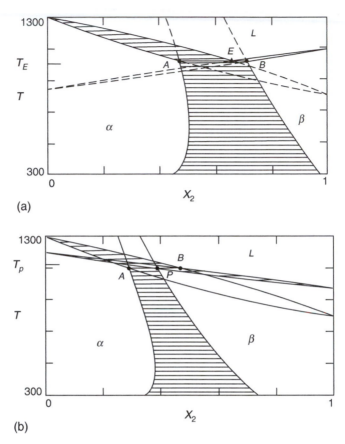

(a)

(b)

FIGURE 10.15 (a) Intersection of three two-phase fields that share the same three phases (α, β, L) may produce (a) a eutectic phase diagram or (b) a peritectic phases diagram.

curve for ($\beta + $L) at B. The points A and B are compositions of α and β that are in equilibrium with the liquid and therefore with each other. Accordingly, the boundaries of the ($\alpha + \beta$) field must also pass through the points A and B. The parts of the ($\alpha + $L) and ($\beta + $L) fields that extend below the eutectic line, shown as dashed lines in Figure 10.15a, are metastable two-phase equilibria, as is the part of the ($\alpha + \beta$) field that extends above the line.

The sequence of (G–X) diagrams that underlies this construction is shown in Figure 10.16. As temperature decreases and the liquid phase becomes progressively less stable with respect to the two solid phases, the liquid phase hanging points and curve moves upward in (G–X) space. The crossing points between α and L and β and L, move toward each other and the associated common tangent lines rotate toward each other. At the eutectic temperature the common tangent lines for ($\alpha + $L) and ($\beta + $L) come into coincidence to satisfy the conditions for the three phase ($\alpha + \beta + $L) equilibrium. The common tangent line for the ($\alpha + \beta$) equilibrium is also coincident with these lines at the eutectic temperature. Below the eutectic

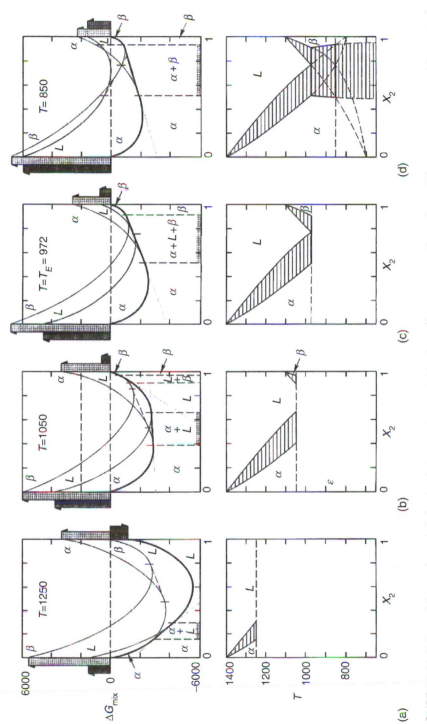

FIGURE 10.16 Pattern of thermodynamic behavior that generates a eutectic phase diagram among three phases. Diagrams are computed from simple regular solution models with $a_0^L = 6000$: $a_0^\alpha = 8000(J/mol)$, and $a_0^\beta = 9000(J/mol)$.

temperature the $(\alpha + L)$ and $(\beta + L)$ common tangent constructions still exist but now the $(\alpha + \beta)$ line has a lower free energy, Figure 10.16d. Thus below T_E the $(\alpha + L)$ and $(\beta + L)$ equilibria are metastable. These metastable equilibria are technologically important because they play a key role in the process by which a eutectic liquid solidifies, determining the rate of the process and the scale of the microstructure that results.

If the $(\alpha + L)$ and $(\beta + L)$ fields slope in the same direction, then a peritectic phase diagram results, Figure 10.15b. Here the point of intersection of the liquidus curves for $(\alpha + L)$ and $(\beta + L)$, labeled B in the diagram, determines the temperature of the three-phase equilibrium. The solid phases in equilibrium with this liquid composition are labeled A and P; these solid phases must be in equilibrium with each other and therefore identify the compositions of the $(\alpha + \beta)$ equilibrium at the peritectic temperature, T_P. The underlying $(G-X)$ relationships are illustrated in Figure 10.17. In this case, as the liquid $(G-X)$ curve moves upward the $(\alpha + L)$ tangent line touches the β curve first at a single point, which identifies the three-phase peritectic equilibrium. The now stable $(\beta + L)$ equilibrium rapidly moves to the side of the diagram and vanishes at the melting point of the β phase; the $(\alpha + \beta)$ field descends to room temperature.

It is possible to produce eutectic and peritectic diagrams from only two-phase forms (α and L) if the α phase contains a miscibility gap. In each case the critical temperature of the miscibility gap must lie at a temperature that is sufficiently high so that the undulating $(G-X)$ curve for the α phase interacts with the liquid curve; the miscibility gap is then metastable above the three-phase line and its maximum is not observed experimentally. The sequence of $(G-X)$ diagrams corresponding to such a eutectic system is shown in Figure 10.18. Development of a similar sequence for a peritectic diagram is left as an exercise for the reader. In order to determine whether a given diagram of these forms is made from three-phase forms, α, β and L, or from only two-phase forms, one of which contributes a miscibility gap, it is only necessary to check the crystal structures of the solid phases of the pure components from which the system is constructed. If they are the same, then the two-phase region connecting the solid phases is a miscibility gap and the underlying thermodynamics is that presented in Figure 10.18. Copper and silver form a simple eutectic phase diagram; however, copper and silver are both face centered cubic crystals. Thus the low temperature two-phase region in this system is in actuality a miscibility gap.

10.1.5 INTERMEDIATE PHASES

Three phases are required to produce an ordinary eutectic or peritectic diagram such as those shown in Figure 10.15 to Figure 10.17. The majority of binary systems exhibit more than just three phases. Those solid solutions that are formed by adding solute to the pure components, i.e., the phases at the sides of the diagram, are called terminal solid solutions. If one of the components has stable allotropic forms, then additional terminal solid solutions exist at the sides of the diagram in the temperature regions where these phase forms exist. In addition, there may exist in

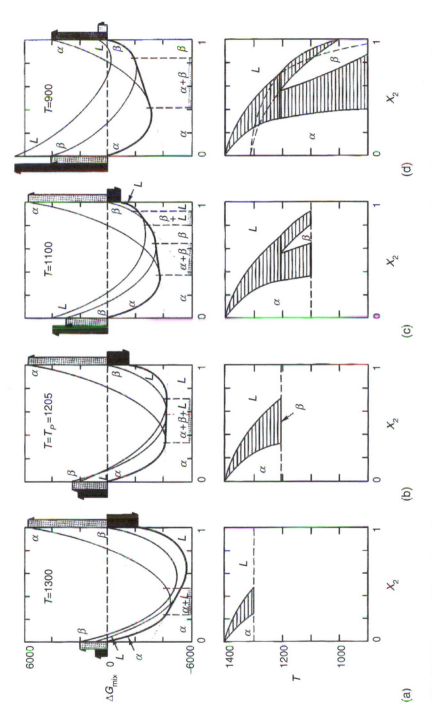

FIGURE 10.17 Pattern of thermodynamic behavior that generates a peritectic phase diagram among three phases. Diagrams are computed from simple regular solution models with $a_0^L = 10,000 : a_0^\alpha = 1500(J/mol)$, and $a_0^\beta = 6000(J/mol)$.

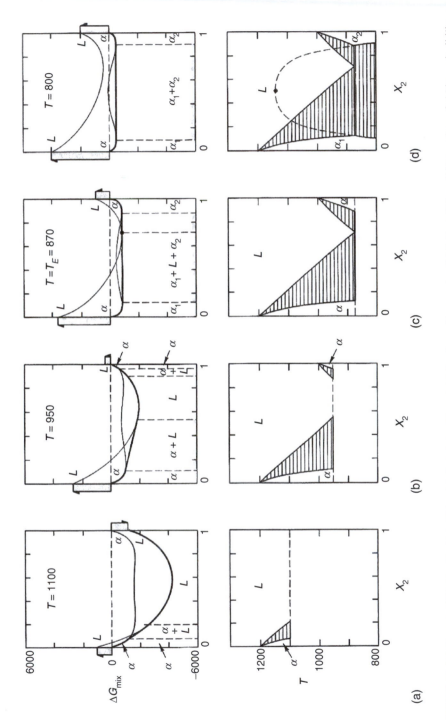

FIGURE 10.18 Pattern of thermodynamic behavior that generates a eutectic phase diagram in which the solid–solid two-phase field is a miscibility gap in the α phase. Regular solution models with $a_0^L = 9000 : a_0^\alpha = 19,000 (J/mol) : T_c = 1143$ K.

the system phase forms that are different from any of those exhibited by the pure components; these are called intermediate phases. Each additional phase form has its own $(G-X)$ diagram which potentially interacts with all of the others to generate the phase diagram.

At any given temperature the sequence of stable one- and two-phase fields that is traversed as X_2 passes from 0 to 1 is determined by the collection of $(G-X)$ diagrams for all of the phases that the system may exhibit, all carefully referred to the same reference states for the pure components. The taut string construction illustrated in Figure 10.7 gives the sequence of alternating one- and two-phase fields that have the minimum free energy and are thus stable. The $(G-X)$ curves for the terminal phases always have a vertical slope at the sides of the diagram. As the temperature changes, the curves move with respect to each other in $(G-X)$ space. Crossing points move; common tangents rotate. Occasionally common tangent lines come into coincidence passing through the unique temperature corresponding to a three-phase field, either eliminating a phase or introducing a new one, to the list of stable phases as the temperature changes.

Figure 10.19 shows a collection of plausible $(G-X)$ diagrams that correspond to a phase diagram with a single intermediate phase, ε. Between T_1 and T_2 the $(\alpha + L)$ and $(L + \varepsilon)$ tangents rotate into coincidence forming the $(\alpha + L + \varepsilon)$ eutectic; the liquid curve lifts off this line leaving the two solid phases. Between T_2 and T_3 the β curve touches the $(\varepsilon + L)$ equilibrium line forming the peritectic $(\varepsilon + \beta + L)$ equilibrium; then the β curve penetrates past the $(\varepsilon + L)$ line to form two-phase fields $(\varepsilon + \beta)$ and $(\beta + L)$. Between T_3 and T_4 the $(\alpha + \varepsilon)$ and $(\varepsilon + \beta)$ tangent lines rotate into coincidence to form the $(\alpha + \varepsilon + \beta)$ eutectoid line. The ε curve lifts off the line leaving the $(\alpha + \beta)$ field at the lowest temperatures.

Intermediate phases frequently have the character of a chemical compound, particularly in ceramic systems. That is, the composition of the phase does not exhibit a significant deviation from some fixed ratio like A_2B or AB_3. On the phase diagram, such an intermediate phase appears as a line compound, Figure 10.20; that is, although the single phase region labeled ε has a structure and a width that is somewhat similar to that shown for the ε phase in Figure 10.19, on the scale at which the diagram is plotted, this width is too small to be resolved. The $(G-X)$ curve for this line compound may be plotted as a single point for practical purposes; this point is labeled P in Figure 10.20. The value of ΔG_{mix} at P may be thought of as the free energy of formation of the line compound from the reference states. Some complicated diagrams with many intermediate phases may be constructed from information about free energies of formation of each of the compounds as a function of temperature.

An important characteristic of the thermodynamics of line compounds is illustrated in Figure 10.20. Small changes in composition across the very limited range of variation available to a line compound are accompanied by large changes in chemical potential of the components. When ε is equilibrated with the α phase its value of X_2 will be slightly less than its stoichiometric value. At that composition, the chemical potential of component 2 in ε is given by the intercept on the two-side of the diagram of the common tangent line connecting α and ε, i.e., by point M. For

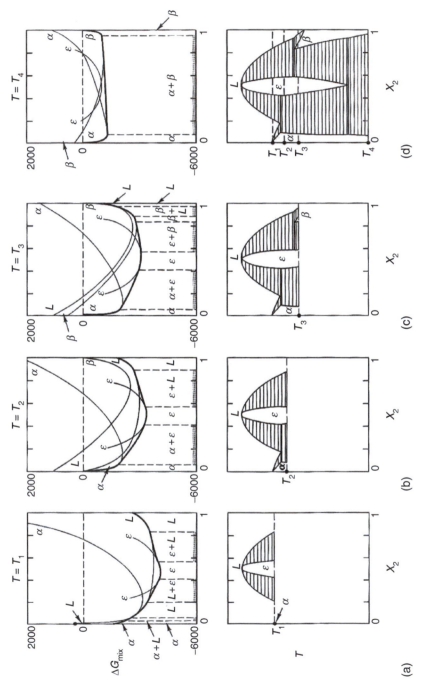

FIGURE 10.19 Free energy composition diagrams that correspond to a phase diagram with a single intermediate phase, ε.

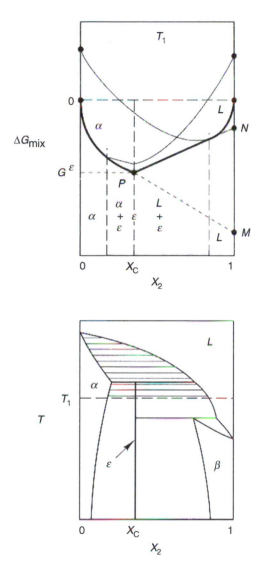

FIGURE 10.20 Free energy composition diagrams that correspond to a phase diagram with an intermediate phase, ε, that forms a line compound.

ε equilibrated with the liquid, the value of X_2 in the ε is only slightly larger than stoichiometric; the chemical potential of component 2 is given by the point N. Thus in line compounds changes in composition that are too small to resolve on the scale of the phase diagram are accompanied by significant changes in chemical potentials.

This behavior reflects the fact that in most solid state line compounds the two components occupy specific sites in the crystal lattice of the phase; each component has its own sublattice. The compound is stoichiometric because the ratios of the two types of sites is fixed by the geometry of the crystal structure. Deviations from this fixed ratio can only occur by introducing defects into the crystal structure. For example, a composition slightly enriched in component 2 may be achieved by leaving some of the sites in the sublattice for component 1 vacant. Conversely, atoms of component 2 may be squeezed into positions between the normal lattice sites, called interstitial positions. Analogous structural defects may accommodate deviations in stoichiometry toward the one-side of the diagram. The energy to form such lattice defects is large on the scale of thermodynamic energies; hence, the large changes in chemical potential are associated with their formation. The thermodynamics of defects in crystals is discussed in detail in Chapter 13.

10.1.6 METASTABLE PHASE DIAGRAMS

Consider a system that may contain p phases, G, L, α, β, ε,..., p. Pairwise intersections between the $(G-X)$ curves for those phases produce two-phase fields. The number of possible two-phase fields that may result is at least $p!/[2!(p-2)!]$ (the number of ways that p items may be arranged in two categories) plus the number of miscibility gaps the system may have. Parts of at least some of these two-phase fields will be stable and appear on the equilibrium phase diagram, defining it, for example, the solid line portions of the two-phase fields in Figure 10.15a,b. Other parts will be metastable, indicated for example by the dashed line portions of the two-phase fields in Figure 10.15, implying that an alternate structure has a lower free energy in that part of the diagram and is the stable equilibrium construction. Some two-phase fields may be metastable over their entire range of existence and thus not appear anywhere on the stable diagram.

Metastable structures frequently play an important role in understanding how microstructures develop in materials science. It has already been pointed out that the metastable extensions of the $(\alpha + L)$ and $(\beta + L)$ equilibria in a eutectic system play a key role in the theory of eutectic solidification and in the control of the resulting microstructure. Formation of solute-rich zones during processing to develop and control precipitation hardening in some systems is frequently associated with the presence of a metastable miscibility gap in the phase diagram. The development of microcrystalline or metallic glass structures in rapid solidification technology is sometimes associated with the presence of a stable phase that is very slow to nucleate or grow.

If during the time of an experiment a particular phase that is known to be stable in the system does not form, then the behavior of the system may be interpreted with the use of a metastable phase diagram. A metastable phase diagram may be developed from information about the stable diagram simply by leaving one or more of the stable phase $(G-X)$ curves out of the construction of the diagram. Figure 10.21 presents an example of the application of this strategy. The stable diagram is shown in Figure 10.21a. In constructing Figure 10.21b, it is assumed that the ε phase does not form. In this example, a simple eutectic phase diagram results when the, phase is deleted from the $(G-X)$ constructions with the result that the liquid phase exists to significantly lower temperatures than in the

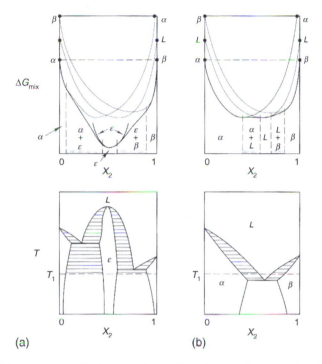

FIGURE 10.21 The metastable phase diagram (b) results from the stable phase diagram (a) if the intermediate phase is left out.

stable diagram, favoring the formation of a low temperature glass phase in this system.

10.2 THERMODYNAMIC MODELS FOR BINARY PHASE DIAGRAMS

Two kinds of information are needed in order to compute a set of $(G-X)$ curves from which to deduce a phase diagram: the thermodynamics of mixing behavior and the relative free energies of the pure components. The ideal and regular solution models and their descendants provide the basis for describing mixing behavior. Relative stabilities of the pure component phase forms requires absolute entropies, heat capacities and entropies of the transformations that may occur for these components. This information may be used to derive expressions that describe the variation of the chemical potentials of the components in each phase at any temperature. Substitution of these model derived expressions for the chemical potentials into the two conditions for chemical equilibrium yields two equations in the unknown phase boundary compositions. These equations may be solved, in principle, to compute $X^{2\alpha}$ and X_2^{β} at that temperature. Repetition of this strategy at a sequence of temperatures generates the $(\alpha + \beta)$ field. Repetition for the other possible combinations of two phases in the system generates the complete set of two-phase fields and their intersections. The stable and metastable parts of the diagram may then be inferred from the known stable parts of the unary phase diagrams, i.e., from their connections to the sides of the diagram. This strategy is first demonstrated for the simplest of models, which assumes that the mixing behavior of each phase is ideal and that the stabilities of the pure components behave in a simple way. The procedure is then illustrated for the simplest regular solution model. Modifications to diagrams brought into play by the existence of intermediate phases that may be treated as line compounds complete this discussion.

10.2.1   IDEAL SOLUTION MODELS FOR PHASE DIAGRAMS

In order to compute the phase boundaries in a two-phase field from a solution model, it is necessary to derive expressions for the chemical potentials of the components in both phases. Consider a system consisting of the α and β phases. Recall the expression for the chemical potential of a component in an ideal solution:

$$\Delta\mu_k^\alpha = \mu_k^\alpha - \mu_k^{0\alpha} = RT \ln X_k^\alpha \tag{8.72}$$

Use this expression to write explicit expressions for the chemical potentials of the components in the α phase in a binary system:

$$\mu_1^\alpha = \mu_1^{0\alpha} + RT \ln X_1^\alpha = G_1^{0\alpha} + RT \ln X_1^\alpha \tag{10.15}$$

$$\mu_2^\alpha = \mu_2^{0\alpha} + RT \ln X_2^\alpha = G_2^{0\alpha} + RT \ln X_2^\alpha \tag{10.16}$$

The analogous expressions may be written for the components in a β phase solution that is ideal:

$$\mu_1^\beta = \mu_1^{0\beta} + RT \ln X_1^\beta = G_1^{0\beta} + RT \ln X_1^\beta \tag{10.17}$$

$$\mu_2^\beta = \mu_2^{0\beta} + RT \ln X_2^\beta = G_2^{0\beta} + RT \ln X_2^\beta \tag{10.18}$$

Choose a value of temperature, T. The conditions for equilibrium of the α and β phases are

$$\mu_1^\alpha = \mu_1^\beta \qquad \text{and} \qquad \mu_2^\alpha = \mu_2^\beta$$

Set Equation 10.15 equal to Equation 10.17:

$$\mu_1^\alpha = G_1^{0\alpha} + RT \ln X_1^\alpha = G_1^{0\beta} + RT \ln X_1^\beta = \mu_1^\beta$$

Choose X_2 as the compositional variable in each phase:

$$G_1^{0\alpha} + RT \ln(1 - X_2^\alpha) = G_1^{0\beta} + RT \ln(1 - X_2^\beta)$$

Rearrange:

$$\frac{1 - X_2^\beta}{1 - X_2^\alpha} = e^{-(\Delta G_1^{0\alpha \to \beta}/RT)} \equiv K_1(T) \tag{10.19}$$

Set Equation 10.16 equal to Equation 10.18:

$$\mu_2^\alpha = G_2^{0\alpha} + RT \ln X_2^\alpha = G_2^{0\beta} + RT \ln X_2^\beta = \mu_2^\beta$$

Rearrange:

$$\frac{X_2^\beta}{X_2^\alpha} = e^{-(\Delta G_2^{0\alpha \to \beta}/RT)} \equiv K_2(T) \tag{10.20}$$

The functions $K_1(T)$ and $K_2(T)$ are implicitly defined in these equations; they contain only information about the relative stabilities of the pure components embodied in $\Delta G_1^0(T)$ and $\Delta G_2^0(T)$.

Equation 10.19 and Equation 10.20 are simply restatements of the conditions for equilibrium that define the $(\alpha + \beta)$ system when both solutions are ideal. They are simultaneous equations that are linear in X_2^α and X_2^β. Their solution is straightforward:

$$X_2^\alpha = \frac{K_1 - 1}{K_1 - K_2} \tag{10.21}$$

$$X_2^\beta = K_2 X_2^\alpha = K_2 \frac{K_1 - 1}{K_1 - K_2} \tag{10.22}$$

Equation 10.21 is the functional form $X_2^\alpha = X_2^\alpha(T^\alpha)$ relating composition to temperature for the phase boundary on the α side of the $(\alpha + \beta)$ field; Equation 10.22 is the analogous function $X_2^\beta = X_2^\beta(T^\beta)$ for the β side. To compute the ideal solution phase diagram it only remains to evaluate ΔG_k^0 for pure 1 and 2 for the α to β transformation as a function of temperature.

An estimate of this temperature function is easily obtained. For any pure component the definitional relationship,

$$\Delta G^0(T) = \Delta H^0(T) - T\Delta S^\circ(T) \tag{10.23}$$

where ΔH^0 and ΔS° represent the enthalpy and entropy for the transformation at any temperature. At constant pressure, these temperature functions are given by

$$\Delta H^0(T) = \Delta H^0(T_0) + \int_{T_0}^{T} \Delta C_P^0(T) dT \tag{10.24}$$

and

$$\Delta S^{\circ}(T) = \Delta S^{\circ}(T_0) + \int_{T_0}^{T} \frac{\Delta C_P^0(T)}{T} \, dT \tag{10.25}$$

where ΔC_P^0 is the difference in heat capacity between the α and β phase forms for the pure component. If T_0 is the equilibrium temperature for α and β for that component, then $\Delta H^0(T_0)$ and $\Delta S^{\circ}(T_0)$ are the enthalpy (heat) and entropy of the transformation for the pure component at that equilibrium temperature. Further, recall that these two quantities are related through Equation 7.17:

$$\Delta H^0(T_0) = T_0 \Delta S^{\circ}(T_0) \tag{7.17}$$

Now assume that the difference in heat capacity between the two-phase forms is sufficiently small so that the integrals in Equation 10.24 and Equation 10.25 may be neglected with respect to the magnitudes of the other terms. This is equivalent to the assumption that the heat and entropy of the transformation are independent of temperature. Equation 10.23 may be written

$$\Delta G^0(T) = \Delta H^0(T_0) - T\Delta S^{\circ}(T_0) = T_0 \Delta S^{\circ}(T_0) - T\Delta S^{\circ}(T_0)$$

$$\Delta G^0(T) = \Delta S^{\circ}(T_0)[T_0 - T] \tag{10.26}$$

Adopt the convention that α is the stable phase at low temperature and β at high temperature; then ΔS° is positive. For $T < T_0$, $\Delta G^0(T)$ is positive for the transformation $\alpha \rightarrow \beta$ and the α form is stable; for $T > T_0$, $\Delta G^0(T)$ is negative and the β phase is stable.

With the temperature dependence of the free energies of the pure components thus estimated to be linear, it becomes possible to evaluate the functions $K_1(T)$ and $K_2(T)$ in Equation 10.21 and Equation 10.22 for the ideal solution model.

$$K_1(T) = e^{-(\Delta S_1^{\circ}(T_{01} - T)/RT)} \tag{10.27}$$

and

$$K_2(T) = e^{-(\Delta S_2^{\circ}(T_{02} - T)/RT)} \tag{10.28}$$

Thus in the ideal solution model of a two-phase field the adjustable parameters are:

$$T_{01}, \ T_{02}, \ \Delta S_1^{\circ}, \ \Delta S_2^{\circ}, \qquad (\text{or } \Delta H_1^0, \ \Delta H_2^0)$$

To make the example more concrete, if α is the solid phase and β the liquid, then the information required is the melting points and entropies of fusion of the two pure components. It is relatively simple to write a mathematical application program to compute and plot an ideal solution model two-phase field. Input to the program are the melting points and entropies of fusion of the two pure components. Increment the temperature over the range between the melting points. For each temperature write relations for each ΔG_k^0 in terms of the entropy and melting temperature; express $K_k(T)$ in terms of ΔG_k^0 for both components. Then put these values into Equation 10.21 and Equation 10.22 to compute and plot $X_2^\alpha(T)$ and $X_2^\beta(T)$.

The ideal solution model is not very flexible because the mixing contribution has no adjustable parameters. Given the two equilibrium temperatures which pin the ends of the two-phase field, the only remaining adjustable parameters are the entropies of the transformation for pure 1 and pure 2. Figure 10.22 shows the limited pattern of two-phase field structures that may be obtained from the ideal solution model for various combinations of the entropy differences. If both transformation entropies are small, the field is narrow and straight. If both are large, the field is broad and straight. If one transformation entropy is large and the other small, the field is narrow opposite the side with the small entropy change and broad opposite the side with the large change in entropy. All fields are monotonic, i.e., the ideal solution model is incapable of yielding a two-phase field with a maximum or minimum.

Consider an ideal solution model for a binary system with three phases, α, β and L. Three two-phase fields are possible: $(\alpha + \beta)$, $(\alpha + L)$ and $(\beta + L)$. If it is assumed that the mixing behavior of all three phases is ideal, then the procedure just developed may be applied separately to compute the three two-phase fields in the system. In this case, it will be necessary to know T_0 and ΔS° values for all three two phase equilibria for each component. Values for these six pieces of information are not independent, however, since, for example,

$$\Delta S_1^{\circ \alpha \to L} = \Delta S_1^{\circ \alpha \to \beta} + \Delta S_1^{\circ \beta \to L} \tag{10.29}$$

A similar relation holds for component 2, as well as for the enthalpies of transformation. Thus, of the six parameters contained in an ideal solution model for a three-phase system, only four are independent. In constructing a model for such a system it is essential to be aware of these constraining relationships. The three two-phase fields may be computed from this information by applying Equation 10.21 and Equation 10.22 with values of K_1 and K_2 appropriately evaluated. The diagrams shown in Figure 10.15a and b are ideal solution phase diagrams computed for a eutectic system and a peritectic system. The rapid broadening evident in the $(\alpha + \beta)$ fields at low temperatures derives from the behavior of K_1 and K_2 as $T \to 0$ K.

10.2.2 REGULAR SOLUTION MODEL FOR PHASE DIAGRAMS

The simplest of the regular solution models, Section 8.7.1, which assumes that

$$\Delta G_{\text{mix}}^{\text{xs}} = \Delta H_{\text{mix}} = a_0 X_1 X_1 \tag{10.30}$$

with the result that

$$\Delta \bar{H}_k = \Delta \bar{G}_k^{\text{xs}} = a_0 (1 - X_k)^2 \qquad (k = 1, 2) \tag{10.31}$$

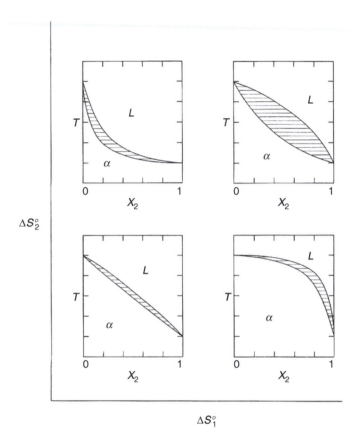

FIGURE 10.22 The ideal solution model can generate a relatively limited pattern of two-phase fields. The four examples shown correspond to different combinations of the entropies of fusion of the pure components, 1 and 2.

provides a much more flexible description of two-phase fields than does the ideal solution model. This model introduces one additional parameter for each phase, a_0^α, a_0^β, etc., which, according to the quasichemical theory discussed in Section 8.7.3, is determined by the relative energies of like and unlike bonds in the phase. Within a single phase, this regular solution model is capable of generating a miscibility gap. In other two-phase equilibria, depending upon the values of the mixing parameters, the model can develop two-phase field shapes that have a maximum or minimum, as well as monotonic structures.

The chemical potential of component k in phase α may in general be written:

$$\mu_k^\alpha = \mu_k^{0\alpha} + \Delta \bar{G}_k^{xs\alpha} + RT \ln X_k^\alpha \tag{10.32}$$

where the excess partial molal Gibbs free energy is a function of composition and temperature. For the simple regular solution model, according to Equation 10.31,

$$\mu_k^\alpha = \mu_k^{0\alpha} + a_0^\alpha (1 - X_k^\alpha)^2 + RT \ln X_k^\alpha \tag{10.33}$$

In deriving the equations that describe a two phase $(\alpha + \beta)$ field, it is first necessary to write out the chemical potentials of both components in both phases:

$$\mu_1^\alpha = \mu_1^{0\alpha} + a_0^\alpha (X_2^\alpha)^2 + RT \ln(1 - X_2^\alpha) \tag{10.34}$$

$$\mu_2^\alpha = \mu_2^{0\alpha} + a_0^\alpha (1 - X_2^\alpha)^2 + RT \ln X_2^\alpha \tag{10.35}$$

$$\mu_1^\beta = \mu_1^{0\beta} + a_0^\beta (X_2^\beta)^2 + RT \ln(1 - X_2^\beta) \tag{10.36}$$

$$\mu_2^\beta = \mu_2^{0\beta} + a_0^\beta (1 - X_2^\beta)^2 + RT \ln X_2^\beta \tag{10.37}$$

Compare these to Equation 10.15 to Equation 10.18 for the ideal solution model. The conditions for equilibrium require that Equation 10.34 be set equal to Equation 10.36 and that Equation 10.35 be equal to Equation 10.37. After some manipulation,

$$\frac{\Delta G_1^{0\alpha \rightarrow \beta}}{RT} + \frac{a_0^\beta}{RT}(X_2^\beta)^2 - \frac{a_0^\alpha}{RT}(X_2^\alpha)^2 + \ln\frac{(1 - X_2^\beta)}{(1 - X_2^\alpha)} = 0 \tag{10.38}$$

$$\frac{\Delta G_2^{0\alpha \rightarrow \beta}}{RT} + \frac{a_0^\beta}{RT}(1 - X_2^\beta)^2 - \frac{a_0^\alpha}{RT}(1 - X_2^\alpha)^2 + \ln\frac{(X_2^\beta)}{(X_2^\alpha)} = 0 \tag{10.39}$$

In a model calculation of a two-phase field, the parameters a_0^α and a_0^β must be input together with the temperature dependence of the ΔG^0 values for the pure components estimated from Equation 10.26 or its more sophisticated versions. These quantities are thus constants at any given temperature. Then Equation 10.38 and Equation 10.39, which are explicit expressions for the conditions for chemical equilibrium, constitute a pair of simultaneous equations in the unknowns X_2^α and X_2^β, which are the phase boundary compositions at that temperature. Unlike the ideal solution model, it is not possible to obtain an analytical solution to this pair of equations; iterative numerical techniques are required. Lupis[1] provides a useful discussion of some of these numerical methods. Such procedures are available in standard mathematical applications programs and may be efficiently computed. To calculate the phase boundaries for the complete two-phase field, choose a temperature, evaluate the ΔG^0 values and solve the equations, plot the result; increment the temperature and repeat the algorithm until the field is complete.

A computer program written to compute phase boundaries for a regular solution model is easily generalized to handle an arbitrary solution model. It is only necessary to input the more general expressions for the partial molal excess free energies in Equation 10.32.

If a_0 is positive for a given phase then at sufficiently low temperatures that phase will exhibit a miscibility gap. This construction was illustrated in Figure 10.11. For the simple regular solution model the excess free energy of mixing is a symmetrical parabola given by Equation 10.30. Since the ideal contribution also has a symmetrical form about $X_2 = 0.5$, ΔG_{mix} is also symmetrical. $(G-X)$ curves computed from this model at several temperatures are shown in Figure 8.5. The phase boundaries may be computed with the numerical algorithm already described.

The boundary of the spinodal region and the critical temperature of the miscibility gap for this model may be derived without resorting to numerical techniques. The spinodal region is defined by the condition that the ΔG_{mix} curve is concave downward. At any given temperature, the limits of this condition are defined by the inflection points in the $(G-X)$ curve where its curvature changes sign. Mathematically, the inflection point on a curve representing a function is described by the condition that the second derivative of the function is zero. In this case the function is

$$\Delta G_{mix} = a_0 X_1 X_2 + RT(X_1 \ln X_1 + X_2 \ln X_2)$$

Take the second derivative, noting again that X_1 and X_2 are related so that this is the total derivative in the context of these two variables:

$$\frac{d^2 \Delta G_{mix}}{dX_2^2} = -2a_0 + \frac{RT}{X_1 X_2}$$

To find the inflection point in the curve, which defines the boundary of the spinodal region at any temperature, set this derivative equal to zero and rearrange the result:

$$X_1 X_2 = \frac{RT}{2a_0} \tag{10.40}$$

This equation describes how the compositions at the boundary of the spinodal region vary with temperature.

The spinodal boundary is thus seen to be a parabola that is symmetrical about $X_2 = 0.5$. As the temperature is increased these inflection points move toward each other; at the critical temperature T_c they meet at $X_2 = 0.5$. Insert this condition into Equation 10.40:

$$(0.5)(0.5) = \frac{RT_c}{2a_0}$$

Solve for the critical temperature:

$$T_c = \frac{a_0}{2R} \tag{10.41}$$

Thus in the simple regular solution model a positive heat of mixing for any phase implies that it exhibits a miscibility gap below a temperature given by Equation 10.41. Whether this miscibility gap is stable, metastable, or interacts with other phases to form additional two-phase fields, is determined by the usual comparisons available through the $(G-X)$ curves for the system. For example, Figure 10.18 illustrates a case in which the critical temperature for a solid state miscibility gap lies in the region in which the liquid phase is thermodynamically stable; interactions between the miscibility gap and the liquid phase produce a eutectic phase diagram.

10.2.3 THE MIDRIB CURVE

Consider the $(G-X)$ curves for two phases at some temperature, T_1. The system will form a two-phase field at the temperature in question if and only if the two curves cross at some point; then a common tangent may be constructed connecting the two curves and, through the conditions for equilibrium, the boundaries of the two-phase field may be defined. The point at which the two $(G-X)$ curves cross necessarily lies within the two-phase field. It is defined by the condition

$$\Delta G^{\alpha}_{\text{mix}}\{\alpha; \beta\} = \Delta G^{\beta}_{\text{mix}}\{\alpha; \beta\} \tag{10.42}$$

where, as usual the reference states must be chosen to be the same for both phases. If the temperature range is scanned to trace out the $(\alpha + \beta)$ field, this crossing point traces out a curve that lies within that field. This midrib curve, though not necessarily at the center of the two-phase field, is guaranteed to lie within it. It provides a useful means of locating the two-phase field in (X_2, T) space and has other applications of technological importance. Properties of midrib curves for the simplest regular solution model are explored in this section.

Focus on an $(\alpha + L)$ two-phase field. Choose α to be the reference state for component 1 and L to be that for component 2. The free energies of mixing of these two phases are given by:

$$\Delta G^{\alpha}_{\text{mix}}\{\alpha; L\} = a^{\alpha}_0 X^{\alpha}_1 X^{\alpha}_2 + RT(X^{\alpha}_1 \ln X^{\alpha}_1 + X^{\alpha}_2 \ln X^{\alpha}_2) - X^{\alpha}_2 \Delta G^{0\alpha \rightarrow L}_2 \tag{10.43}$$

$$\Delta G^{L}_{\text{mix}}\{\alpha; L\} = a^{L}_0 X^{L}_1 X^{L}_2 + RT(X^{L}_1 \ln X^{L}_1 + X^{L}_2 \ln X^{L}_2) + X^{L}_1 \Delta G^{0\alpha \rightarrow L}_1 \tag{10.44}$$

The last term in each equation accommodates the change in reference state required. At any temperature T^* the midrib point is given by the point of intersection of these curves, Equation 10.42. Note also that at this point of intersection the value of X_2 is the same for both phases; thus

$$X^{\alpha}_2 = X^{L}_2 = X^* \qquad \text{and} \qquad X^{\alpha}_1 = X^{L}_1 = (1 - X^*)$$

When Equation 10.43 and Equation 10.44 are substituted into Equation 10.42 the ideal mixing terms cancel:

$$a^{\alpha}_0 X^*(1 - X^*) - X^* \Delta G^{0\alpha \rightarrow L}_2 = a^{L}_0 X^*(1 - X^*) - (1 - X^*)\Delta G^{0\alpha \rightarrow L}_1$$

Rearrange:

$$(a^{\alpha}_0 - a^{L}_0)X^*(1 - X^*) = (1 - X^*)\Delta G^{0\alpha \rightarrow L}_1 + X^* \Delta G^{0\alpha \rightarrow L}_2$$

Apply Equation 10.26 to estimate the free energies of the pure components:

$$(a_0^\alpha - a_0^L)X^*(1 - X^*) = (1 - X^*)\Delta S_1^{\circ\alpha\to L}(T_{01} - T^*) + X^*\Delta S_2^{\circ\alpha\to L}(T_{02} - T^*)$$

Introduce $\Delta a_0 = a_0^L - a_0^\alpha$ and simplify the notation for the entropy factors:

$$-\Delta a_0 X^*(1 - X^*) - (1 - X^*)T_{01}\Delta S_1^\circ - X^*T_{02}\Delta S_2^\circ = -T^*[(1 - X^*)\Delta S_1^\circ + X^*\Delta S_2^\circ]$$

Solve for T^* to obtain the equation for the midrib curve:

$$T^* = \frac{\Delta a_0 X^*(1 - X^*) + (1 - X^*)T_{01}\Delta S_1^\circ + X^*T_{02}\Delta S_2^\circ}{(1 - X^*)\Delta S_1^\circ + X^*\Delta S_2^\circ} \tag{10.45}$$

As a check, examine the behavior at $X^* \to 0$. Terms with X^* as a factor are zero and $(1 - X^*) \to 1$; $T^* \to T_{01}$ as $X^* \to 0$, as required by the phase diagram. Similarly, at $X^* \to 1$, $T^* \to T_{02}$. Thus the midrib curve passes through the melting points (or more generally for any $\alpha + \beta$ field, the transformation temperatures for the pure components) at the sides of the diagram.

In order to identify the limiting conditions for formation of a two-phase field with a maximum or minimum, it is useful to examine the slope of the midrib curve at the sides of the phase diagram. The derivative of Equation 10.45 with respect to X^* yields

$$\frac{dT^*}{dX^*} = \frac{\Delta a_0[(1 - X^*)^2\Delta S_1^\circ - (X^*)^2\Delta S_2^\circ] + [T_{02} - T_{01}]\Delta S_1^\circ \Delta S_2^\circ}{[(1 - X^*)\Delta S_1^\circ + X^*\Delta S_2^\circ]} \tag{10.46}$$

The slope of the $T^*(X^*)$ curve at the sides of the diagram may be found by examining the limiting cases. As $X^* \to 0$, $T^* \to T_{01}$, the melting point on the pure 1 side of the diagram. Equation 10.46 simplifies for this limiting case:

$$\frac{dT^*}{dX^*} = \frac{\Delta a_0 + \Delta S_2^\circ \Delta T_0}{\Delta S_1^\circ} \tag{10.47}$$

where $\Delta T_0 = T_{01} - T_{02}$, i.e., the difference between the melting points of the pure components. Similarly, in the limit as $X^* \to 1$, $T^* \to T_{02}$ and the slope is given by

$$\frac{dT^*}{dX^*} = \frac{-\Delta a_0 + \Delta S_1^\circ \Delta T_0}{\Delta S_2^\circ} \tag{10.48}$$

Retain the convention that the component with the higher melting point is selected to be component 1. If the slope of the midrib curve at $X^* = 0$ is negative, then the two-phase field is either monotonic or has a minimum. However, if the slope at $X^* = 0$ is positive, then the resulting two-phase field must exhibit a maximum. Thus the condition that defines a two-phase field with a maximum in this model is

$$\Delta a_0 = a_0^L - a_0^\alpha > \Delta S_2^\circ \Delta T_0 \tag{10.49}$$

Examine the sign of the slope of the midrib curve at $X^* = 1$. If this slope is negative, then the two-phase field may either be monotonic or exhibit a maximum. However, if the slope is positive at $X^* = 1$, then the two-phase field must exhibit a minimum. The numerator in Equation 10.48 must be positive, or

$$\Delta a_0 = a_0^L - a_0^\alpha < -\Delta S_1^\circ \Delta T_0 \qquad (10.50)$$

It is concluded that, at least in the simplest regular solution model, a comparison of the difference in solution parameter values with the entropies of fusion of the pure components determines whether a two-phase field is monotonic or exhibits an extremum. Figure 10.10 is an example of a phase diagram computed from this model with parameters that satisfy inequality Equation 10.50.

The midrib curve also comes into play in interpreting some kinetic phenomena. Focus on the portion of a two-phase field and its $(G-X)$ curves sketched in Figure 10.23. A sample of composition X' is initially heated to T_a and brought to equilibrium as the stable β phase. During processing this sample is quenched to T_1 and held at that temperature. When it arrives at T_1, still in the β phase form, its thermodynamic condition is given by the point P on Figure 10.23. The equilibrium structure at T_1 is the two-phase mixture given by the point R. However, it is possible for the metastable condition, labeled Q, to form spontaneously from the starting point at P, since Q has a lower free energy than P. Further, since the two-phase mixture is made up of two solutions with compositions different from X', formation of this state requires diffusion of the components through the solid phases to change their compositions. Solid-state diffusion is a relatively slow process. In contrast, formation of the state labeled Q requires no composition change, but only a change in crystal structure from β to α. Depending upon the kinetics of this process, it may thus be possible to form α of the same composition as β; the final structure that may then form from this intermediate state with a microstructure that is entirely different from that which may form by the direct decomposition of β.

In contrast, if the starting alloy composition were X'', on the other side of the midrib point, the as-quenched condition at T_1 is labeled M. The α phase of the same composition as this quenched-in β phase has a higher free energy. Thus α of the same composition has a free energy given by L, which lies above M. Evidently α of the same composition, X'', cannot form directly from β for this composition and the transformation can only proceed directly to the two-phase mixture labeled N. It is evident that the midrib separates a two-phase field into two regions, one in which a structural change without a composition change may occur on the way to the stable structure and a second region in which this possibility is ruled out.

The midrib curve locates the two-phase field in (X_2, T) space, gives its curvature and reveals whether it is monotonic or has a maximum or minimum. Equation 10.45, which describes this curve for the simplest regular solution model, can be derived without resorting to numerical analysis. This result is easily generalized to include more sophisticated solution models.

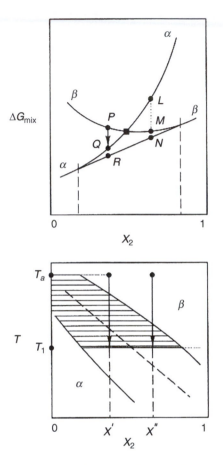

FIGURE 10.23 Role played by the crossing point of two $G-X$ curves in delimiting possible transformation sequences in processing.

10.2.4 PATTERN OF REGULAR SOLUTION PHASE DIAGRAMS WITH TWO PHASES

The pattern of phase diagrams that may be generated from two phases, α and L, in the simplest regular solution model is laid out in Figure 10.24. All of the diagrams in this figure have the same melting points and entropies of fusion for the pure components. The pattern that develops is entirely due to variations in the mixing parameters, a_0^α and a_0^L. The top row of diagrams has a fixed positive value of a_0^α; each diagram thus has a miscibility gap in the solid solution with a known critical temperature T_c. The bottom row of diagrams are calculated with a small value of a_0^α. The value of Δa_0 chosen for the first column of diagrams satisfies inequality Equation 10.50 and produces an $(\alpha + L)$ field with a minimum. The middle column has Δa_0 values that produce monotonic $(\alpha + L)$ fields and the last column has a Δa_0 value that produces a maximum in the $(\alpha + L)$ field. The requirement that $a_0^L > a_0^\alpha$ in diagram (c) implies that the liquid phase also has a miscibility gap with a critical

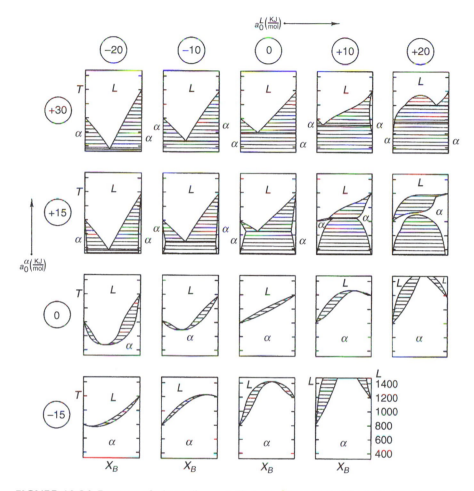

FIGURE 10.24 Patterns of phase diagrams that can be generated by only two phases, α and L, with the simplest regular solution model.

temperature above that for the solid phase. It is evident that a wide variety of phase diagrams may be produced with only two phases and their miscibility gaps by this simplest regular solution model.

10.2.5 DIAGRAMS WITH THREE OR MORE PHASES

Because this model is capable of developing miscibility gaps and two-phase fields with extrema, the variety of phase diagrams that may be produced when a third phase is added to the model is too broad to be discussed systematically in this context. As in the case of an ideal solution model, the strategy for calculating a diagram with three or more phases is based upon the computation of all of the possible two-phase fields that may appear in the structure applying Equation 10.38 and Equation 10.39. Intersections of these two-phase fields produce the three-phase invariant equilibrium lines in the diagram. The problem of sorting out which of these equilibria are stable and which are metastable is nontrivial and increases in complexity as the number of phases in the system increases. In computer calculations of phase diagrams, this problem is circumvented by applying a procedure that is based directly upon the principle of minimization of Gibbs free energy rather than upon the conditions for equilibrium that are derived from it. This strategy is discussed in Section 10.6.

An important and challenging problem appears when additional phases must be incorporated into phase diagram models. Because it is essential that the reference states for each component must be the same in all phases, it is necessary to obtain estimates of the stabilities of all of the phase forms in the system relative to the reference phase for each pure component. Consider a system that contains four phases: liquid L, face centered cubic α, body centered cubic β and hexagonal ε. Suppose, at a given temperature, pure 1 in the α (FCC) phase form is chosen as the reference state for component 1. In order to develop a $(G-X)$ diagram at the given temperature, it is necessary to obtain estimates of the free energies of pure component 1 for L, β and ε relative to the α phase. The same information must be estimated for pure component 2. This estimate is straightforward for the liquid phase, applying the approach leading to Equation 10.26. However, it may be that pure component 1 does not exist in the β (BCC) or ε (HCP) forms at any temperature and pressure. Consider, for example, the aluminum–zinc system. Pure aluminum is FCC (α); the terminal phase on the zinc-rich side is HCP (ε). Calculation of the $(G-X)$ diagram for the HCP (ε) phase at any temperature T requires an estimate of the free energy difference between hexagonal and face centered forms of pure aluminum at T. However, aluminum does not exist in the hexagonal form at any temperature and pressure. Thus an extrapolation from a region in which pure aluminum is stable in this phase form is not possible. Alternate strategies must be devised to obtain this information. The most common approach employs experimental measurements of phase boundaries and inverts the model calculation to obtain information about $\Delta G_k^{\,0}$ and ΔS_k° and T_{0k} for transformations that cannot be made to occur in pure k. An alternative approach involves the calculation of energies of formation of crystal structures from "first principles" calculations. This approach is under development in computational materials science, and uses energetic models for the pairwise interactions of individual atoms arranged in the crystal structure of interest to compute these properties.

EXAMPLE 10.1

Use of phase boundary information to estimate $\Delta G_{Zn}^{0\alpha \to \varepsilon}$ at 1200 K. Figure 10.25 shows a portion of the copper-zinc phase diagram. Assume for the purpose of this example that both the liquid and α phases form ideal solutions. Given the following information:

$$\text{Cu}: \ T_m = 1357 \text{ K}; \qquad \Delta S_{Cu}^{\circ \alpha \to L} = 9.60 \text{ J/mol K}$$

$$\text{Zn}: \ T_m = 693 \text{ K}; \qquad \Delta S_{Zn}^{\circ \varepsilon \to L} = 10.51 \text{ J/mol K}$$

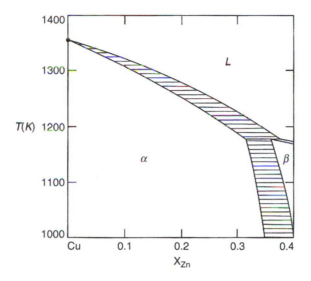

FIGURE 10.25 Portion of the copper-rich side of the copper–zinc phase diagram at high temperatures.

At 1200 K, the compositions read from the phase diagram are $X_{Zn}^{\alpha} = 0.27$; $X_{Zn}^{L} = 0.34$. In the ideal solution model the ratio of these two compositions is simply related to $K_{Zn}(1200\,K)$, (see Equation 10.20):

$$\frac{X_{Zn}^{L}}{X_{Zn}^{\alpha}} = K_{Zn} = e^{-(\Delta G_{Zn}^{0\alpha \to L}/RT)}$$

Solve for

$$\Delta G_{Zn}^{0\alpha \to L} = -8.314 \times 1200 \ln\left(\frac{0.34}{0.27}\right) = -2300\,J/mol$$

Apply Equation 10.26 to estimate $\Delta G_{Zn}^{0\varepsilon \to L}$ at 1200 K:

$$\Delta G_{Zn}^{0\varepsilon \to L} = \Delta S_{Zn}^{0\varepsilon \to L}(T_{mZn} - T) = 9.64(692 - 1200) = -4900\,J/mol$$

Note that

$$\Delta G_{Zn}^{0\alpha \to \varepsilon} = G_{Zn}^{0\varepsilon} - G_{Zn}^{0\alpha}$$
$$= G_{Zn}^{0\varepsilon} - G_{Zn}^{0L} + G_{Zn}^{0L} - G_{Zn}^{0\alpha}$$
$$= -\Delta G_{Zn}^{0\varepsilon \to L} + \Delta G_{Zn}^{0\alpha \to L}$$
$$\Delta G_{Zn}^{0\alpha \to \varepsilon} = -(-5300) + (-2300) = +3000\,J/mol$$

which provides an estimate of the Gibbs free energy of FCC zinc relative to HCP zinc, although FCC zinc never forms in the laboratory.

This example is based upon the simplest of phase diagram models in which both solutions are assumed to be ideal. Evidently, this strategy can be applied through more sophisticated models for the phases involved and invoking more points on the two phase $(\alpha + L)$ field to provide a statistically based estimate of this quantity and others like it. Once a valid estimate of the energetics of FCC zinc is obtained from this diagram and perhaps some others in which zinc also dissolves in FCC phases, then this result may be used in all phase diagrams that have zinc as a component and contain an FCC phase, including ternary and higher order systems.

10.2.6 MODELING PHASE DIAGRAMS WITH LINE COMPOUNDS

Figure 10.19 to Figure 10.21 illustrate the role played by intermediate phases in determining the structure of phase diagrams. The composition range of such phases is frequently too narrow to be resolved on a phase diagram; the single-phase field for such a phase plots as a vertical line on the phase diagram. In this case, it is usually sufficient to represent the $(G-X)$ curve for the line compound by a single point which plots the free energy of formation of the compound from the reference states for the components in the diagram at the composition of the line compound, Figure 10.20.

It is crucial to note that in the remainder of the $G-X$ curve calculations in the system, the free energy is given in units of energy per gram atom of solution. In tabulated information, free energies of formation are generally reported as energy per mole of the compound formed. If the compound has the formula M_uX_v, then 1 mol of the compound contains $(u + v)$ gram atoms of M and X atoms. Thus the free energy value required for representation on a $G-X$ diagram is the free energy of formation computed from tabulated values divided by $(u + v)$. The value of G^ε corresponding to the point P in Figure 10.20 is determined from the formation reaction for that compound:

$$2A + B = A_2B \qquad \Delta G_f^0$$

$$G^\varepsilon = \frac{\Delta G_f^0}{2 + 1} = \frac{\Delta G_f^0}{3}$$

As an illustration of the treatment of line compounds, suppose that the terminal α phases in the system shown in Figure 10.20 is dilute and obeys the simple regular solution model. The chemical potentials of the components in a dilute solution are given by:

$$\text{Raoult's Law:}\quad \Delta\bar{\mu}_1^\alpha = RT \ln(1 - X_2^\alpha) \tag{10.51}$$

$$\text{Henry's Law:}\quad \Delta\bar{\mu}_2^\alpha = RT \ln \gamma_2^{0\alpha} + RT \ln X_2^\alpha \tag{10.52}$$

where X_2 is the mole fraction of component 2 and γ^0 is the Henry's law constant for the solution. In the simple regular solution model it has been shown that $RT \ln \gamma_2^0 = a_0$, (see Equation 8.119). Thus, in this model Henry's law becomes:

$$\text{Henry's Law:}\quad \Delta\bar{\mu}_2^\alpha = a_0 + RT \ln X_2^\alpha \tag{10.53}$$

Equation 10.51 and Equation 10.53 identify the intercepts at $X_2 = 0$ and $X_2 = 1$ of a line that is tangent to the $(G-X)$ curve for the α phase at any composition X_2 in its dilute range. The equation of the tangent line to the $(G–X)$ curve for the α phase may be obtained by recalling from analytical geometry the equation of a line that passes through two given points, (x_1, y_1) and (x_2, y_2):

$$(y - y_1) = \frac{y_2 - y_1}{x_2 - x_1}(x - x_1)$$

For the tangent line with intercepts given by Equation 10.51 and Equation 10.53, $x_1 = 0$, $x_2 = 1$ and y_1 and y_2 are given by these equations. Substitute these values into the equation for the straight line and rearrange:

$$y = \lfloor a_0 + RT \ln X_2 - RT \ln(1 - X_2) \rfloor x + RT \ln(1 - X_2)$$

Since the solution is assumed to be dilute, $X_2 \ll 1$, $(1 - X_2) \cong 1$ and $\ln(1 - X_2) \cong \ln(1) = 0$. Thus,

$$y = \lfloor a_0 + RT \ln X_2 \rfloor x \tag{10.54}$$

This is the equation of a line that is tangent to the $(G–X)$ curve for the α phase in its dilute range, touching the curve at any composition X_2.

The value of $X_2 = X_2{}^\alpha$ that describes the phase boundary on the α side of the $(\alpha + \varepsilon)$ field in Figure 10.20 lies on the particular tangent that also passes through the point P, which corresponds to $y = G^\varepsilon$ at $x = X_c$, the mole fraction of component 2 in the ε phase. Substitute these values for y and x in Equation 10.54:

$$G^\varepsilon = \lfloor a_0 + RT \ln X_2^\alpha \rfloor X^\varepsilon$$

Solve for the composition of the phase boundary:

$$\ln X_2^\alpha = \frac{1}{RT}\left(\frac{G^\varepsilon}{X^\varepsilon} - a_0\right) \tag{10.55}$$

This is the equation for the phase boundary on the α side of the $(\alpha + \varepsilon)$ field, i.e., the solvus line for the terminal phase α.

Apply the definitional relationship to the free energy of formation of the compound:

$$\Delta G_f^\varepsilon = \Delta H_f^\varepsilon - T\Delta S_f^\varepsilon$$

where ΔH_f^ε is the heat of formation of the compound and ΔS_f^ε is the entropy of formation of 1 mol of the compound. The corresponding values per gram atom of ε formed,

$$G^\varepsilon = H^\varepsilon - TS^\varepsilon$$

are obtained by dividing ΔH_f and ΔS_f by the number of gram atoms in 1 mol of the compound. Then, the equation for the phase boundary may be written:

$$\ln X_2^\alpha = \frac{1}{RT}\left(\frac{H^\varepsilon - TS^\varepsilon}{X^\varepsilon} - a_0\right) = \left(\frac{H^\varepsilon - a_0 X^\varepsilon}{X^\varepsilon R}\right)\frac{1}{T} - \frac{S^\varepsilon}{X^\varepsilon R}$$

where H^ε and S^ε are values per gram atom of compound formed. The solubility limit in the α phase thus varies with temperature according to the relation

$$X_2^\alpha = A e^{(B/RT)} \tag{10.56}$$

where

$$A = e^{-(S^\varepsilon/X^\varepsilon R)} \qquad \text{and} \qquad B = \frac{H^\varepsilon - a_0 X^\varepsilon}{X^\varepsilon}$$

Heats of formation for most compounds are negative; the associated reaction is exothermic. Thus B is negative in Equation 10.56 and the solubility, X_2^α, increases with temperature. Further, the more stable is the compound, i.e., the more negative is its free energy of formation, the smaller will be the solubility limit in the terminal α phase.

As the temperature is raised an intermediate compound may either melt congruently, as does the ε phase in Figure 10.21, or may decompose into two other phases, as in Figure 10.20. The condition that defines a congruent melting point is depicted in Figure 10.26. With increasing temperature the free energy of formation of the compound becomes progressively less negative; simultaneously, the liquid $(G{-}X)$ curve descends in the diagram as the liquid phase increases its stability. At the congruent melting point, Figure 10.26b, the free energy of formation of the compound is equal to the free energy of mixing of the liquid phase:

$$\Delta G_{\text{mix}}^{\text{L}} = G^\varepsilon = H^\varepsilon - T_{\text{mc}} S^\varepsilon$$

$$\Delta G_{\text{mix}}{}^{\text{Lxs}} + RT_{\text{mc}}[(1 - X^\varepsilon)\ln(1 - X^\varepsilon) + X^\varepsilon \ln X^\varepsilon] = H^\varepsilon - T_{\text{mc}} S^\varepsilon$$

This assumes that the reference state for the evaluation of the heat and entropy of formation of the compound are the pure liquid phases for the components. If these

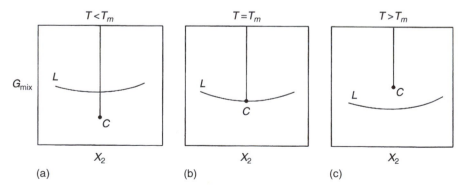

FIGURE 10.26 Thermodynamic condition for congruent melting of an intermediate line compound.

quantities are referred to the pure solid states, then additional terms must be added to bring the reference states for the components into agreement as presented in Section 10.1.1. Assume that the excess free energy of mixing is not a function of temperature. Solve for the melting point:

$$T_{mc} = \frac{H^{\varepsilon} - \Delta G_{mix}^{Lxs}}{R[(1 - X^{\varepsilon})\ln(1 - X^{\varepsilon}) + X^{\varepsilon}\ln X^{\varepsilon}] + S^{\varepsilon}} \qquad (10.57)$$

The excess free energy in the liquid phase may be positive or negative; the remaining terms in this equation are negative. One conclusion readily obtained from this result is: the larger (more negative) is the heat of formation of the compound, the higher will be its congruent melting point.

Figure 10.27 shows a phase diagram for a system in which the terminal solid solutions are dilute and all of the intermediate phases appear as line compounds, along with a sketch of a $(G-X)$ diagram constructed at T_1. The entire diagram may

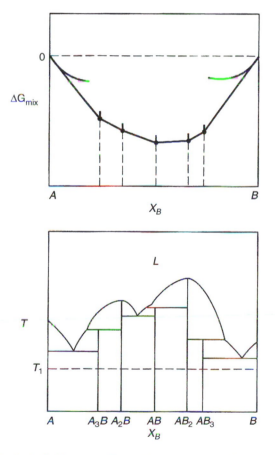

FIGURE 10.27 Typical $G-X$ curve for a binary system with very limited terminal solubilities and a number of intermediate line compounds.

be generated from a model for the liquid phase and enthalpies and entropies of the compounds, so that ΔG_f may be estimated as a function of temperature for each phase. This kind of phase diagram is typical of systems treated in chemistry where the participating phases are essentially stoichiometric in composition.

10.3 THERMODYNAMIC MODELS FOR THREE COMPONENT SYSTEMS

The strategies developed for modeling binary phase diagrams extend directly to the description of three component systems. Four variables (T, P, X_2 and X_3, for example) are required to describe the state of a single phase. Thus the $B–X$ curve that represents the mixing behavior of binary systems at some chosen temperature and pressure expands to a $B–X_2–X_3$ surface for a phase with three components, Figure 10.28. The typical $B–X–X$ surface is a distorted parabaloid hanging over the Gibbs triangle from three points on the axes representing the free energies of the pure components relative to their chose reference states. Curves of intersection with

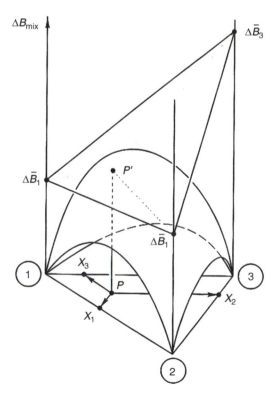

FIGURE 10.28 Mixing behavior of ternary solutions is represented by a surface over the Gibbs triangle. Tangent plane-intercepts construction yields the partial molal properties of the three components corresponding to any composition P in the three-component system, an extension of the construction presented for binary systems in Figure 8.1.

the sides of the prism in Figure 10.28 are $B-X$ curves for the corresponding binary systems.

The tangent line-intercept construction shown in Figure 8.1 gives a graphical visualization of the partial molal properties for the components in a binary system at any composition. An analogous construction applies to surface plots of ΔB_{mix} vs. composition in three component systems, Figure 10.28. To find the partial molal properties of each of the components for any composition P construct the tangent plane to the surface at that composition, P'. The intercepts of that tangent plane on the three axes are the corresponding partial molal properties.

The competition between two different phase forms for stability can be represented by superimposing the two $G-X-X$ surfaces characteristic of these phases with the reference states for the three components carefully chosen to be the same in both phase forms, Figure 10.29. The conditions for equilibrium between two phases in a three-component system require that

$$\mu_1^\alpha = \mu_1^\beta; \qquad \mu_2^\alpha = \mu_2^\beta; \qquad \mu_3^\alpha = \mu_3^\beta \qquad (10.58)$$

If the reference states are consistently chosen, the conditions may be restated

$$\Delta\mu_1^\alpha = \Delta\mu_1^\beta; \qquad \Delta\mu_2^\alpha = \Delta\mu_2^\beta; \qquad \Delta\mu_3^\alpha = \Delta\mu_3^\beta \qquad (10.59)$$

Thus if P^α represents a point on the α surface and P^β one on the β surface, this pair of points will represent a pair of compositions that may exist in equilibrium if and only if the tangent plane at P^α and that at P^β are the same plane. If this is true for the pair of points, then the intercepts on the three axes will be the same for P^α as for P^β and the conditions for two-phase equilibrium will be satisfied. Thus the common tangent construction devised for binary systems extends to ternary two-phase equilibria; however, the tangent becomes a tangent plane that touches both surfaces. The pair of points P^α and P^β are ends of a tie line on the Gibbs triangle. As the common tangent plane rolls over the two surfaces always maintaining double contact, it identifies a continuous series of pairs of tangent points which project to form the two-phase field on the composition plane.

Because the constructions involved are three-dimensional, this graphical visualization of the competition between phase forms for stability in ternary systems is not as useful as is the $G-X$ construction for binary systems. Nonetheless, the construction is available and may provide a useful thinking tool.

Equation 10.58 forms the basis for calculating two-phase fields in ternary systems from solution models for the phases involved. Since, for any component in a multicomponent system,

$$\mu_k^I = G_k^{0I} + \Delta\bar{G}_k^{xsI} + RT \ln X_k^I \qquad (I = \alpha, \beta; \qquad k = 1, 2, 3) \qquad (10.60)$$

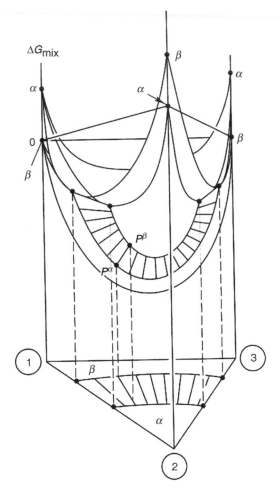

FIGURE 10.29 The common tangent plane construction describes a pair of points, P^α and P^β, on two different $G-X-X$ surfaces (α phase and β phase) that satisfy the conditions for two-phase equilibrium in a three-component system. As the common tangent plane rolls over the surfaces this set of point pairs project to the Gibbs triangle to generate the two phase $\alpha + \beta$ field.

The conditions for equilibrium yield three equations of the form:

$$\Delta G_k^{\alpha \to \beta} + [\Delta \bar{G}_k^{xs\beta} - \Delta \bar{G}_k^{xs\alpha}] + RT \ln \left[\frac{X_k^\beta}{X_k^\alpha} \right] = 0 \qquad (k = 1, 2, 3) \qquad (10.61)$$

The solution model may be used to evaluate the excess free energies in terms of composition and temperature. The free energy change for the phase transformation for the pure components may be computed for each of the unary systems as in Chapter 7. This system of three equations relates four variables, $X_2^\alpha, X_3^\alpha, X_2^\beta$ and X_3^β;

there thus remains one degree of freedom. In order to compute the set of composition pairs that delineate the $\alpha + \beta$ equilibria, it is necessary to choose a value for one of these four variables, say X_2^α. Then compute values for the other three corresponding to the chosen value of X_2^α. An iterative equation solver routine available in standard mathematical applications software may be used for this purpose. The value of X_2^α is then incremented and the calculation repeated until the full collection of tie lines that constitute the two-phase field is generated.

The principle of minimization of the Gibbs free energy to determine the domains of stability of the phases also applies in ternary systems. The taut string construction is replaced by a taut elastic sheet construction. The free energy surfaces are considered to be rigid surfaces hanging from appropriate fixed points on the unary axes. An elastic sheet is attached at the three corners of the diagram and pulled taut against the free energy surfaces. Where it comes into contact with the surface of a particular phase, that phase has the minimum free energy and is stable. Between the surfaces, the elastic sheet traces out the common tangent plane construction that represents two-phase equilibria. Where the sheet is stretched between three free energy surfaces, a three-phase tie triangle represents the condition for equilibrium. This visualization is also useful in systems with line compounds so that the one-phase regions essentially are limited to points on the Gibbs triangle, Figure 10.30.

10.4 CALCULATION OF PHASE DIAGRAMS IN POTENTIAL SPACE

If the coordinates used to represent the states of the phases in a system are chosen from the thermodynamic potentials, then the resulting phase diagram is a simple cell structure.

The counterpart to the common binary diagram plotted in (X_2, T) space is the isobaric diagram in (a_2, T) space. The thermodynamic database information necessary to compute a binary diagram in potential space is identical with that required for the more familiar (X_2, T) diagrams modeled in Section 10.2. Since these calculations require models for the relation between chemical potential and composition, corresponding relations between composition and activity are implicit. No new information is required to compute an (a_2, T) diagram.

In (a_2, T) space, every two-phase field is represented by a single curve that sets the limits of stability for the phases it separates. The quantitative form of that curve depends upon the choice of the reference state used in defining the activity of component 2. Figure 10.31 illustrates this point for a simple ideal solution phase diagram model. Figure 10.31a shows the diagram in (X_2, T) space.

Consider first the case in which the α phase form is chosen as the reference state for component 2. Then, in the α phase solid solution region, according to the ideal solution model, $a_2^\alpha = X_2^\alpha$. The identical simple relation will not hold for the L phase because, in order for a_2 to be equal to X_2 in an ideal solution, the reference state for component 2 must be the same phase form as the solution, i.e., L in this case. The equation describing the α boundary of the $(\alpha + L)$ field has the form $X_2^\alpha = X_2^\alpha(T^\alpha)$ given by Equation 10.21. Since for this choice of reference state $a_2^\alpha = X_2^\alpha$, the same

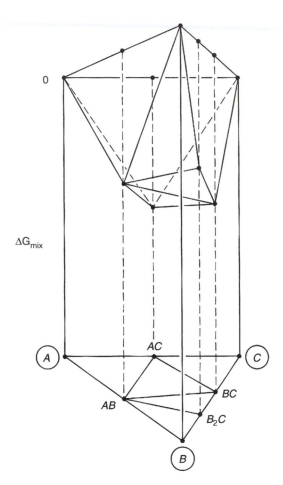

FIGURE 10.30 A taut elastic sheet construction gives the domains of stability of one-, two- and three-phase combinations illustrated here for a system that has line compounds at stoichiometric compositions AB, AC, BC and B_2C.

equation describes the condition for $(\alpha + L)$ equilibrium in the (a_2, T) phase diagram shown in Figure 10.31b. Thus the curve representing the $(\alpha + L)$ equilibrium for this choice of reference state for component 2 for this ideal solution model coincides with the curve on the α side of the $(\alpha + L)$ field in Figure 10.31a.

The same argument demonstrates that if the L phase form is chosen as reference state for component 2 at all temperatures, the two-phase field in the (a_2, T) diagram, Figure 10.31c, is identical with the curve on the L side of the $(\alpha + L)$ field in Figure 10.31a.

These simple identities between an appropriate phase boundary on the (X_2, T) diagram and the two-phase field on the (a_2, T) diagram exist in this case because the system is composed of ideal solutions for which, with appropriate choice of

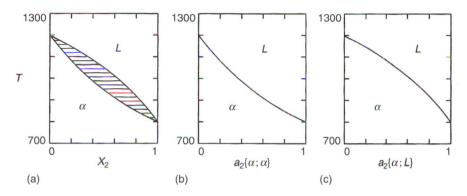

FIGURE 10.31 Ideal solution model of a two-phase (α + L) field in a binary phase diagram (a) in (X_2, T) space and (b), (c) in (a_2, T) space. Reference states for a_2 are the α phase form in (b) and the L phase form in (c).

reference state, the activity is equal to the atom fraction. More generally, a_2 and X_2 are functionally related within each phase but are not simply equal to each other. Perhaps the most straightforward strategy for obtaining a two-phase field curve in an (a_2, T) diagram is to compute the composition-temperature diagram as outlined in Section 10.2. Then use the model relationship between activity and composition to convert one of the $X_2(T)$ curves to the corresponding $a_2(T)$.

In constructing phase diagrams in potential space it is important to be explicit about the choice of the reference state for the component represented on the activity axis and to maintain that choice in modeling the mixing behavior of all of the phases on the diagram.

Since three-phase equilibria are formed by the intersection of three two-phase equilibria, in (a_2, T) space they appear as the point of intersection of the three curves representing the two-phase equilibria involved. This is further evident from the observation that when three phases coexist, one of the conditions for equilibrium requires that the chemical potential of component 2 be the same in all three phases. Since the reference states are consistently chosen to be the same for all of the phases, this condition implies that the activity of component 2 will be the same in all three phases at the triple point. Thus three-phase equilibria appear as triple points where the three related two-phase equilibrium curves intersect.

10.5   COMPUTER CALCULATIONS OF PHASE DIAGRAMS

The thermodynamic underpinnings of phase diagrams developed in this chapter and in Chapter 9 provide the basis for comprehensive computer calculations of binary, ternary and higher-order phase diagrams. This subject is so central to the field of materials science that a monthly journal, *CALPHAD*, focused on computer calculations of phase diagrams, has been published since the late 1970s. A variety of compilations of phase diagrams[2-8] exist for metallic and ceramic systems. Early versions of these compilations were based entirely on direct observation of the phases that exist at equilibrium in samples of the material prepared and equilibrated at a sequence of temperatures and compositions that blanket the $(T - X_k)$ space. More recent compilations are based on CALPHAD-derived computer calculations that combine database information from all sources: direct observation of equilibrated structures, thermodynamic measurements of phase stabilities, thermo-dynamic information about solution behavior, and so on, to arrive at an optimized phase diagram. These programs have the capability to handle multicomponent systems, which are of central interest in most industrial applications. Components of these programs include:

1. Comprehensive and expanding databases, compiled from the world literature, which are used to compute the Gibbs free energy of the phases in a system as a function of temperature and composition.
2. Optimizer programs that are used to assess the literature information, weighting results from different methods and different sources to yield a statistically best set of data, and also to optimize the phase diagram produced at the end of the process.
3. A computation engine that finds the minimum in the Gibbs free energy function over the composition domain at each temperature thus identifying the domains of stability of one, two, three and more phases.
4. An engine that generates graphical displays of the resulting equilibrium maps, as well as a wide variety of related information.

The output is a diagram for any given system that provides a best fit with all the available data. As experience and assessments accumulate, this information is being incorporated into comprehensive and assessed databases that strive to develop consistent descriptions for the thermodynamics of the elements, solid and liquid solutions, and intermediate compounds. For example, as discussed in Section 7.1.4, the thermodynamic information for pure FCC chromium that is part of the input in the calculation of the iron–chromium phase diagram must be the same information that provides part of the input to compute the gold–chromium diagram or the copper–chromium diagram, etc. (Gold and copper are also FCC structures.) Similarly, the same assessed model that produces the binary copper–nickel phase diagram and is thus a part of the input to compute the Cu–Ni–Au ternary diagram must also form the basis for computing the Cu–Ni–Fe system.

An early comprehensive study of phase diagrams that applied computer calculations based upon thermodynamic models was carried out by Kaufmann[9] in the late 1960s. His text provides a comprehensive introduction to the subject, carrying the reader systematically through unary systems, binary systems of increasing complexity, including line compounds, and ultimately to ternary systems. Equation 10.26 was assumed to describe the behavior of the pure components. The single parameter regular solution model described mixing behavior so that Equation 10.38 and Equation 10.39 define the phase boundaries for each two-phase field. An algorithm based upon atom sizes and electronegativities was devised to estimate values for the mixing parameters for each type of solution. Typical results of this study are displayed in Figure 10.32. A pair of elements that are adjacent to each other in the periodic table produce a simple diagram that may be reasonably represented in this model. As the elements become more separated in the periodic table, the diagram increases in complexity as intermediate phases appear and the description with this simple model becomes unsatisfactory. The comprehensive and sophisticated CALPHAD approach grew from these simple models.

The Gibbs free energy function lies at the heart of CALPHAD software. By general agreement, the reference state for each element is taken to be the pure element at one atmosphere pressure, 298 K, and in the phase form that is stable for that element at these conditions. Thus Gibbs free energy values in every system in any phase form at any temperature, pressure and composition must be calculated from this choice of reference states for the participating elements. The generic equation is

$$G_m^p = G^{0p} + G^{p\ xs} + G^{p\ id} \tag{10.62}$$

Here

$$G^{0p} = \sum_{j=1}^{e} X_k^{\,p} \Delta G_j^{0\alpha \to p}(T) \tag{10.63}$$

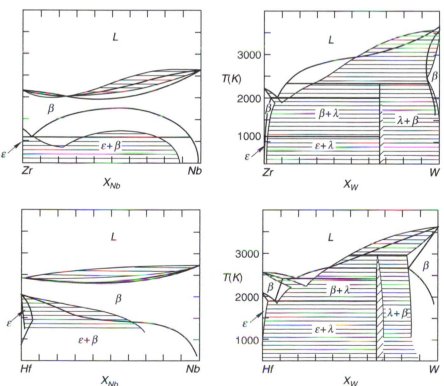

FIGURE 10.32 Representative examples of phase diagrams computed with Kaufman's strategies, comparing experimentally determined (lightly sketched, with filled two-phase fields) and calculated (heavily drawn lines) diagrams.

where X_k^p is the atom fraction of component k in the solution p and $\Delta G_j^{0\alpha \rightarrow p}(T)$ is the Gibbs free energy of element j in the phase form p at T relative to the reference form α at 298 K (the stable phase form for j at 298 K). Procedures for computing these phase stabilities of the elements were reviewed in Section 7.1.4. The second term is the excess Gibbs free energy of forming the p phase solution from the pure elements in the p phase at the temperature T:

$$G_m^{p\ xs} = \sum_{k=1}^{c} X_k^p \overline{\Delta G_k^{p\ xs}} \tag{10.64}$$

Computation of the excess Gibbs energy of mixing is the goal of all of the solution models reviewed in Section 8.7. The third term is the ideal component of the Gibbs free energy of mixing, given by

$$G^{p\ id} = RT \sum_{k=1}^{c} X_k \ln X_k \tag{10.65}$$

An equivalent formulation in terms of site fractions is used where the description for the phase p requires a sublattice model.

In addition to producing phase diagrams, modules in these programs can yield other kinds of equilibrium maps (Figure 1.4) including predominance diagrams (Figure 1.7 and Chapter 12) and Pourbaix electrochemical diagrams (Figure 1.9 and Chapter 15).

Besides furnishing a comprehensive methodology for generating and evaluating equilibrium maps, successful CALPHAD models provide by extrapolation thermodynamic information about the phases in the system outside their domains of stability. This kind of information is crucial to the development of kinetic descriptions of behavior in processing materials. The Gibbs free energy of a metastable or unstable phase relative to the value for the stable state of that system provides a measure of how far the system is from the equilibrium state. Theories of rates at which processes occur invariably contain a factor that makes use of some form of this free energy difference, universally (though without precision) referred to as the "driving force" for the process. Prediction of the rates at which microstructures evolve is crucial to the design of materials processes in industry.

10.6 SUMMARY

Free energy–composition $(G-X)$ diagrams provide a graphical basis for visualizing the connection between the thermodynamics of the phases in a system and the phase diagram that it exhibits.

The common tangent construction is a graphical representation of the conditions for equilibrium in a two-phase system, expressed either as equality of the chemical potentials or as a minimization of the Gibbs free energy function.

Relative Gibbs free energies of the pure components combine with mixing behavior of the solutions to determine the location and shape of two-phase fields.

Two-phase fields identify the boundaries of single-phase regions and intersect to form three-phase fields; determination of the two-phase fields in a system is sufficient to determine the phase diagram.

At sufficiently low temperatures, any phase that exhibits a positive excess free energy will exhibit a miscibility gap.

The spinodal region within a miscibility gap, between the inflection points in the $(G-X)$ curve, is a region in which nonuniform systems are more stable than uniform ones. This attribute provides the basis for spinodal decomposition.

The incorporation of line compounds as intermediate phases may be adequately described by their free energy of formation (per gram atom of compound) as a function of temperature.

The ideal solution model can only generate monotonic two-phase fields; the pattern of shapes is very limited and determined by the entropies of transformation of the pure components.

Even the simplest, one parameter regular solution model is capable of producing the full variety of phase diagram configurations. Additional terms, including a temperature dependence in the excess free energy, may be added to provide quantitative agreement with measured diagrams.

Extensive computer programs incorporating databases, a variety of solution models, statistical bases for model parameter selection and free energy minimization algorithms have been devised to compute binary, ternary and even higher-order phase diagrams.

The establishment of firm connections between the thermodynamics of solutions and phase diagrams, implemented by computer calculations, yields information about metastable phases and equilibria that form the basis for the analysis, and eventually the prediction, of the processes in materials science that control microstructure and properties.

HOMEWORK PROBLEMS

Problem 10.1. Replot the $G-X$ curves shown in Figure 10.2 for the following choices of reference states for components 1 and 2:
   a. $\{L; L\}$.
   b. $\{L; \alpha\}$.
   c. $\{\alpha; \alpha\}$.
Curves are computed with the following information: solution models are simple regular solutions. $T_{m1} = 1500$ K; $T_{m2} = 850$ K; $\Delta S_{m1} = 9(\text{J}/\text{mol K})$; $\Delta S_{m2} = 7(\text{J}/\text{mol K})$; $a_0^\alpha = 8400(\text{J}/\text{mol})$; $a_0^L = 10,500$ (J/mol).

Problem 10.2. The A–B system forms regular solutions in both the β and liquid phase forms. Parameters for the heat of mixing for these phases are $a_0^\alpha = -8200$ J/mol and $a_0^L = -10,500$.

    a. Compute and plot curves for ΔG_{mix} for the β and liquid phases for these models at 800 K.Pure A melts at 1050 K with a heat of fusion of 8200 J/mol. Pure B melts at 660 K with a heat of fusion of 6800 J/mol.

    b. Change the reference states appropriately to plot and compare $\Delta G_{mix}^{\beta}\{\beta; L\}$ and $\Delta G_{mix}^{L}\{\beta; L\}$.

Neglect differences in heat capacities for the phases for the pure components.

Problem 10.3.   The free energy of mixing (J/mol) of a liquid solution is modeled by

$$\Delta G_{mix}\{L; L\} = 8400 X_A X_B + RT[X_A \ln X_A + X_B \ln X_B]$$

    a. Compute and plot a_B as a function of composition at 600 K for this solution.

    b. The free energy of fusion of pure B at 600 K is computed to be -1200 J/mol. Compute and plot the activity of B as a function of composition for the choice of reference states $\{L; \alpha\}$.

Problem 10.4.   Use the taut string construction given in Figure 10.7 to sketch a plot of the chemical potential of component 2 as a function of composition for this system.

Problem 10.5.   Consider the (G/X) curve plotted in Figure 10.5. A mechanical mixture is formed by taking quantities of two solutions with compositions given by the points A and B in this figure so as to give a system with an average composition X_2^0. Prove that the Gibbs free energy of mixing for this two-solution mixture is given by the point C.

Problem 10.6.   It has been argued that, if the (G/X) curves for two different phases cross within the diagram, then it is always possible to construct a common tangent line that straddles the crossing point. What aspect of the behavior of (G/X) curves guarantees that both tangent points will in fact lie within the composition range, $0 < X_2 < 1$?

Problem 10.7.   Prove that a change in the reference states of either or both of the (G/X) curves producing a common tangent construction at some temperature will not alter the compositions at which the tangents occur.

Problem 10.8.   Sketch a sequence of (G/X) curves that would produce a maximum in an $(\alpha + L)$ two-phase field.

Problem 10.9.   Consider an α solid solution that obeys the "simplest regular solution" model with the interaction parameter $a_0 = 12,700$ (J/mol). This solution exhibits a miscibility gap below $T_c = 746\ K$. At 650 K,

    a. Compute the compositions of the limits of the spinodal region.

    b. Compute and plot $\Delta \mu_B^{\alpha} = \mu_B^{\alpha} - \mu_B^{0\alpha}$ as a function of composition.

   c. Illustrate the peculiar behavior inside the spinodal region by computing
      and plotting

$$\frac{d\Delta\mu_B^\alpha}{dX_2}$$

   as a function of composition.

Problem 10.10. Sketch a series of (G/X) curves that will produce a peritectic phase diagram in which both solid phases are the same phase form, α, so that the solid–solid two-phase field is a miscibility gap.

Problem 10.11. Look up the phase diagram for the BaO_2–TiO_2 system. Sketch a plausible (G/X) diagram for this system at 1400°C.

Problem 10.12. In Figure 10.27, one of the intermediate compounds forms at the composition AB_2. Sketch the phase diagram in the vicinity of this composition, expanding the composition axis by a factor of, say, 1000, so that the departures from stoichiometry may be displayed.

Problem 10.13. Refer again to the phase diagram for the BaO_2–TiO_2 system obtained for Problem 10.11. Suppose that under practical solidification conditions in this system, the $BaTi_3O_7$ phase does not form.
   a. Estimate the composition and temperature of the last liquid to solidify in
      this system if this phase does not form.
   b. Suppose also that the $BaTi_2O_5$ phase is also sluggish to form; then what is
      the temperature and composition of the last liquid to solidify?

Problem 10.14. Use a mathematics application software package to compute and plot a simple ideal solution phase diagram. Write the program so that:
   a. Input is T_1^0, T_2^0, ΔS_1° and ΔS_2°.
   b. Output is a plot of the two phase field.

Fix the melting points. Try several combinations of values of ΔS_1° and ΔS_2°, and explore the kinds of diagrams that may be developed. Values of these parameters range around 1 (J/mol K) for solid–solid transformations, 8 for melting, and 90 for vaporization.

Problem 10.15. Use the computer program developed in Problem 10.14 as a basis to compute and plot a phase diagram for a system involving three phases, α, β and L. Identify the stable and metastable portions of this diagram.

Problem 10.16. The system A–B obeys the simple regular solution model. A melts at 1352 K with an entropy of fusion of 6.9 (J/mol K); B melts at 1148 K with an entropy of fusion of 8.5 (J/mol K). Neglect differences in heat capacities of the pure solid and liquid phases. For the liquid phase, $a_0^l = -5600$ (J/mol); for the α phase $a_0^\alpha = -11,400$ (J/mol). Find the compositions of the boundaries of the two phase $(\alpha + L)$ field at 1300 K. (Note: this will require solution of two simultaneous nonlinear equations; standard mathematics applications packages have this feature.)

Problem 10.17. Compute and plot the phase boundaries and the spinodal boundaries for a solution that obeys the model

$$\Delta G_{mix}^\alpha = 10,600X_1X_2 + RT[X_1\ln X_1 + X_2\ln X_2]$$

Problem 10.18. Find the critical temperature for the miscibilty gap that will be found in a regular solution with

$$\Delta H_{mix} = X_1X_2[4850X_1 + 12,300X_2]$$

Problem 10.19. Compute and plot the midrib curves for the $(\alpha + L)$ fields for two systems with the following properties:

a.	T_{0k} (K)	ΔS_k° (J/mol K)
Component 1	1283	8.8
Component 2	942	6.3

$$a_0^\alpha = 7280(J/mol); \qquad a_0^L = -2100(J/mol)$$

b.	T_{0k} (K)	ΔS_k° (J/mol K)
Component 1	1283	8.8
Component 2	942	6.3

$$a_0^\alpha = -4800 \ (J/mol); \qquad a_0^L = 5200(J/mol)$$

Problem 10.20. At 1550 K the solubility of oxygen in zirconium is estimated to be 120 ppm. The Zr–O system forms a very stable compound, ZrO_2, an important ceramic refractory. No other oxides are stable at 1550 K. Estimate the Henry's law coefficient for oxygen in zirconium at 1550 K.

Problem 10.21. Assuming that the liquid solution of oxygen in silicon is an ideal solution, estimate the melting point of quartz, a crystal form of silica (SiO_2). The free energy of formation of quartz is:

$$\Delta G_f^0 = -910,900 + 182.33T \text{ J/mol}$$

Note carefully that the reference states in this relation are crystal silicon and gaseous oxygen. Compare your estimate with the observed melting point of quartz.

Problem 10.22. Consider the potential–composition diagram for the Fe–Cr–O system shown in Figure 9.16. Replot this diagram on the Gibbs triangle. Plot the tie lines in the two-phase fields quantitatively.

REFERENCES

1. Lupis, C.H.P., *Chemical Thermodynamics of Materials*, Elsevier Science Publishing Co., Inc., New York, pp. 219–229, 1983.
2. Hansen, M., *Constitution of Binary Alloys*, 2nd ed., McGraw-Hill Book Co., New York, 1958.
3. Elliott, R.P., *Constitution of Binary Alloys, First Supplement*, McGraw-Hill Book Co., New York, 1965.
4. Shunk, F.A., *Constitution of Binary Alloys, Second Supplement*, McGraw-Hill Book Co., New York, 1969.
5. Hultgren, R., Desai, P.D., Hawkins, D.T., Gleiser, M., and Kelley, K.K., *Selected Values of the Thermodynamic Properties of Binary Alloys*, ASM, Materials Park, OH, 1973.
6. Levin, E.M., Robbins, C.R., and McMurdie, H.F., *Phase Diagrams for Ceramists*, American Ceramic Society, Columbus, OH, 1964.
7. Levin, E.M., Robbins, C.R., and McMurdie, H.F., *Phase Diagrams for Ceramists, 1969 Supplement*, American Ceramic Society, Columbus, OH, 1969.
8. *Bulletin of Alloy Phase Diagrams*, ASM, Materials Park, OH, Journal published bimonthly since 1980 in collaboration with the National Institute for Standards and Technology.
9. Kaufman, L. and Bernstein, H., *Computer Calculations of Phase Diagrams*, Academic Press, New York, 1970.

11 Multicomponent Multiphase Reacting Systems

CONTENTS

At the heart of the idea of a chemical reaction is the visualization of the chemical molecule. A molecule is an arrangement of atoms of some of the elements in the system into a specific geometric and energetic configuration. This configuration is so specific that it can be represented by a chemical formula: CO_2, H_2O, H_2, CH_4, Al_2O_3, HNO_3; which describes succinctly the number of atoms of each element in the molecule.

In a system consisting of a mixture of molecular types the atoms may spontaneously redistribute themselves among the various molecules that are present. In a closed system this rearrangement necessarily occurs without changing the total number of atoms of each element in the system. Such a rearrangement is commonly called a chemical reaction; a system that is capable of this kind of change is classified as a reacting system. The notion that such a process is a rearrangement of the atoms present among the molecular types in the system, requiring conservation of the number of atoms of each element, is succinctly expressed by statements that take the form:

$$C + O_2 = CO_2$$

or

$$2H_2 + O_2 = 2H_2O$$

These expressions, which are usually also referred to as chemical reactions, are in essence statements of the conservation of the atoms of the elements in such systems. Equivalent statements may be obtained by dividing all of the coefficients by a constant value, e.g.,

$$H_2 + \frac{1}{2}O_2 = H_2O$$

is an equally valid statement of the conservation of hydrogen and oxygen atoms in a system that contains these three components.

If a system consists of e elements and c components, some of which are molecules, then the number r of independent conservation statements or reactions that may be written for the system is

$$r = c - e \qquad (11.1)$$

Thus a system that contains the elements C and O ($e = 2$) and is made up of the molecular species O_2, CO and CO_2 ($c = 3$) exhibits a single independent chemical reaction:

$$2CO + O_2 = 2CO_2$$

Such a system is termed a univariant reacting system.

If the system also contains elemental carbon as a component, C, so that $c = 4$: C, O_2, CO and CO_2, then the number of independent chemical reactions is $r = 4 - 2 = 2$:[1]

$$C + O_2 = CO_2 \qquad [11.1]$$

$$2C + O_2 = 2CO \qquad [11.2]$$

Other reactions may be written in this system; e.g.,

$$C + CO_2 = 2CO \qquad [11.3]$$

but this statement is not independent of the first two expressions of elemental conservation. If statements [11.1] and [11.2] are true for the system, then [11.3] must also be true because it is a linear combination of the first two conservation

---

[1] Equation numbers representing chemical reactions are enclosed in brackets, [ ], to distinguish them from numbers for mathematical equations, which are enclosed in parentheses, ( ).

statements: $[11.3] = [11.2] - [11.1]$. To demonstrate this relationship, reverse the statement of reaction $[11.1]$ and add it to reaction $[11.2]$:

$$CO_2 = C + O_2 \qquad\qquad [11.1a]$$

$$2C + O_2 = 2CO \qquad\qquad [11.2]$$

Combine these statements:

$$CO_2 + 2C + O_2 = C + O_2 = 2CO$$

eliminate redundant terms (O_2 appears on both sides, as does C) to obtain statement $[11.3]$:

$$CO_2 + C = 2CO \qquad\qquad [11.3]$$

Thus this system is a bivariant reacting system. Systems for which $r = (c - e) > 1$ are multivariant reacting systems.

The thermodynamic apparatus necessary to handle multivariant reacting systems is the subject of this chapter. The treatment first focuses upon reactions in the gas phase where the components unambiguously exist in molecular form. The general strategy for finding conditions for equilibrium is applied first to a univariant system to identify clearly how a reacting system differs from a nonreacting system in this context. The derivation leads to the familiar law of mass action and the definition of the equilibrium constant for the reaction. Treatment of a bivariant system shows how these results generalize when more than one independent reaction may occur. Further generalization from the bivariant to the multivariant case is then straightforward. These results can be conveniently summarized in an equilibrium map referred to in Figure 1.4 as a gas phase equilibrium map, such as that shown in Figure 1.6.

The treatment of multiphase systems parallels that developed for describing reactions in the gas phase. In solids and liquids, even in line compounds, which are practically stoichiometric, the chemical concept of the molecule loses the clear meaning that it has for the gas phase. Crystals that are composed of ordered arrays of molecular units such as O_2 or CO_2 are relatively rare. Most solid compounds are crystals with components of ionic, covalent and sometimes metallic bonding. Molecular formulas such as SiO_2, Al_2O_3, NaCl and FeO arise from the nature of the bonding and the geometry of the crystal structure. All such compounds exhibit defects, which support deviations from the strict molecular ratio that characterizes a molecule (see Chapter 13). Rigorous treatment of such systems may be best formulated in terms of their phase diagrams, as developed in Chapter 10, which allow incorporation of such deviations in composition. However, the treatment of such systems can be greatly simplified if the fiction of a molecular formula is assumed. This simplification is important because it makes analysis of such complex systems more practical. These points are developed in detail in Section 11.2.

Treatment of multicomponent multiphase systems as reacting systems leads to a representation of the competition between the components in the form of predominance diagrams first referred to in Figure 1.4 and illustrated in Figure 1.7. Strategies for developing such diagrams and their relation to the phase diagram for the same system are presented in Section 11.3.

11.1 REACTIONS IN THE GAS PHASE

Application of the general strategy for finding conditions for equilibrium to a reacting system yields new relationships among the chemical potentials, which are superimposed upon the set of conditions for equilibrium already derived for multicomponent, multiphase nonreacting systems. These relationships derive directly from the conservation of elements, which accompanies chemical reactions and is succinctly represented in the statement of chemical equations. The simplest illustration of this connection is embodied in the description of a univariant reacting system in which all of the components are gases; this class of system is treated first.

11.1.1 UNIVARIANT REACTIONS IN THE GAS PHASE

Consider a gas mixture that contains the three components O_2, CO and CO_2. Since $r = c - e = 3 - 2 = 1$, this is a univariant system. Only one chemical reaction may be written among these components:

$$2CO + O_2 = 2CO_2$$

To find the conditions for equilibrium in such a system, first write an expression for the change in entropy that it may experience. The combined statement of the first and second laws for this multicomponent single-phase system is given in Chapter 8, Equation 8.49

$$dU' = TdS' - P\,dV' + \sum_{k=1}^{c} \mu_k dn_k \qquad (11.2)$$

Rearrange this equation to express the change in entropy:

$$dS' = \frac{1}{T}dU' + \frac{P}{T}dV' - \frac{1}{T}\sum_{k=1}^{c} \mu_k dn_k$$

Written explicitly in terms of the three components in this application,

$$dS' = \frac{1}{T}dU' + \frac{P}{T}dV' - \frac{1}{T}[\mu_{CO}dn_{CO} + \mu_{O_2}dn_{O_2} + \mu_{CO_2}dn_{CO_2}] \qquad (11.3)$$

The general criterion states that the entropy is a maximum in an isolated system. If a system is isolated from its surroundings, then:

Its internal energy cannot change:

$$dU' = 0 \qquad (11.4)$$

Its volume cannot change:

$$dV' = 0 \qquad (11.5)$$

In systems considered up to now, the third isolation constraint expresses the conservation of the components in the system since, if a system is isolated, no matter crosses its boundary. This constraint had the form

$$dn_k = 0 \qquad (k = 1, 2, 3, ..., c) \qquad (11.6)$$

However, if the system is capable of chemical reactions, then this set of conditions no longer holds in an isolated system; even if matter cannot cross the boundary of the system, the number of moles of each of the components may still change. The atoms contained in the system may rearrange themselves over the molecular species in the process, decreasing the number of molecules of some components and increasing the number of others. Thus in an isolated system capable of chemical reactions,

$$dn_k \neq 0 \qquad (k = 1, 2, 3, ..., c) \qquad (11.7)$$

However, it will be true that the number of gram atoms of each of the elements in the system cannot change no matter how they rearrange themselves among the molecules; atoms cannot be created or destroyed and, in an isolated system, are not exchanged with the surroundings. Thus, the isolation constraint that applies to a reacting system may be stated:

$$dm_i = 0 \qquad (i = 1, 2, 3, ..., e) \qquad (11.8)$$

where m_i is the total number of gram atoms of element i in the system. This condition applies separately to each element in the system. This reformulation of the isolation constraint is the essential difference between reacting and nonreacting systems. The consequences of this difference will now be developed.

The system under study contains two elements, carbon (C) and oxygen (O). It is easy to compute the number of gram atoms of carbon in the system since each molecule of carbon monoxide has one carbon atom and each molecule of carbon dioxide also has one carbon atom. Thus the total number of gram atoms of carbon atoms, m_C, is

$$m_C = n_{CO_2} + n_{CO} \qquad (11.9)$$

A count of the oxygen atoms in the system gives m_O:

$$m_O = n_{CO} + 2n_{CO_2} + 2n_{O_2} \qquad (11.10)$$

The coefficient of each term on the right sides of these expressions corresponds to the number of atoms of the element in question contained in the corresponding molecular formula. No matter what reactions occur within the system, if the system is isolated, the number of carbon and oxygen atoms cannot change. The isolation constraints may be obtained by taking the differentials of Equation 11.9 and Equation 11.10 and setting the results equal to zero:

$$dm_C = 0 = dn_{CO} + dn_{CO_2} \tag{11.11}$$

$$dm_C = 0 = dn_{CO} + 2dn_{CO_2} + 2dn_{O_2} \tag{11.12}$$

Equation 11.11 implies that in an isolated system

$$dn_{CO} = -dn_{CO_2} \tag{11.13}$$

While Equation 11.12 gives

$$dn_{O_2} = -\frac{1}{2}[dn_{CO} + 2dn_{CO_2}]$$

Insert Equation 11.13 and simplify:

$$dn_{O_2} = -\frac{1}{2}dn_{CO_2} \tag{11.14}$$

These relationships are embodied in the chemical equation for the reaction which states that for every mole of CO_2 formed, 1 mol of CO and one half mole of O_2 must be consumed. Equation 11.13 and Equation 11.14 are essentially restatements of the isolation constraints, Equation 11.11 and Equation 11.12. These conservation equations demonstrate that, *although there are three components in this system, the number of moles of only one of them may be varied independently.*

The change in entropy that this system can experience when it is isolated from its surroundings is obtained by substituting the isolation constraints, Equation 11.4, Equation 11.5, Equation 11.13 and Equation 11.14 into the expression for the entropy, Equation 11.3

$$dS'_{iso} = \frac{1}{T}(0) + \frac{P}{T}(0) - \frac{1}{T}\left[\mu_{CO}(-dn_{CO_2}) + \mu_{O_2}\left(-\frac{1}{2}dn_{CO_2}\right) + \mu_{CO_2}dn_{CO_2}\right]$$

which may be simplified to:

$$dS'_{iso} = -\frac{1}{T}\left[\mu_{CO_2} - \left(\mu_{CO} + \frac{1}{2}\mu_{O_2}\right)\right]dn_{CO_2} \tag{11.15}$$

The linear combination of chemical potentials contained in brackets in this expression is defined to be the affinity for the reaction:

$$A = \left[\mu_{CO_2} - \left(\mu_{CO} + \frac{1}{2}\mu_{O_2}\right)\right] \tag{11.16}$$

This property of the system, which through the chemical potentials depends upon the temperature, pressure and composition of the system, may be thought of as "the chemical potential of the products minus the chemical potential of the reactants" for the reaction

$$CO + \frac{1}{2}O_2 = CO_2 \qquad [11.4]$$

With introduction of this notation Equation 11.15 becomes

$$dS'_{iso} = -\frac{1}{T}A\, dn_{CO_2} \qquad (11.17)$$

If the temperature, pressure and composition of the gas mixture are known, then the chemical potentials of the components and hence the affinity, may be computed. Suppose for a given state this computation gives a value for A that is negative, i.e., $A < 0$. This implies that, for the given state the chemical potential of the reactants is higher than that of the products. Examine Equation 11.17, keeping in mind that the second law requires that in an isolated system entropy can only increase. If A is negative, then, in order for dS to be positive, dn_{CO_2} must be positive. Thus, if the chemical potential of the reactants is higher than that of the products, then the only process that is possible is an increase in the number of moles of products: products form. If, on the other hand, for the given composition, calculation yields $A > 0$, implying the products have a higher chemical potential than the reactants, then in order for dS to be positive, dn_{CO_2} must be negative: products decompose.

The condition for equilibrium in this system corresponding to the maximum in its entropy is obtained by setting the coefficient of dn_{CO_2} equal to zero

$$A = \left[\mu_{CO_2} - \left(\mu_{CO} + \frac{1}{2}\mu_{O_2} \right) \right] = 0 \qquad (11.18)$$

Thus the system attains equilibrium when its composition arrives at the state for which the chemical potential of the reactants equals that of the products. For compositions on one side of this balance point $\mu_{reactants} > \mu_{products}$ and products form; if $\mu_{reactants} < \mu_{products}$, then products decompose. This relationship constitutes the additional thermodynamic apparatus needed to treat reacting systems.

In order to demonstrate the generality of this result, consider a system consisting of two elements, M and X and three components: the molecule M_aX_b, X_2 and the molecule M_rX_s. If a system composed of these three components is isolated from its surroundings, the change in entropy for this system for any internal process is:

$$dS'_{iso} = -\frac{1}{T}[\mu_{M_aX_b}dn_{M_aX_b} + \mu_{X_2}dn_{X_2} + \mu_{M_rX_s}dn_{M_rX_s}] \qquad (11.19)$$

The number of gram atoms of each of the elements may be expressed in terms of the number of moles of the components at any instant in time:

$$m_M = an_{M_aX_b} + (0)n_{X_2} + rn_{M_rX_s}$$

and

$$m_X = bn_{M_aX_b} + 2n_{X_2} + sn_{M_rX_s}$$

where the coefficients in each case are given by the corresponding formula for the molecular component. The isolation constraints yield:

$$dm_M = 0 = a\, dn_{M_aX_b} + r\, dn_{M_rX_s} \Rightarrow dn_{M_aX_b} = -\frac{r}{a} dn_{M_rX_s} \qquad (11.20)$$

and

$$dm_X = 0 = b\, dn_{M_aX_b} + 2\, dn_{X_2} + s\, dn_{M_rX_s}$$

$$\Rightarrow dn_X = -\frac{1}{2}(b\, dn_{M_aX_b} + s\, dn_{M_rX_s}) = -\frac{1}{2}\left(-\frac{br}{a} dn_{M_rX_s} + s\, dn_{M_rX_s}\right)$$

$$dn_{X_2} = -\frac{(as-br)}{2a} dn_{M_rX_s} \qquad (11.21)$$

Substitute these relationships into Equation 11.19:

$$dS'_{iso} = -\frac{1}{T}\left[\mu_{M_aX_b}\left(-\frac{r}{a} dn_{M_rX_s}\right) + \mu_{X_2}\left(-\frac{as-br}{2a} dn_{M_rX_s}\right) + \mu_{M_rX_s} dn_{M_rX_s}\right]$$

$$dS'_{iso} = -\frac{1}{T}\left[\mu_{M_rX_s} - \left(\frac{r}{a}\mu_{M_aX_b} + \frac{as-br}{2a}\mu_{X_2}\right)\right]dn_{M_rX_s} \qquad (11.22)$$

which may be written as Equation 11.17 with the affinity

$$A = \left[\mu_{M_rX_s} - \left(\frac{r}{a}\mu_{M_aX_b} + \frac{as-br}{2a}\mu_{X_2}\right)\right] \qquad (11.23)$$

The first term represents the chemical potential of the products and the terms in brackets that for the reactants for the balanced reaction:

$$\frac{r}{a}M_aX_b + \frac{as-br}{2a}X_2 = M_rX_s \qquad [11.5]$$

The reader may verify that this equation is balanced by showing that the number of M atoms is the same on both sides, as is the number of X atoms.

Equation 11.22 has the generic form

$$dS'_{iso} = -\frac{1}{T}A\, dn_{M_rX_s} \qquad (11.24)$$

The condition for equilibrium is

$$A = \left[\mu_{M_r X_s} - \left(\frac{r}{a} \mu_{M_a X_b} + \frac{as - br}{2a} \mu_{X_2} \right) \right] = 0 \qquad (11.25)$$

If for a given composition $A > 0$, then $dn_{M_r X_s}$ must be negative so that the entropy change is positive: products decompose. If $A < 0$, products form spontaneously.

It is evident that these results may be extended to any chemical reaction

$$lL + mM = rR + sS \qquad [11.6]$$

for which

$$A = \mu_{products} - \mu_{reactants} \qquad (11.26)$$

$$A = (r\mu_R + s\mu_S) - (l\mu_L + m\mu_M) \qquad (11.27)$$

The condition for equilibrium corresponding to the maximum in the entropy is $A = 0$ and the direction of the reaction otherwise is determined by the sign of the affinity.

Recall that the chemical potentials of components are not usually reported directly in experimental studies. This aspect of solution behavior of a component is normally supplied in terms of the activity or, equivalently, the activity coefficient of the component. Since the affinity evidently plays a key role in the analysis of reacting systems, it is useful to rewrite the affinity, given in its general form in Equation 11.27, in terms of the activities of the components.

Recall the definition of activity a_k of component k in a solution, Equation 8.66:

$$\mu_k = \mu_k^0 + RT \ln a_k = G_k^0 + RT \ln a_k \qquad (11.28)$$

where G_k^0 is the Gibbs free energy per mole of component k when it is in its standard or reference state. An expression like this applies to each component in Equation 11.27. These substitutions lead to an equivalent expression for the affinity of the reaction.

$$A = [r(G_R^0 + RT \ln a_R) + s(G_S^0 + RT \ln a_S)] - [l(G_L^0 + RT \ln a_L) + m(G_M^0 + RT \ln a_M)]$$

Collect terms that are similar:

$$A = [(rG_R^0 + sG_S^0) - (lG_L^0 + mG_M^0)] + RT[(r \ln a_R + s \ln a_S) - (l \ln a_L + m \ln a_M)]$$

The first set of brackets is the change in Gibbs free energy that would accompany the complete conversion of l moles of L and m moles of M when they are in their standard states to form r moles of R and s moles of S in their standard states. This quantity is called the standard free energy change for the reaction with the notation

$$\Delta G^0 \equiv \lfloor (rG_R^0 + sG_S^0) - (lG_L^0 + mG_M^0) \rfloor \qquad (11.29)$$

The four terms in the second set of brackets may be written as a single term by employing the properties of the logarithm function, $x \ln y = \ln y^x$, and $\ln x + \ln y = \ln xy$. The expression for the affinity becomes

$$A = \Delta G^0 + RT \ln \left[\frac{a_R{}^r a_S{}^s}{a_L{}^l a_M{}^m} \right] \qquad (11.30)$$

The notation may be further simplified by introducing a quantity called the proper quotient of activities for the reaction, defined by

$$Q \equiv \frac{a_R^r a_S^s}{a_L^l a_M^m} \qquad (11.31)$$

This quantity may be thought of as the ratio of "the activity of the products" to "the activity of the reactants", always keeping in mind that these expressions refer to products of the activities of the participating components raised to powers corresponding to the coefficients in the stoichiometric equation.

With these definitions the affinity for the reaction may be written:

$$A = \Delta G^0 + RT \ln Q \qquad (11.32)$$

If the relations between the activities of the components and the composition of the mixture are known, then, for a given composition, a_L, a_M, a_R and a_S may be evaluated. The proper quotient of activities, Q, thus has a particular value for that composition, which may be computed from this information.

Whatever its initial state, the system will spontaneously evolve until it achieves the specific composition that is its equilibrium state. At that composition Q takes on a particular value which could be written Q_{equil}; the symbol uniformly adopted to represent this quantity is K, called the equilibrium constant for the reaction. Thus

$$K \equiv Q_{equil} = \left[\frac{a_R^r a_S^s}{a_L^l a_M^m} \right]_{equil} \qquad (11.33)$$

K is the value that the proper quotient of activities takes on when the composition of the system achieves its equilibrium distribution. Recall that the condition for equilibrium corresponds to $A = 0$, at which $Q = K$. Put these conditions into Equation 11.32 to obtain the familiar law of mass action:

$$A = 0 = \Delta G^0 + RT \ln K$$

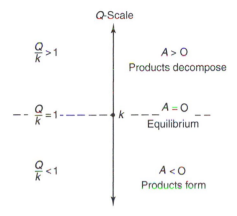

FIGURE 11.1 Sketch for the scales for affinity and Q/K showing the ranges of spontaneous decomposition and of formation of the products in a chemical reaction.

Thus the condition for equilibrium in a univariant reacting system may be written

$$\Delta G^0 = -RT \ln K \tag{11.34}$$

This is the working equation most widely used in solving practical problems that deal with chemical reactions.

The expression for the affinity at any state, Equation 11.32, may be rewritten by inserting Equation 11.34 for ΔG^0:

$$A = -RT \ln K + RT \ln Q$$

Again applying the properties of logarithms,

$$A = RT \ln \frac{Q}{K} \tag{11.35}$$

This expression for the affinity is entirely equivalent to Equation 11.27 or Equation 11.32. No simplifying assumptions have been introduced, only simplifying notation. With this form of the affinity expression the conditions for spontaneous change may be reinterpreted. Recall that in the range of compositions for which the affinity is negative, products will form. From Equation 11.35, A will be less than zero if $(Q/K) < 1$: products form. For the range in which A is positive, products will decompose; this corresponds to the range of values for which $(Q/K) > 1$: products decompose. For the unique composition at which $(Q/K) = 1$, i.e., $Q = K$, the affinity is zero and the system has the equilibrium composition. This situation is represented in Figure 11.1.

EXAMPLE 11.1.

A gas mixture at 1 atm total pressure and at the temperature 700°C has the following composition:

Component	H_2	O_2	H_2O
Mole fraction	0.01	0.03	0.96

At 700°C, for the reaction

$$2H_2 + O_2 = 2H_2O \qquad \Delta G^0 = -440 \text{ kJ}$$

Determine the direction of spontaneous change for this system.
The equilibrium constant for this reaction at 700°C (973 K) may be computed by rearranging Equation 11.34:

$$K = e^{-(\Delta G^0/RT)} = e^{-(-440,000/8.314 \times 973)} = 4.2 \times 10^{23}$$

Note that equilibrium constants are unitless quantities. A gas mixture may be considered to be an ideal solution; thus the activities of the components are given by their mole fractions in the mixture. The proper quotient of activities for the mixture given is

$$Q = \frac{a_{H_2O}^2}{a_{H_2}^2 a_{O_2}} = \frac{X_{H_2O}^2}{X_{H_2}^2 X_{O_2}} = \frac{(0.96)^2}{(0.01)^2(0.03)} = 3.1 \times 10^5$$

Thus $Q/K = (3.1 \times 10^5)/(4.2 \times 10^{23}) = 7.2 \times 10^{-19} \ll 1$. According to Figure 11.1, there is a strong tendency for products (H_2O) to form in this system.

EXAMPLE 11.2.

What will be the equilibrium composition for the system described in Example 11.1? From a practical point of view, the magnitude of the equilibrium constant means that at equilibrium the numerator in the proper quotient of activities is 23 orders of magnitude larger than the denominator. This suggests that the system will maximize the H_2O content relative to the other components. However, if all of the hydrogen present is converted to water, not all of the oxygen will be consumed. Conversion of the 0.01 mol of H_2 to water will consume only 0.005 mol of O_2; in the process, 0.01 mol of additional water will be formed. For each mole of the initial mixture in the system the final mixture will contain negligible hydrogen, $0.96 + 0.01 = 0.97$ mol of H_2O and $0.03 - 0.005 = 0.025$ mol of excess oxygen. The total number of moles will be reduced from 1.00 to $0.97 + 0.025 = 0.995$. Expressed in mole fractions, the final composition will be:

Component	H_2	O_2	H_2O
Mole fraction	negligible	$\dfrac{0.025}{0.995} = 0.025$	$\dfrac{0.970}{0.995} = 0.975$

The form of Equation 11.34, which expresses the condition for equilibrium, yields an exponential relationship between the equilibrium constant and the standard free energy change for a reaction. Measured values for ΔG^0 for chemical reactions of technological interest span the range from about $+100$ to -1000 kJ. As a consequence values of K range over many orders of magnitude. The value of Q that may be obtained for any composition for a system that might be of interest in a practical problem may also range over many orders of magnitude. Thus it is usually very likely that the value of Q computed for a given composition of practical interest will differ from the value of K representing the equilibrium composition for the system by many orders of magnitude, as was the case in Example 11.1. The qualitative answer to the generic question posed in Example 11.1, namely, "Given the composition of the system, which way lies equilibrium?", is usually very easily obtained. Except for systems with initial compositions that happen to lie near the equilibrium composition, Q/K is orders of magnitude greater or less than 1; in this case neither Q nor K need to be known with accuracy to answer this qualitative question.

In contrast, the generic question posed by Example 11.2, namely, "Given the initial composition of a system, what will be its composition at equilibrium?", requires accurate knowledge of the equilibrium constant and computations that incorporate conservation of the elements in the system. A general strategy for treating such problems is presented in the next section.

11.1.2 MULTIVARIANT REACTIONS IN THE GAS PHASE

A system with two independent chemical reactions is the simplest example of a multivariant reacting system. Consider again a system with two elements, carbon (C) and oxygen (O). In addition to the three components considered in Section 11.1.1 (CO, O_2, and CO_2) this system may exhibit measurable quantities of a fourth component, carbon vapor (C(g)). The extent to which this component may be neglected in treating gas mixtures containing carbon and oxygen can only be assessed by including this component in the analysis. In this system $e = 2$, $c = 4$ and $r = 4 - 2 = 2$. It is a bivariant reacting system.

Expressions for the number of gram atoms of carbon and oxygen take the form:

$$m_C = (1)n_{C(g)} + (0)n_{O_2} + (1)n_{CO_2} + (1)n_{CO} \tag{11.36}$$

$$m_O = (0)n_{C(g)} + (2)n_{O_2} + (2)n_{CO_2} + (1)n_{CO} \tag{11.37}$$

The compositional isolation constraints become

$$dm_C = dn_{C(g)} + dn_{CO_2} + dn_{CO} = 0 \tag{11.38}$$

$$dm_O = 2\,dn_{O_2} + 2n_{CO_2} + dn_{CO} = 0 \tag{11.39}$$

Thus the four compositional variables are related by two linear equations. This system of equations has two degrees of freedom. Choose dn_{CO} and dn_{CO_2} as independent variables. From Equation 11.38,

$$dn_{C(g)} = -(dn_{CO_2} + dn_{CO})$$ (11.40)

From Equation 11.39,

$$dn_{O_2} = -\left(dn_{CO_2} + \frac{1}{2}dn_{CO}\right)$$ (11.41)

The expression for the change in entropy for this isolated system contains four terms, one for each component:

$$dS'_{iso} = -\frac{1}{T}[\mu_{C(g)}dn_{C(g)} + \mu_{O_2}dn_{O_2} + \mu_{CO_2}dn_{CO_2} + \mu_{CO}dn_{CO}]$$

Use Equation 11.40 and Equation 11.41 to eliminate the dependent variables:

$$dS'_{iso} = -\frac{1}{T}\left[\mu_{C(g)}(-dn_{CO_2} - dn_{CO}) + \mu_{O_2}\left(-dn_{CO_2} - \frac{1}{2}dn_{CO}\right)\right.$$
$$\left. +\mu_{CO_2}dn_{CO_2} + \mu_{CO}dn_{CO}\right]$$

Collect terms in dn_{CO} and dn_{CO_2}:

$$dS'_{iso} = -\frac{1}{T}\left\{\left[\mu_{CO} - \left(\mu_{C(g)} + \frac{1}{2}\mu_{O_2}\right)\right]dn_{CO}\right.$$
$$\left. +[\mu_{CO_2} - (\mu_{C(g)} + \mu_{O_2})] + dn_{CO_2}\right\}$$ (11.42)

Inspection of this equation reveals that the coefficient of dn_{CO} is the affinity, $A_{[CO]}$ for the reaction

$$C(g) + \frac{1}{2}O_2 = CO$$

while the coefficient of dn_{CO_2} is the affinity $A_{[CO_2]}$ for the reaction

$$C(g) + O_2 = CO_2$$

Thus Equation 11.42 may be written

$$dS'_{iso} = -\frac{1}{T}[A_{[CO]}dn_{CO} + A_{[CO_2]}dn_{CO_2}]$$ (11.43)

This is a general expression for the change in entropy that this system may exhibit when it is isolated from its surroundings, expressed in terms of the two independent variables in the system. The constrained maximum in S may be found by setting the coefficients of the differentials in this expression equal to zero. Thus in this bivariant system the conditions for equilibrium are:

$$A_{[CO]} = \mu_{CO} - \left(\mu_{C(g)} + \frac{1}{2}\mu_{O_2}\right) = 0 \tag{11.44}$$

and

$$A_{[CO_2]} = \mu_{CO_2} - (\mu_{C(g)} + \mu_{O_2}) = 0 \tag{11.45}$$

The same strategy that converts the expression for the affinities in Equation 11.27 to that in Equation 11.32 and leads to the condition for equilibrium in a univariant reacting system given in Equation 11.34 may be applied separately to each of these equations. The conditions for equilibrium in this bivariant system may be written:

$$\Delta G^0_{[CO]} = -RT \ln K_{[CO]} \tag{11.46}$$

and

$$\Delta G^0_{[CO_2]} = -RT \ln K_{[CO_2]} \tag{11.47}$$

The standard free energy changes and equilibrium constants are those that apply to the CO and CO_2 formation reactions [CO] and [CO_2] that describe this system.

While the derivation of the conditions for equilibrium is straightforward for this bivariant case, the conditions for spontaneous change, which is based upon the requirement that the entropy increase in Equation 11.43, are not so straightforward. For example, even if both affinities have the same sign it is possible for dn_{CO} and dn_{CO_2} to have opposite signs and the sum still yield a positive change in entropy. The directions of the reactions that increase the entropy in an isolated bivariant reacting system depend explicitly upon the values of the quantities involved; no general statement may be made. This situation evidently becomes more complicated as additional independent reactions are added in a multivariant system.

A reacting system is bivariant if it is characterized by two independent chemical reactions. Other reactions among the components of the system may be written but these are not independent: they are linear combinations of the two independent reactions. In the introduction it was shown that the reaction

$$C(g) + CO_2 = 2CO \tag{11.7}$$

can be obtained by subtracting reaction [CO_2] from reaction [CO] after doubling its coefficients. Thus, as these reactions are written, [11.7] = 2[CO] − [CO_2]. The stoichiometric statement embodied in [11.7] is a linear combination of those in [CO_2] and [CO] and is thus not an independent statement. It is also possible to write the reaction

$$2CO + O_2 = 2CO_2 \tag{11.8}$$

For this reaction, [11.8] = 2[CO] − 2[CO_2].

It is easy to show that the affinities for reactions [11.7] and [11.8] are the analogous linear combinations of the affinities for reactions [CO] and [CO_2].

The affinity for reaction [11.7] is defined as

$$A_{[11.7]} = 2\mu_{CO} - (\mu_{C(g)} + \mu_{CO_2}) \tag{11.48}$$

The linear combination of $A_{[CO]}$ and $A_{[CO_2]}$ indicated is

$$A_{[11.7]} = 2A_{[CO]} - A_{[CO_2]} = 2\left[\mu_{CO} - \left(\mu_{C(g)} + \frac{1}{2}\mu_{O_2}\right)\right] - [\mu_{CO_2} - (\mu_{C(g)} + \mu_{O_2})]$$

$$= 2\mu_{CO} - 2\mu_{C(g)} - \mu_{O_2} - \mu_{CO_2} + \mu_{C(g)} + \mu_{O_2}$$

$$2A_{[CO]} - A_{[CO_2]} = 2\mu_{CO} - (\mu_{C(g)} + \mu_{CO_2}) = A_{[11.7]} \tag{11.49}$$

The demonstration that $A_{[11.8]} = 2A_{[CO_2]} - 2A_{[CO]}$ is left as an exercise to the reader.

The relationships between the affinities also imply relations between the standard free energy changes and the equilibrium constants for these reactions. For example, given that

$$A_{[k]} = mA_{[i]} - nA_{[j]} \tag{11.50}$$

Substitute expression 11.32 for each of the affinities in this expression:

$$\Delta G^0_{[k]} + RT \ln K_{[k]} = m(\Delta G^0_{[i]} + RT \ln K_{[i]}) - n(\Delta G^0_{[j]} + RT \ln K_{[j]})$$

$$\Delta G^0_{[k]} + RT \ln K_{[k]} = (m\Delta G^0_{[i]} - n\Delta G^0_{[j]}) + RT \ln \frac{K_{[i]}{}^m}{K_{[j]}{}^n}$$

Compare corresponding terms:

$$\Delta G^0_{[k]} = m\Delta G^0_{[i]} - n\Delta G^0_{[j]} \tag{11.51}$$

and

$$K_{[k]} = \frac{K_{[i]}{}^m}{K_{[j]}{}^n} \tag{11.52}$$

Thus any reaction that is a linear combination of other reactions has a standard free energy change that is the same linear combination of the ΔG^0 values for the contributing reactions. The equilibrium constant for such a reaction is a product of K values for the contributing reactions raised to powers that correspond to coefficients in the linear combination relating the reactions.

Incidentally, the description of the system emerged in terms of reactions [CO] and [CO$_2$] in this derivation because the affinities for these reactions appeared in Equation 11.42. There is nothing special about this result; it may be traced to the choice that was made for the independent composition variables in the conservation Equation 11.40 and Equation 11.41. If n_C or n_{O_2} had been chosen as independent variables, then Equation [11.7] and Equation [11.8] would have appeared as the appropriate description of the system.

Since at equilibrium $A_{[CO]}$ and $A_{[CO_2]}$ are independently zero (Equation 11.44 and Equation 11.45) and the affinity for reaction [11.7] is a linear combination of these values, Equation 11.50, then when the system reaches its equilibrium state,

$$A_{[11.7]} = 0 \qquad\qquad (11.53)$$

Similarly, because $A_{[11.8]}$ is a linear combination of $A_{[CO]}$ and $A_{[CO_2]}$, at equilibrium the affinity for reaction [11.8] must also be zero:

$$A_{[11.8]} = 0 \qquad\qquad (11.54)$$

Thus the working equations that describe the equilibrium state for the system are:

$$\Delta G_{[j]}{}^0 = -RT \ln K_{[j]} \qquad [j = 1, 2, 3, 4] \qquad (11.55)$$

Only two of these four equations are independent.

These arguments extend directly to multivariant reacting systems with e elements forming c components. In a system with a large number of components, there are $r = (c - e)$ independent equations of the form of Equation 11.55 that describe the equilibrium state. However, as the preceding argument demonstrates, at equilibrium it is possible to write an equation like Equation 11.55 for every possible reaction that can be written among the components in the system; only r of these equations will be independent.

The general strategy for determining the equilibrium composition in a multivariant reacting system makes use of the independent conditions for chemical equilibrium and the equations that require that the number of gram atoms of the elements in the system be conserved. If a single-phase system contains c components, the equilibrium state will be specified by assigning values to the mole fractions of each of these c components. If the components are built from e elements, then there are e conservation of atoms equations, one for each element, of the form:

$$m_j = \sum_{k=1}^{c} b_{jk} n_k \qquad (j = 1, 2, ..., e) \qquad (11.56)$$

where the coefficient b_{jk} is the number of atoms of element j contained in a molecule of k. (If k does not contain the element j, then $b_{jk} = 0$ for that component.) Note that these equations are linear in the unknown n_k values. There are $r = (c - e)$ independent reactions in this system, each with its condition for equilibrium, Equation 11.55 and corresponding value for its equilibrium constant. Each of these equilibrium constants is related to the composition of the system through an equation like Equation 11.33; these relationships are normally not linear because the activity values involved have exponents. Adding these $r = (c - e)$ equilibrium constant relations to the set of e conservation of atoms equations yields a total of $[e + (c - e)] = c$ equations in c unknowns; the state of the system is therefore mathematically fully determined.

The equilibrium composition of the mixture is determined by the initial composition when the system is considered to be isolated. Further analysis shows that the final composition is actually a function only of the *concentrations of the elements* in the system, no matter how they are distributed among the molecular species in the system. The crucial chemical variables in the analysis is thus the set of mole fractions of the elements given by

$$X_j = \frac{m_j}{\displaystyle\sum_{j=1}^{e} m_j} \qquad (j = 1, 2, ..., e) \tag{11.57}$$

Further, since these elemental fractions add to one, there are $(e - 1)$ independent chemical variables that actually determine the final equilibrium composition of the system. Different molecular compositions that happen to have the same set of elemental fractions will produce the same final equilibrium composition.

EXAMPLE 11.3.

A gas mixture has following composition:

Component	H_2	O_2	H_2O	CO	CO_2	CH_4
Mole fraction	0.05	0.05	0.15	0.25	0.40	0.10

Find the equilibrium composition of this mixture at 600°C.
The set of conservation equations corresponding to Equation 11.56 for this system is

$$m_C = (0)n_{H_2} + (0)n_{O_2} + (0)n_{H_2O} + (1)n_{CO} + (1)n_{CO_2} + (1)n_{CH_4}$$

$$m_O = (0)n_{H_2} + (2)n_{O_2} + (1)n_{H_2O} + (1)n_{CO} + (2)n_{CO_2} + (0)n_{CH_4}$$

$$m_H = (2)n_{H_2} + (0)n_{O_2} + (2)n_{H_2O} + (0)n_{CO} + (0)n_{CO_2} + (4)n_{CH_4}$$

The total number of moles in the system at any instant is the sum of the number of moles of each of the components;

$$n_T = n_{H_2} + n_{O_2} + n_{H_2O} + n_{CO} + n_{CO_2} + n_{CH_4}$$

For this example, $c = 6$ and $e = 3$; there are three independent reactions. Find three nonredundant reactions that incorporate all six components:

$$2H_2 + O_2 = 2H_2O \tag{1}$$

$$2CO + O_2 = 2CO_2 \tag{2}$$

$$CH_4 + 2O_2 = 2H_2O + CO_2 \tag{3}$$

The standard free energy changes for these reactions at 600°C are:

$$\Delta G^0_{[1]} = -406,200; \qquad \Delta G^0_{[2]} = -414,500; \qquad \Delta G^0_{[3]} = -797,900$$

The corresponding equilibrium constants, assuming the gas mixture is ideal so that the activities are equal to the mole fractions, are:

$$K_{[1]} = \frac{X^2_{H_2O}}{X^2_{H_2} X_{O_2}} = 2.02 \times 10^{24} \qquad K_{[2]} = \frac{X^2_{CO_2}}{X^2_{CO} X_{O_2}} = 6.34 \times 10^{24}$$

$$K_{[3]} = \frac{X^2_{H_2O} X_{CO_2}}{X_{CH_4} X^2_{O_2}} = 5.53 \times 10^{47}$$

since for each component $X_k = n_k/n_T$. There are thus seven equations relating the seven variables, (six n_k values and n_T). This system of equations can be solved using a math applications software iterative solver program for the final equilibrium composition that will be obtained for a system that has the initial composition given at the beginning of the example.

The equilibrium composition of the gas phase is computed from these six simultaneous equations to be:

Component	H_2	H_2O	CO	CO_2	CH_4	O_2
Mole fraction	0.136	0.144	0.231	0.445	0.052	1.6×10^{-24}

The problem posed in Example 11.3 is typical of this class of problems. An initial composition of the system is known; this permits calculation of the number of gram atoms, m_j, of each of the elements. The molecular formulas of the components give the coefficients in the set of equations corresponding to the conservation of the elements in the system. A set of $r = (c - e)$ independent chemical reactions may then be formulated. Values for ΔG^0 are then obtained from an appropriate database for these reactions. For many common molecular components, values for the reaction that forms the component from its elements, ΔH_f^0 and ΔS^0_f are tabulated; (see, for example Refs. [1–4] or Appendix E). Since $\Delta G^0 = \Delta H^0 - T\Delta S^0$, this information permits computation of ΔG_f^0 at any temperature, for any formation reaction involving these components.[2] The equilibrium constant, K, may be computed for each of these reactions from the condition for equilibrium, Equation 11.55. Of the resulting c equations in c unknowns, e are linear, being derived from the conservation of the elements. The remaining $(c - e)$ equations come from the equilibrium constants and are

---

[2] Note that this calculation of ΔG^0 assumes that ΔH^0 and ΔS^0 are each independent of temperature; this assumption is a good approximation for most chemical reactions because ΔH^0 and ΔS^0 are large in comparison to heat capacity effects.

nonlinear equations. Mathematics applications packages, such as TK-Solver™ and MathCad™, provide a framework for solving such sets of equations.

More explicit application software, such as the program SOLGASMIX,[5] and later ChemSage, has been developed that permit solution of this class of problem for a system with an arbitrary number of components. In its most complete form this software incorporates a database with enthalpies and entropies of formation of a large number of molecular components, which may normally form in the gas phase. Input to the program is a list of the elements in the system and a range of temperatures that are of interest in the problem. The software:

1. Develops a list of the molecular components in its database that form from the elements that are input.
2. Generates a list of independent balanced chemical reactions among these components.
3. Looks up enthalpies and entropies for these reactions in its database.
4. Computes ΔG_j^0 values for each reaction at a temperature of interest.
5. Computes the corresponding equilibrium constant, K_j.
6. Writes the e conservation of element equations.
7. Writes the $(r - e)$ equilibrium constant relations.
8. Solves this set of equations for the compositions of each of the components at equilibrium.

Output is a list of the equilibrium concentrations of all of the molecular components considered in the program. Typically, concentrations of perhaps 20 to 30 components may be included in the output. The vast majority of these will be small enough to be neglected for most practical purposes and the composition of the gas at equilibrium may be usefully specified in terms of a handful of components. Variation of the gas composition with temperature may be explored by incrementing T and repeating the algorithm. Figure 11.2 shows a typical output of the SOLGASMIX program.

It is frequently true that in a particular practical problem, the focus is upon the concentration of one of the components. For example, if oxidation of a part that will be heat treated in a gas atmosphere is an issue, then the focal point of the thermodynamics of the atmosphere will be the partial pressure of oxygen, P_{O_2}. This gives rise to an application of this kind of software to produce the pattern of oxygen partial pressures as a function of the gas composition. Such an equilibrium gas composition map is shown in Figure 1.6, which plots contours of the partial pressure of oxygen on a ternary composition triangle with the elements carbon, hydrogen and oxygen at the corners. For any initial molecular composition use Equation 11.56 to compute the number of moles of each of the elements and Equation 11.57 to compute the elemental atom fractions. Plot these on the composition triangle and read the oxygen potential. This kind of diagram can be generated for other combinations of elements (e.g., H, N and O) or for quaternary mixtures as a function of temperature. Alternatively, they may plot partial pressure contours for any component in the system, e.g., P_{H_2}, P_{CO}, P_{H_2O}, etc. Such maps can be very useful in designing furnace atmospheres to achieve specific combinations of partial pressures of the reactive components.

TEMPERATURE = 1000 K;
TOTAL PRESSURE = 1.000e + 00 ATM;

INITIAL COMPOSITION;

CO(g)	CO2(g)	SO2
3.12E-01	5.09E-01	1.79E-01

EQUILIBRIUM COMPOSITION:

1 COS (g)	p = 9.25E-03
2 S_2O (g)	p = 2.34E-04
3 SO_3 (g)	p = 6.03E-10
4 SO_2 (g)	p = 3.22E-02
5 SO (g)	p = 5.19E-06
6 CS_2 (g)	p = 1.13E-05
7 CS (g)	p = 6.23E-10
8 S_8 (g)	p = 3.44E-08
9 S_7 (g)	p = 4.47E-07
10 S_6 (g)	p = 6.84E-06
11 S_5 (g)	p = 2.75E-05
12 S_4 (g)	p = 6.60E-06
13 S_3 (g)	p = 2.15E-03
14 S_2 (g)	p = 7.30E-02
15 S (g)	p = 1.39E-09
16 C_3O_2 (g)	p = 4.68E-23
17 CO_2 (g)	p = 8.78E-01
18 CO (g)	p = 5.30E-03
19 O_3 (g)	p = 8.78E-36
20 O_2 (g)	p = 1.00E-16
21 O (g)	p = 1.56E-18
22 C_5 (g)	p = 3.39E-64
23 C_4 (g)	p = 1.75E-59
24 C_3 (g)	p = 2.39E-46
25 C_2 (g)	p = 1.37E-43

FIGURE 11.2 An example of the equilibrium composition of a gas mixture with a given initial composition at 1000 K computed by SOLGASMIX.[5]

11.2   REACTIONS IN MULTIPHASE SYSTEMS

It is not unusual in materials science for a system of practical interest to be multicomponent, multiphase and capable of exhibiting chemical reactions. For example, the oxidation of a metal involves three phases: the metal, the ceramic oxide and the gas phase, which is the source of the oxygen. The description of such a system begins with a list of the components in each phase. The construction of a valid list is a nontrivial task; within a given phase the components are those chemical species whose composition can be varied independently when the system is not at equilibrium.

In the gas phase the list includes, in principle, all of the known molecular species that the elements in the system may combine to form. If electrical effects are important in the problem at hand, this list will be extended to include ionic species as well (see Chapter 15). In metallic phases the list normally incorporates the elements in the system since molecular combinations separate into their elemental components upon dissolution in metals. Ceramic phases include the vast range of ionic, covalent and polar solid phases normally described in chemistry as chemical compounds, such as oxides, carbides, nitrides, sulfides, sulfates, etc. If variations from the stoichiometric composition are of interest in the problem then variations in the number of moles of the elemental components must be included in the description of a compound phase; if electrical fields are involved, the components may be usefully taken to be the ionic species that make up the ceramic phase. Ionic species will also play a key role in behavior if one of the phases is an electrolyte.

As an example, consider the oxidation of copper metal. Three phases will normally participate: the metallic phase (α), the gas phase (g) and the ceramic phase, copper oxide (ε), nominally Cu_2O. In a restricted range of oxygen potential and temperature, CuO may also form. At equilibrium all three of these phases are solutions: some oxygen atoms will dissolve in copper metal, some copper vapor will exist in the gas along with the oxygen, the oxide phase will depart from its stoichiometric composition. Some components exist in one phase but not in others. The gas phase contains the molecular form of oxygen, O_2. The oxygen in the copper metal is monatomic; in order to dissolve oxygen in copper, the oxygen molecule must dissociate into atoms at the metal–gas interface. The oxygen in the oxide is present as oxygen ions, the copper as copper ions. These ions do not exist either in the metal or the vapor. Other ionic species may exist as defects in the oxide.

The system may be visualized as consisting of three phases: the metallic phase (α), the ceramic solution (ε) and the gas phase (g). The components may be taken to be: Cu^α, O^α, $Cu^{++\varepsilon}$, $O^{--\varepsilon}$, Cu^g and O_2^g. This list is not exhaustive. Spectroscopic analysis of the gas phase will reveal monatomic oxygen O_1, ozone O_3 and a variety of copper and oxygen complexes, some of which are ions, present in very small but detectable quantities.

In most practical applications the list of components that is considered to be sufficient to describe the behavior of the system is much shorter than the exhaustive list contemplated in the last paragraph. Components may be deleted from the description of the behavior of a system if it is known from independent information that the variation in their number of moles is negligible in the context of the problem. The simplest treatment of the copper–oxygen system visualizes only three components: $Cu(\alpha)$; $CuO(\varepsilon)$; and $O_2(g)$. Each of the phases in the system (α, ε, g) is also viewed as a component (Cu, CuO and O_2) because the compositions of the phases are approximated to be invariant. At this level of sophistication there are two elements and three components in this three-phase system. Since $r = (c - e) = (3 - 2) = 1$, it is a univariant system.

The strategy for finding the conditions for equilibrium may now be applied to this three-phase, three-component system. The number of copper atoms in the system at any instant in time is:

$$m_{Cu} = n_{Cu}^\alpha + n_{CuO}^\varepsilon \tag{11.58}$$

The oxygen atoms:

$$m_O = 2n_{O_2}^g + n_{CuO}^\varepsilon \tag{11.59}$$

In an isolated system the number of atoms of each element is conserved:

$$dm_{Cu} = 0 = dn_{Cu}^\alpha + dn_{CuO}^\varepsilon \Rightarrow dn_{Cu}^\alpha = -dn_{CuO}^\varepsilon$$

$$dm_O = 0 = 2dn_{O_2}^g + dn_{CuO}^\varepsilon \Rightarrow dn_{O_2}^g = -\frac{1}{2}dn_{CuO}^\varepsilon$$

The expression for the change in entropy of this three-phase system will involve terms in U' and V' for each of the phases; when these terms are combined with the internal energy and volume isolation constraints they yield the conditions for thermal $(T^\alpha = T^\varepsilon = T^g)$ and mechanical $(P^\alpha = P^\varepsilon = P^g)$ equilibrium. Since the current focus is upon the chemical effects in the system, the entropy may be written:

$$dS'_{iso} = [[\ ]] - \frac{1}{T}[\mu_{Cu}^\alpha dn_{Cu}^\alpha + \mu_{CuO}^\varepsilon dn_{CuO}^\varepsilon + \mu_{O_2}{}^g dn_{O_2}^g] \tag{11.60}$$

where the notation $[[\ ]]$ signifies the four noncompositional terms in the expression. Apply the conservation equations:

$$dS'_{iso} = [[\ ]] - \frac{1}{T}\left[\mu_{Cu}^\alpha(-dn_{CuO}^\varepsilon) + \mu_{CuO}^\varepsilon dn_{CuO}^\varepsilon + \mu_{O_2}^g\left(-\frac{1}{2}dn_{CuO}^\varepsilon\right)\right]$$

$$dS'_{iso} = [[\ ]] - \frac{1}{T}\left[\mu_{CuO}^\varepsilon - \left(\mu_{Cu}^\alpha + \frac{1}{2}\mu_{O_2}^g\right)\right](dn_{CuO}^\varepsilon) \tag{11.61}$$

To find the conditions for equilibrium, set the coefficients of the differentials in this expression equal to zero. Applying that tactic to the terms in $[[\ ]]$ yields the conditions for thermal and mechanical equilibrium, as previously noted. The coefficient of the compositional variable may be recognized as the affinity for the reaction:

$$Cu(\alpha) + \frac{1}{2}O_2(g) = CuO(\varepsilon)$$

defined to be

$$A_{CuO} = \mu_{CuO}^\varepsilon - \left(\mu_{Cu}^\alpha + \frac{1}{2}\mu_{O_2}^g\right) \tag{11.62}$$

The condition for chemical equilibrium in this system requires that the coefficient of dn_{CuO} in Equation 11.61 be equal to zero:

$$A_{CuO} = 0 \tag{11.63}$$

Express the chemical potentials in terms of the corresponding activities and collect like terms to derive

$$\Delta G^0_{CuO} = -RT \ln K_{CuO} \qquad (11.64)$$

where

$$\Delta G^0_{CuO} = G^0_{CuO} - \left(G^0_{Cu} + \frac{1}{2}G^0_{O_2}\right) \qquad (11.65)$$

and

$$K_{CuO} = \frac{a_{CuO}}{a_{Cu}a_{O_2}^{1/2}} \qquad (11.66)$$

Thus, the condition for equilibrium in this multiphase reacting system is formally identical with that derived for reactions in the gas phase.

EXAMPLE 11.4.

Find the partial pressure of oxygen that exists in a system in which pure copper is equilibrated with CuO at 900°C. The standard free energy for the formation of CuO from copper and oxygen at 900°C is

$$\Delta G^0_{900°C} = -52.7 \text{ kJ}$$

The equilibrium constant for the reaction is

$$K = e^{-(-52,700/8.314 \times 1173)} = 220$$

The proper quotient of activities may be computed by noting that $a_{Cu} = 1$ and $a_{CuO} = 1$ since they are in their reference states for the given conditions. The activity of oxygen is the ratio of its partial pressure in the system to the pressure in the reference state:

$$a_{O_2} = \frac{P_{O_2}}{P^0_{O_2}} = P_{O_2}$$

Since the reference state used in evaluating ΔG^0 for the reaction is 1 atm, the activity of oxygen is numerically equal to its partial pressure, P_{O_2}. The equilibrium constant, Equation 11.66, is thus

$$K_{CuO} = \frac{a_{CuO}}{a_{Cu}a_{O_2}^{1/2}} = \frac{1}{1P_{O_2}^{1/2}} = 220$$

Solve for the partial pressure of oxygen:

$$P_{O_2} = \left(\frac{1}{220}\right)^2 = 2.0 \times 10^{-5} \text{ (atm)}$$

A reacting system is multivariant if $r = c - e > 1$. In multiphase systems, the extension from univariant to multivariant reactions follows the same logical sequence that was developed for the single gas phase in Section 11.1.2. The result is the same. The conditions for chemical equilibrium correspond to a set of equations of the form:

$$A_{[j]} = 0 \qquad [j = 1, 2, ..., r] \tag{11.67}$$

and the equivalent working equations

$$\Delta G^0_{[j]} = -RT \ln K_{[j]} \qquad [j = 1, 2, ..., r] \tag{11.68}$$

where r is the number of independent chemical reactions for the system. It is further true that, since any chemical reaction that may be written among the components in the system may be constructed from a linear combination of the independent reactions, Equation 11.67 and Equation 11.68 apply to every chemical reaction that may be written among the components in the system.

In treating practical problems involving multicomponent, multiphase reacting systems it is frequently convenient to limit consideration to a single equilibrium that exists between a few of the components that happen to be of interest in a particular application. This tactic is possible because an equation of the form 11.68 holds for every reaction. Caution must be applied in making such a simplification; the equilibrium situation thus deduced may be metastable. One or more of the components left out of consideration in making such a simplification may in fact be the predominant specie in the actual stable equilibrium state. Deductions based upon the simplified analysis may therefore be misleading. A representation of multivariant equilibria that incorporates all of the components in a system, called the predominance diagram for the system, is presented in Section 11.4.

11.3 PATTERNS OF BEHAVIOR IN COMMON REACTING SYSTEMS

It is clear that evaluation of the equilibrium constant for a reaction is the key to the solution of practical problems in such systems. The equilibrium constant is, in turn, most frequently computed from database information about the standard free energy change for the reaction, ΔG^0, through Equation 11.68. The variation of the equilibrium state with temperature and pressure may be traced to the variation of ΔG^0. Since the standard free energy change is defined in terms of the free energy of the components in their reference states and these are normally defined to be the pure components, computation of the variation of ΔG^0 with temperature and pressure employs familiar thermodynamic relationships derived in earlier chapters for simple one-component systems.

11.3.1 RICHARDSON–ELLINGHAM CHARTS FOR OXIDATION

The standard free energy change for any reaction may be expressed in terms of the standard enthalpy and entropy of the reaction through the definitional relationship:

$$\Delta G^0 = \Delta H^0 - T\Delta S^\circ \tag{11.69}$$

In most studies of reacting systems in materials science the variation of the behavior of the system with temperature at 1 atm pressure is most important. The variation of the enthalpy with temperature at constant pressure is given by:

$$\Delta H^0(T) = \Delta H^0(T_0) + \int_{T_0}^{T} \Delta C_P^0(T)dT \qquad (11.70)$$

where ΔC_P^0 is the heat capacity of the product pure components minus the heat capacity of the reactants, with coefficients given in the balanced chemical equation. On the scale of energies associated with thermodynamic processes, heats of reactions are usually very large numbers, typically a few hundred kilojoules. Heat exchanges associated with altering the temperature of reactants or products more typically range below a few tens of kilojoules. The difference in these heating effects for reactants and products is yet an order of magnitude smaller. The second term in Equation 11.70 therefore is negligible for most practical purposes and the heat of reaction is independent of temperature. A similar observation holds for the standard entropy of a reaction: in the relation

$$\Delta S^\circ(T) = \Delta S^\circ(T_0) + \int_{T_0}^{T} \frac{\Delta C_P^0(T)}{T}dT$$

the second term may be neglected. Thus the enthalpy and entropy terms in Equation 11.69 may be treated as constants, i.e., independent of temperature.

With these approximations a plot of ΔG^0 for a reaction vs. temperature is based simply on Equation 11.69 with ΔH^0 and ΔS° as constant. Accordingly, $\Delta G^0(T)$ is expected to be linear with a slope equal to $(-\Delta S^\circ)$ and an intercept at $T = 0$ K equal to the standard enthalpy change for the reaction, ΔH^0. Figure 11.3 shows such a plot for the formation of nickel oxide. Deviations from linearity are measurable but not important in most practical applications. The curve has three discontinuities in its slope. These are associated with phase changes of the components in the reaction. For example, at 1450°C nickel metal melts. Below that temperature the reference state for nickel is pure crystalline (FCC) nickel, with the reaction

$$2Ni(c) + O_2 = 2NiO$$

Above the melting point of nickel the stable form and thus the reference state for the component in the reaction, is pure liquid nickel. The ΔG^0 plot represents the reaction

$$2Ni(l) + O_2 = 2NiO(c)$$

The enthalpies and entropies of these reactions differ by amounts that correspond to the difference in reference states for the nickel:

$$2Ni(c) = 2Ni(l)$$

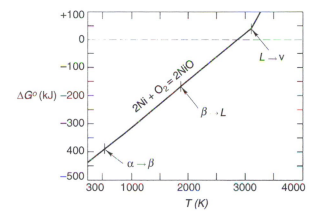

FIGURE 11.3 Standard free energy change for the formation of nickel oxide (NiO) plotted as a function of temperature illustrates the points that the temperature dependence of ΔH^0 and ΔS° can be neglected.

i.e., by twice the heat and entropy of fusion of nickel. Thus at the melting point the slope of the ΔG^0 plot changes by $2\Delta S_f^\circ$ for pure nickel. The intercept at $T = 0$ K changes by $2\Delta H_f^0$. The more drastic change in slope and intercept that occurs at 3380°C is associated with the vaporization of nickel; entropies of vaporization are typically an order of magnitude larger than entropies of fusion. A break in the curve, which decreases its slope, may be traced to a phase change for the product, in this case NiO.

Thus a plot of ΔG^0 vs. temperature is typically a broken line with straight segments; the discontinuities in slope correspond to phase changes that occur in the components of the reaction.

It is frequently useful to compare the stabilities of families of compounds in seeking answers to practical questions such as "Which of these metals is more resistant to oxidation at 1000°C?" or "Which of these nitrides is more stable, i.e., less likely to dissociate, at 1400°C?" Such comparisons are facilitated by the existence of patterns of behavior within classes of similar compounds. In developing a comparison of the oxides or the nitrides or the carbides, it is useful to formulate the problem in terms of reactions that are written so as to involve the same number of moles of the component that is common in all of the reactions. Thus to compare oxidation behavior, write and balance all of the reactions to be considered in terms of the consumption of 1 mol of oxygen.

Figure 11.4 shows a plot that is formulated in this way, known as a Richardson–Ellingham Chart, introduced independently in the 1940s by these scientists. Standard free energy changes for a collection of oxidation reactions are plotted as a function of temperature on this chart. The curve representing a given reaction is a broken line, with changes in slope occurring at phase changes for the metal or oxide. Each line segment is described by Equation 11.69 and has a slope equal to $-\Delta S^\circ$ for the reaction and an intercept at $T = 0$ K equal to ΔH^0. At low temperatures where both metal and oxide are solid phases, the lines all have essentially the same slope.

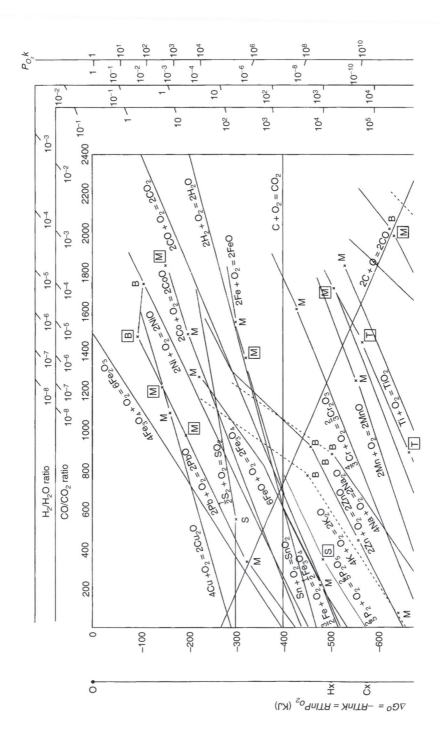

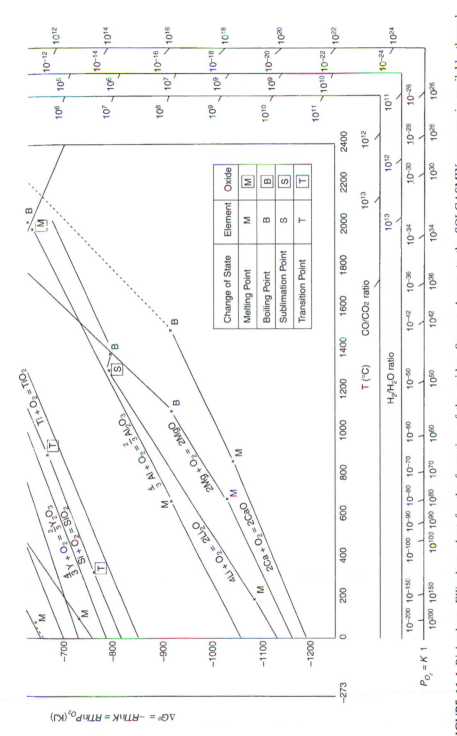

FIGURE 11.4 Richardson–Ellingham chart for the formation of the oxides. *Source:* Access to the SOLGASMIX program is available through F*A*C*T* (Facility for the Analysis of Chemical Thermodynamics) Ecole Polytechnique, CRCT, Montreal, Quebec, Canada.

Evidently the standard entropy changes for these reactions are nearly identical. This reflects the fact that the primary contribution to the change in entropy for these reactions comes from the change in volume associated with the contraction of 1 mol of the gas phase for each unit of reaction that occurs. Differences between entropies of the solid phases exist but are small in comparison with this effect. In support of this point note that the reaction

$$C + O_2 = CO_2$$

is essentially horizontal on this plot. In this case 1 mol of gas forms 1 mol of gas; the volume change is essentially zero and the change in entropy associated with the volume change is zero. Similarly, the reaction

$$2C + O_2 = 2CO$$

has a negative slope because the system expands from 1 to 2 mol of gas for each unit of reaction that occurs.

Since the entropies of the reactions that transform a metal to its oxide have similar values, the major differences between the curves are contained in the heats of reaction, displayed as the intercept for each curve on the $T = 0$ K ordinate. Since the temperature scale plotted on the diagram is in degrees Celsius the intercept at 0 K corresponds to an intercept on the line at $-273°C$ to the left of the ordinate. The order of the reaction lines from the top to the bottom of the chart is primarily determined by the corresponding heat of reaction per mole of oxygen consumed.

Each of the oxidation reactions represented in this chart is written on the basis of 1 mol of oxygen consumed. If the oxide has the formula M_uO_v, the equation for the reaction reads:

$$\frac{2u}{v}M + O_2 = \frac{2}{v}M_uO_v \tag{11.71}$$

One consequence of this tactic is that the equilibrium constant for all of the reactions has the form

$$K = \frac{a_{M_uO_v}^{(2/v)}}{a_M^{(2u/v)}P_{O_2}} \tag{11.72}$$

In most applications the departure of the composition of the oxide from its reference state is negligible and the activity of the oxide may be taken to be 1. If the metal in the problem is pure or the solvent in a dilute solution, its activity may also be taken to be 1. With these two assumptions, Equation 11.72 may be written

$$K = \frac{1}{P_{O_2}} \tag{11.73}$$

In this case the value of P_{O_2} determined from the chart is the partial pressure of oxygen that is in equilibrium with the pure metal and its oxide. This value of P_{O_2} is frequently referred to as the dissociation pressure of the oxide because it reports the limit of stability of the oxide in question as the oxygen pressure in the system is decreased. If the value of P_{O_2} for a given atmosphere under consideration lies below the pressure that is in equilibrium with the oxide, the oxide will spontaneously decompose or dissociate into the metal and oxygen.

It is very useful to construct some additional scales on this chart that make it possible to read equilibrium constants and dissociation pressures graphically. Because for any reaction $\Delta G^0 = -RT \ln K = (-R \ln K)T$, any combination of values $(\Delta G_0, T)$, i.e., any point on the chart, may be viewed as representing a particular value of K. The locus of points in $(\Delta G_0, T)$ space that all share the same value of K is a straight line through the origin with a slope equal to $(-R \ln K)$. The value of the equilibrium constant corresponding to any point on the chart could be read directly if an envelope of such labeled straight lines were superimposed as shown in Figure 11.5. The same information could be obtained simply by constructing the scale at the right and bottom perimeter of this graph with points labeled that correspond to the intercept of a line with a fixed K value on that scale. With this scale the value of K corresponding to the combination of ΔG_0 and T labeled P can be obtained simply by overlaying a straight edge from the origin at O through P; the intercept on the K scale, properly interpolated, gives the corresponding value for the equilibrium constant.

EXAMPLE 11.5.
Find the equilibrium constant for the oxidation of zinc at 700°C.
Find the zinc oxide reaction line on the chart. Read the value of ΔG^0 for this reaction at 700°C:

$$\Delta G^0_{700°C} = -500 \text{ (KJ)}$$

The equilibrium constant corresponding to this condition may be calculated algebraically from this value:

$$K = e^{-(-500,000/8.134 \times 973)} = 7.0 \times 10^{26}$$

To solve the problem using the K scale on the chart, find the point on the zinc oxide line at 700°C. With a straight edge, lay a line between this point and the origin. Read the intercept of this line on the K scale; interpolate, visualizing a log scale, between 10^{26} and 10^{27}.

If the approximations that lead to Equation 11.73 are valid, i.e., if the metal and oxide activities are approximately 1, then the value of K read from the K scale is simply the reciprocal of the dissociation pressure for the oxide in the reaction under consideration. An oxygen pressure scale constructed with these assumptions is incorporated into the Richardson–Ellingham chart, Figure 11.4. Use of this scale allows quick answers to a variety of questions that may be asked about the reactions represented in the chart. Points on the K scale are given by the reciprocal of the value on the P_{O_2} scale.

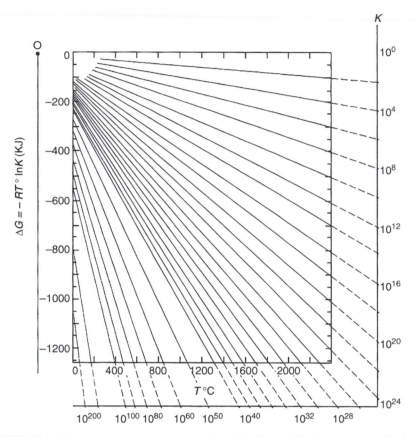

FIGURE 11.5 Lines that represent constant values for the equilibrium constant, K, plotted in $\Delta G^0 - T$ space corresponding to the Richardson–Ellingham chart.

EXAMPLE 11.6.

Evaluate the dissociation pressure of zinc oxide at 700°C.

The equilibrium constant for this reaction at 700°C was computed in Example 11.5 to be 7.0×10^{26}. According to Equation 11.73,

$$P_{O_2} = \frac{1}{K} = 1.4 \times 10^{-27} \text{ atm}$$

The problem may be solved graphically with the P_{O_2} scale in Figure 11.4. On the zinc oxide reaction line mark the point at 700°C. Lay a straight edge from this point through the origin. Read the intercept on the P_{O_2} scale, interpolating logarithmically between 10^{-27} and 10^{-26}.

If the assumptions leading to Equation 11.73 are not valid, then the activity of the metal in the system, e.g., as a solute in a dilute solution, must be measured or modeled and input to the calculation. If the solvent in this binary metallic system is noble, i.e., does not form an oxide spontaneously, then the K scale may be used to compute the equilibrium constant and the more general form, Equation 11.72, applied with the activity of the oxide taken as 1. If the second element in the alloy may also react to form an oxide, then a more sophisticated treatment is required.

EXAMPLE 11.7.

Compute the partial pressure of oxygen in an atmosphere that is equilibrated at 700°C with zinc oxide and a gold–zinc alloy with $X_{Zn} = 0.005$. Assume the Henry's law constant for this dilute solution is 8.5 at 700°C. Except for its effect upon the activity of zinc in solution, the gold may be treated as inert in this system.
The activity of zinc in the alloy is

$$a_{Zn} = \gamma_{Zn}^0 X_{Zn} = 8.5(0.005) = 0.043$$

Insert this value in the equilibrium constant read from the K scale on the chart:

$$K = \frac{1}{a_{Zn}{}^2 P_{O_2}} = 7.0 \times 10^{26}$$

Solve for P_{O_2}:

$$P_{O_2} = \frac{1}{(0.043)^2 7.0 \times 10^{26}} = 7.9 \times 10^{-24} \text{ atm}$$

The partial pressure of oxygen in a system is frequently referred to as its oxygen potential. This term derives from the fact that the chemical potential of oxygen in the system is functionally determined by the oxygen partial pressure since

$$\mu_{O_2} = \mu_{O_2}^0 + RT \ln a_{O_2} = \mu_{O_2}^0 + RT \ln \frac{P_{O_2}}{P_{O_2}{}^0} = \mu_{O_2}^0 + RT \ln P_{O_2} \quad (11.74)$$

since the oxygen pressure in the reference state in most databases is 1 atm. If the oxygen potential in a system lies above the equilibrium value at any temperature, then the metal will oxidize and the oxide is stable. If the potential lies below the equilibrium value at the temperature of the system, the oxide is unstable and will dissociate. This situation is illustrated in Figure 11.6. The standard free energy change for the oxidation of zinc at 700°C is given by the point B on this diagram; the corresponding equilibrium oxygen potential is labeled C. Consider a system with the oxygen potential labeled E on the P_{O_2} scale. From the construction that produced the P_{O_2} scale, it may be deduced that the point labeled D has a y value, in units of ΔG, given by $RT \ln(P_{O_2})_E$. The length (B−D) on the graph is thus:

$$B-D = RT \ln(P_{O_2})_{eq} - RT \ln(P_{O_2})_E = RT \ln \frac{(P_{O_2})_{eq}}{(P_{O_2})_E}$$

This difference is precisely the affinity for the reaction when the partial pressure is $(P_{O_2})_E$. Recall Equation 11.35

$$A = RT \ln \frac{Q}{K}$$

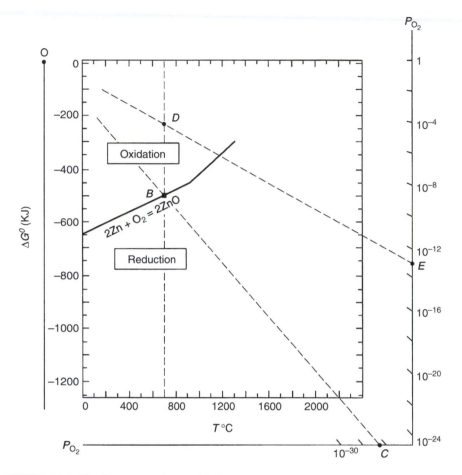

FIGURE 11.6 The line responding equilibrium states between Zn, ZnO and O_2 divides the graph into two domains. Combinations of oxygen potential and temperature that lie above the line oxidize the metal; combinations below the line reduce the oxide.

If the activity of metal and oxide are taken to be 1, then Q is $1/(P_{O_2})_D$ and K is $1/(P_{O_2})_{eq}$ and

$$A = RT \ln \frac{(P_{O_2})_{eq}}{(P_{O_2})_E} = B - D \qquad (11.75)$$

Thus the vertical distance from a point on the chart that represents any given nonequilibrium oxygen potential to the equilibrium line is equal to the affinity for the reaction for that composition. If D lies above the line the affinity is negative and, by arguments previously developed, products (the oxide) spontaneously form. In the context of the Richardson–Ellingham chart points in the domain above, the equilibrium line have affinities that correspond to oxidation of the metal or the formation of a stable oxide. Points lying below the line yield a positive affinity and products decompose; in this region the oxide is unstable and dissociates.

EXAMPLE 11.8.

Find the affinities for the oxidation reactions for copper, nickel, zinc, titanium and aluminum in a system with an oxygen potential of 10^{-16} atm at 1000°C. The standard free energies for these reactions at 1000°C may be read from the chart

	Cu_2O	NiO	ZnO	TiO_2	Al_2O_3
ΔG^0_{1000} (kJ)	− 175	− 255	− 425	− 680	− 855

The value of $RT \ln P_{O_2}$ corresponding to 10^{-16} atm at 1000°C is

$$D = (8.314)(1273)\ln(10^{-16}) = -390,000 \text{ J} = -390 \text{ kJ}$$

The values for the affinities are obtained by subtracting D from ΔG^0 for each reaction:

	Cu_2O	NiO	ZnO	TiO_2	Al_2O_3
A_{1000} (kJ)	+215	+135	− 35	− 290	− 465

Recall that the affinity may be thought of as $\mu_{prod} - \mu_{react}$. If $A > 0$, $\mu_{prod} > \mu_{react}$ and products decompose. For $A < 0$, $\mu_{prod} < \mu_{react}$ and products form. It is concluded that this atmosphere will reduce Cu_2O and NiO, while ZnO, TiO_2, and Al_2O_3 are stable in this system.

The demonstration that the affinity for a reaction for any given oxygen potential can be visualized as the vertical distance from the equilibrium line in the chart provides a convenient tool for deducing patterns of behavior. For example, consider a system with a fixed oxygen pressure containing a metal M and its oxide M_uO_v. Figure 11.6 serves to illustrate this situation for Zn and ZnO but the strategy is general. The point E locates the line through the origin that represents the locus of points on the chart that have the oxygen potential given by E. As the temperature of the system is changed for this fixed atmosphere, visualize the point D moving along the line OE. This line crosses the equilibrium reaction line for M at some temperature T_{eq} for which P_{O_2} at the point E is the dissociation pressure. At lower temperatures, D lies above the line and the M oxidizes; above T_{eq} this atmosphere is reducing to M_uO_v. Thus if the oxide is heated in this atmosphere it will be stable up to T_{eq} and will then begin to decompose with further increase in temperature.

A comparison of the relative stabilities of the oxides may be visualized with the same construction. Consider any temperature, T, on the chart. The oxygen potential for any atmosphere represented by a point D located on that vertical constant temperature line may be obtained from the P_{O_2} scale. Oxides with equilibrium lines that lie above D will be reduced, i.e., are unstable, in that atmosphere; those below are stable. Evidently the order of stability of the oxides at any given temperature is given by the order in which their equilibrium lines cross that temperature line.

This observation is the basis for one strategy for preventing the oxidation of a particular metal at high temperatures in the laboratory. For example, the oxidation of a nickel sample may be prevented by encapsulating it with titanium chips, usually prepared by simply turning a titanium rod in a lathe and collecting the chips. The formation of TiO_2, by far the more stable oxide in this system, establishes the oxygen potential in the system. This potential is well below the dissociation pressure of NiO and a nickel sample will remain clean. The titanium is said to be a "getter" of oxygen. This strategy is also used in high vacuum systems where a coiled filament of titanium is heated by passing an electric current once a moderate vacuum is established by a vacuum or mercury pump. The hot titanium filament "gets" oxygen molecules in the system and dissolves them in the titanium, which also happens to have an unusually high solubility for oxygen.

The arguments just presented for evaluating the affinity extend in a straightforward way to systems in which the metal and oxide are not in their pure reference states so that their activities differ from 1. In this case, the point B in Figure 11.6 has an equilibrium constant, K, which may be read from the K scale on the chart. A system not in equilibrium that has an activity quotient Q above K on that scale may be represented by the corresponding point D at the temperature T. The affinity for the reaction in this state is given by Equation 11.35 and again corresponds to the vertical distance (B–D) on the chart. If D lies above B, the metal will oxidize and oxide is stable; if D is below B, the oxide dissociates.

11.3.2   OXIDATION IN CO/CO_2 AND H_2/H_2O MIXTURES

In Section 11.3.1 it is assumed that the only component present in the gas phase is oxygen; thus the only means of controlling the oxygen partial pressure is through a reduction in the total pressure of the system. Since the best vacuum attainable in the laboratory is about 10^{-10} atm, which is well above most dissociation pressures, this approach to controlling oxidation is both inflexible and very limited. The most convenient means for controlling the oxygen potential of an atmosphere and thus the oxidation reactions that occur in its presence is through control of the chemical composition of the gas phase. The simplest atmospheres that provide this control incorporate mixtures of CO/CO_2, H_2/H_2O or both. The development required to understand the behavior of these two classes of atmospheres is identical.

Focus upon oxidation behavior in an atmosphere containing CO and CO_2. At any temperature the oxidation potential, i.e., the partial pressure of oxygen in the equilibrated atmosphere, is controlled by the ratio of the partial pressures of CO and CO_2. This relationship derives from the condition for equilibrium that applies to an atmosphere containing these three components. The reaction that describes this univariant system is

$$2CO + O_2 = 2CO_2$$

The standard free energy change for this reaction is plotted as a function of temperature on the Richardson–Ellingham chart. The equilibrium constant for this reaction, assuming the gas mixture to be ideal, is

$$K = \frac{P_{CO_2}^2}{P_{CO}^2} \frac{1}{P_{O_2}} \qquad (11.76)$$

which again makes use of the fact that the reference state partial pressures for any component X, P_X^0 is 1 atm so that $a_X = P_X/P_X{}^0 = P_X$, i.e., the activity is numerically equal to the partial pressure of X in the atmosphere. At any temperature ΔG^0 for this reaction may be read from the chart and the numerical value of K can be computed or read from the K scale. At that temperature for any value of the partial pressure of oxygen there is a unique value for the ratio (P_{CO_2}/P_{CO}):

$$P_{O_2} = \frac{P_{CO_2}^2}{P_{CO}^2} \frac{1}{K} \qquad (11.77)$$

A high ratio of CO_2 to CO in the atmosphere yields a high oxygen potential; to decrease the oxygen potential, increase the relative concentration of the reducing gas, CO.

The construction in Figure 11.5 demonstrates that the ordinate on the chart may be interpreted as a $[RT \ln P_{O_2}]$ scale. With this interpretation in mind, consider the locus of points in this graph that correspond to a fixed ratio of $[P_{CO_2}/P_{CO}]$. The connection is established through the condition for equilibrium related to Equation 11.76:

$$\Delta G_{[CO_2]}^0 = \Delta H_{[CO_2]}^0 - T\Delta S_{[CO_2]}^\circ = -RT \ln\left(\frac{P_{CO_2}^2}{P_{CO}^2} \frac{1}{P_{O_2}} \right)$$

which may be written

$$\Delta H_{[CO_2]}^0 - T\Delta S_{[CO_2]}^\circ = -RT \ln\left(\frac{P_{CO_2}}{P_{CO}} \right)^2 - RT \ln \frac{1}{P_{O_2}}$$

$$RT \ln P_{O_2} = \Delta H_{[CO_2]}^0 + T\left[R \ln\left(\frac{P_{CO_2}}{P_{CO}} \right)^2 - \Delta S_{[CO_2]}^\circ \right] \qquad (11.78)$$

For a constant ratio of $[P_{CO_2}/P_{CO}]$ a plot of this relationship on a $[RT \ln P_{O_2}]$ vs. T chart is a straight line with an intercept at $T = 0$ K of $\Delta H_{CO_2}^0$ and a slope equal to the quantity in brackets in Equation 11.78. Figure 11.7 visualizes the envelope of straight lines, all passing through the point C on the $T = 0$ K line, that represent constant (CO/CO_2) ratios. This envelope can be replaced by constructing a (CO/CO_2) scale around the perimeter of the chart in the same manner as the K and P_{O_2} scales already discussed.

To calculate the oxygen potential of any given CO/CO_2 mixture at a particular temperature use the K scale to determine the equilibrium constant for the CO/CO_2 reaction, substitute into Equation 11.77 and compute the partial pressure of oxygen in equilibrium in the given atmosphere and temperature. To determine the oxygen potential graphically using Figure 11.7, lay a straight edge on the chart connecting the C point with the given value on the (CO/CO_2) scale, labeled A in the example in Figure 11.7. Mark the point P at which this line intersects the temperature of interest. The oxygen partial pressure corresponding to the point P is found as before: lay a straight edge from O through P and read its intersection on the P_{O_2} scale, labeled B in Figure 11.7. This construction may be used to read oxygen potentials for any combination of (CO/CO_2) ratio and temperature on the chart.

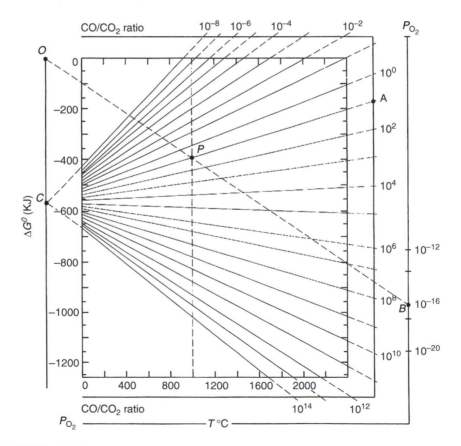

FIGURE 11.7 Locus of points (oxygen potential and temperature) on the Richardson–Ellingham chart that corresponds to fixed values of the ratio CO/CO_2 in the gas phase.

The procedure for relating (H_2/H_2O) ratios to oxygen partial pressures is identical except that the intercept of the envelope of radiating lines is $\Delta H^0_{H_2O}$, labeled H on the chart, and the (H_2/H_2O) scale applies. All four scales, K, P_{O_2}, (CO/CO_2) and (H_2/H_2O), are constructed in Figure 11.4.

EXAMPLE 11.9.

Estimate the oxygen potential for a gas mixture with a CO_2/CO ratio of 10^{-4} at 1200°C. The standard free energy change for the CO/CO_2 reaction at 1200°C may be read from Figure 11.4: $\Delta G^0 = -310$ kJ. The corresponding equilibrium constant is

$$K = e^{-(-310,000/8.314 \times 1473)} = 9.8 \times 10^{10}$$

The corresponding oxygen partial pressure is given by Equation 11.77:

$$P_{O_2} = \left(\frac{P_{CO_2}}{P_{CO}}\right)^2 \frac{1}{K} = (10^{-4})^2 \frac{1}{9.8 \times 10^{10}} = 1.0 \times 10^{-19} \text{ atm}$$

To obtain the same result graphically, lay a straight edge on the chart connecting the point C on the ordinate and the value 10^{+4} on the (CO/CO_2) scale. (Caution: the scale is constructed in terms of the ratio of CO to CO_2; the input information in this example gives the value for the ratio of CO_2 to CO.) Mark the point at which this line crosses the 1200°C temperature line; the coordinates of this point are (1200, -530). Now lay a straight edge from the O point on the chart through this point and read to P_{O_2} scale:

$$P_{O_2} \cong 10^{-19} \text{ atm}$$

EXAMPLE 11.10.

What is the maximum water content that could be tolerated in a hydrogen atmosphere used to prevent the oxidation of copper samples annealed at 900°C?

Use the H_2/H_2O scale to find the ratio in equilibrium with copper and its oxide at 900°C. Lay a straight edge from the H point on the ordinate through the point on the diagram where the ΔG^0 curve for Cu and Cu_2O crosses the 900°C line; the coordinates of this point are (900, -184). Read the corresponding equilibrium ratio on the (H_2/H_2O) scale:

$$\frac{P_{H_2}}{P_{H_2O}} \cong \frac{1}{10^3} \cong 10^{-3}$$

The corresponding oxygen partial pressure is about 10^{-8} atm and may be neglected in computing the composition of the atmosphere. Thus so long as the partial pressure of water vapor is kept below 10^{-3} or about 0.1%, the formation of copper oxide at 900°C may be prevented. The composition of water vapor may be monitored by measuring the dew point of the atmosphere.

The arguments presented in Section 11.3.1 that demonstrate the application of the Richardson–Ellingham chart to compute the affinity for a system that is not at equilibrium may also be applied to the (CO/CO_2) and (H_2/H_2O) scales on the chart. For example, any combination of an atmosphere composition given by a (CO/CO_2) ratio and a temperature plots as a point D on the chart, as shown in Figure 11.6.

Every such point has a corresponding oxygen potential. The vertical distance from that point D to a point B on an equilibrium line that represents oxidation of some metal M of interest is the affinity for the reaction

$$\frac{2u}{v}M + 2CO_2 = \frac{2}{v}M_uO_v + 2CO \tag{11.79}$$

If D lies above B, the atmosphere oxidizes M; if D is below B, the oxide dissociates.

11.4   PREDOMINANCE DIAGRAMS AND MULTIVARIANT EQUILIBRIA

Consider a system with c components made from e elements and exhibiting $r = (c - e)$ independent reactions. The conditions for chemical equilibrium in this system may be described by a system of equations that have the form given by Equation 11.66:

$$\Delta G^0_{[j]} = -RT \ln K_{[j]} \qquad [j = 1, 2, ..., r] \tag{11.66}$$

It was emphasized that an equation of this form may be written for every chemical reaction that may be devised among the components in the system, although only r of these equations are independent. The rest of the equations describing the system may be deduced from these independent equations. As the number of components increases, the calculation and representation of the equilibrium state becomes progressively less tractable.

A convenient representation of the behavior of a multivariant system is embodied in a predominance diagram that may be constructed to represent its complex behavior as a reacting system. Of the collection of independent potentials necessary to describe the state of the system, two that are of particular interest in a given application are chosen as variables to be explored in describing the behavior of the system. All but one of the remaining potentials is fixed. This permits display of the rivalry between the components that are competing for predominance in the system on a two-dimensional graph. Since the variables involved are all thermodynamic potentials, such predominance diagrams take the form of a cell structure with areas that represent domains of predominance of a particular component separated by lines that are *limits of* predominance of competing components; these lines meet at triple points. Predominance diagrams have an appearance that is similar to phase diagrams plotted in potential space, see (Sections 10.3 and 10.4 in Chapter 10). However, as will be demonstrated, while a predominance diagram may provide an approximate representation of such a phase diagram, they are not identical with a phase diagram. Domains of predominance displayed in these diagrams are not quantitavely identical with the domains of stability normally represented in phase diagrams.

11.4.1 POURBAIX HIGH TEMPERATURE OXIDATION DIAGRAMS

This form of predominance diagram chooses as independent variables the oxygen potential and temperature and displays the domains of predominance of the various oxides of a metal that exhibits more than one stable oxide. Consider a metal M that is found to exist in two different oxidation states yielding oxides with the formulas MO and MO_2. This system of four components (M, O_2, MO and MO_2) and two elements (M and O) has $r = 4 - 2 = 2$ independent reactions; it is bivariant. A total of three reactions may be written among the components; as in the development of the Richardson–Ellingham chart, it is convenient to write these reactions in terms of 1 mol of O_2 consumed in each case:

$$2M + O_2 = 2MO \qquad [11.9]$$

$$M + O_2 = MO_2 \qquad [11.10]$$

$$2MO + O_2 = 2MO_2 \qquad [11.11]$$

Let R denote the component other than O_2 on the reactant side of each equation and P denote the product component. Then each of these reactions has the generic form

$$xR + O_2 = yP \qquad [11.12]$$

The condition for equilibrium associated with each reaction has the form:

$$\Delta G^0 = -RT \ln K = -RT \ln \frac{a_P{}^y}{a_R{}^x} \frac{1}{P_{O_2}} \qquad (11.80)$$

which may also be written

$$\Delta H^0 - T\Delta S° = -RT \ln f + RT \ln P_{O_2}$$

where f may be defined to be the predominance ratio for the reaction:

$$f \equiv \frac{a_P{}^y}{a_R{}^x} \qquad (11.81)$$

If $f \gg 1$, the product component in the reaction predominates at equilibrium; if $f \ll 1$, the reactant predominates. Solve for

$$\ln P_{O_2} = \frac{\Delta H^0}{RT} + \left[\ln f - \frac{\Delta S°}{R} \right] \qquad (11.82)$$

In the practical application of this equation it is convenient to convert from natural to common logarithms by dividing through by ln 10 = 2.303:

$$\log P_{O_2} = \frac{\Delta H^0}{R'T} + \left[\log f - \frac{\Delta S^\circ}{R'} \right] \qquad (11.83)$$

where $R' = 2.303$, $R = 19.147$ (J/mol K). This is the working equation involved in constructing these predominance diagrams. It applies to all oxidation reactions written so that O_2 is a reactant with a coefficient of 1; it is a rewritten form of Equation 11.80, the condition for equilibrium, with no simplifying assumptions.

Assume that ΔH^0 and ΔS° are independent of temperature, i.e., that the integrals involving ΔC_P for the reaction are negligible, see Equation 11.70. For a fixed value of f, the predominance ratio, a plot of $\log P_{O_2}$ vs. $(1/T)$ for any given equilibrium is a straight line with slope equal to $(\Delta H^0/R')$ and intercept given by $[\log f - \Delta S^\circ/R']$. It is conventional to graph the $(1/T)$ scale in reverse so that the temperature qualitatively increases from left to right; this convention changes the sign of the slope on the graph to $(-\Delta H^0/R')$. Since oxidation reactions are exothermic (heat is given off in the reaction so that the heat absorbed by the system is negative) slopes of lines on these plots are positive.

Figure 11.8 sketches a plot of Equation 11.82 for a reaction similar to [11.9]. In this representation the predominance ratio, f, is treated as a parameter and allowed to vary from 10^{-4} to 10^{+4}. Since the scale is logarithmic, each order of magnitude change in f displaces the intercept by 1 unit. Below the line, where $f \ll 1$, the reactants predominate at equilibrium; above the line where $f \gg 1$, the product component is the predominant specie at equilibrium. In the range of values of f near $f = 1$ neither reactants nor products may be said to predominate. It is convenient to define the condition $f = 1$ to be the *limit of predominance* with respect to the competing components. The line drawn with the condition $f = 1$ serves as a boundary between domains of predominance of the components involved.

The competition between three of the components, such as is represented by Equation [11.9] to Equation [11.11] at the beginning of this section, may be represented by three envelopes of parallel lines similar to those shown for a single reaction in Figure 11.8. Note that the values of ΔH^0 and ΔS° for these three reactions are not independent; reaction [11.11] is a linear combination of reactions [11.9] and [11.10]. For the particular stoichiometry in this example, [11.11] = 2[11.10] − [11.9]. Figure 11.9(a) illustrates the juxtaposition of three competing reaction lines. Figure 11.9(b) shows details of the structure in the region around their common intersection. Focus upon the solid lines representing the limits of predominance, $f = 1$, in each case. That these three lines intersect at a single point can be derived algebraically from the relationships between their slopes and intercepts dictated by the linear relation among the three chemical reactions they represent. Segments of these three intersecting lines bound domains of predominance of each of the competing components. Extensions of these lines beyond the triple point are dashed because in each case they represent competition between two components in a domain in which the third component has independently

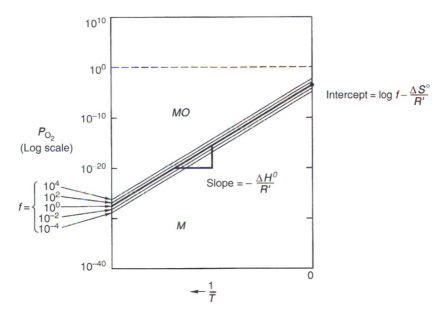

FIGURE 11.8 Plot of equilibrium oxygen potential for reaction [11.9] as a function of temperature for a set of values of the predominance ratio, f.

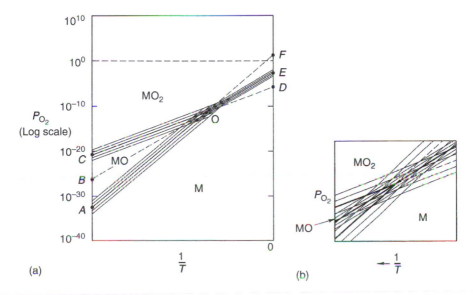

FIGURE 11.9 (a) Predominance diagram for a system with three competing components M, MO and MO_2. (b) The pattern of behavior near the triple point. Darker lines correspond to the limits of predominance for which $f = 1$.

been shown to be predominant. Below the line labeled OA in Figure 11.9(a), M predominates with respect to MO; above the line OC, MO_2 predominates at equilibrium relative to MO. Thus the dashed segment OB represents the competition between M and MO_2 in a region in which MO has been shown to be the predominant component with respect to both M and MO_2. The same argument may be applied to the extensions labeled OD and OF.

In order to generate a high temperature oxidation predominance diagram it is necessary to list all of the oxide forms known to exist for the given metal. Heats and entropies of formation of each of the oxide forms must be obtained from a database as input. Next, an exhaustive list of all of the possible reactions between metal and oxides, as well as between the oxides, is assembled. Values of ΔH^0 and ΔS° may be computed for all of these reactions from the information about the formation reactions; no additional information is required. The generic form of an oxidation reaction on these diagrams is

$$xM_aO_b + O_2 = yM_uO_v$$

It is easy to show that the coefficients required to balance these equations depend upon the chemical formulas of the components as

$$x = \frac{2u}{va - ub} \qquad y = \frac{2a}{va - ub} \qquad (11.84)$$

Reactions involving the pure metal M may be included in this compilation simply by setting $b = 0$. Enthalpies and entropies of all reactions are related to the corresponding properties in the formation reactions:

$$\Delta H^0_{[J]} = y\Delta H_{f,M_uO_v^0} - x\Delta H_{f,M_aO_b^0} \qquad (11.85)$$

and similarly for the entropy change.

An efficient approach for generating such a diagram focuses upon computation of the triple points formed by the intersection of triplets of predominance limit lines. Begin with a list of the components other than oxygen, i.e., the metal and all of its oxides. Make an exhaustive list of the arrangements of these components taken three at a time; this represents the possible triple points that may exist in the system. The coordinates of any triple point may be computed from the intersection of any two of the three lines that meet at that point. Consider, for example, the triple point formed in the limits of predominance for three oxides, M_2O, MO and MO_2. The corresponding oxidation reactions, based on 1 mol of oxygen, may be written with coefficients determined by inspection or computed from Equation 11.84:

$$2M_2O + O_2 = 4MO \qquad\qquad [A]$$

$$\frac{2}{3}M_2O + O_2 = \frac{4}{3}MO_2 \qquad\qquad [B]$$

$$2MO + O_2 = 2MO_2 \qquad\qquad [C]$$

From heats and entropies of formation of these compounds, compute ΔH^0 and ΔS° for each of these reactions. Note that [C] is a linear combination of [A] and [B]. Expressions based on Equation 11.83 give equations for the limits of predominance for reactions [A] and [B]:

$$\log P_{O_2} = \frac{\Delta H^0_{[A]}}{R'T} - \frac{\Delta S^\circ_{[A]}}{R'}$$

$$\log P_{O_2} = \frac{\Delta H^0_{[B]}}{R'T} - \frac{\Delta S^\circ_{[B]}}{R'}$$

$$\log P_{O_2} = \frac{\Delta H^0_{[C]}}{R'T} - \frac{\Delta S^\circ_{[C]}}{R'}$$

Since the slopes and intercepts of these lines are related, it can be shown that these lines intersect at a single point. The coordinates of this triple point in $(1/T, \log P_{O_2})$ space may be found by solving any two of these equations simultaneously. The resulting relations are:

$$\frac{1}{T_{ABC}} = \frac{\Delta S^\circ_{[A]} - \Delta S^\circ_{[B]}}{\Delta H^0_{[A]} - \Delta H^0_{[B]}} \tag{11.86}$$

$$\log P_{O_2,ABC} = \frac{\Delta H^0_{[B]}\Delta S^\circ_{[A]} - \Delta H^0_{[A]}\Delta S^\circ_{[B]}}{R'(\Delta H^0_{[A]} - \Delta H^0_{[B]})} \tag{11.87}$$

Coordinates of all of the triple points that exist among the components taken three at a time may be computed from these equations. Triple points that lie outside the domain of the diagram (e.g., for which $(1/T)$ is negative or $P_{O_2} > 10^{10}$) are deleted from the list. A further subset of triple points may be deleted from consideration because they lie in a region of predominance of a component that is not involved in forming the triple point. This subset may be determined from an analysis of the predominance ratios for all of the reactions that contain the three components in a given triple point. The remaining triple points all appear on the finished predominance diagram. To complete the diagram, construct straight lines that represent limits of predominance between pairs of triple points that share a common pair of components. Finally, label the fields thus enclosed with the component that is predominant in that area.

EXAMPLE 11.11.
Construct a predominance diagram for manganese and its oxides:

Component	ΔH^0_f (kJ)	ΔS°_f (J/K)
Mn	0	0
MnO	−385	−73
Mn_3O_4	−1387	−357
MnO_2	−521	−184

(Mn$_2$O$_3$ may be observed under some conditions but is neglected in this example.) First write down all of the balanced reactions that combine these oxides with each other or with the metal Mn. There are 4!/(2!2!) or six such reactions. Compute the values of ΔH^0 and ΔS° for all six reactions; they are linear combinations of the formation quantities given above. Use these values to compute slopes and intercepts and plot the limits of predominance given by setting $f = 1$ in Equation 11.83. Next write out the possible combinations of these four components taken three at a time: 4!/(3!2!) or four possible triple points. Use Equation 11.86 and Equation 11.87 to compute the triple points in the diagram:

Triple Point	$(1/T)(K - 1)$	$\log P_{O_2}$
[Mn, MnO, Mn$_3$O$_4$]	-4.24×10^{-4}	24.21
[Mn, MnO, MnO$_2$]	-1.53×10^{-4}	13.76
[Mn, Mn$_3$O$_4$, MnO$_2$]	-3.19×10^{-4}	10.48
[MnO, Mn$_3$O$_4$, MnO$_2$]	2.81×10^{-4}	11.04

In this case three of the triple points yield negative values for $1/T$. This is mathematically valid, but not thermodynamically so. Only the [MnO, Mn$_3$O$_4$, MnO$_2$] lies within the diagram; however, even this point occurs at a very high oxygen potential corresponding to a partial pressure of oxygen of about 10^{11} atm. Connect the dots to form the diagram (Figure 11.10).

11.4.2 Predominance Diagrams with Two Compositional Axes

Consider the reaction of a metal M with an atmosphere that contains two independent compositional variables. The most common example of this class of

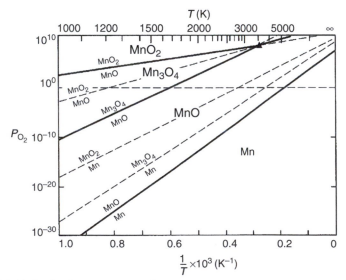

FIGURE 11.10 Predominance diagram for the manganese–oxygen system constructed by computing triple points and connecting them appropriately.

diagram displays domains of predominance of the various compounds that may be formed in the presence of an atmosphere containing two active molecular species, e.g., oxygen and sulfur, O_2 and S_2. The axes of the diagram are taken to be the potentials $\log P_{O_2}$ and $\log P_{S_2}$. In order to plot such a diagram the temperature must be fixed. Competing components now involve oxides, sulfides and components with M, S and O in their formulas. Three types of generic equations may be constructed:

$$\text{No } S_2 : x_1 M_a S_b O_c + O_2 = y_1 M_u S_v O_w \qquad [11.13]$$

$$\text{No } O_2 : x_2 M_a S_b O_c + S_2 = y_2 M_u S_v O_w \qquad [11.14]$$

$$\text{Both } S_2 \text{ and } O_2 : x_3 M_a S_b O_c + O_2 + m S_2 = y_3 M_u S_v O_w \qquad [11.15]$$

Normally equations of type [11.13] are simple oxidation equations, generically identical with that which leads to Equation 11.84 for the stoichiometric coefficients. However, the components may contain sulfur provided that $b/a = v/u$. A similar statement may be made for the sulfidation equation of type [11.14]; Equation 11.84 also yields the stoichiometric coefficients required to balance this class of equations. Conservation of M, S and O atoms for this class of equations yields three equations for the coefficients for equations in class [11.15]:

$$x_3 = \frac{2u}{aw - cu}; \qquad y_3 = \frac{2a}{aw - cu}; \qquad m = \frac{av - ub}{aw - cu} \qquad (11.88)$$

EXAMPLE 11.12.
Construct a balanced equation with the components MS_2, O_2, S_2, and $M_2S_3O_5$. Comparison with the generic form given in Equation [11.15] above gives:

$$M_a S_b O_c \text{ is } MS_2 \Rightarrow a = 1; \qquad b = 2; \qquad c = 0$$

$$M_u S_v O_w \text{ is } M_2 S_3 O_5 \Rightarrow u = 2; \qquad v = 3; \qquad w = 5$$

so that

$$(aw - cu) = (1 \times 5 - 0 \times 2) = 5$$

and

$$(av - ub) = (1 \times 3 - 2 \times 2) = -1$$

The corresponding coefficients are:

$$x_3 = \frac{(2 \times 2)}{5} = \frac{4}{5}; \qquad y_3 = \frac{(2 \times 1)}{5} = \frac{2}{5}; \qquad m = -\frac{1}{5}$$

The balanced equation reads:

$$\frac{4}{5}MS_2 + O_2 = \frac{1}{5}S_2 + \frac{2}{5}M_2S_3O_5$$

The negative result obtained for m simply places S_2 on the right side of the equation. To check the balance note that there are 4/5 mol of M atoms, 2 mol of O and 8/5 mol of S on both sides of the equation.

A list of the components considered to exist in the system yields an exhaustive compilation of the reactions between competing components. The condition for equilibrium for any of these reactions may be written:

$$\Delta G^0_{[j]} = -RT \ln K_{[j]} = R'T \log K_{[j]} \qquad [j = 1, 2, ..., r] \qquad (11.89)$$

For reactions of type [11.13] that do not involve S_2:

$$K_{[11.13]} = \frac{f_{[11.13]}}{P_{O_2}}$$

where $f_{[11.13]}$ is the predominance ratio for the nonoxygen components. Equation 11.89 becomes

$$\Delta G^0_{[11.13]} = -R'T \log f_{[11.13]} + R'T \log P_{O_2}$$

At the limit of predominance, which defines the boundary between regions of predominance of the two nonoxygen components in this case, $f_{[11.13]} = 1$ and $\log f_{[11.13]} = 0$. Thus the equation defining the limit of predominance for this class of reactions is

$$\log P_{O_2} = \frac{\Delta G^0_{[11.13]}}{R'T} \qquad (11.90)$$

i.e., for the fixed temperature of the system, $\log P_{O_2}$ is a constant. A similar argument applied to equations of type [11.14] yields

$$\log P_{S_2} = \frac{\Delta G^0_{[11.14]}}{R'T} \qquad (11.91)$$

for the limit of predominance. The equilibrium constant for reactions that involve both S_2 and O_2 is

$$K_{[11.15]} = \frac{f_{[11.15]}}{P_{O_2} P_{S_2}{}^m}$$

Equation 11.89 becomes

$$\Delta G^0_{[11.15]} = -R'T[\log f_{[11.15]} - \log P_{O_2} - m \log P_{S_2}]$$

The limit of predominance for this class of reactions is obtained by setting $f_{[11.15]} = 1$ or $\log f_{[11.15]} = 0$. Then

$$\log P_{O_2} = -m \log P_{S_2} + \frac{\Delta G^0_{[11.15]}}{R'T} \tag{11.92}$$

Expression 11.90 to Expression 11.92 are the equations for the limits of predominance for each of the three classes of reactions. When plotted on coordinates of $\log P_{O_2}$ vs. $\log P_{S_2}$ as in Figure 11.11, these equations yield simple results. On this scale Equation 11.90 simply reads "$\log P_{O_2}$ is a constant" and plots as a horizontal line. This reflects the fact that S_2 is not involved in class [11.13] reactions so that their competition is independent of the S_2 content in the atmosphere. Equation 11.91 reads "$\log P_{S_2}$ is a constant and plots as a vertical line." Competitions between components that involve both O_2 and S_2, class [11.15],

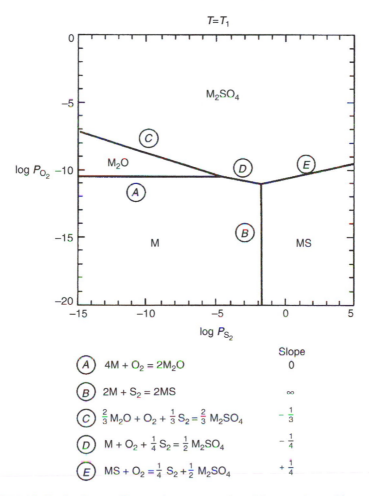

FIGURE 11.11 Predominance diagram for a system describing reactions with a gas phase containing both oxygen and sulfur.

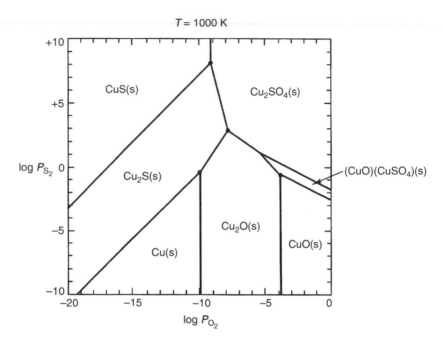

FIGURE 11.12 Predominance diagram calculated for the system Cu–O–S at 1000 K.

described by Equation 11.92 plot as a straight line with slope equal to $-m$ (which is determined by the stoichiometric formulas for the compounds involved and is given by Equation 11.88) and an intercept (at log $P_{S_2} = 0$ or $P_{S_2} = 1$) on the log P_{O_2} scale at $\Delta G_{[11.15]}^{0}/R'T$. An example of the predominance diagram for copper in oxygen–sulfur atmospheres at 1000 K is shown in Figure 11.12.

11.4.3 INTERPRETATION OF PREDOMINANCE DIAGRAMS

Predominance diagrams provide a convenient means for displaying the behavior of multivariant reacting systems. Their construction assumes that the system has reached equilibrium and also that heats and entropies of reactions are essentially independent of temperature. It is important that the distinction be established between a predominance diagram for a system and the phase diagram for the same system. Figure 11.13 compares the predominance diagram and the phase diagram for the iron–oxygen system. They are similar but not identical in appearance; both form simple cell structures with areas bounded by curves and triple points.

An area in a phase diagram plotted on potential coordinates is a domain of stability of a single phase. The phases shown in this diagram are: α and γ iron-rich terminal phases, the liquid and the three intermediate phases, two line compounds (at least at low temperatures), Fe_2O_3 and Fe_3O_4 and wustite, which is near (but not on) the composition FeO and exhibits significant deviations from that stoichiometric

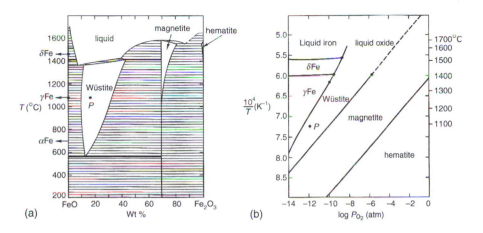

FIGURE 11.13 Comparison of (a) the phase diagram with (b) the analogous predominance diagram in ($\log P_{O_2} - 1/T$) space for the Fe–O system.

composition. At the composition and temperature labeled P in Figure 11.13(a) the system will consist of a single phase, wustite, at equilibrium.

In contrast, the predominance diagram treats the three intermediate compounds, Fe_2O_3, Fe_3O_4, and FeO, as components rather than phases. At the corresponding point P the predominance diagram in Figure 11.13(b) the system is considered to consist of a mixture of all of the components Fe, O_2, FeO, Fe_2O_3 and Fe_3O_4. In this view the component designated by the molecular formula FeO predominates over all of the remaining components at the point P. At equilibrium the system is predicted to consist mostly of the chemical component FeO.

It will be observed that from a practical point of view these two results are essentially equivalent; Figure 11.13(a) and (b) appear very similar. Regions of phase stability correspond reasonably well with regions of predominance of the component that has the nominal composition of that phase. From a fundamental point of view, these descriptions are distinctly different. However, it can be demonstrated with rigor that with careful manipulation these different descriptions of system behavior can be transformed into one another. Specifically, the true phase diagram for a system may be derived from a description of the system in terms of components that involve assumed intermediate molecular compositions such as M_2O or M_2O_5 and not just M and O. However, in this transformed version of predominance diagrams, the limits of stability of the phases do not correspond precisely with the limits of predominance which, after all, are given by the condition $f_{[j]} = 1$, chosen because it simplifies the computations involved. Nonetheless, predominance diagrams provide a useful basis for displaying the pattern of compound formation as a function of the applied thermodynamic potentials. They may be used to place limits on tolerable atmosphere compositions, form a basis for the design of atmospheres, support suggestions for candidate materials for use in a given atmosphere and suggest possible chemical and thermal histories that a failed material may have experienced.

11.5 COMPOUNDS AS COMPONENTS IN PHASE DIAGRAMS

It is common practice in the chemical literature, particularly in ceramic materials, to choose chemical compound compositions as the components in displaying a phase diagram. Where this is justified the description of behavior of the systems may be greatly simplified. It is not always justified.

Figure 11.14 presents the phase diagram for the alumina–silica system. The components in this diagram are taken to be Al_2O_3 and SiO_2. The four phases that exist in this system (alumina, mullite, silica and liquid) are solutions. These solutions are viewed as mixtures, not of the components Al, Si and O, but of the components Al_2O_3 and SiO_2. The free energies of mixing of these phases are each expressed in terms of the mole fractions of the components Al_2O_3 and SiO_2. Chemical potentials and activities in the system are properties of the components Al_2O_3 and SiO_2.

$$\mu_{Al_2O_3} \equiv \left(\frac{\partial G^\alpha}{\partial n_{Al_2O_3}} \right)_{T,P,n_{SiO_2}}$$

The conditions for equilibrium that determine the positions of the phase boundaries are also familiar; for example, for the $(\alpha + \varepsilon)$ [alumina + mullite] two-phase equilibrium,

$$\mu_{Al_2O_3}^\alpha = \mu_{Al_2O_3}^\varepsilon; \qquad \mu_{SiO_2}^\alpha = \mu_{SiO_2}^\varepsilon$$

The calculation of this phase diagram may be formulated in terms of solution models for the phases, viewed as solutions of Al_2O_3 and SiO_2 and information about the relative stability of the phase forms for pure alumina and silica as presented in detail in

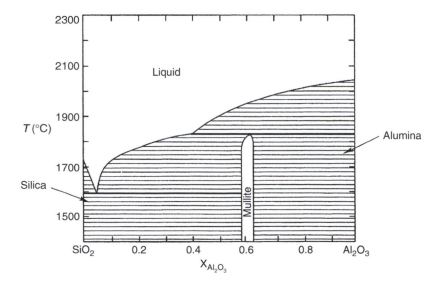

FIGURE 11.14 Alumina–silica phase diagram in which molecular compounds are treated as components.

Chapter 10. For many purposes this representation is adequate to understand the behavior of the system.

On a more sophisticated level, it is understood that this diagram is actually a pseudobinary section through the ternary Al–Si–O phase diagram. The sequence of compositions used to designate the abscissa in Figure 11.14 lies along the straight line connecting the compositions corresponding to Al_2O_3 on the Al–O binary side of the diagram with the composition SiO_2 on the Si–O binary. The single-phase regions that are intersected by that line may exhibit small but finite composition ranges, departing from stoichiometric ratios in directions, which lie outside the pseudobinary composition line. Since these small deviations are associated with defects in the structure of the phases, they may exert influences on properties that are significant. Whether the system can be treated as a binary alumina–silica system or as the ternary Al–Si–O system depends upon the application.

In many cases it is not possible to represent the phase relationships in a system made from two compounds on a binary phase diagram. Figure 11.15 shows the ternary isotherm for a system A–B–C. This system has intermediate compounds on the binaries at AB, AC and BC. The compounds AB and AC lie at ends of a two-phase field in the ternary and, at least at the temperature shown, may be treated as forming a binary system. A binary may also be constructed

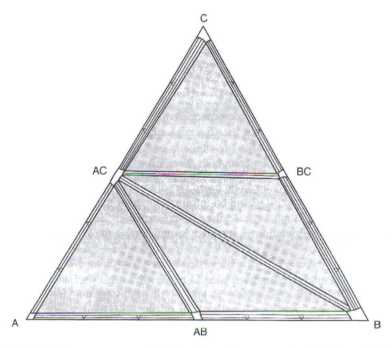

FIGURE 11.15 The ternary system ABC has three line compounds: AB, AC and BC. These compounds may be used in plotting the binary ceramic systems (AB–AC) and (AC–BC). However the (BC–AB) section cuts across two and three phase fields, a "binary" phase diagram with AB and BC as components cannot be constructed.

using the compounds AC and BC as components. However, an attempt to construct a binary phase diagram using the two compounds AB and BC would be disastrous. Equilibria that are sampled by the sequence of compositions along the line joining these compound compositions involve tie lines and tie triangles connecting compositions that are far outside the line. The behavior of the system in this region cannot be described by a binary phase diagram involving compound components.

11.6 SUMMARY

A system is classified as reacting if the number of components, c, is larger than the number of elements, e. The number of independent chemical reactions a system may exhibit, r, is equal to $c - e$.

In an isolated reacting system the equations conserving the number of moles of each component must be replace by constraints that conserve the number of gram atoms of each element. These constraints give rise to relations between changes in molar concentration that may occur, which in turn produce relations between the chemical potentials of those components when the system attains equilibrium.

The affinity for a reaction is defined to be the difference between the chemical potential of the products and reactants. For

$$lL + mM = rR + sS$$

$$A \equiv (r\mu_R + s\mu_S) - (l\mu_L + m\mu_M)$$

The affinity is zero when the system reaches equilibrium. If, for a given nonequilibrium state,

$$A < 0 \qquad \text{Products Form}$$
$$A > 0 \qquad \text{Products Decompose}$$

The condition for equilibrium in a reacting system is

$$A = 0$$

Substitution of the definition of activity for each chemical potential converts this equation to

$$\Delta G^0 = -RT \ln K$$

where ΔG^0 is the standard free energy for the reaction,

$$\Delta G^0 \equiv (rG_R^0 + sG_S^0) - (lG_L^0 + mG_M^0)$$

and K is the equilibrium constant:

$$K \equiv \frac{a_R{}^r a_S{}^s}{a_L{}^l a_M{}^m}$$

For a system that has not reached equilibrium the proper quotient of activities, Q, has the same form as K but the activities in this ratio correspond to some state of interest other than the equilibrium state. The affinity may be written:

$$A = RT \ln \frac{Q}{K}$$

and the direction of spontaneous change for the given state depends upon the value of Q/K:

$$\frac{Q}{K} < 1 \Rightarrow A < 0 \qquad \text{Products Form}$$

$$\frac{Q}{K} > 1 \Rightarrow A > 0 \qquad \text{Products Decompose}$$

For a multivariate reacting system for which $r = (c - e) > 1$, there are $(c - e)$ independent conditions for equilibrium that have the form

$$\Delta G^0_{[j]} = -RT \ln K_{[j]} \qquad [j = 1, 2, ..., (c - e)]$$

Since it can be shown that all reactions that may be written among the components are linear combinations of this set of independent reactions, an equation of this form must be satisfied for every chemical reaction that can be written for the system.

If the system involves condensed phases then some components may exhibit departures from the simple stoichiometric ratios of the elements characteristic of the gas phase. To the extent that departures from stoichiometry may be neglected, multiphase systems may be treated as reacting systems and the conditions for equilibrium described above may be applied. Where departures from stoichiometry are important in the application at hand, it is necessary to recognize these compositional variations and treat the problem as a competition for stability between phases rather than as a competition for predominance between components.

The Richardson–Ellingham chart for the oxides is a succinct summary of the thermodynamic behavior of oxide-forming reactions. Especially constructed scales on the diagram permit easy estimation of the equilibrium oxygen potential and the equivalent CO/CO_2 ratio and H_2/H_2O ratio for any metal and its oxide presented on the chart. Similar charts may be constructed for other classes of compounds such as sulfides, chlorides, nitrides and carbides.

If departures from stoichiometry are not critical, intermediate phases may be treated as components in a multivariate reacting system. The equilibrium condition of the system may then be adequately represented on a predominance diagram. The typical predominance diagram is plotted in potential space, with axes either temperature or the chemical potential of governing components. The lines on the diagram are limits of predominance that arise in the comparison of the reacting components two at a time. Domains enclosed within the lines are regions in which, at equilibrium, one component has a concentration that is larger than that of every other component in the competition.

HOMEWORK PROBLEMS

Problem 11.1.   Consider a gas mixture at a high temperature with the following components:

$$H_2, H_2O, O_2, SiO(g), SiH_4, Si(g)$$

a. List the elements in the system.
b. How many independent chemical reactions does this system have?
c. Make a complete list of all of the possible reactions.
d. From this list choose a set of independent reactions.
e. Select one of the remaining reactions and show that it is a linear combination of independent reactions.

Problem 11.2.   A gas mixture at 1200 K has the following composition

Component	CO	CO_2	O_2
Mole fraction	0.25	0.60	0.15

Compute the affinity for the reaction

$$CO + \frac{1}{2}O_2 = CO_2$$

for this mixture. In what direction will the reaction go?

Problem 11.3.   Consider the following gas mixture at 900 K .

Component	H_2	H_2O	O_2	CH_4	CO	CO_2
Mole fraction	0.19	0.05	1×10^{-5}	0.30	0.05	0.50

a. Write a set of independent reactions for this system.
b. Obtain equilibrium constants for these reactions.
c. Evaluate the affinity for each of these reactions.
d. What can you conclude about the direction of spontaneous change for this system?

Problem 11.4.   A gas mixture initially has the composition:

Component	CO	O_2	CO_2
Mole fraction	0.50	0.12	0.38

    a. Write out the conservation equations for atoms of carbon and oxygen in terms of the three n_k values.

    b. Write an expression for the total number of moles, n_T in the system.

    c. Write the expression for the equilibrium constant in terms of the n_k values and n_T.

    d. Solve this set of equations to find the equilibrium composition for this reaction at 1000 K.

Problem 11.5. A gas mixture of nitrogen and oxygen may contain the following components:

$$N_2, O_2, NO, NO_2, N_2O_4, N_2O, N_2O_5$$

    a. Write out the set of formation reactions for this mixture.

    b. Write all of the reactions of the form

$$nA + O_2 = mB$$

    that may be constructed for this system.

    c. From this list, how many sets of three equations that each share in three components (other than oxygen) may be written?

    d. List the triplets of components involved.

Problem 11.6. Nickel forms NiO upon exposure to an atmosphere containing oxygen.

    a. Find the standard free energy of formation of NiO as a function of temperature.

    b. Evaluate and plot the equilibrium constant for this reaction as a function of temperature.

    c. Compute the equilibrium oxygen partial pressure at 962 K, assuming the nickel and NiO are pure.

    d. Compute the affinity for a system with pure nickel and pure NiO in air at 962 K.

Problem 11.7. Consider the oxidation of silicon with water vapor at 800°C.

    a. Find the standard free energy of formation of SiO_2 and H_2O.

    b. Combine them to compute ΔG^0 for the reaction

$$Si + 2H_2O = SiO_2 + 2H_2$$

    c. Evaluate the equilibrium constant and the (H_2/H_2O) ratio in equilibrium with silicon and its oxide at 800°C.

    d. Compute the equilibrium partial pressure of oxygen in a silicon–silica system at 800°C.

e. Compute the equilibrium partial pressure of oxygen for a gas mixture at 800°C with an (H_2/H_2O) ratio computed in part (c).
f. Compare oxygen pressure values in parts (d) and (e).

Problem 11.8.   Use the Richardson–Ellingham chart to verify the results of the calculation in Problem 11.7.

Problem 11.9.   Use the Richardson–Ellingham chart for oxides to find the following:
   a. The dissociation pressure of CoO at 1000°C.
   b. The equilibrium constant for the formation of SiO_2 at 800°C.
   c. The (CO/CO_2) ratio in equilibrium with Na and Na_2O at 600°C.
   d. Will an atmosphere with an H_2/H_2O ratio of $10^5/1$ prevent the oxidation of chromium at 1200°C?
   e. Find the composition of an H_2/H_2O atmosphere that has the same oxygen potential as an atmosphere with a CO/CO_2 ratio of 10^3 at 900°C.
   f. Find the oxygen potential in part (e).

Problem 11.10.   From the Richardson–Ellingham chart, will:
   a. Aluminum reduce Fe_2O_3 to iron at 1200°C?
   b. An atmosphere with a CO/CO_2 ratio of 0.06 oxidize nickel at 800°C?
   c. Silica oxidize liquid zinc at 700°C?
   d. Sodium reduce water to oxygen and hydrogen at 500°C?

Problem 11.11.   Compute the affinities for the following reactions under the following conditions. Assume nongaseous components are pure.

Reaction	T (°C)	P_k (atm)
$4Cu + O_2 = 2Cu_2O$	800	$P_{O_2} = 2 \times 10^{-8}$
$2Ca + O_2 = 2CaO$	1200	$P_{O_2} = 2 \times 10^{-16}$
$2Ni + S_2 = 2NiS$	700	$P_{S_2} = 2 \times 10^{-2}$
$Ca + Cl_2 = CaCl_2$	600	$P_{Cl_2} = 3 \times 10^{-6}$

Problem 11.12.   Compute and plot a Pourbaix high temperature diagram for copper and its oxides, CuO and Cu_2O.

Problem 11.13.   Compute and plot a predominance diagram for nitrogen and its oxides, NO, NO_2, N_2O_4, N_2O and N_2O_5. Plot only the limits of predominance.

Problem 11.14.   Write and balance chemical equations relating the following components:
   a. Na_2O and Na_2SO_3.
   b. Nb_2O_5 and NbS_2O_5.

   c. Cu_2O and $CuSiO_4$.
   d. Cu_2S and $CuSO_4$.

Problem 11.15. List the possible triple points that may exist in a log P_{O_2} − log P_{S_2} predominance diagram involving the following components:

$$Fe, FeO, Fe_2O_3, Fe_3O_4, FeS_2, FeS, FeSO_4$$

Problem 11.16. Can you find a CO/CO_2 atmosphere that will both prevent the oxidation of an iron-carbon alloy with $a_C = 0.48$ at 1200 K and at the same time carburize it?

Problem 11.17. Initially the composition of a system is

Component	H_2	H_2O	CO	CO_2	O_2
Mole fraction	0.40	0.10	0.40	0.08	0.02

   a. Find the equilibrium composition for this system at 800°C.
   b. Compute the affinities for the reactions from this starting point at 800°C.

REFERENCES

1. Lide, D.R., Editor in chief, *CRC Handbook of Physics and Chemistry*, 71st ed., CRC Press, Boca Raton, FL, 1990.
2. Cox, J.D., Wagman, D.D., and Medvedev, V.A., *CODATA Key Values of Thermodynamics*, Hempshire Publishing Corporation, New York, 1989.
3. Barin, I., Knacke, O., and Kubaschewski, O., *Thermochemical Properties of Inorganic Substances Supplement*, Springer, New York, 1977.
4. Chase, M.W. et al., JANAF thermodynamic tables, *J. Phys. Chem. Ref. Data*, 11(Suppl. 2) 1982.
5. Access to the SOLGASMIX program is available through F*A*C*T* (Facility for the Analysis of Chemical Thermodynamics), Ecole Polytechnique, CRCT, Montreal, Quebec, Canada.

12 Capillarity Effects in Thermodynamics

CONTENTS

Thermodynamic systems that consist of more than one phase necessarily have internal interfaces, i.e., regions in which the transition from the intensive properties of one phase to those of a neighboring phase is accommodated. Since this transition is accommodated over distances that are of the order of a few atoms, these regions are nearly two-dimensional in their geometry and are commonly called surfaces or interfaces between the phases.

Effects associated with these zones of transition and their geometry are unusually important in materials science. Solid materials have an internal microstructure; the geometry of microstructure plays a key role in the behavior of materials. Properties are altered and controlled by processing the material, e.g., heat treating it, hot pressing it, extruding it, etc., under conditions that change the internal microstructure. Since the features in the microstructure are defined by the interfaces that bound them, the thermodynamic behavior of interfaces plays a central role in both the processing of materials and in their performance in service.

Atoms residing in the transition zone between two phases necessarily have patterns of neighbors that are not characteristic of either phase. Each element of surface of an interface thus has an extra energy associated with it that may be called its surface or interfacial energy. A rigorous definition of this property is given in Section 12.2. This surface energy operates to influence properties of the adjacent phases through the geometry of the surface. The geometric property that operates in these relationships is the local curvature of the surface or interface. Section 12.1 develops the geometric concepts associated with the curvature of surfaces and the relation between curvature and other geometric properties of the interface.

Effects upon thermodynamic properties that derive from the curvature of interfaces in a system are commonly called capillarity effects because the initial documentation of these relationships was studied experimentally with fine glass tubes, called capillaries, which could be made with reproducible internal radii or curvature. These capillarity effects derive from the conditions for equilibrium in a multicomponent two-phase system with curved interfaces. The general criterion for equilibrium, and the strategy for deriving the governing equations at equilibrium, shown in the vision of the structure of thermodynamics in Figure 4.1, will be applied to this class of systems in Section 12.3. It is shown that, in the presence of curved interfaces, the condition for mechanical equilibrium is altered. A purely mechanical development of this result is presented in Section 12.4. The consequences of this result are deduced in Section 12.5.

The anisotropy of surface energy in crystalline materials and its role in determining the equilibrium shape of crystals is developed in Section 12.6.

A second geometric feature found in most microstructures results from the intersection of three interfaces to form a line, called a triple line in the structure. The conditions for equilibrium at a triple line are developed in Section 12.7. Application to wetting of one phase by another, penetration of a liquid into a solid along its internal grain boundaries, and the development of the geometry of microstructures illustrate these results.

Owing to the arrangement of atoms in the transition zone between phases being different from either phase, the properties of that region will be different from either adjacent phase. In particular, the composition of the components may be significantly different in the transition zone than in either adjacent phase when the system attains equilibrium. This phenomenon, called adsorption, is dealt with in Section 12.8.

12.1 THE GEOMETRY OF SURFACES

In order to develop the relations between the thermodynamic properties of surfaces and the properties of the adjacent phases it is first necessary to introduce the concepts required to describe the geometry in the vicinity of a point P on a surface. The quantities required, called the principal normal curvatures at P, are defined in this section. In addition, it is shown that if a curved surface moves in space, e.g., as the result of kinetic processes that occur in the adjacent phases, the change in area of the surface is related to the volume swept through by this displacement through the local curvature of the surface.

Consider first the definition of the curvature at a point P on a curved line in two-dimensional space, Figure 12.1. In this construction two points, A and B, on the curve are selected. A circle may be constructed through any three points in the plane; construct the circle that passes through the points A, P and B. This circle has a center and a radius. Now let A and B approach P. In the limit as A and B arrive at P a unique circle is approached in this construction. This circle, called the "osculating circle at P" because it just kisses the curve at this point, has a center, O in Figure 12.1, called the center of curvature for P and a radius r, called the radius of curvature at P. The reciprocal of the radius,

$$k = \frac{1}{r} \tag{12.1}$$

is called the local curvature of the curve at the point P. In general, k varies continuously as P moves along the curve.

Now consider an element of surface on a smoothly curved surface embedded in three-dimensional space, Figure 12.2. There exists a tangent plane at any point P and a unit vector **N** perpendicular to the tangent plane called the normal vector at P. A plane that intersects this surface at P and contains the normal, i.e., a plane that is locally perpendicular to the tangent plane, Figure 12.2a, produces a curve of intersection with the surface that is conceptually identical to the plane curve shown in Figure 12.1. Thus the curvature of this line of intersection through P has a value given by the construction outlined for Figure 12.1.

If the intersecting plane is rotated to another orientation that still contains the normal, the intersecting curve in general will be different. The local curvature at P will vary as the intersecting plane rotates about the normal through P, Figure 12.2b. In differential geometry it is shown that there are two orientations of the normal plane for which the curvature of the intersecting line has a maximum and a minimum value; these are called the principal directions at P on the surface. It can be shown that these directions are perpendicular to each other.

These observations lead to the construction shown in Figure 12.3. **N** is the normal vector at P, and $\hat{u}$ and $\hat{v}$ are unit vectors in the principal directions. r_1 and r_2 are the principal radii of curvature at P; $\kappa_1 = 1/r_1$ and $\kappa_2 = 1/r_2$ are the principal normal curvatures at P. The latter two quantities are the most convenient descriptors of the local surface geometry.

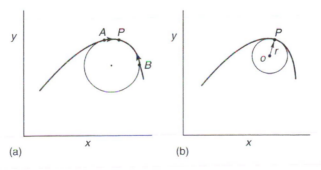

FIGURE 12.1 Definition of the curvature at a point P on a line in two dimensions.

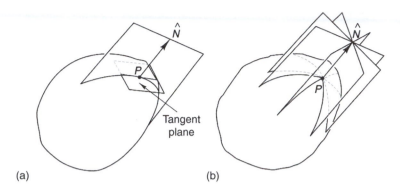

FIGURE 12.2 (a) A plane through a point P on a surface and containing the local surface normal, N, intersects the surface with a plane curve that can be analyzed as in Figure 12.1. (b) the local curvature of each of these curves of intersection varies with the orientation of the intersecting plane.

Two geometric properties related to the local curvatures find useful application:

$$H \equiv \frac{1}{2}(\kappa_1 + \kappa_2) \tag{12.2}$$

is the local mean curvature of the surface at P. The product of the curvatures,

$$K \equiv \kappa_1 \kappa_2 \tag{12.3}$$

is sometimes called the total curvature at P. It will be shown that thermodynamics operates in the geometry of a microstructure through H, the local mean curvature.

Consider a smooth closed surface, i.e., one that encloses a microstructural feature such as a particle of the β phase. Define the curvature to be positive if the vector at P that points toward its center of curvature points into the β phase and negative if it points outward. Three kinds of surface elements may be found on a smooth surface that encloses a feature or a phase, Figure 12.4:

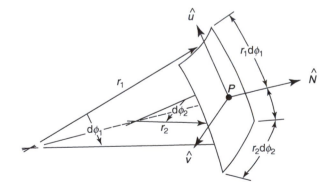

FIGURE 12.3 The normal N and principle directions u and v define two planes (N, u) and (N, v) that are normal to the surface at P. The lengths r_1 and r_2 are radii of curvatures in these planes. The reciprocals of these radii are the *principle normal curvatures* at P.

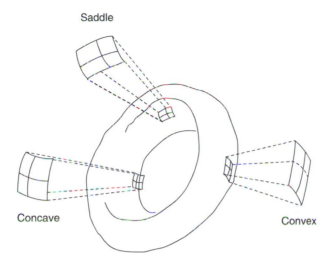

FIGURE 12.4 Three classes of surfaces are illustrated for a bowl: (a) convex, (b) concave, and (c) saddle.

a. Convex surface, for which both vectors point inward.
b. Concave surface, with both vectors pointing outward.
c. Saddle surface, for which the two vectors point in opposite directions.

The local mean curvature, H, is always positive for convex surface elements, negative for concave elements, and may be either positive or negative for saddle surface, depending upon the relative values of the curvatures.

Figure 12.3 shows an element of surface near P with principal radii of curvature r_1 and r_2 generating the element by rotating through infinitesimal arcs $d\phi_1$ and $d\phi_2$. The dimensions of the element are $r_1 d\phi_1$ by $r_2 d\phi_2$. The area of the element

$$dA_0 = r_1 d\phi_1 r_2 d\phi_2 \qquad (12.4)$$

Now suppose as a result of processes that occur in the adjacent phases the interface moves a distance δn along its normal, Figure 12.5. The volume swept out by this displacement is

$$\delta V = \delta n dA_0 = \delta n r_1 d\phi_1 r_2 d\phi_2 \qquad (12.5)$$

The radii of curvature change to $(r_1 + \delta n)$ and $(r_2 + \delta n)$. The area of the element after displacement is

$$dA_1 = (r_1 + \delta n)(r_2 + \delta n) d\phi_1 d\phi_2 = [r_1 r_2 + (r_1 + r_2)\delta n + \delta n^2] d\phi_1 d\phi_2$$

The term δn^2 may be neglected as a higher order differential. The change in area associated with the displacement of the interfacial element is

$$\delta A = dA_1 - dA_0 = (r_1 + r_2)\delta n d\phi_1 d\phi_2$$

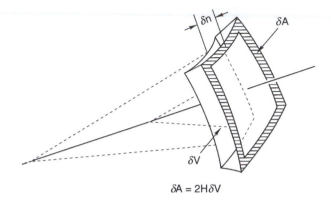

FIGURE 12.5 The displacement of a surface element a distance δn along its normal sweeps through a volume δV and changes the area of the element by an amount δA.

Multiply and divide by $r_1 r_2$:

$$\delta A = \frac{(r_1 + r_2)}{r_1 r_2} \, \delta n r_1 r_2 \mathrm{d}\phi_1 \mathrm{d}\phi_2 \tag{12.6}$$

Note that the ratio may be written

$$\frac{r_1 + r_2}{r_1 r_2} = \frac{1}{r_1} + \frac{1}{r_2} = \kappa_1 + \kappa_2 = 2H$$

Further, the remaining product of factors in Equation 12.6, $\delta n r_1 r_2 \mathrm{d}\phi_1 \mathrm{d}\phi_2$, may be recognized as the volume swept through, δV (see Equation 12.5). Thus the change in surface area of a curved surface as it migrates is related to the volume swept through and the local mean curvature:

$$\delta A = 2H \delta V \tag{12.7}$$

This relationship constitutes a constraint upon the variables that describe the system; it plays a key role in the derivation of the conditions for equilibrium in systems with curved internal interfaces.

12.2   SURFACE EXCESS PROPERTIES

This section presents the apparatus developed by J. Willard Gibbs for ascribing an appropriate part of the total value of an extensive property of a system to the presence of internal interfaces. Gibbs realized that the zone of transition between two phases is too thin to admit direct measurement of its properties. (That assumption is being tested today.). He proposed a strategy for evaluating the contribution due to surfaces in terms of the measurable properties of the macroscopic system. This strategy leads to the definition of surface excess thermodynamic properties.

Figure 12.6 illustrates a system composed of two phases, α and β, and the transition zone, labeled σ, between them. The boundaries of the transition zone need not be precisely defined so long as they enclose the region in which occurs all of the changes in intensive properties from their values in α to their values in β. Gibbs referred to the transition zone σ as the "physical surface of discontinuity," recognizing that it is not truly a surface but is a slab of small but finite thickness.

Let B' be any extensive thermodynamic property. A corresponding local density of B may be defined at any point P in the system. Consider a small volume V' that has the value B' for the property of interest. The local density of the property B at the point P is given by

$$\lim_{V' \to 0} \frac{B'}{V'} \equiv b_v \tag{12.8}$$

Since it has a value defined at each point in the system, b_v is an intensive property. The value of the property B' contained in a volume element dv neighboring P is $b_v dv$. The total value of B' for the system is evidently the integral of the density b_v over the volume of the system:

$$B' = \iiint_{V'} b_v dv \tag{12.9}$$

If the system is in internal equilibrium and no external fields operate, the intensive properties within each phase do not vary with position outside of the physical surface of discontinuity. Figure 12.6b sketches the variation of b_v with position through the physical surface of discontinuity. Suppose the system under consideration has a constant cross-sectional area. Then the total value of B'_{sys} for the system may be visualized in Figure 12.6b as proportional to the shaded area under the curve.

Gibbs next introduced a true two-dimensional geometric surface, which he called the "dividing surface". The location of the dividing surface is arbitrary except that it lies within the physical surface of discontinuity, Figure 12.6c. It will be shown

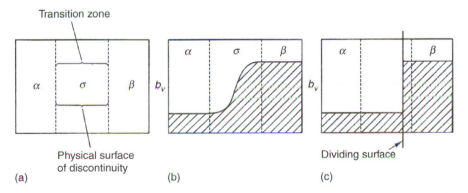

FIGURE 12.6 (a) Visualization of the zone of transition, σ, between two phases defining the *physical surface of discontinuity*. (b) Sketch of the variation of the density of a thermodynamic property, b_v, through the zone of transition. (c) The hypothetical system introducing the *dividing surface* with the step function variation of the density, b_v.

that some choices for the actual location in a given application are more convenient than others. Once the location has been chosen in conjunction with one of the properties of the system, it must be retained for all of the others.

The surface excess of B, i.e., the part of the total value of B' for the system that may be associated with the presence of interface, is assigned by imagining a hypothetical system in which the densities b_v^α and b_v^β in α and β are maintained up to the dividing surface, at which there is a discontinuity in b_v, Figure 12.6c. The total value of B'_{hyp} for this hypothetical system may be visualized as being proportional to the shaded area under the density curve in Figure 12.6c; this quantity may be calculated from the known densities, b_v^α and b_v^β, in the parts of the system away from the interface where these quantities are constant, and the location of the dividing surface.

The surface excess of B for the system as a whole is defined to be:

$$B'^s \equiv B'_{sys} - B'_{hyp} \tag{12.10}$$

That is, B'^s is the difference in B evaluated for the actual system and that computed from the intensive properties b_v^α and b_v^β and the location of the dividing surface. This quantity may be visualized as being proportional to the difference between the area under the curve in Figure 12.6b and under the step function in Figure 12.6c. Let A be the area of the interface in the system. Then the specific surface excess of B is given by

$$B^s \equiv \frac{B'^s}{A} \tag{12.11}$$

This quantity, which has units of "B per unit area," may be defined for any extensive thermodynamic property. Thus S^s is the specific surface excess entropy for the α/β interface, G^s is the specific surface excess Gibbs free energy, and so on. Of particular interest in Section 12.7 is the specific surface excess number of moles of component k, which is the central notion in the description of adsorption phenomena. Specific surface excess properties are intensive variables; this observation is important because it implies that they are functions of the intensive properties of the system but not the extensive properties.

Equation 12.10 and Equation 12.11 may be combined with minor rearrangement to give

$$B'_{sys} = B'_{hyp} + B'^s = B'_{hyp} + B^s A \tag{12.12}$$

Since the hypothetical system consists of an α and a β part, this expression may be written more explicitly as

$$B'_{sys} = B'^\alpha_{hyp} + B'^\beta_{hyp} + B^s A \tag{12.13}$$

An arbitrary change in the state of this system must now explicitly include the possibility that the area of the interface may change. Thus for an arbitrary change,

$$dB'_{sys} = dB'^\alpha_{hyp} + dB'^\beta_{hyp} + B^s dA \tag{12.14}$$

12.3 CONDITIONS FOR EQUILIBRIUM IN SYSTEMS WITH CURVED INTERFACES

The general strategy for finding conditions for equilibrium is applied to a system consisting of α and β phases each containing c components and separated by curved interfaces. First, write an expression for the change in entropy of the system. According to Equation 12.14, this will include contributions from two phases and the surface excess term:

$$dS'_{sys} = dS'^{\alpha}_{hyp} + dS'^{\beta}_{hyp} + S^s dA \tag{12.15}$$

where S^s is the specific superficial excess entropy for the system. The hypothetical subsystems, with entropy densities of the phases extending to the dividing surface, have intensive properties that are identical with those of the α and β phases outside the zone of transition. Thus,

$$dS'_{sys} = \left[\frac{1}{T^{\alpha}} dU'^{\alpha}_{hyp} + \frac{P^{\alpha}}{T^{\alpha}} dV'^{\alpha}_{hyp} - \sum_{k=1}^{c} \frac{\mu_k^{\alpha}}{T^{\alpha}} dn_{k,hyp}^{\alpha} \right]$$
$$+ \left[\frac{1}{T^{\beta}} dU'^{\beta}_{hyp} + \frac{P^{\beta}}{T^{\beta}} dV'^{\beta}_{hyp} - \sum_{k=1}^{c} \frac{\mu_k^{\beta}}{T^{\beta}} dn_{k,hyp}^{\beta} \right] + S^s dA \tag{12.16}$$

The extensive properties in the right-hand side of this relation are understood to be those of the hypothetical system.

Next, write out the isolation constraints. The internal energy is constant in an isolated system with internal interfaces may be written:

$$dU'_{sys} = 0 = dU'^{\alpha}_{hyp} + dU'^{\beta}_{hyp} + U^s dA$$

or

$$dU'^{\beta}_{hyp} = -(dU'^{\alpha}_{hyp} + U^s dA) \tag{12.17}$$

The volume is constant:

$$dV'_{sys} = 0 = dV'^{\alpha}_{hyp} + dV'^{\beta}_{hyp} + V^s dA$$

The superficial excess volume, V^s, is zero since the construction that leads to the definition of superficial excess properties assumes that the total volume of the hypothetical system is the same as the actual system. Thus this constraint yields simply:

$$dV'^{\alpha}_{hyp} = -dV'^{\beta}_{hyp} \tag{12.18}$$

The total number of moles of each component is constant in the system:

$$dn_{k,sys} = 0 = dn_{k,hyp}^{\alpha} + dn_{k,hyp}^{\beta} + n_k^s dA$$

or

$$dn^\beta_{k,\text{hyp}} = -(dn^\alpha_{k,\text{hyp}} + n^s_k dA) \tag{12.19}$$

Substitute the isolation constraints, Equation 12.17 to Equation 12.19 into the entropy expression Equation 12.16 to eliminate the dependent variables in an isolated system.

$$dS'_{\text{sys,iso}} = \frac{1}{T^\alpha}dU'^\alpha_{\text{hyp}} + \frac{P^\alpha}{T^\alpha}dV'^\alpha_{\text{hyp}} - \sum_{k=1}^{c}\frac{\mu^\alpha_k}{T^\alpha}dn^\alpha_{k,\text{hyp}} + \frac{1}{T^\beta}(-dU'^\alpha_{\text{hyp}} - U^s dA)$$

$$+ \frac{P^\beta}{T^\beta}(-dV'^\alpha_{\text{hyp}}) - \sum_{k=1}^{c}\frac{\mu^\beta_k}{T^\beta}(-dn^\alpha_{k,\text{hyp}} - n^s_k dA) + S^s dA$$

Collect like terms:

$$dS'_{\text{sys,iso}} = \left[\frac{1}{T^\alpha} - \frac{1}{T^\beta}\right]dU'^\alpha_{\text{hyp}} + \left[\frac{P^\alpha}{T^\alpha} - \frac{P^\beta}{T^\beta}\right]dV'^\alpha_{\text{hyp}}$$

$$- \sum_{k=1}^{c}\left[\frac{\mu^\alpha_k}{T^\alpha} - \frac{\mu^\beta_k}{T^\beta}\right]dn^\alpha_{k,\text{hyp}} + \left[S^s - \frac{1}{T^\beta}U^s + \sum_{k=1}^{c}\frac{\mu^\beta_k}{T^\beta}n^s_k\right]dA \tag{12.20}$$

The variables in this equation are still not all independent: there exists a geometric relationship between the area and volume variables derived in Equation 12.7. In order to apply this relationship it is first necessary to choose a convention for the sign of the surface curvature. Let the curvature be defined to be positive when the surface is convex relative to the β phase. Then the volume enclosed by the surface is that of the β phase, and Equation 12.7 must be written

$$dA = 2H dV^\beta_{\text{hyp}} = -2H dV^\alpha_{\text{hyp}} \tag{12.21}$$

Substitute this result into Equation 12.20 to eliminate dA:

$$dS'_{\text{sys,iso}} = \left[\frac{1}{T^\alpha} - \frac{1}{T^\beta}\right]dU'^\alpha_{\text{hyp}}$$

$$+ \left[\left(\frac{P^\alpha}{T^\alpha} - \frac{P^\beta}{T^\beta}\right) - \left(S^s - \frac{1}{T^\beta}U^s + \sum_{k=1}^{c}\frac{\mu^\beta_k}{T^\beta}n^s_k\right)2H\right]dV'^\alpha_{\text{hyp}}$$

$$- \sum_{k=1}^{c}\left[\frac{\mu^\alpha_k}{T^\alpha} - \frac{\mu^\beta_k}{T^\beta}\right]dn^\alpha_{k,\text{hyp}} \tag{12.22}$$

The conditions for equilibrium in this system may now be determined by setting all of the coefficients of the differentials in this equation simultaneously equal to zero.

The coefficient of the internal energy yields the condition for thermal equilibrium:

$$\frac{1}{T^\alpha} - \frac{1}{T^\beta} = 0 \quad \Rightarrow \quad T^\alpha = T^\beta \tag{12.23}$$

The coefficient of dn_k yields the condition for chemical equilibrium:

$$\frac{\mu_k^\alpha}{T^\alpha} - \frac{\mu_k^\beta}{T^\beta} = 0 \quad \Rightarrow \quad \mu_k^\alpha = \mu_k^\beta \quad (k = 1, 2, ..., c) \tag{12.24}$$

These results are identical in form with those obtained for a system without curved interfaces. The condition for mechanical equilibrium is not the same:

$$\left(\frac{P^\alpha}{T^\alpha} - \frac{P^\beta}{T^\beta} \right) - \left(S^s - \frac{1}{T^\beta} U^s + \sum_{k=1}^{c} \frac{\mu_k^\beta}{T^\beta} n_k^s \right) 2H = 0 \tag{12.25}$$

Since by Equation 12.23 the temperatures are equal, they may be eliminated from this equation by multiplying by T.

To simplify the application of this result, introduce the definition,

$$\gamma \equiv U^s - TS^s - \sum_{k=1}^{c} \mu_k n_k^s \tag{12.26}$$

It will subsequently be shown that, in simple systems, γ is the specific interfacial free energy. Then the condition for mechanical equilibrium in Equation 12.25 may be written

$$P^\beta - P^\alpha = 2\gamma H \tag{12.27}$$

Thus, at equilibrium in a two-phase system with curved interfaces, the pressures in the two phases are not equal. It can be shown that γ is always positive. Whether the pressure in the β phase is larger or smaller than in the α phase depends upon the sign of the mean curvature, H. Since, by the choice of convention made earlier, H is positive for surface elements that are convex relative to the β phase, the pressure inside a simple closed particle of the β phase is higher than that outside it at equilibrium. The magnitude of the pressure difference increases as the particle size gets smaller or, more generally, as H increases. The influences exerted by curved surfaces upon the thermodynamic behavior of systems developed in the remainder of this chapter are all directly traceable to this condition for mechanical equilibrium.

Alternate visualizations of γ may be obtained by developing the combined statement of the first and second laws for two-phase systems with curved interfaces. From the definition of surface excess quantities, the change in internal energy experienced by such a system may be written:

$$dU'_{\text{sys}} = dU'^{\alpha}_{\text{hyp}} + dU'^{\beta}_{\text{hyp}} + U^s dA \tag{12.28}$$

Application of the combined statements of the first and second laws to the uniform hypothetical systems gives

$$dU'_{\text{sys}} = \left[T^{\alpha} dS'^{\alpha}_{\text{hyp}} - P^{\alpha} dV'^{\alpha}_{\text{hyp}} + \sum_{k=1}^{c} \mu^{\alpha}_k dn^{\alpha}_k \right]$$
$$+ \left[T^{\beta} dS'^{\beta}_{\text{hyp}} - P^{\beta} dV'^{\beta}_{\text{hyp}} + \sum_{k=1}^{c} \mu^{\beta}_k dn^{\beta}_k \right] + U^s dA \tag{12.29}$$

If this two-phase system remains in equilibrium as its state is changed, Equation 12.23 and Equation 12.24 permit replacement of T^{α} and T^{β} with a single system temperature, T, and μ^{α}_k and μ^{β}_k with μ_k for the system. Equation 12.29 may be written,

$$dU'_{\text{sys}} = T(dS'^{\alpha}_{\text{hyp}} + dS'^{\beta}_{\text{hyp}}) - P^{\alpha} dV'^{\alpha}_{\text{hyp}} - P^{\beta} dV'^{\beta}_{\text{hyp}}$$
$$+ \sum_{k=1}^{c} \mu_k (dn^{\alpha}_{k,\text{hyp}} + dn^{\beta}_{k,\text{hyp}}) + U^s dA \tag{12.30}$$

Use Equation 12.14 to rewrite the terms in brackets in terms of the superficial excess properties:

$$dU'_{\text{sys}} = T(dS'_{\text{sys}} - S^s dA) - P^{\alpha} dV'^{\alpha}_{\text{hyp}} - P^{\beta} dV'^{\beta}_{\text{hyp}} + \sum_{k=1}^{c} \mu_k (dn_{\text{sys}} - n^s_k dA)$$
$$+ U^s dA \tag{12.31}$$

Collect like terms:

$$dU'_{\text{sys}} = T dS'_{\text{sys}} - P^{\alpha} dV'^{\alpha}_{\text{hyp}} - P^{\beta} dV'^{\beta}_{\text{hyp}} + \sum_{k=1}^{c} \mu_k dn_{\text{sys}} + \left(U^s - T S^s - \sum_{k=1}^{c} \mu_k n^s_k \right) dA$$

The coefficient of dA is precisely γ: thus

$$dU'_{\text{sys}} = T dS'_{\text{sys}} - P^{\alpha} dV'^{\alpha}_{\text{hyp}} - P^{\beta} dV'^{\beta}_{\text{hyp}} + \sum_{k=1}^{c} \mu_k dn_{\text{sys}} + \gamma dA \tag{12.32}$$

By applying a coefficient relation to dA in this equation γ may be appropriately interpreted in terms of the change in internal energy of a two-phase system with change in internal area,

$$\gamma = \left(\frac{\partial U'}{\partial A} \right)_{S', V'^{\alpha}, V'^{\beta}, n_k} \tag{12.33}$$

For a unary system ($c = 1$) it is possible to choose the location of the dividing surface so that n^s is zero. For this case Equation 12.26 becomes simply

$$\gamma = U^s - TS^s = F^s = G^s \qquad (12.34)$$

That is, γ is the specific interfacial free energy. (The Gibbs and Helmholtz functions are identical in this case because V^s is zero.) This property of the interface has units of energy per unit area and is usually reported in units of J/m^2. The value of γ depends upon the nature of both phases involved. For crystalline materials the value of γ also depends upon the orientation of the crystal plane(s) at the interface and the orientation of the interface itself. Table 12.1 presents typical values of γ measured

TABLE 12.1
Typical Values for Surface Energies of Selected Interfaces[3,4]

Material	γ (ergs/cm^2) (T °C)	Material	γ (ergs/cm^2) (T °C)
	Liquid metals at their melting point		
Cesium	60 (mp)	Gold	1140 (mp)
Lead	450 (mp)	Copper	1300 (mp)
Aluminum	866 (mp)	Nickel	1780 (mp)
Silicon	730 (mp)	Rhenium	2700 (mp)
	Solid–vapor surfaces of pure metals		
Bismuth	550 (250)	Copper	1780 (925)
Aluminum	980 (450)	Nickel	2280 (1060)
Gold	1400 (1100)	Tungsten	2800 (2000)
	Solid–liquid interfaces for pure metals		
Sodium	20	Gold	132
Lithium	30	Copper	177
Lead	33	Platinum	240
	Grain boundaries in pure metals[a]		
Aluminum	324 (450)	Copper	625 (925)
Iron (δ phase)	468 (1450)	Nickel	866 (1060)
Iron (γ phase)	756 (1350)	Tungsten	1080 (2000)
	Compounds		
Water (liquid)	72 (25)	MgO	1000 (25)
NaCl (100) face	300 (25)	TiC	1190 (1100)
Be$_2$O$_3$ (liquid)	80 (900)	CaF$_2$ (111)	450 (25)
Al$_2$O$_3$ (liquid)	700 (2080)	CaCO$_3$ (1010)	230 (25)
Al$_2$O$_3$ (solid)	905 (1850)	LiF (100)	340 (25)

[a] Grain boundary energies depend very strongly upon the character of the boundary.

Source: Murr, L.E., *Interfacial Phenomena in Metals and Alloys*, Addison-Wesley, Reading, MA, p. 101, 1975. With permission. Shewmon, P.G. and Robertson, W.M., *Metal Surfaces, Structure, Energetics and Kinetics*, ASM, Materials Park, OH, 1963.

for a variety of surfaces and interfaces. Values of γ for additional systems are collected in Appendix F.

EXAMPLE 12.1

Compute the pressure in a 1 μm diameter water droplet formed in supercooled water vapor at 80°C. The vapor pressure of water at 80°C is 0.52 atm. The specific interfacial free energy of water may be taken to be 80 erg/cm^2.
For a spherical particle the mean curvature, $H = 1/r$. From Equation 12.27,

$$P^{\mathrm{L}} = P^{\mathrm{V}} + 2\gamma\left(\frac{1}{r}\right) = 0.52 \text{ (atm)} + 2\left(80\frac{\text{ergs}}{\text{cm}^2}\right)\left(\frac{1}{0.5 \times 10^{-4}\text{cm}}\right)\left(\frac{1 \text{ J}}{10^7\text{ergs}}\right)$$

$$\times\left(\frac{82.06 \text{ cc}-\text{atm}}{8.314 \text{ J}}\right)$$

$$P^{\mathrm{L}} = 0.52 \text{ (atm)} + 0.10 \text{ (atm)} = 0.62 \text{ (atm)}$$

12.4 SURFACE TENSION: THE MECHANICAL ANALOGUE OF SURFACE FREE ENERGY

The mechanical effects derived from the surface free energy may also be interpreted as arising from a force acting tangentially in the physical surface of discontinuity. Figure 12.7 shows an element of area with arbitrary principal curvatures. If this element were cut out of the surface it would shrink and disappear. In order to prevent the shrinkage of the element it would be necessary to apply a tensile force tangentially around its perimeter.

Define the surface tension σ of the surface to be a normalized expression of this force with units of force per unit length, usually reported in units of dyn/cm. The force acting on the edge AB arising from surface tension is $F_1 = \sigma r_1 d\phi_1$ since $r_1 d\phi_1$ is the length over which the normalized force σ acts. Similarly, the force acting over the line CD is also $F_1 = \sigma r_1 d\phi_1$. The force acting on the segment BC and DA are each $F_2 = \sigma r_2 d\phi_2$. These force vectors act in the surface, Figure 12.7. Since the surface is curved, these forces are not quite perpendicular to the local surface normal, N. They make small angles $(1/2)d\phi_1$ and $(1/2)d\phi_2$ with vectors perpendicular to the normal. The component of the force acting along the segment AB that acts normal to the surface is

$$F_1 \sin\left(\frac{1}{2}d\phi_2\right) \cong F_1\left(\frac{1}{2}d\phi_2\right) = \sigma r_1 d\phi_1\left(\frac{1}{2}d\phi_2\right) \qquad (12.35)$$

since in the limit of small angles $\sin(x) \to x$. The normal component derived from the force acting along CD has the same value. The normal component due the force

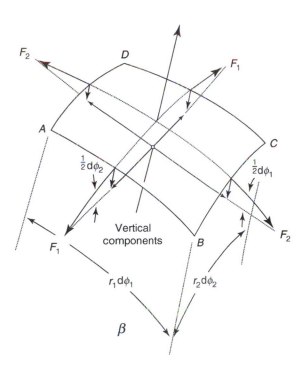

FIGURE 12.7 Forces acting on a surface element derived from the surface tension. If the surface is curved the forces have component that acts normal to the surface.

acting along BC is the same as that along DA and is given by

$$F_2 \sin\left(\frac{1}{2}d\phi_1\right) \cong F_2\left(\frac{1}{2}d\phi_1\right) = \sigma r_2 d\phi_2\left(\frac{1}{2}d\phi_1\right) \qquad (12.36)$$

The total force acting in the direction normal to the surface derived from the surface tension is the sum of these four components:

$$F_{surf} = 2\left[\sigma r_1 d\phi_1\left(\frac{1}{2}d\phi_2\right)\right] + 2\left[\sigma r_2 d\phi_2\left(\frac{1}{2}d\phi_1\right)\right] \qquad (12.37)$$

The normal force acting upon the β phase adjacent to the element of surface in Figure 12.7 is the sum of the force exerted by the α phase and that exerted by the surface tension acting through the curvature of the local surface element:

$$F^\beta = F^\alpha + F_{surf} \qquad (12.38)$$

The force exerted by the α phase may be expressed in terms of the pressure in the α phase acting over the area of the surface element:

$$F^\alpha = P^\alpha r_1 r_2 d\phi_1 d\phi_2 \tag{12.39}$$

The force acting on the β phase may be expressed in terms of the pressure in the β phase:

$$F^\beta = P^\beta r_1 r_2 d\phi_1 d\phi_2 \tag{12.40}$$

Substitute Equation 12.37, Equation 12.39, and Equation 12.40 into Equation 12.38:

$$P^\beta r_1 r_2 d\phi_1 d\phi_2 = P^\alpha r_1 r_2 d\phi_1 d\phi_2 + \sigma[r_1 + r_2]d\phi_1 d\phi_2$$

Divide through by $[r_1 r_2 d\phi_1 d\phi_2]$:

$$P^\beta = P^\alpha + \sigma \frac{r_1 + r_2}{r_1 r_2} \tag{12.41}$$

Note that

$$\frac{r_1 + r_2}{r_1 r_2} = \frac{1}{r_1} + \frac{1}{r_2} = 2H \tag{12.42}$$

So that Equation 12.41 may be written

$$P^\beta = P^\alpha + 2\sigma H \tag{12.43}$$

The equation is identical with Equation 12.27 with the specific interfacial free energy, γ, replaced by the surface tension σ.

Thus for the simplest case of a unary two-phase system in which both phases are isotropic fluids, the surface tension, which is a mechanical force acting in the surface, and the surface free energy, which is an excess energy associated with the presence of the interface, are equal. Their units (dyn/cm and erg/cm^2 = (dyn–cm)/cm^2) are convertible. However, the situation increases significantly in complexity if one or both phases are crystalline. The specific interfacial free energy is a scalar, i.e., a property of the system that has a value at each point on the surface, and varies with the orientation(s) of the crystal(s) involved. The surface tension passes to a surface stress in this more general case and is a tensor, as are all other local mechanical conditions in the solid state. Treatment of this more sophisticated case is beyond the scope of this introductory text. Comprehensive and understandable presentations of these ideas may be found in Refs. [1,2].

12.5 CAPILLARITY EFFECTS ON PHASE DIAGRAMS

The influence of curved interfaces upon the behavior of materials systems is manifested primarily through the shift of phase boundaries on phase diagrams derived from the altered condition for mechanical equilibrium. The melting point of a phase is altered if the solid phase is present as fine particles. The vapor pressure in equilibrium with liquid is raised if the liquid is present as a dispersion of droplets. Phase boundaries in a binary system are slightly shifted if the minor phase is finely divided. Although these effects are usually small unless the particulate phase is extremely fine, they play a profound role in determining the course of many microstructural processes that occur in materials science.

12.5.1 PHASE BOUNDARY SHIFTS IN UNARY SYSTEMS

Consider a unary two phase system, $(\alpha + \beta)$, in which it is explicitly recognized that the curvature of the $\alpha\beta$ boundary plays a role in the conditions for equilibrium:

$$T^\beta = T^\alpha \tag{12.44}$$

$$P^\beta = P^\alpha + 2\gamma H \tag{12.45}$$

$$\mu^\beta = \mu^\alpha \tag{12.46}$$

Recall that Equation 12.45 presumes that H is positive for surface elements that are convex relative to the β phase. The following development is an incremental generalization of the Clausius–Clapeyron equation derived in Chapter 7.

In considering possible variations in the state of this system it is necessary to include not only temperature and pressure but also explicitly to include variations that may arise from changes in the state of subdivision of the system reported in the geometric factor H. If an element of volume of the α phase is taken through an arbitrary change in state, the change in chemical potential is related to changes in temperature and pressure it may experience by the equation:

$$d\mu^\alpha = -S^\alpha dT^\alpha + V^\alpha dP^\alpha \tag{12.47}$$

Similarly, for a sample of the β phase,

$$d\mu^\beta = -S^\beta dT^\beta + V^\beta dP^\beta \tag{12.48}$$

If throughout these changes in state the α and β phases are maintained in thermodynamic equilibrium, then Equation 12.44 to Equation 12.46 must hold. Then

$$dT^\beta = dT^\alpha = dT \tag{12.49}$$

$$dP^\beta = dP^\alpha + 2\gamma dH \tag{12.50}$$

$$d\mu^\beta = d\mu^\alpha = d\mu \tag{12.51}$$

Equation 12.50 recognizes that the pressure in the β phase may be changed as a result of two independent influences: the pressure in the α phase and the curvature of the $\alpha\beta$ interface.

These last five equations may be combined into a single expression:

$$d\mu^\beta = -S^\beta dT + V^\beta(dP^\alpha + 2\gamma dH) = d\mu^\alpha = -S^\alpha dT + V^\alpha dP^\alpha$$

Combine like terms:

$$(S^\alpha - S^\beta)dT - (V^\alpha - V^\beta)dP^\alpha + 2\gamma V^\beta dH = 0 \qquad (12.52)$$

Define, as in Chapter 7,

$$\Delta S \equiv S^\alpha - S^\beta$$
$$\Delta V \equiv V^\alpha - V^\beta$$

Since by convention the β phase is the reference phase for the definition of sign of the curvatures in the system, these definitions define changes from the reference (β) phase to the α phase. This convention is maintained throughout this development and must be used in applying the results to practical problems. With these definitions Equation 12.52 may be written:

$$\Delta S dT - \Delta V dP^\alpha + 2\gamma V^\beta dH = 0 \qquad (12.53)$$

The first two terms are identical with those obtained in the derivation of the Clausius–Clapeyron Equation 7.14 in which variations due to geometric effects were not considered; the third term contains the effect of capillarity influences upon the equilibrium conditions.

12.5.2 Vapor Pressure in Equilibrium with Curved Surfaces

Systematic exploration of Equation 12.53 is most readily carried out by comparing equilibrium states in systems in which one of the three variables is fixed. To derive an expression for the effect of curvature on vapor pressure, compare two liquid/vapor systems that are at the same temperature but differ infinitesimally in their state of subdivision reported by a difference in mean curvature, dH, Figure 12.8. In this comparison the β phase is the liquid, α is the vapor and $dT = 0$; Equation 12.53 becomes

$$-\Delta V dP + 2\gamma V^L dH = 0 \qquad (12.54)$$

If it is assumed that the vapor behaves like an ideal gas,

$$\Delta V = V^G - V^L \cong V^G = \frac{RT}{P}$$

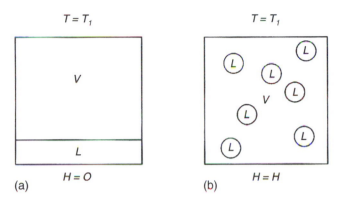

FIGURE 12.8 Compare two unary liquid plus vapor systems that have the same temperature: (a) a bulk system with flat interfaces and mean curvature $H = 0$; (b) a system in which the liquid is finely divided in droplet form with a mean curvatures everywhere equal to H.

since the volume of one mole of liquid is very small compared with that occupied by one mole of gas. Equation 12.54 may be written

$$-\frac{RT}{P}dP + 2\gamma V^L dH = 0$$

or

$$d(\ln P) = \frac{2\gamma V^L}{RT}dH \qquad (12.55)$$

Since γ and V^L are not significant functions of the curvature of the interface in the system, this equation may be integrated from conditions corresponding to the bulk liquid with a flat interface and $H = 0$ to a system with the liquid surface exhibiting an arbitrary value of H:

$$\int_{H=0}^{H} d(\ln P) = \int_{H=0}^{H} \frac{2\gamma V^L}{RT}dH$$

$$\ln\frac{P(H)}{P(H=0)} = \frac{2\gamma V^L}{RT}(H - 0)$$

$$P(H) = P(H = 0)e^{(2\gamma V^L/RT)H} \qquad (12.56)$$

where $P(H)$ is the vapor pressure in a system with liquid droplets that have surface mean curvature H and $P(H = 0)$ is the equilibrium vapor pressure over the bulk liquid at the same temperature. Since H has units of length^{-1}, the coefficient of H in the exponent must have units of length. This quantity may be called the capillarity length scale for liquid–vapor systems, written λ_v:

$$\lambda_v \equiv \frac{2\gamma V^L}{RT} \tag{12.57}$$

With this definition Equation 12.56 may be written

$$P(H) = P(H = 0)e^{\lambda_v H} \tag{12.58}$$

This equation is frequently used in an approximate form obtained by expanding the exponential in a series and neglecting terms beyond the linear term:

$$P(H) = P(H = 0)[1 + \lambda_v H] \tag{12.59}$$

This approximation is valid so long as $\lambda_v H \ll 1$. It is concluded that the vapor pressure in a system containing liquid droplets is higher than the value found over a bulk liquid at the same temperature. In order for the effect of curvature to be significant, the product $\lambda_v H$ must be larger than about 0.01. Table 12.2 gives typical values of the capillarity length scale for a variety of liquids at their melting points. Evidently the effect of capillarity becomes significant for droplets with a radius of curvature smaller than about 1 μm (10^{-6} m or 10^{-4} cm).

EXAMPLE 12.2

Calculate and plot as a function of radius the vapor pressure in a system with liquid zinc droplets suspended in zinc vapor at 900 K. The vapor pressure of zinc over bulk liquid at this temperature is 0.0263×10^{-2} atm; the molar volume of the liquid is 9.5 cc/mol and $\gamma = 768$ erg/cm^2.

The capillarity length scale for this system is

$$\lambda_v = \frac{2\left(768 \frac{\text{ergs}}{\text{cm}^2}\right)\left(9.94 \frac{\text{cc}}{\text{mole}}\right)}{\left(8.314 \frac{\text{J}}{\text{mole K}}\right)\left(10^7 \frac{\text{ergs}}{\text{J}}\right)900 \text{ K}} = 2.0 \times 10^{-7} \text{ cm}$$

TABLE 12.2
Typical Values of Capillary Length Scales

	Liquid–vapor γ_v (nm)	Solid–liquid λ_m (nm)
Water	0.8	0.2
Metals	1	0.4
Salts	2	—
Oxides	0.8	—

Insert this value into Equation 12.58 to compute the vapor pressure in this system:

$$P(r) = (0.0263 \text{ atm}) \, e^{2.0 \times 10^{-7} (\text{cm})/r(\text{cm})}$$

This result is plotted in Figure 12.9. The computed vapor pressure begins to deviate measurably from the bulk value for droplets below 1 μm in size.

Equation 12.59 describes the shift in vapor pressure as a function of curvature at each temperature along the liquid–vapor curve. For a given value of H this effect may be visualized as a shift in the vapor pressure curve on the (P, T) diagram, Figure 12.10. Note that this capillarity shift may also be visualized as a depression of the vaporization temperature at a fixed pressure. This latter result could have been derived from Equation 12.53 by comparing the equilibrium temperature of two systems that are constrained to have the same vapor pressure, but different states of subdivision. This application of the strategy will be explored for the solid–liquid equilibrium in the next section.

12.5.3 EFFECT OF CURVATURE UPON THE MELTING TEMPERATURE

In order to examine the effect of curvature upon solid–liquid equilibria compare the systems shown in Figure 12.11. In this comparison the external pressure, which is also the pressure in the liquid phase, is the same in the two systems; they differ in the value of H that is characteristic of each system. Equation 12.53 forms the basis for determining the difference in temperature that these two systems will exhibit when

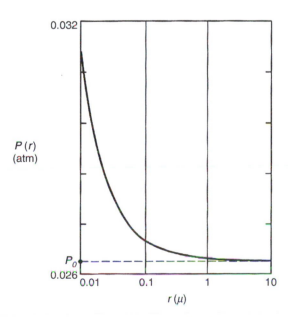

FIGURE 12.9 Plot of the vapor pressure of zinc as a function of droplet radius at 900 K.

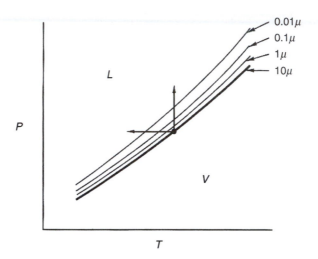

FIGURE 12.10 Shift in the vapor pressure curve with size of the liquid droplets in a L + V system.

they come to equilibrium. In the convention used in deriving Equation 12.53 the β phase is the finely divided solid phase, s, and α is the liquid. Since, in this comparison $dP = 0$, Equation 12.53 may be written

$$\Delta S dT + 2\gamma V^S dH = 0 \qquad (12.60)$$

Rearrange:

$$dT = -\frac{2\gamma V^S}{\Delta S} dH$$

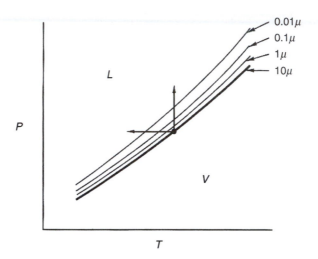

FIGURE 12.11 Compare two unary solid plus liquid systems that have the same external pressure: (a) a bulk system with a flat interface and $H = 0$; (b) a system in which the solid is finely divided as particles with a mean curvature everywhere equal to H.

Integrate:

$$\int_{T(H=0)}^{T(H)} dT = - \int_{H=0}^{H} \frac{2\gamma V^S}{\Delta S} dH = - \frac{2\gamma V^S}{\Delta S} H$$

since the terms in the integrand are not functions of curvature. Thus,

$$T(H) = T(H = 0) - \frac{2\gamma V^S}{\Delta S} H \qquad (12.61)$$

where $T(H)$ is the equilibrium melting temperature in the system with solid particles that have mean curvature H and $T(H = 0)$ is the bulk melting temperature at the same pressure. This equation may also be written

$$T(H) = T(H = 0)[1 - \lambda_m H] \qquad (12.62)$$

where the coefficient λ_m has units of length and may be viewed as a capillarity length scale for melting, given by

$$\lambda_m = \frac{2\gamma V^S}{T(H = 0)\Delta S} = \frac{2\gamma V^S}{\Delta H} \qquad (12.63)$$

since the heat of fusion, $\Delta H = T(H = 0)\Delta S$. Because $\Delta S = S^L - S^S$, the entropy of fusion in the system is positive; it is concluded that increasing the curvature of the solid particles in a solid–liquid system lowers the melting point. Application of this result to each value of P along the melting curve produces a shift in that phase boundary to lower temperatures, Figure 12.12. Typical values of capillarity length scales for solid–liquid systems are given in Table 12.2.

EXAMPLE 12.3
When a liquid is cooled below its bulk melting point the solid phase forms and grows as dendrites. The scale of the resulting microstructure, i.e., whether it is fine grained or a coarse structure and the rate at which it solidifies is strongly influenced by the size of the tips of these tree-like structures as they grow. The tip radius is in turn determined by the temperature difference between the liquid adjacent to the tip and that of the supercooled liquid surrounding it. Calculate this temperature difference for a silicon dendrite with a tip radius of 0.5 μm growing into a surrounding liquid that is undercooled 5°C below the bulk melting point of silicon. The surface energy of the liquid–solid interface may be taken to be 1050 erg/cm^2.

The melting point of bulk silicon is 1685 K; the heat of fusion is 50.7 KJ/mol; the molar volume of solid silicon is 12 cc/mol. The capillarity length scale for this system may be

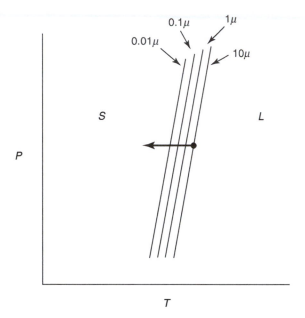

FIGURE 12.12 Shift in the melting temperature with the size of the particles of the solid phase suspended in the liquid.

calculated from Equation 12.63:

$$\lambda_m = \frac{2\left(1050\,\frac{\text{ergs}}{\text{cm}^2}\right)\left(10^{-7}\,\frac{\text{J}}{\text{erg}}\right)\left(12\,\frac{\text{cc}}{\text{mole}}\right)}{\left(50,700\,\frac{\text{J}}{\text{mole}}\right)} = 5.0 \times 10^{-8}\ \text{cm}$$

Assuming that the liquid and solid in a volume element at the tip of the dendrite are in local equilibrium, the temperature of the liquid at the interface is given by Equation 12.62:

$$T = 1685\ \text{K}\left[1 - (5.0 \times 10^{-8}\ \text{cm})\,\frac{1}{0.5\ \mu\text{m}\left(\dfrac{10^{-4}\ \text{cm}}{1\ \mu\text{m}}\right)}\right]$$

$$T = 1683\ \text{K}$$

The temperature of the surrounding supercooled liquid is

$$T^L = 1685\ \text{K} - 5\ \text{K} = 1680\ \text{K}$$

The temperature difference between that of the liquid at the tip and in the bulk liquid is $+3$ K. Thus the temperature at the tip lies above that of the surrounding liquid. Heat will flow from the tip into the liquid, promoting solidification and the solid silicon dendrite will grow in these surroundings.

EXAMPLE 12.4

Assume an array of dendrites with a distribution of tip size exists in the supercooled liquid silicon described in Example 12.3. If the temperature of a tip lies above that of the surrounding supercooled liquid, heat will flow away from the tip into the surroundings and the tip will grow. Conversely, if a tip has a temperature below T^L, heat will flow from the surrounding liquid toward the tip, melting it. Compute the radius of the tip that will neither grow nor melt.

If the tip radius is such that its equilibrium melting temperature equals that of the supercooled surroundings, 1680 K in this case, the tip will remain stationary because it is in thermal equilibrium with its surroundings. The tip radius at which this occurs is found by setting $T(H)$ equal to 1680 K and solving for $H = 1/r$:

$$1680 \text{ K} = 1685 \text{ K}\left[1 - (5.0 \times 10^{-8} \text{ cm})\frac{1}{r} \right]$$

Solve for the tip radius:

$$r = \frac{(5.0 \times 10^{-8}(\text{cm}))(1685 \text{ K})}{1685 \text{ K} - 1680 \text{ K}} = 1.7 \times 10^{-5}(\text{cm}) = 0.17(\mu\text{m})$$

This value for r is the critical radius for growth; tips smaller than 0.17 μm will melt under these conditions while those that are larger than this radius will grow.

12.5.4 EFFECT OF CURVATURE ON CHEMICAL POTENTIAL IN UNARY SYSTEMS

In the bulk system shown in Figure 12.8a the chemical potential of the vapor and liquid phases are equal at equilibrium. Equation 12.46 demonstrates that this is also true in a system with curved interfaces, Figure 12.8b. In each case the chemical potential is uniform in each system at equilibrium. However, the value of the chemical potential in the two phases in the bulk system, Figure 12.8a, is different from the value of the chemical potential in the system with curved interfaces, Figure 12.8b. The section derives a relation between the chemical potential of the two phases in a bulk system and the chemical potential in a two-phase system, $\alpha + \beta$, in which the $\alpha\beta$ interface has a mean curvature, H[1].

Equation 12.47 and Equation 12.48 provide generic expressions for the dependence of the chemical potentials of the α and β phases upon temperature and pressure. Focus upon a comparison of flat and curved interface systems at the same temperature. Then the chemical potentials of the two phases depend only upon their

---

[1] A common error in the literature makes the interpretation that the dependence of chemical potential upon curvature translates as a difference in chemical potential across the interface that may be applied as a driving force for interface motion, for example, in theories of grain growth. As the derivation of the conditions for equilibrium shows, the chemical potential across a curved interface in local equilibrium is zero. The difference arises in a comparison between a bulk system and a system with curved interfaces, not between the phases across the interface in a single system.

pressures:

$$d\mu_T^\alpha = V^\alpha dP^\alpha; \qquad d\mu_T^\beta = V^\beta dP^\beta \tag{12.64}$$

Substitute for dP^β from the condition for mechanical equilibrium, Equation 12.50:

$$d\mu_T^\alpha = V^\alpha dP^\alpha; \qquad d\mu_T^\beta = V^\beta(dP^\alpha + 2\gamma dH) \tag{12.65}$$

Apply the condition for chemical equilibrium, Equation 12.51

$$d\mu_T^\alpha = d\mu_T^\beta; \qquad V^\alpha dP^\alpha = V^\beta(dP^\alpha + 2\gamma dH) \tag{12.66}$$

Solve this equation for dP^α, which is the change in pressure when the curvature changes by an amount dH:

$$dP^\alpha = \frac{2\gamma V^\beta}{V^\alpha - V^\beta} dH \tag{12.67}$$

The chemical potential change of the α phase accompanying such a change in curvature of the $\alpha\beta$ interface is

$$d\mu_T^\alpha = V^\alpha dP^\alpha = V^\alpha \frac{2\gamma V^\beta}{V^\alpha - V^\beta} dH \tag{12.68}$$

A comparison between the value of μ^α in the bulk phase and that in a system with interface curvature H is obtained by integrating between these two states:

$$\int_{H=0}^{H} d\mu_T^\alpha = \int_{H=0}^{H} \frac{2\gamma V^\alpha V^\beta}{V^\alpha - V^\beta} dH \tag{12.69}$$

Assuming that values of γ and the molar volumes do not vary significantly with H (which translates as a small variation in pressure), integration gives the dependence of the chemical potential of the α phase upon curvature:

$$\mu_T^\alpha(H) = \mu_T^\alpha(H=0) + \frac{2\gamma V^\alpha V^\beta}{V^\alpha - V^\beta} H \tag{12.70}$$

which is equal to the $\mu^\beta(H)$ in the system through the condition for chemical equilibrium.

If the α phase is the vapor, then $V^\alpha \gg V^\beta$ so that $V^\alpha - V^\beta \approx V^\alpha$ and this result simplifies to:

$$\mu_T^\alpha(H) = \mu_T^\alpha(H=0) + 2\gamma V^\beta H \tag{12.71}$$

This form of the relationship of chemical potential to curvature is sometimes erroneously applied to the case in which both phases are condensed phases. It is only valid if one of the phases is the vapor phase. If both phases are condensed phases then Equation 12.70 applies.

If one phase is the vapor, then the direction of change in chemical potential of the system depends only upon the sign of the curvature. If both are condensed, the direction of change also depends upon the sign of $V^\alpha - V^\beta$, the difference in molar volumes. Note in this equation that the β phase is the reference phase for defining the signs of the principal normal curvatures. Equation 12.70 and Equation 12.71 differ by the factor $V^\alpha/(V^\alpha - V^\beta)$ which may add a factor of about ten to the effect of curvature since molar volume differences tend to be small in comparison with their values.

12.5.5   PHASE BOUNDARY SHIFTS IN BINARY SYSTEMS

The altered condition for mechanical equilibrium that characterizes capillarity effects also operates on the boundaries of an $\alpha + \beta$ field in a binary system, producing shifts that are similar to those just derived for unary systems. In order to develop expressions for these capillarity shifts it is necessary to follow a strategy that is analogous to that leading to Equation 12.53 for a unary system. The following development paraphrases that strategy.

Equation 12.23, Equation 12.24, and Equation 12.27 present the conditions for equilibrium in a two-phase multicomponent system. For a binary system $c = 2$ and these equations may be written:

$$T^\beta = T^\alpha; \qquad \mu_1^\beta = \mu_1^\alpha; \qquad \mu_2^\beta = \mu_2^\alpha; \qquad P^\beta = P^\alpha + 2\gamma H$$

Focus for the moment upon the behavior of component 1 in the α phase. Since the chemical potential is an intensive property, it is a function of the temperature, pressure and composition of the α phase:

$$d\mu_1^\alpha = -\bar{S}_1^\alpha dT^\alpha + \bar{V}_1^\alpha dP^\alpha + \mu_{12}^\alpha dX_2^\alpha \tag{12.72}$$

where the mole fraction of component 2 has been chosen as the independent composition variable. $\bar{S}_1^\alpha$ and $\bar{V}_1^\alpha$ are the partial molal entropy and volume for component 1. The coefficient,

$$\mu_{12}^\alpha \equiv \left(\frac{\partial \mu_1}{\partial X_2}\right)^\alpha_{T,P} \tag{12.73}$$

reports the change in chemical potential of component 1 with composition in the α phase. This term can be evaluated from a solution model for the α phase. Analogous expressions may be written for changes with composition of the other three chemical potentials in the system, μ_{22}^α, μ_{12}^β and μ_{22}^β.

If α and β are maintained in equilibrium as the state of the $(\alpha + \beta)$ system is altered the conditions for equilibrium require that

$$dT^\beta = dT^\alpha = dT \tag{12.74}$$

$$dP^\beta = dP^\alpha + 2\gamma dH \tag{12.75}$$

$$d\mu_1^\beta = d\mu_1^\alpha \tag{12.76}$$

$$d\mu_2^\beta = d\mu_2^\alpha \tag{12.77}$$

Expressions like Equation 12.72 may be substituted into Equation 12.76 to give

$$d\mu_1^\alpha = -\bar{S}_1^\alpha dT^\alpha + \bar{V}_1^\alpha dP^\alpha + \mu_{12}^\alpha dX_2^\alpha = d\mu_1^\beta = -\bar{S}_1^\beta dT^\beta + \bar{V}_1^\beta dP^\beta + \mu_{12}^\beta dX_2^\beta$$

Use the condition for thermal equilibrium to replace dT^α and dT^β with dT. Write dP^β in terms of dP^α and dH with the condition for mechanical equilibrium:

$$d\mu_1^\alpha = -\bar{S}_1^\alpha dT^\alpha + \bar{V}_1^\alpha dP^\alpha + \mu_{12}^\alpha dX_2^\alpha = d\mu_1^\beta$$
$$= -\bar{S}_1^\beta dT^\beta + \bar{V}_1^\beta [dP^\alpha + 2\gamma dH] + \mu_{12}^\beta dX_2^\beta$$

Substitution for dP^β introduces the mean curvature into the relation. Collect like terms:

$$-(\bar{S}_1^\alpha - \bar{S}_1^\beta)dT + (\bar{V}_1^\alpha - \bar{V}_1^\beta)dP^\alpha - 2\gamma\bar{V}_1^\beta dH + \mu_{12}^\alpha dX_2^\alpha - \mu_{12}^\beta dX_2^\beta = 0$$

Define

$$\Delta\bar{S}_1 \equiv \bar{S}_1^\alpha - \bar{S}_1^\beta \tag{12.78}$$

$$\Delta\bar{V}_1 \equiv \bar{V}_1^\alpha - \bar{V}_1^\beta \tag{12.79}$$

Note that these quantities are differences in partial molal properties for component 1 as distinguished from the corresponding total properties which appear in Equation 12.53 for a unary system. With these definitions,

$$-\Delta\bar{S}_1 dT + \Delta\bar{V}_1 dP^\alpha - 2\gamma\bar{V}_1^\beta dH + \mu_{12}^\alpha dX_2^\alpha - \mu_{12}^\beta dX_2^\beta = 0 \tag{12.80}$$

The same derivation applied to Equation 12.77 gives the corresponding equation for component 2:

$$-\Delta\bar{S}_2 dT + \Delta\bar{V}_2 dP^\alpha - 2\gamma\bar{V}_2^\beta dH + \mu_{22}^\alpha dX_2^\alpha - \mu_{22}^\beta dX_2^\beta = 0 \tag{12.81}$$

Equation 12.80 and Equation 12.81 are generalizations of Equation 12.53 derived for a unary system. They represent two equations relating the five variables, T, P, H, X_2^α and X_2^β. In order to explore the effect of H upon phase boundary compositions, it is necessary to fix temperature and pressure. With $dT = 0$ and $dP^\alpha = 0$, the analysis focuses upon the variation of the compositions of the phases with curvature. With these constraints Equation 12.80 and Equation 12.81 become:

$$-2\gamma\bar{V}_1^\beta dH + \mu_{12}^\alpha dX_2^\alpha - \mu_{12}^\beta dX_2^\beta = 0 \tag{12.82}$$

$$-2\gamma\bar{V}_2^\beta dH + \mu_{22}^\alpha dX_2^\alpha - \mu_{22}^\beta dX_2^\beta = 0 \tag{12.83}$$

The chemical potential derivatives that appear as coefficients in these equations, the μ_{kj}'s, are related within each phase by the Gibbs–Duhem equation. For any given phase,

$$X_1 d\mu_1 + X_2 d\mu_2 = 0$$

For a fixed temperature and pressure $d\mu_1$ may be evaluated from Equation 12.73:

$$(d\mu_1)_{T,P} = \mu_{12} dX_2$$

Similarly, for μ_2:

$$(d\mu_2)_{T,P} = \mu_{22} dX_2$$

Insert these evaluations into the Gibbs–Duhem equation:

$$X_1 \mu_{12} dX_2 + X_2 \mu_{22} dX_2 = 0$$

Thus the compositional derivative of μ_2 is related to that for μ_1:

$$\mu_{22} = -\frac{X_1}{X_2}\mu_{12} \tag{12.84}$$

This equation applies separately to the α or β phase.
With these relationships Equation 12.82 and Equation 12.83 may be rewritten

$$\mu_{12}^\alpha\left(\frac{dX_2^\alpha}{dH}\right) + \mu_{12}^\beta\left(\frac{dX_2^\beta}{dH}\right) = 2\gamma\bar{V}_1^\beta \tag{12.85}$$

$$-\frac{X_1^\alpha}{X_2^\alpha}\mu_{12}^\alpha\left(\frac{dX_2^\alpha}{dH}\right) - \frac{X_1^\beta}{X_2^\beta}\mu_{12}^\beta\left(\frac{dX_2^\beta}{dH}\right) = 2\gamma\bar{V}_2^\beta \tag{12.86}$$

The derivatives in these equations are the rates of change of the $(\alpha + \beta)$ phase boundary compositions with curvature. These equations form two simultaneous linear equations in these derivatives. They may be solved by substitution or determinants to give

$$\left(\frac{dX_2^\alpha}{dH}\right) = 2\gamma[X_1^\beta\bar{V}_1^\beta + X_2^\beta\bar{V}_2^\beta]\frac{X_2^\alpha}{\mu_{12}^\alpha(X_2^\alpha - X_2^\beta)} \tag{12.87}$$

$$\left(\frac{dX_2^\beta}{dH}\right) = 2\gamma[X_1^\alpha \bar{V}_1^\beta + X_2^\alpha \bar{V}_2^\beta]\frac{X_2^\beta}{\mu_{12}^\beta(X_2^\alpha - X_2^\beta)} \tag{12.88}$$

Note that the quantity in brackets in Equation 12.87 is simply the molar volume of the β phase, V^β; the corresponding quantity in Equation 12.88 has units of volume per mole but is not equal to the molar volume of either phase.

The chemical potential of component 1 generally decreases with increasing component 2 (except within a miscibility gap) so that the μ_{12}'s are typically negative in these equations. Choose component 2 so that it is rich in the β phase and the solute in α, i.e., so that $X_2^\beta > X_2^\alpha$. Then both derivatives are positive. This shift in the compositions at the phase boundaries of an $(\alpha + \beta)$ field is sketched in Figure 12.13; with increasing curvature of the $\alpha + \beta$ phase boundaries for both phases become richer in component 2. The magnitude of the shift is particularly sensitive to the values of the μ_{12} factors and to the width of the two-phase field given by $(X_2^\beta - X_2^\alpha)$. Evidently, the narrower the two-phase field, the larger will be the magnitude of the shift.

These results, although general, are most frequently applied to the case in which both the α and β phases are dilute solutions. The α phase is dilute in component 2 and the β phase dilute in component 1. The chemical potential of component 1 in the

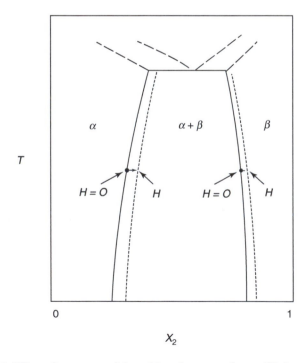

FIGURE 12.13 Effect of curvature of the $\alpha\beta$ interface upon the equilibrium composition of the α and β phases.

α phase is given by Raoult's law for the solvent:

$$\mu_1^\alpha = \mu_1^{o\alpha} + RT \ln X_1^\alpha$$

Evaluate

$$\mu_{12}^\alpha \equiv \left(\frac{d\mu_1^\alpha}{dX_2^\alpha}\right)_{T,P} = \left(\frac{d\mu_1^\alpha}{dX_1^\alpha}\right)_{T,P}\frac{dX_1^\alpha}{dX_2^\alpha} = -\frac{RT}{X_1^\alpha} \cong -RT$$

since in dilute α, $X_1^\alpha \cong 1$. In the β phase, where component 1 is the solute, Henry's law describes its chemical potential:

$$\mu_1^\beta = \mu_1^{\beta o} + RT \ln\gamma_1^o + RT \ln X_1^\beta$$

where γ_1^o, the Henry's law coefficient, is independent of composition. The required derivative,

$$\mu_{12}^\beta \equiv \left(\frac{d\mu_1^\beta}{dX_2^\beta}\right)_{T,P} = \left(\frac{d\mu_1^\beta}{dX_1^\beta}\right)_{T,P}\frac{dX_1^\beta}{dX_2^\beta} = -\frac{RT}{X_1^\beta}$$

Further, since both α and β are dilute solutions $(X_2^\alpha - X_2^\beta) \cong (-X_2^\beta) \cong -1$. With these evaluations, Equation 12.87 and Equation 12.88 become

$$\left(\frac{dX_2^\alpha}{dH}\right) = 2\gamma V^\beta \frac{X_2^\alpha}{-RT(-1)} = \frac{2\gamma V^\beta}{RT}X_2^\alpha \tag{12.89}$$

$$\left(\frac{dX_2^\beta}{dH}\right) = 2\gamma \bar{V}_1^\beta \frac{1}{-\left(\dfrac{RT}{X_1^\beta}\right)(-1)} = \frac{2\gamma \bar{V}_1^\beta}{RT}X_1^\beta \tag{12.90}$$

Note that the volume factor in Equation 12.88 simplifies to the partial molal volume of component 1 in the β phase.

Equation 12.89 may be integrated after separating the variables:

$$\int_{X_2^\alpha(H=0)}^{X_2^\alpha(H)} \frac{dX_2^\alpha}{X_2^\alpha} = \int_{H=0}^{H} \frac{2\gamma V^\beta}{RT}dH$$

where $X_2^\alpha(H = 0)$ is the composition of the α side of the two-phase field in a system with flat interfaces (a bulk system) and $X_2^\alpha(H)$ is the composition of α in equilibrium with β when the $\alpha\beta$ interface has a mean curvature H. Since the integrand on the right side is independent of the curvature of the interface, integration gives

$$\ln\left[\frac{X_2^\alpha(H)}{X_2^\alpha(H=0)}\right] = \frac{2\gamma V^\beta}{RT}H \tag{12.91}$$

The coefficient of H has the same form as that obtained for the shift in vapor pressure in a unary system, Equation 12.56. It may be defined as a capillarity length scale for the composition shift:

$$\lambda_c^\alpha \equiv \frac{2\gamma V^\beta}{RT} \tag{12.92}$$

With this substitution and some rearrangement Equation 12.91 may be written

$$X_2^\alpha(H) = X_2^\alpha(H = 0)e^{\lambda_c^\alpha H} \tag{12.93}$$

An approximate version of this equation is usually applied in practical situations, obtained by expanding the exponential and neglecting terms beyond the first order in H:

$$X_2^\alpha(H) = X_2^\alpha(H = 0)[1 + \lambda_c^\alpha H] \tag{12.94}$$

Thus the concentration of component 2 in the α phase, which is the solute in this case, in equilibrium with the β phase in a system in which the $\alpha\beta$ interface has mean curvature H is larger than the concentration in the analogous bulk system, i.e., one for which the interface is flat. This increase in solubility is proportional to the mean curvature H of the interface.

A similar result may be obtained for the composition of the β phase in this system by integrating Equation 12.90:

$$X_2^\beta(H) = X_2^\beta(H = 0)e^{\lambda_c^\beta H} \tag{12.95}$$

where the capillarity length scale for the β phase is given by

$$\lambda_c^\alpha \equiv \frac{2\gamma \bar{V}_1^\beta}{RT} \tag{12.96}$$

If $\lambda_c^\beta H$ is small compared to 1,

$$X_2^\beta(H) = X_2^\beta(H = 0)[1 + \lambda_c^\beta H] \tag{12.97}$$

Thus in the β phase the equilibrium concentration of component 2, which is the solvent in this case, is increased in proportion to the mean curvature of the interface.

Recall that the convention for choosing the sign of H assumed that the mean curvature is positive for surface elements that are convex relative to the β phase. If in the system under consideration the α phase is present as particles in a matrix of β, then, by this convention, H is negative in this system. The equations describing the capillarity shift in such a system are exactly those given here, unchanged. Within these equations, however, H is negative; thus the phase boundary shifts will be in the opposite direction on the composition axis.

The shifts in composition at constant temperature and pressure given in Equation 12.94 and Equation 12.97 may be computed at each temperature along a phase boundary. The result may be viewed as a shift in the phase boundaries of the α and β phases, Figure 12.13. The effects are qualitatively similar if the α and β phases are not dilute so that the general Equation 12.87 and Equation 12.88 must be applied.

12.5.6 LOCAL EQUILIBRIUM AND THE APPLICATION OF CAPILLARITY SHIFTS

The development presented in Section 12.5 visualizes a system composed of two phases with an interface on which the mean curvature is constant. The simplest example of such a microstructure would consist of a collection of spherical particles of the β phase which all have the same radius, r, so that $H = 1/r$. In real microstructures there always exists a distribution of mean curvature values over the interfaces in the system. Even in the simplest case of spherical particles, real microstructures exhibit a distribution of sizes. Thus the structures that are visualized in Figure 12.8 and Figure 12.11 never exist in real structures.

The results derived in that development are generally used in practice by applying the principle of local equilibrium at an element of interface in the two phase structure. Each interfacial element has its value of mean curvature, H. It is assumed that volume elements in the α and β phases that are adjacent to the interface are *locally* in equilibrium. Explicitly, it is assumed that the conditions for equilibrium, equality of temperature, and chemical potentials, and the condition for mechanical equilibrium that gives rise to a pressure difference determined by the local curvature of the interface, all hold for the α and β volume elements in contact at the interface. The results derived from these conditions, contained for example in Equation 12.94 and Equation 12.97, also hold for these volume elements.

If the principle of local equilibrium is applied at each interfacial element in a microstructure, the deductions from that assumption give rise to effects that may be used to understand microstructural changes that may be ongoing. Figure 12.14a sketches a microstructure consisting of a dispersion of β particles in an α matrix. Prolonged exposure to high temperature causes such a microstructure to coarsen to

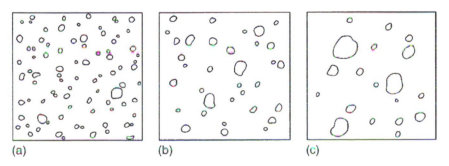

(a) (b) (c)

FIGURE 12.14 Three stages of a coarsening process in which a dispersion of fine β particles in an α matrix evolves toward a microstructure with the same total quantity of β particles, but fewer in number and larger in size.

the structure sketched in Figure 12.14b. This process is called coarsening or Ostwald ripening. Small particles shrink and disappear while large particles grow in this process; at a later time the same volume of β phase is distributed over fewer larger particles. An understanding of how this process occurs may be obtained by focusing upon the interaction between two neighboring particles, Figure 12.15. According to Equation 12.94 the concentration of solute in α adjacent to the small particle will be shifted more by capillarity than that adjacent to the large particle, Figure 12.15. The concentration difference thus created in the alpha phase produces a diffusion flux of solute from the region of high concentration (by the small particle) to the region of lower concentration (adjacent to the large particle). As a result the small particle shrinks as it supplies the solute to this diffusion process and the large particle grows as it accumulates solute. Repetition of this interaction between pairs of interfacial elements throughout the microstructure yields the coarsening process.

Application of capillarity shift effects on phase diagrams through the principle of local equilibrium is invoked in the analysis of a variety of processes that produce changes in microstructure including coarsening, dendrite growth, eutectic and eutectoid transformations, powder processing in ceramics and powder metallurgy, grain growth and nucleation. This section has provided the fundamental thermodynamic background underlying these applications.

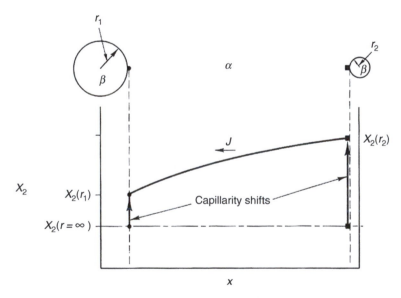

FIGURE 12.15 The principle of local equilibrium implies capillary shifts in the matrix composition at the $\alpha\beta$ phase boundaries which sets up conditions for diffusion of solute through the α matrix from smaller particles (with higher mean curvature and larger shifts) toward larger particles (with smaller curvature and smaller compositional shifts). The diffusional flow is away from small particles (which shrink) toward large particles (which grow).

12.6 THE EQUILIBRIUM SHAPE OF CRYSTALS: THE GIBBS–WULFF CONSTRUCTION

The developments in Section 12.5 assume that the value of the specific interfacial free energy is isotropic; that is, it does not depend upon the orientation of the element of interface under consideration. If one or both of the phases involved is crystalline, then this assumption may be a poor approximation. It may be readily visualized that different crystal faces exposed to their equilibrium vapor will exhibit differences in surface energy. The atoms in a close packed plane thus exposed will have a configuration of neighbors that is more like that in the bulk crystal than atoms lying on an irrational crystal plane at the surface. In general the value of γ varies significantly with the orientation of the crystal plane exposed to its vapor.

Visualize a plot $\gamma(\theta, \phi)$ over the sphere of orientation, Figure 12.16a. Orientations corresponding to close packed planes may exhibit local minima in the form of cusps in this γ plot because their surface energies are significantly smaller than neighboring orientations. The crystal may minimize its surface energy by favoring the formation of surface facets with these low energy orientations. The resulting equilibrium shape may be a polyhedron completely bounded by facets, or may consist of facets connected by smoothly curved surfaces, Figure 12.16b and c. The shape of the crystal at equilibrium can be modeled from such a γ plot by applying once more the strategy for finding conditions for equilibrium. In order to develop this strategy it is first necessary to review some aspects of the geometry of polyhedra.

The volume of a polyhedron may be related to its surface area and its radius in much the same way as a sphere. Label the faces on the polyhedron shown in Figure 12.17a from 1 to F. The point O is the centroid (center of mass) of the polyhedron. Let A_j be the area of the jth face. Define λ_j to be the perpendicular distance from the centroid to the jth face; this property is called the pedal function for the jth face. Draw lines from the centroid O to the corners of the jth face. This construction establishes a pyramid with the jth face as its base, O as its apex and λ_j as its altitude. The volume of this pyramid is $(1/3)A_j\lambda_j$. The volume of the polyhedron

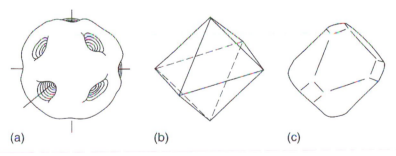

(a)   (b)   (c)

FIGURE 12.16 A plot of specific surface free energy in spherical coordinates (a) may yield an equilibrium crystal shape composed entirely of flat facets (b), or a faceted shape with rounded corners (c).

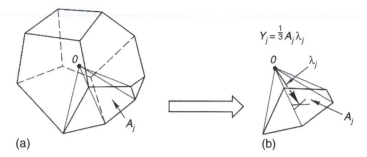

FIGURE 12.17 The volume of a flat faced polyhedron (a) can be computed (b) from the areas, A_j, of the individual facets and their pedal functions, λ_j.

is obtained by summing over the F faces:

$$V^c = \sum_{j=1}^{F} \frac{1}{3} A_j \lambda_j \tag{12.98}$$

Now compare the geometry of this polyhedron with one that is obtained from it by displacing each of the facets a distance $d\lambda_j$, which may be different for each face. The jth face sweeps through a volume given by $A_j d\lambda_j$. The total change in volume of the polyhedron is the sum of the contributions from the faces:

$$dV^c = \sum_{j=1}^{F} A_j d\lambda_j \tag{12.99}$$

In this evaluation of dV^c contributions along the edges and at the corners are neglected; however, since these are of a higher order in the differential $d\lambda_j$, in the limit they are in fact negligible.

This volume change may also be computed by differentiating Equation 12.98:

$$dV^c = \sum_{j=1}^{F} \frac{1}{3} [A_j d\lambda_j + \lambda_j dA_j] \tag{12.100}$$

Set these two expressions for dV^c equal to each other and simplify:

$$\sum_{j=1}^{F} A_j d\lambda_j = \frac{1}{2} \sum_{j=1}^{F} \lambda_j dA_j \tag{12.101}$$

Substitute this result into Equation 12.99 to obtain a relation between the volume change of a faceted crystal and the changes in areas of its facets:

$$dV^c = \frac{1}{2} \sum_{j=1}^{F} \lambda_j dA_j \tag{12.102}$$

   This expression is the analog to Equation 12.7 for smooth surfaces, connecting the change in volume of the crystal to changes in its area.

   Next apply the general strategy for finding conditions for equilibrium. The sequence of relationships that result in this application is essentially equivalent to that developed in Section 12.3 for curved interfaces except that each term involving the change in area in Equation 12.15 to Equation 12.20 is replaced by the corresponding summation over the facets on the crystal surface. With α designated as the vapor (v) and β the crystal (c) (12.21) becomes

$$dS'_{sys,iso} = \left(\frac{1}{T^v} - \frac{1}{T^c}\right)dU''_{hyp} + \left(\frac{P^v}{T^v} - \frac{P^c}{T^c}\right)dV''_{hyp}$$
$$- \sum_{k=1}^{c}\left(\frac{\mu_k^v}{T^v} - \frac{\mu_k^c}{T^c}\right)dn_{k,hyp}^v - \sum_{j=1}^{F}\frac{\gamma_j}{T^c}dA_j \qquad (12.103)$$

where

$$\gamma_j = U_j^s - TS_j^s - \sum_{k=1}^{c}\mu_k n_{k,j}^s \qquad (12.104)$$

for the jth face. The differential in the second term in Equation 12.103 may be converted to dV'^c_{hyp} since, in this isolated system, $dV'^v_{hyp} = -dV'^c_{hyp}$. Apply Equation 12.102 to relate this volume change to the changes in facet areas:

$$dS'_{sys,iso} = \left(\frac{1}{T^v} - \frac{1}{T^c}\right)dU'^v + \left(\frac{P^v}{T^v} - \frac{P^c}{T^c}\right)\left[-\frac{1}{2}\sum_{j=1}^{F}\gamma_j dA_j\right]$$
$$- \sum_{k=1}^{c}\left(\frac{\mu_k^v}{T^v} - \frac{\mu_k^c}{T^c}\right)dn_k^v + \sum_{j=1}^{F}\frac{\gamma_j}{T^c}dA_j \qquad (12.105)$$

Collect like terms:

$$dS'_{sys,iso} = \left(\frac{1}{T^v} - \frac{1}{T^c}\right)dU'^v_{hyp} - \sum_{k=1}^{c}\left(\frac{\mu_k^v}{T^v} - \frac{\mu_k^c}{T^c}\right)dn_{k,hyp}^v$$
$$+ \sum_{j=1}^{F}\left[\left(\frac{P^c}{T^c} - \frac{P^v}{T^v}\right)\lambda_j - \frac{\gamma_j}{T^c}\right]dA_j$$

With the entropy of the isolated system thus expressed in terms of independent variables the conditions for equilibrium are found by setting the coefficients equal to zero. The first $(c + 1)$ terms in this equation yield the usual conditions for thermal and chemical equilibrium. The remaining F terms are each of the form

$$\left(\frac{P^c}{T^c} - \frac{P^v}{T^v}\right)\frac{1}{2}\lambda_j - \frac{\gamma_j}{T} = 0$$

which, given that $T^c = T^v = T$, may be rearranged as

$$P^c = P^v + 2\frac{\gamma_j}{\lambda_j} \qquad (j = 1, 2, ..., F) \tag{12.106}$$

This condition for mechanical equilibrium is analogous to Equation 12.27 for smoothly curved surfaces.

The shape of the equilibrium crystal may be inferred from the observation that in Equation 12.106 the pressures are values for the crystal and vapor phases and are thus the same for all of the exposed faces. Consequently, at equilibrium the ratio

$$\frac{\gamma_j}{\lambda_j} = \text{constant} \tag{12.107}$$

for all of the faces. This implies that the pedal function for each face, i.e., the perpendicular distance from the centroid of the crystal to each facet, is proportional to the surface energy of that facet. Facets with a low surface energy will take on a position closer to the centroid of the crystal than those with higher surface energy. As a consequence these facets will occupy more of the surface area of the resulting crystal.

The relationship given in Equation 12.107 provides the basis for determining the quantitative shape of a crystal at equilibrium embodied in the Gibbs–Wulff construction. Figure 12.18 shows a two-dimensional polar plot of λ vs. orientation that illustrates this construction. For any given orientation the radius vector of the polar plot, r, is proportional to the surface energy. In simple terms, Equation 12.107 states that the pedal function for each face is proportional to the value of γ for that face and thus the radius vector on the polar plot. This geometric relation may be visualized by constructing the plane that is perpendicular to r at the tip of each radius vector on the γ polar plot. Repetition of this construction for the set of all radius vectors over the sphere of orientation yields a collection of overlapping planes which envelop and delineate the shape that satisfies the conditions for mechanical

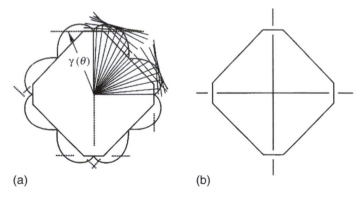

(a)   (b)

FIGURE 12.18 (a) Polar plot of γ vs. orientation for a two-dimensional crystal and the corresponding Gibbs–Wulff construction yielding (b) the equilibrium two-dimensional crystal shape.

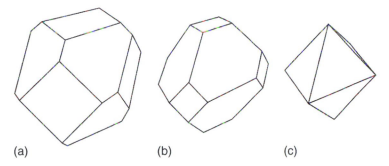

FIGURE 12.19 Equilibrium crystal shapes derived from a variety of γ polar plots.

equilibrium for the crystal that has the given polar plot. Equilibrium shapes for a variety of polar plots are sketched in Figure 12.19.

This condition for mechanical equilibrium for fine crystals may be translated into capillarity shifts of vapor pressure and equilibrium compositions by paraphrasing the strategies that yield Equation 12.56, Equation 12.62, Equation 12.87 and Equation 12.88, in each case replacing the product γH with the ratio γ_j/λ_j

12.7 EQUILIBRIUM AT TRIPLE LINES

Consider the rudimentary three phase $(\alpha + \beta + \varepsilon,)$ system shown in Figure 12.20. These phases meet pair-wise at $\alpha\beta$, $\alpha\varepsilon$ and $\beta\varepsilon$, interfaces; each interface has its value of specific interfacial free energy designated, respectively, $\gamma_{\alpha\beta}$, $\gamma_{\beta\varepsilon}$, and $\gamma_{\alpha\varepsilon}$. In the development presented in this section it is assumed that these surface energies are isotropic. The three phases meet along a line designated $\alpha\beta\varepsilon$, and called a triple line. The three interfaces also meet along this triple line. In general, the triple line is not straight or even a plane curve; it is a space curve. Figure 12.21 shows a plane

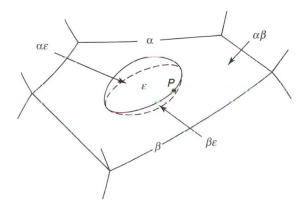

FIGURE 12.20 A triple line $(\alpha\beta\varepsilon)$ in a three phase microstructure. A particle of the ε phase lies on the interface between α and β phases. The line in the $\alpha\beta$ interface where all three phases meet is an $\alpha\beta\varepsilon$ triple line.

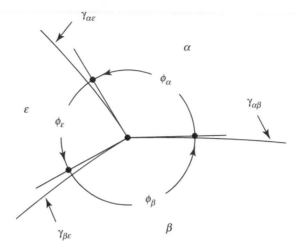

FIGURE 12.21 View of the local microstructure on a plane perpendicular to the triple line in Figure 12.20 at the point P.

section through this system at the point P, constructed so that it is locally perpendicular to the triple line. The traces of the three interfaces form interior angles ϕ_α, ϕ_β, and ϕ_ε as shown in this figure.

The condition for equilibrium in this system may be found by applying the general strategy used repeatedly in this presentation. Development of this application is cumbersome because the system consists of three phases and three interfaces. The change in entropy is given by

$$dS'_{sys} = dS'^\alpha_{hyp} + dS'^\beta_{hyp} + dS'^\varepsilon_{hyp} + S^s_{\alpha\beta}dA_{\alpha\beta} + S^s_{\beta\varepsilon}dA_{\beta\varepsilon} + S^s_{\alpha\varepsilon}dA_{\alpha\varepsilon} \quad (12.108)$$

where S^s_{ij} is the specific superficial excess entropy for the ij interface. The isolation constraints are

$$dU'_{sys} = 0 = dU'^\alpha_{hyp} + dU'^\beta_{hyp} + dU'^\varepsilon_{hyp} + U^s_{\alpha\beta}dA_{\alpha\beta}$$
$$+ U^s_{\beta\varepsilon}dA_{\beta\varepsilon} + U^s_{\alpha\varepsilon}dA_{\alpha\varepsilon} \quad (12.109)$$

$$dV'_{sys} = 0 = dV'^\alpha_{hyp} + dV'^\beta_{hyp} + dV'^\varepsilon_{hyp} \quad (12.110)$$

$$dn_{k,sys} = 0 = dn^\alpha_{k,hyp} + dn^\beta_{k,hyp} + dn^\varepsilon_{k,hyp} + n^s_{k,\alpha\beta}dA_{\alpha\beta} + n^s_{k,\beta\varepsilon}dA_{\beta\varepsilon}$$
$$+ n^s_{k,\alpha\varepsilon}dA_{\alpha\varepsilon} \quad (k = 1, 2, ..., c) \quad (12.111)$$

Use of the isolation constraints to eliminate dependent variables in the expression for the entropy yields the familiar terms that lead to conditions for thermal, mechanical, and chemical equilibrium among the three phases along with three additional terms:

$$dS'_{sys,iso} = [...2(c + 2)\text{terms}...] + \frac{\gamma_{\alpha\beta}}{T}dA_{\alpha\beta} + \frac{\gamma_{\beta\varepsilon}}{T}dA_{\beta\varepsilon} + \frac{\gamma_{\alpha\varepsilon}}{T}dA_{\alpha\varepsilon} \quad (12.112)$$

These three changes in area of the interfaces are not independent. It will be necessary to derive the relationships between these area changes associated with an arbitrary displacement of the triple line.

A displacement $d\lambda$ of a triple line that is arbitrary in both magnitude and direction is shown in Figure 12.22. The area of each of the three interfaces is changed as a result of this displacement by amounts that are determined by the length changes $dl_{\alpha\beta}$, $dl_{\beta\varepsilon}$ and $dl_{\alpha\varepsilon}$. For an element of length dL along the triple line, the change in area of each of the boundaries is:

$$dA_{\alpha\beta} = -dL \cdot dl_{\alpha\beta} = -dL \cdot d\lambda [\cos\theta]$$

$$dA_{\alpha\varepsilon} = -dL \cdot dl_{\alpha\varepsilon} = -dL \cdot d\lambda [\cos\phi_\alpha \cos\theta + \sin\phi_\alpha \sin\theta]$$

$$dA_{\beta\varepsilon} = -dL \cdot dl_{\beta\varepsilon} = -dL \cdot d\lambda [-\cos\phi_\beta \cos\theta + \sin\phi_\beta \sin\theta]$$

where θ gives the arbitrary direction of displacement of the triple line. Substitute these results into the expression for the entropy, Equation 12.112:

$$dS'_{\text{sys,iso}} = \ldots + \frac{1}{T}[\gamma_{\alpha\beta}(-dL \cdot d\lambda)(\cos\theta)] + \frac{1}{T}[\gamma_{\alpha\varepsilon}(-dL \cdot d\lambda)(\cos\phi_\alpha \cos\theta + \sin\phi_\alpha \sin\theta)]$$

$$+ \frac{1}{T}[\gamma_{\beta\varepsilon}(-dL \cdot d\lambda)(-\cos\phi_\beta \cos\theta + \sin\phi_\beta \sin\theta)]$$

$$dS'_{\text{sys,iso}} = [\ldots] + \frac{1}{T}(-dL \cdot d\lambda) \times [\gamma_{\alpha\beta}(\cos\theta) + \gamma_{\alpha\varepsilon}(\cos\theta\cos\phi_\alpha + \sin\theta\sin\phi_\alpha)$$

$$+ \gamma_{\beta\varepsilon}(-\cos\theta\cos\phi_\beta + \sin\theta\sin\phi_\beta)]$$

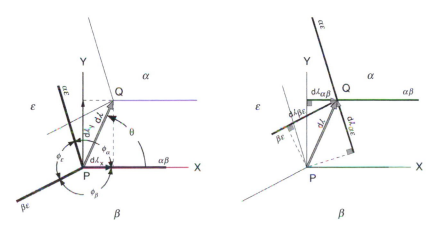

FIGURE 12.22 (a) The arbitrary displacement $d\lambda$ of a triple point P on a triple line to new triple point location Q can be described in terms of its components, $d\lambda_x$ and $d\lambda_y$. (b) As a result the lengths of the traces of $\alpha\beta$, $\alpha\varepsilon$ and $\varepsilon\beta$ interfaces change by amounts $dl_{\alpha\beta}$ $dl_{\alpha\varepsilon}$ and $dl_{\beta\varepsilon}$.

Collect like terms in $\cos\theta$ and $\sin\theta$:

$$dS'_{\text{sys,iso}} = [\ldots] + \frac{1}{T}[-dL]\cdot[\gamma_{\alpha\beta} + \gamma_{\alpha\varepsilon}\cos\phi_\alpha - \gamma_{\beta\varepsilon}\cos\phi_\beta]\cdot dl\cos\theta + \frac{1}{T}[-dL][\gamma_{\alpha\varepsilon}\sin\phi_\alpha$$

$$+ \gamma_{\beta\varepsilon}\sin\phi_\beta]\cdot dl\sin\theta$$

$$dS'_{\text{sys,iso}} = [\ldots] + \frac{1}{T}[-dL]\cdot[\gamma_{\alpha\beta}\cos\phi_\alpha + \gamma_{\beta\varepsilon}\cos\phi_\varepsilon + \gamma_{\alpha\varepsilon}]\cdot dl_x$$

$$+ \frac{1}{T}[-dL]\cdot[\gamma_{\alpha\beta}\sin\phi_\alpha - \gamma_{\beta\varepsilon}\sin\phi_\varepsilon]dl_y \qquad (12.113)$$

where dl_x and dl_y are the independent components of the displacement of the triple line.

Equation 12.113 thus expresses the change in entropy for an isolated three-phase system with interfaces and triple line in terms of variables that are now independent. The conditions for equilibrium are found by setting all of the coefficients of the independent differentials in this equation equal to zero. The terms included in the brackets, $\{\ldots\}$, give the usual conditions for thermal, mechanical and chemical equilibrium. The remaining two coefficients, treated explicitly in Equation 12.105, yield

$$\gamma_{\alpha\beta}\cos\phi_\alpha + \gamma_{\beta\varepsilon}\cos\phi_\varepsilon + \gamma_{\alpha\varepsilon} = 0 \qquad (12.114)$$

and

$$\gamma_{\alpha\beta}\sin\phi_\alpha - \gamma_{\beta\varepsilon}\sin\phi_\varepsilon = 0 \qquad (12.115)$$

It may be noted that these equations are identical to those that would be derived from a mechanical balance of three force vectors of magnitudes $\gamma_{\alpha\beta}$, $\gamma_{\beta\varepsilon}$ and $\gamma_{\alpha\varepsilon}$, acting at the triple point. In this force balance Equation 12.114 represents the sum of the forces in the x direction and Equation 12.115 is the sum in the y direction.

These two equations may also be written:

$$\frac{\gamma_{\alpha\beta}}{\sin\phi_\varepsilon} = \frac{\gamma_{\beta\varepsilon}}{\sin\phi_\alpha} = \frac{\gamma_{\alpha\varepsilon}}{\sin\phi_\beta} \qquad (12.116)$$

by applying some trigonometric identities (see Problem 12.13). These conditions for equilibrium at a triple line form the basis for determining relative interfacial energies of the intersecting surfaces since, with minor rearrangement,

$$\frac{\gamma_{\beta\varepsilon}}{\gamma_{\alpha\beta}} = \frac{\sin\phi_\alpha}{\sin\phi_\varepsilon} \qquad (12.117)$$

$$\frac{\gamma_{\alpha\varepsilon}}{\gamma_{\alpha\beta}} = \frac{\sin\phi_\beta}{\sin\phi_\varepsilon} \qquad (12.118)$$

Thus if the absolute value of any one of the three surface energies has been measured independently, the values of the other two may be determined simply by measuring the angles between the traces of the surfaces on a section that is perpendicular to the triple line.

EXAMPLE 12.5

A nickel bicrystal is equilibrated with its vapor. Where the grain boundary intersects the free surface, a groove is formed as shown in Figure 12.23. The surface energy of a nickel–vapor interface is nearly isotropic at 1400 K and may be taken to be 2040 erg/cm^2 = 2.04 J/m^2 (Appendix F). Determine the interfacial energy of this grain boundary.

Measure the dihedral angle (the angle between the surface normals) at the root of the groove. Since the grain boundary energy is significantly smaller than the surface energy, this angle will be near 180°. The most accurate method for determining such angles is based upon sighting vertically down on the groove with an interference microscope and analyzing the interference fringes. For this illustrative example, measurement of the angle between tangent lines drawn at the root of the groove yields $\phi_\alpha = 168°$. The remaining angles are equal to each other and thus equal to $(360° - \phi_\alpha)/2$. Equation 12.117 becomes

$$\frac{\gamma_{gb}}{\gamma_{sv}} = \frac{\sin\phi_\alpha}{\sin\left(\dfrac{360° - \phi_\alpha}{2}\right)} = \frac{\sin(168°)}{\sin(96°)} = \frac{0.208}{0.994} = 0.209$$

$$\gamma_{gb} = 0.209\,\gamma_{sv} = 0.209 \times 2.04\,\frac{J}{m^2} = 0.42\,\frac{J}{m^2}$$

Equation 12.112 and Equation 12.113 assume that the surface energies of the three interfaces involved are not functions of orientation. If one or more of the three phases is crystalline, this assumption will not be valid. A more general set of conditions for equilibrium, which considers variations in the energies of the boundaries due to rotations as well as translations, introduces additional "torque" terms into these equations that derive from the changes in the γ values with orientation. While such effects have been shown to be important in the analysis of

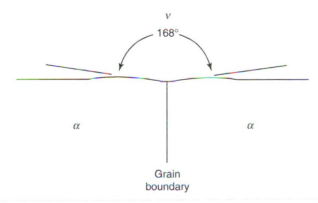

FIGURE 12.23 Configuration commonly observed after annealing where a grain boundary meets a free surface. The angle at the root of the groove yields the ratio of the grain boundary energy to the surface energy for the material.

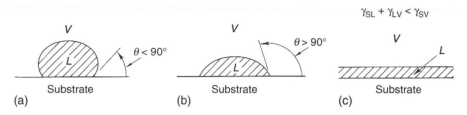

FIGURE 12.24 The angle measurement in a sessile drop experiment provides a relationship between the surface energies of the three interfaces that meet at the triple line; this angle reports the tendency for the liquid phase to wet (spread over) the substrate surface.

low angle grain boundaries, it is usually not necessary to appeal to this level of sophistication in the description of the behavior of materials systems. See Ref. [5] for more details.

The sessile drop experiment is sometimes used to assess surface energies, Figure 12.24. A droplet of a liquid is placed on an inert substrate and allowed to equilibrate with its vapor. For metals, glasses and ceramics this experiment is carried out in a muffle furnace since the system must be above the melting point of the droplet. A telescope is sighted on the droplet and a shadow image of the droplet is recorded. An analysis of the droplet shape that incorporates both gravitational and capillarity effects yields the liquid–vapor surface energy and the dihedral angle θ shown in Figure 12.24. Since the substrate is inert, it remains flat and equilibrium is not obtained in the vertical direction; i.e., Equation 12.115 is not realized. Equation 12.114 holds with α, β and ε, becoming the vapor (v), liquid (l) and solid (s) phases. Then, $\phi_\alpha = \phi_v = \theta$; $\phi_\beta = \phi_l = \pi - \theta$; and $\phi_\varepsilon = \phi_s = \pi$. Equation 12.114 becomes

$$\gamma_{vl}\cos\theta + \gamma_{ls}\cos(\pi) + \gamma_{vs} = 0 \tag{12.119}$$

So that

$$\gamma_{vl}\cos\theta - \gamma_{ls} + \gamma_{vs} = 0$$

or

$$\cos\theta = \frac{\gamma_{ls} - \gamma_{vs}}{\gamma_{vl}} \tag{12.120}$$

The dihedral angle θ is taken as a measure of the tendency of the liquid to wet the substrate. In Figure 12.24 it is shown that for values of θ less than 90° the liquid equilibrates as a bead resting on the substrate. As θ approaches 0° a given volume of droplet spreads over an increasing area of the substrate. The limiting case, $\theta = 180°$, $\cos\theta = -1$, corresponds to the condition

$$\gamma_{vs} = \gamma_{lv} + \gamma_{ls} \tag{12.121}$$

If the liquid–substrate and the liquid–vapor interfaces have a combined energy that is lower than the substrate–vapor interface energy, the system can minimize its surface energy by replacing the high energy (sv) interface completely with the low energy (ls) interface, Figure 12.24c. The liquid completely covers the solid substrate with a film and is said to wet the solid.

These same phenomena operate to influence the geometry of microstructures of materials, sometimes with disastrous effects. Figure 12.25 shows a particle of the β phase at a grain boundary in the matrix phase α. If it is assumed that $\gamma_{\alpha\beta}$ is isotropic, two of the interfaces at this $\alpha\beta\alpha$ triple line have the same energy; the third, $\alpha\alpha$, is a grain boundary in the matrix phase. Equation 12.114 becomes

$$\gamma_{\alpha\alpha} = 2\gamma_{\alpha\beta}\cos\frac{\theta}{2} \qquad (12.122)$$

If $\theta = 0°$, corresponding to $\gamma_{\alpha\alpha} = 2\gamma_{\alpha\beta}$, the β phase wets the grain boundary completely. For values of $\gamma_{\alpha\beta}$ smaller than this limiting value the system can minimize its surface energy by completely replacing the relatively high energy grain boundary with two $\alpha\beta$ interfaces; at equilibrium the β phase exists as a grain boundary film, Figure 12.25b. If during processing of this material the system is taken to a temperature in which the β phase is liquid, all of the grain boundaries in the structure are replaced by a liquid film and the material disintegrates into separate grains. This phenomenon is occasionally used in research to disintegrate a polycrystal into separate grains so that the distribution of shapes and sizes of grains may be examined. If the same phenomenon occurs during hot working an alloy ingot, the disintegration of the grain structure may be disastrous.

The equilibrium angle at triple lines determined by the relative energies of the three interfaces that meet there plays a role in a significant number of phenomena that influence the evolution of microstructures in materials science. The presence of a small amount of liquid phase accelerates sintering in the processing of some important ceramics; if the liquid is retained as a thin glassy film in the final product, the mechanical properties may be significantly degraded. The shape of porosity in the late stages of sintering of ceramics may be significantly influenced by the dihedral angle developed at grain boundaries that intersect the surface of internal porosity. Grain growth in polycrystalline metallic and ceramic systems results from a continuing attempt of the system to establish the equilibrium angles at the grain edges and corners. Large grain sizes are detrimental to mechanical properties.

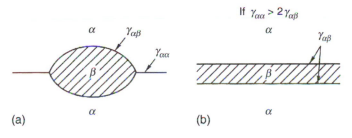

FIGURE 12.25 A particle of the β phase may tend to nucleate and grow at a grain boundary in the matrix phase. If $\gamma_{\alpha\alpha} > 2\ \gamma_{\alpha\beta}$, the β phase wets the grain boundary and may form a continuous grain boundary film at equilibrium.

Failures in thin film stripes that connect active devices on integrated circuits may be associated with the development of pinholes initiated at triple lines in the grain boundary network of the film. Wetting tendencies are crucial in the spreading of adhesives and solders in joining of materials as well as the application of coatings that protect them. The nucleation of particles of a newly forming phase is strongly favored if the particles form heterogeneously on a substrate that the new phase tends to wet; the energy required to form the surface of the new small particle, which is the primary barrier to nucleation, is thus significantly reduced. The distribution of second-phase particles in a microstructure at grain faces, triple lines, or quadruple points in the grain boundary network, or dispersed within the grains, is largely determined by the competing energetics of formation of nuclei at these various sites.

12.8   ADSORPTION AT SURFACES

Every extensive thermodynamic property has a specific surface excess contribution associated with interfaces in the system. In particular, the number of moles of any component may be expected to exhibit such a surface excess, which may be positive or negative. Component k is said to be adsorbed at the surface or interface in an equilibrated system. The general strategy underlying the description of adsorption of components at interfaces is presented in this section. The treatment is confined to liquid–vapor and solid–vapor surfaces because this permits certain simplifications but still allows development of the strategy. For a more general treatment of the thermodynamics of adsorption see Ref. [6].

12.8.1   MEASURES OF ADSORPTION

Figure 12.26a illustrates the variation of concentration of component k through a system containing a flat surface between the α and vapor phases, where α may be liquid or solid. In general, except near the critical point, the concentration of any component in the vapor phase may be treated as negligible in comparison with the α phase because the molar volume of the vapor is very large in comparison with V^{α}. Figure 12.26b illustrates the corresponding hypothetical system used in defining surface excess quantities with C_k^{α} constant up to the position of the dividing surface at x_s. It has been shown in Section 12.2 that the superficial excess of any extensive property may be visualized as the difference between the area under the curve in Figure 12.26a and the area under the step function defined by the hypothetical system. For component k, this difference is proportional to the shaded area in Figures. 12.26b and c. Mathematically,

$$n_k^s = n_k - C_k^{\alpha} A x_s = \int_a^b C_k A \, dx - C_k^{\alpha} A x_s \qquad (12.123)$$

Figure 12.26c shows an alternative choice for the position of the dividing surface at x_s' with the corresponding shaded area visualizing the surface excess of component k. This shaded area is roughly twice that shown in Figure 12.26b. Evidently, the value of the surface excess of k may be very sensitive to the choice of

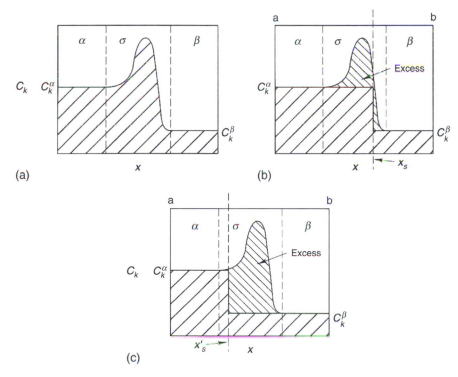

FIGURE 12.26 (a) Illustration of the distribution of component k through the physical surface of discontinuity between the α and β phases. (b) The corresponding hypothetical system with the dividing surface located at x_s. (c) An alternate choice for the position x_s.

the position of the dividing surface. Since the total thickness of the surface of discontinuity is about one atom, displacement of the dividing surface by a fraction of an angstrom may produce a significant change in the computed value for the surface excess of component k or any other surface excess property.

This impractical situation may be avoided by following yet another strategy due to Gibbs, defining reduced measures of adsorption that are independent of the choice of the position of the dividing surface. Consider the surface excess of the total number of moles of all of the components defined for the dividing surface at x_s:

$$n_T^S = n_T - C_T^\alpha A x_s = \int_a^b C_T A dx - C_T^\alpha A x_s \qquad (12.124)$$

where

$$C_T = \sum_{k=1}^c C_k$$

at each point in the system. Eliminate the position of the dividing surface in Equation 12.123 and Equation 12.124. First, solve Equation 12.124 for the product Ax_s:

$$Ax_s = \frac{1}{C_T^\alpha}\left[\int_a^b C_T(x)A\,dx - n_T^s\right]$$

Substitute this result into Equation 12.123:

$$n_k^s = \int_a^b C_k A\,dx - C_k^\alpha \frac{1}{C_T^\alpha}\left[\int_a^b C_T(x)A\,dx - n_T^s\right]$$

Collect the surface excess terms on the left side of the equation:

$$n_k^s - \frac{C_k^\alpha}{C_T^\alpha}n_T^s = \int_a^b C_k A\,dx - \frac{C_k^\alpha}{C_T^\alpha}\int_a^b C_T(x)A\,dx \qquad (12.125)$$

Evaluation of the integrals on the right side of this equation does not require a definition of the dividing surface; x_s has been eliminated from the equation. Thus the quantity on the left side of this equation is independent of the position of the dividing surface. These integrals are illustrated in Figure 12.27. The first is the total area

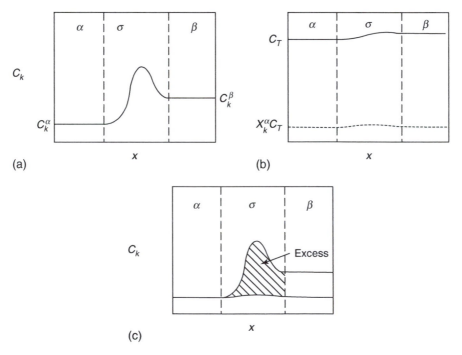

(a)

(b)

(c)

FIGURE 12.27 (a) Sketch of the variation of the concentration of k through the physical surface of discontinuity. (b) Variation of the total concentration, C_T, with position; dashed line is $X_k^\circ C_T$. (c) The shaded area is the *reduced surface excess* of component k.

under the curve for component k, Figure 12.27a. The second integral is the area under the curve for the total number of moles of all of the components, Figure 12.27b. The second term on the right side of Equation 12.125 multiplies this by $c_k^\alpha/c_T^\alpha = X_k^\alpha$, the mole fraction of k in the α phase; the dashed line in Figure 12.27b represents this term. The cross-hatched area in Figure 12.27c is thus a representation of the quantity defined on the left-hand side of Equation 12.125. This area reports the total number of atoms of component k in the surface of discontinuity, reduced in proportion to the fraction of the total number of atoms in this region that would be component k if this component were present in the same proportion as in the α phase. The quantity defined on the left side of Equation 12.125 is called the reduced surface excess of component k.

The corresponding specific reduced surface excess of component k, written $\bar{\Gamma}_k$ may be obtained by dividing the surface excess by the area of the dividing surface:

$$\bar{\Gamma}_k \equiv \frac{n_k^s - X_k^\alpha n_T^s}{A} \tag{12.126}$$

Evidently this property is also independent of the choice of location of the dividing surface.

12.8.2 THE GIBBS ADSORPTION EQUATION

A strategy based upon a generalization of the Gibbs–Duhem equation leads to the working equation for interpreting adsorption phenomena. Consider a two-phase multicomponent system with flat internal interfaces. If the system is in internal equilibrium, then the temperature, pressure and chemical potentials each have a single value for the system. The combined statement of the first and second laws for a two-phase system with flat internal interfaces may be written:

$$dU'_{sys} = TdS'_{sys} - PdV'_{sys} + \sum_{k=1}^{c} \mu_k dn_{k,sys} + \gamma dA \tag{12.127}$$

It is not necessary to write out separate terms for the two phases since the intensive properties are the same in both phases. Consider the formation of this two-phase system by adding quantities of the components to an initially empty system at constant T, P, and composition. The total internal energy of this system may be computed by integrating Equation 12.127 for this process:

$$U'_{sys} = TS'_{sys} - PV'_{sys} + \sum_{k=1}^{c} \mu_k n_{k,sys} + \gamma A \tag{12.128}$$

Take the total differential of this expression for an arbitrary change in state of the system:

$$dU'_{sys} = TdS'_{sys} + S'_{sys}dT - PdV'_{sys} - V'_{sys}dP + \sum_{k=1}^{c} \mu_k dn_{k,sys}$$

$$+ \sum_{k=1}^{c} n_{k,sys}d\mu_k + \gamma dA + Ad\gamma \tag{12.129}$$

Comparison of this result with Equation 12.127 yields a generalized form of the Gibbs–Duhem equation:

$$S'_{sys}dT - V'_{sys}dP + \sum_{k=1}^{c} n_{k,sys}d\mu_k + Ad\gamma = 0 \tag{12.130}$$

Each of the system properties may be expressed in terms of the corresponding surface excess property by applying Equation 12.13:

$$[S'^{\alpha}_{hyp} + S'^{\beta}_{hyp} + S^s]dT - [V'^{\alpha}_{hyp} + V'^{\beta}_{hyp}]dP + \sum_{k=1}^{c} [n^{\alpha}_{k,hyp} + n^{\beta}_{k,hyp} + n^s_k]dn_k$$

$$+ Ad\gamma = 0 \tag{12.131}$$

A separate Gibbs–Duhem equation holds for each of the uniform hypothetical parts of the system: for the α and β parts,

$$S'^J_{hyp}dT - V'^J_{hyp}dP + \sum_{k=1}^{c} [n^J_{k,hyp}d\mu_k = 0 \qquad (J = \alpha, \beta) \tag{12.132}$$

Removal of these terms from Equation 12.131 leaves the relation

$$S^s dT + \sum_{k=1}^{c} n^s_k d\mu_k + Ad\gamma = 0 \tag{12.133}$$

Solve for $d\gamma$:

$$d\gamma = -\frac{S^s}{A} - \sum_{k=1}^{c} \frac{n^s_k}{A}d\mu_k = -s^s dT - \sum_{k=1}^{c} \Gamma_k d\mu_k \tag{12.134}$$

where s^s and Γ_k are specific interfacial excess properties. This result is one form of the Gibbs adsorption equation.

With some manipulation the Gibbs adsorption equation may be expressed in terms of reduced specific interfacial excess properties:

$$d\gamma = -\overline{S^s}dT - \sum_{k=1}^{c} \overline{\Gamma}_k d\mu_k \tag{12.135}$$

It is most useful in this form because the quantities contained are independent of the choice of the dividing surface. This equation is usually applied to the evaluation of the variation of the adsorption of a particular component with composition of the α

phase, at constant temperature. With $dT = 0$, Equation 12.135 reads

$$d\gamma = -\sum_{k=1}^{c} \bar{\Gamma}_k d\mu_k \qquad (12.136)$$

For a binary system the summation in this equation has two terms. With proper choice of the dividing surface the surface excess of component 1 may be set to zero. Then Equation 12.136 may be written:

$$d\gamma = -\bar{\Gamma}_2 d\mu_2$$

and the reduced specific surface excess may be computed from

$$\bar{\Gamma}_2 = -\frac{d\gamma}{d\mu_2} \qquad (12.137)$$

If component 2 is the solute in a dilute solution, Henry's law relates μ_2 to X_2, the solute concentration:

$$d\mu_2 = RTd \ln X_2$$

and the adsorption equation becomes:

$$\bar{\Gamma}_2 = -\frac{d\gamma}{RTd \ln X_2} = -\frac{X_2}{RT}\frac{d\gamma}{dX_2} \qquad (12.138)$$

In a dilute binary solution if the addition of solute lowers the surface free energy (the derivative in Equation 12.138 is negative), then the surface excess for that component is positive: solute adsorbs on the surface. If the surface energy is raised by the addition of the solute, then the solute will be depleted at the surface at equilibrium.

12.9 SUMMARY

The superficial excess contribution to any extensive thermodynamic property may be visualized by comparing the value for a real system to that obtained for a hypothetical system in which intensive properties of each phase have a discontinuity at a dividing surface.

Application of the general strategy for finding conditions for equilibrium to a multicomponent two-phase system with a curved interface yields the usual conditions for thermal and chemical equilibrium but the condition for mechanical equilibrium is found to be

$$P^\beta = P^\alpha + 2\gamma H$$

where H is the local mean curvature of the surface and

$$\gamma = U^s - TS^s - \sum_{k=1}^{c} \mu_k n_k^s$$

is, in simple cases, the specific interfacial free energy of the interface. A mechanical analog yields the same relationship between the pressures if both phases are fluids and demonstrates that for such cases the surface tension and specific interfacial free energy are identical.

The altered condition for mechanical equilibrium associated with curved interfaces may be combined with the conditions for thermal and mechanical equilibrium to compute shifts of phase boundaries associated with the curvature of internal interfaces. For a unary system, the vapor pressure is shifted by

$$P(H) = P(H = 0)e^{(2\gamma V^L /RT)H} = P(H = 0)e^{\lambda_v H}$$

where λ_v is the capillarity length scale for the liquid–vapor equilibrium. The melting point is decreased in proportion to the curvature:

$$T(H) = T(H = 0) - \frac{2\gamma V^L}{\Delta S}H$$

where ΔS is the entropy of fusion.

The phase boundaries of a two-phase field in a binary system are also shifted if the interface between the phases is curved. For dilute terminal phases,

$$X_2^\alpha(H) = X_2^\alpha(H = 0)e^{(2\gamma V^\beta/RT)H}$$

$$X_2^\beta(H) = X_2^\beta(H = 0)e^{(2\gamma \overline{V_1^\beta}/RT)H}$$

These concepts are frequently applied locally in microstructures in which the curvature varies with position, invoking the assumption of local equilibrium.

For crystalline solids γ varies with orientation. The condition for equilibrium yields an equilibrium shape for the crystal defined by the condition

$$P^c = P^v + 2\frac{\gamma_j}{\lambda_j}$$

where γ_j is the interfacial energy of the jth face and λ_j is its pedal function. The Gibbs–Wulff construction yields the equilibrium polyhedral shape for any crystal, given a polar plot of γ over the sphere of orientation.

Where three phases meet at a triple line, the equilibrium configuration of the mating surfaces is given by

$$\frac{\gamma_{\alpha\beta}}{\sin\phi_\varepsilon} = \frac{\gamma_{\beta\varepsilon}}{\sin\phi_\alpha} = \frac{\gamma_{\alpha\varepsilon}}{\sin\phi_\beta}$$

This condition is equivalent to a mechanical force balance at the triple line. Relative values of these angles may play a key role in wetting of one phase on another, microstructural shapes, nucleation processes, and in thin film device failures.

Components tend to adsorb at interfaces at equilibrium. In a dilute binary system the reduced surface excess of the solute is given by

$$\bar{\Gamma}_2 = -\frac{X_2}{RT}\frac{d\gamma}{dX_2}$$

Adsorption is positive if component 2 lowers the surface energy; component 2 is reduced at the surface if it increases γ.

HOMEWORK PROBLEMS

Problem 12.1. Sketch a familiar object that has convex, concave, and saddle surface elements. Draw an exploded view of the object that illustrates each of these classes of features.

Problem 12.2. Write out an explicit expression for the change in internal energy for a two-phase, two-component system, including surface terms.

Problem 12.3. It has been asserted that V^s, the "specific superficial excess volume," is zero for any interface. Prove this assertion.

Problem 12.4. Paraphrase the development leading to Equation 12.33 to show that

$$\gamma = \left(\frac{\partial F'}{\partial A}\right)_{T,V^\alpha,V^\beta,n_k}$$

Problem 12.5. A thin wire of radius r and length l will tend to shorten under the action of surface tension, unless a weight F is hung on it sufficient to balance the contracting forces:
   a. Write a force balance that relates the weight F to the surface tension σ and the dimensions of the wire.
   b. Write an expression for the change in energy associated with an incremental lengthening dl of the rod in terms of γ and the dimensions of the rod.
   c. Interpret the energy change in terms of the work of displacing the force F through a distance dl.
   d. Equate the two alternate evaluations of F to show that $\sigma = \gamma$.

Problem 12.6. Use Equation 12.53 to sketch a surface that represents the variation of vapor pressure with temperature and curvature.

Problem 12.7. Compute the vapor pressure of liquid copper over a flat surface at 1400 K. Compute the equilibrium vapor pressure inside an 0.5 μm diameter bubble of copper vapor suspended in liquid copper at 1400 K.

Problem 12.8. It is asserted that the capillarity shift of the melting point of a material is significantly smaller than the shift expected for an allotropic transformation. Justify this conjecture.

Problem 12.9. Below 892 K the phase diagram for the system A–B consists of dilute terminal solid solutions α and β with no intermediate phases. At 680 K the solubility limits are $X_B^\alpha = 0.025$ and $X_B^\beta = 0.967$. The molar volume of β is 9.5 cc/mol and the partial molal volume of component 1 in β is 11.2 cc/mol. The interfacial free energy is 500 erg/cm^2:
   a. Compute the capillarity length scales for the α and β phases.
   b. Compute and plot the equilibrium interface compositions as a function of particle size. Assume the particles are spheres.

Problem 12.10. The ε to β transformation occurs at 1155 K in pure titanium. The element B is a β stabilizer when added to titanium. Assume for this problem that the β and ε phases are ideal solutions. The properties of the system are:

Component	$T_k^{\varepsilon \rightarrow \beta}$(K)	$\Delta S_k^{o\varepsilon \rightarrow \beta}$(J/mol K)	V^β (cc/mol)
Titanium	1155	4.2	11.5
B	830	5.2	9.7
$\gamma = 470$ erg/cm^2			

   a. Compute the bulk compositions, X_B^ε and X_B^β, at 1100 K.
   b. An alloy with $X_B = 0.12$ is quenched from the β phase to 1100 K where ε, nucleates and grows. Compute and plot the interface composition in the β phase as a function of particle radius.
   c. Sketch the capillarity shift on the phase diagram.

Problem 12.11. Sketch the microstructure at the advancing ($\alpha\beta$/L) interface in a solidifying eutectic. How does the capillarity shift alter the composition in the liquid at the curved interface
   a. In front of the α phase?
   b. In front of the β phase?
   Illustrate your answer by sketching shifted liquidus curves on the phase diagram near the eutectic point.

Problem 12.12. The tetrakiadecahedron is a polyhedron with six {100} faces and eight {111} faces. (Tetrakiadeca is Greek for 14.) On a regular tetrakiadecahedron all of the edges have the same length. Calculate the ratio of $\gamma_{\{111\}}$ to $\gamma_{\{100\}}$ that would be required to produce this shape.

Problem 12.13. Show that Equation 12.108 are consistent with Equation 12.106 and Equation 12.107.

Problem 12.14. The surface energy of the interface between nickel and its vapor is estimated to be 1580 (erg/cm^2) at 1100 K. The average dihedral angle measured for grain boundaries intersecting the free surface is 151°. Thoria dispersed nickel alloys are made by dispersing fine particles of ThO_2 in nickel powder and consolidating the aggregate. The particles are left at the grain boundaries in the nickel matrix. Prolonged heating at elevated temperatures gives the particles their equilibrium shape. The average dihedral angle measured inside the particle is found to be 157°. Estimate the specific interfacial energy of the thoria–nickel interface.

Problem 12.15. Nuclear reactors used in submarine power plants circulate liquid sodium as a coolant in the reactor core. Stainless steel piping is proposed for use in the heat exchangers. The average grain boundary energy in stainless steel is 250 (erg/cm^2), while that of the solid liquid interface is 110 (erg/cm^2). Will liquid sodium significantly penetrate the grain boundaries in stainless steel?

Problem 12.16. Visualize a gold thin film stripe as a connector in a microelectronic chip. Suppose the stripe is 0.1 μm thick and has a bamboo grain structure. (This means the grain boundaries run laterally completely across the stripe.) Take the grain boundary energy of gold at 600 K to be 420 (erg/cm^2); its surface energy is 1440 (erg/cm^2)
    a. Compute the dihedral angle where a grain boundary meets the external surface.
    b. Find a critical grain boundary spacing, s_c, for which the equilibrium grain shape will produce a hole in the film.

Problem 12.17. Show that the specific interfacial excess properties associated with a grain boundary are independent of the choice of the position of the dividing surface.

Problem 12.18. Sketch a concentration profile crossing an interface between α and β phases. Sketch a plot of the variation of the specific interfacial excess of the component as the choice of the position of the dividing surface moves from one side of the physical surface of discontinuity to the other. Pay particular attention to the form of the curve.

REFERENCES

1. Herring, C., *Structure and Properties of Solid Surfaces*, Gomer, R. and Smith, C.S., eds., University of Chicago Press, Chicago, IL, 1952.
2. Murr, L.E., *Interfacial Phenomena in Metals and Alloys*, Addison-Wesley, Reading, MA, 1975.
3. Murr, L.E., *Interfacial Phenomena in Metals and Alloys*, Addison-Wesley, Reading, MA, p. 101, 1975.

4. Kingery, W.D., Bowen, H.K., and Uhlmann, D.R., *Introduction to Ceramics*, 2nd ed., Wiley, New York, p. 183, 1976.
5. Shewmon, P.G. and Robertson, W.M., *Metal Surfaces, Structure, Energetics and Kinetics*, ASM, Materials Park, OH, 1963.
6. Lupis, C.H.P., *Chemical Thermodynamics of Materials*, Elsevier Science Publishing Company, New York, pp. 389–430, 1983.

13 Defects in Crystals

CONTENTS

Stable solids are generally crystalline. Atoms of the components vibrate about well-defined positions in space relative to one another. This local pattern repeats indefinitely in three dimensions. This characteristic periodic structure plays a key role in determining most of the physical, electronic, mechanical and chemical properties of solids.

Solid crystals are not perfect. These imperfections in the crystalline arrangement of the atoms in space occur as isolated points, along lines or as surfaces in the structure. The thermodynamics of surfaces was dealt with in Chapter 12. Line defects, known as dislocations, dominate the mechanical behavior of ductile crystals and are important defects in electronic materials; however, an understanding of their behavior is not traditionally formulated in thermodynamic terms. In contrast, most of the elements of the description and behavior of point defects are formulated thermodynamically.

Perhaps the most important role played by point defects in the behavior of solids is in diffusion, i.e., the atom-by-atom transport of components through the crystal lattice. Most processes in materials science that produce changes in microstructure involve diffusion. The elemental step in diffusion is the motion of an atom from a normal crystal site into an adjacent point defect in the crystal. Processes such as precipitation, phase changes, sintering, oxidation, solid state bonding and some forms of creep depend upon the presence of point defects in the system. Point defects influence the resistivity of conductors, the losses in insulators and the conductivity of semiconductors. Their influence is particularly important in stoichiometric and nonstoichiometric ceramic and intermetallic compounds where the number and distribution of a variety of defect types control performance.

This chapter begins by developing a thermodynamic description of the behavior of point defects in elemental crystals, applying one more time the general strategy for finding conditions for equilibrium previewed in Figure 1.4. This approach is then extended to stoichiometric binary compounds, which by definition are constrained to a fixed ratio of the two elements comprising the system. These results set the groundwork for the treatment of the more general case of nonstoichiometric compounds. Analogous concepts are applied to the description of electronic defects to establish the connection between phenomenological thermodynamics and the electron theory of the behavior of matter.

13.1   POINT DEFECTS IN ELEMENTAL CRYSTALS

The concept of a point defect presumes the existence of a periodic lattice of sites that are normally occupied by atoms in a crystal. Two primary classes of point defects are found to exist in elemental crystals: vacancies and interstitials.[1] A vacancy exists in a crystal where a normal lattice site is unoccupied (Figure 13.1a). An interstitial defect occurs when an atom occupies a position in the crystal other than a normal lattice site (Figure 13.1b). While each defect contributes an increase to the energy of the crystal, each defect also increases the entropy of the crystal. These effects combine to guarantee that at equilibrium a crystal will contain some point defects: a crystal is not perfect at equilibrium. The concentration of point defects in an elemental crystal is normally very small: even where it is highest, near the melting point, defects occur at only about one in 10,000 sites. Concentrations of interstitials in elemental crystals are expected to be much smaller than vacancies under the same conditions. Nonetheless, this small fraction of defect sites plays a crucial role in materials science.

---

[1] Every crystal lattice has a set of normal sites and a set of interstitial sites which lie between the normal sites. If atoms of a second component added to a crystal are small enough, they will tend to occupy interstitial sites; hydrogen, carbon, nitrogen and sometimes oxygen dissolve in most metals as such interstitial components. Solute atoms that normally are found in normal lattice sites are called substitutional components. A small atom of an interstitial component occupying an interstitial site is not normally considered to be a defect in the crystal. However, if an atom of a substitutional component occupies an interstitial site, the configuration is called an interstitial defect.

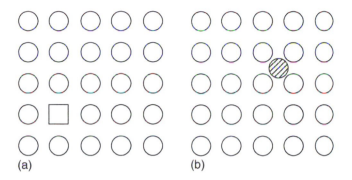

FIGURE 13.1 Two classes of point defects in an elemental crystal: (a) vacancy and (b) interstitial.

13.1.1 Conditions for Equilibrium in a Crystal with Vacant Lattice Sites

To find the concentration of defects at equilibrium, apply the familiar strategy. Consider a system composed of a homogeneous crystalline phase (α) and its vapor (g). The combined statement of the first and second laws must explicitly recognize that the crystal may change its internal energy not only by changing entropy, volume and the number of moles of each component, but also by changing the number of vacancies in the system.

$$U'^\alpha = U'^\alpha(S'^\alpha, V'^\alpha, n_1^\alpha, n_2^\alpha, \dots, n_c^\alpha, n_v^\alpha) \tag{13.1}$$

where n_v is the number of vacant lattice sites in the crystal. The change in internal energy for the crystal when taken through an arbitrary change in state is

$$dU'^\alpha = T^\alpha dS'^\alpha - P^\alpha dV'^\alpha + \sum_{k=1}^{c} \mu_k^\alpha dn_k^\alpha + \mu_v^\alpha dn_v^\alpha \tag{13.2}$$

This equation implicitly contains as a coefficient relationship the definition of μ_v, the chemical potential of a vacancy in the crystal. The analogous expression for the vapor phase is familiar:

$$dU'^g = T^g dS'^g - P^g dV'^g + \sum_{k=1}^{c} \mu_k^g dn_k^g \tag{13.3}$$

Rearrange each of these equations to solve for the change in entropy for each phase. Then combine them to yield an expression for the change in entropy of the system:

$$dS_{sys} = \frac{1}{T^\alpha} dU'^\alpha + \frac{P^\alpha}{T^\alpha} dV'^\alpha - \frac{1}{T^\alpha} \sum_{k=1}^{c} \mu_k^\alpha dn_k^\alpha - \frac{1}{T^\alpha} \mu_v^\alpha dn_v^\alpha$$
$$+ \frac{1}{T^g} dU'^g + \frac{P^g}{T^g} dV'^g - \frac{1}{T^g} \sum_{k=1}^{c} \mu_k^g dn_k^g \tag{13.4}$$

In deriving the conditions for equilibrium attention is focused on an isolated system. Because the system is isolated from its surroundings,

$$dU'_{sys} = 0 = dU'^\alpha + dU'^g \Rightarrow dU'^g = -dU'^\alpha \tag{13.5}$$

$$dV'_{sys} = 0 = dV'^\alpha + dV'^g \Rightarrow dV'^g = -dV'^\alpha \tag{13.6}$$

$$dn_{k,sys} = 0 = dn_k^\alpha + dn_k^g \Rightarrow dn_k^g = -dn_k^\alpha \tag{13.7}$$

Isolation of the system from its surroundings does not put any limitation upon the number of lattice sites in the crystal. Lattice sites may be created or destroyed by moving an atom from the interior of the crystal to its surface and vice versa. Thus, the number of vacancies n_v is not constrained in an isolated system. Insert these isolation constraints into Equation 13.4 and collect terms:

$$dS_{sys,iso} = \left(\frac{1}{T^\alpha} - \frac{1}{T^g}\right)dU'^\alpha + \left(\frac{P^\alpha}{T^\alpha} - \frac{P^g}{T^g}\right)dV'^\alpha$$
$$- \sum_{k=1}^{c}\left(\frac{\mu_k^\alpha}{T^\alpha} - \frac{\mu_k^g}{T^g}\right)dn_k^\alpha - \frac{\mu_v^\alpha}{T^\alpha}dn_v^\alpha \tag{13.8}$$

In this equation n_v is an independent variable: n_v may vary independently in an isolated system because lattice sites may be created or annihilated with no other changes in the system. The maximum in entropy that corresponds to the equilibrium state for the system is found by setting all of the coefficients of the differentials of the independent variables in the system equal to zero. This yields the familiar conditions for thermal, mechanical and chemical equilibrium in this solid–vapor system. In addition, it yields the condition:

$$\mu_v^\alpha = 0 \tag{13.9}$$

Thus in a crystal at equilibrium, the chemical potential of vacancies is zero.

13.1.2 THE CONCENTRATION OF VACANCIES IN AN ELEMENTAL CRYSTAL AT EQUILIBRIUM

A mixture of vacant sites with sites occupied by normal atoms may be considered to be a dilute solution of vacancies and normal atoms. The chemical potential of any component, including vacancies, is identical with the partial molal Gibbs free energy. Apply a definitional relationship the partial molal Gibbs free energy:

$$\mu_V = \bar{G}_V = \bar{H}_V - T\bar{S}_V = \bar{H}_V - T(\bar{S}_V^{xs} + \bar{S}_V^{id}) \tag{13.10}$$

where $\bar{H}_V$ is the partial molal enthalpy of vacancies in the crystal, $\bar{S}_V^{xs}$ is the partial molal excess entropy and $\bar{S}_V^{id}$ is the ideal entropy term, $-k \ln X_V$:

$$\mu_v^\alpha = [\bar{H}_v - T\bar{S}_v^{xs}] + kT \ln X_v^\alpha \tag{13.11}$$

where X_v^α is the atom fraction of vacant sites in the crystal and $\bar{H}_v$ and $\bar{S}_v^{xs}$ are partial molal properties associated with vacancies. According to Equation 13.9, at equilibrium this quantity is zero:

$$\mu_v^\alpha = 0 = \lfloor \bar{H}_v - T\bar{S}_v^{xs} \rfloor + kT \ln X_v^\alpha \qquad (13.12)$$

Thus, the equilibrium fraction of vacant sites in a crystal is:

$$X_v = e^{\bar{S}_v^{xs}/k} e^{-\bar{H}_v/kT} \qquad (13.13)$$

In this equation $\bar{H}_v$ may be thought of as the enthalpy of formation of a vacancy from a perfect crystal. $\bar{S}_v^{xs}$ is the excess entropy associated with this process, physically associated with changes in the vibrational behavior of atoms surrounding the vacant site. In many texts this quantity is written $\bar{S}_v^{vib}$. Since the enthalpy of formation of a vacancy is positive, Equation 13.13 demonstrates that the concentration of vacancies increases with temperature.

Experimental tests of these relations are usually indirect. For example, defects act as scattering centers in the flow of electrons through a conductor. In a dilute solution of vacancies each vacancy makes the same contribution to scattering the flow of electrons; hence the electrical resistivity is proportional to the vacancy concentration. Changes in electrical resistivity may thus be used to monitor changes in vacancy concentrations, provided that other influences may be eliminated.[2] Careful thermal analysis may also be used to monitor changes in vacancy concentration; measurement of the power required to heat two samples, which are identical except for defect concentrations, has been used to follow interstitial and vacancy annihilation processes during annealing after deformation or neutron irradiation.

In a few cases direct measurements of the equilibrium vacancy concentration as a function of temperature have been carried out (Figure 13.2). In one sample the

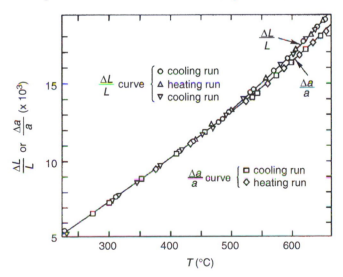

FIGURE 13.2 Comparison of molar volume of a crystal computed from dilatometric (linear expansion) measurements and x-ray lattice parameter measurements. The difference between these curves at any temperature is the volume of vacancies in the crystal. *Source:* Simmons, R. and Balluffi, R., *Phys. Rev.*, 117, 52, 1960.

length was measured carefully as a function of temperature in a sensitive dilatometer. These length measurements could be converted to changes in the sample volume. The volume increases partly because the average distance between atoms expands with temperature (normal thermal expansion) but also as a result of an increase in the number of vacant lattice sites with temperature. X-ray measurements on the same sample at a set of temperatures permits evaluation of the lattice parameters of the crystal and thus the mean interatomic distance. The volume computed for 1 mole of atoms from this x-ray measurement is that of a defect-free crystal. Subtraction of the volume calculated from the x-ray measurements from the total volume measured directly gives the volume of sites in the crystal that are vacant and thus the number of vacancies. Equation 13.13 suggests that a plot of the logarithm of the measured vacancy concentration vs. $1/T$ should be linear with a slope equal to H_v/k and an intercept (at $1/T = 0$) equal to S_v^{xs}/k. Such a plot is shown in Figure 13.3. Table 13.1 presents properties of vacancies for a few metallic systems.[3]

13.1.3 INTERSTITIAL DEFECTS AND DIVACANCIES

These arguments easily extend to other defects in elemental crystals with analogous results. At equilibrium, the chemical potential of an interstitial defect is zero, implying a relation between defect concentration and temperature of the general form:

$$X_D = f_D e^{\bar{S}_D^{xs}/k} e^{-\bar{H}_D/kT} \tag{13.14}$$

where f_D is the ratio of the number of interstitial sites to the number of normal lattice sites in the crystal.

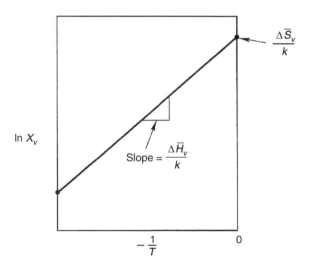

FIGURE 13.3 Arrhenius plot of atom fraction of vacancies in a crystal vs. the reciprocal of the temperature; the slope is proportional to the enthalpy of formation of a vacancy and the intercept gives the vibrational entropy of a vacancy.

TABLE 13.1
Properties of Vacancies for Some Typical Metals

Metal	$X_v (T_m)$	ΔS_v (J/mole K)	ΔH_v (J/mole)
Aluminum	9.0×10^{-4}	19	74
Copper	1.9×10^{-4}	12	113
Gold	7.2×10^{-4}	10	92
Lead	2.0×10^{-4}	21	56
Platinum	6.0×10^{-4}	9	135
Silver	1.7×10^{-4}	12	101

Source: Mukherjee, K., *Trans. AIME*, 6, 1324, 1966.

Theoretical calculations have demonstrated that the enthalpy to form a vacancy is significantly lower than that associated with the formation of an interstitial defect. At the same temperature it may be expected that the concentration of interstitial defects is very much smaller (usually by several orders of magnitude) than that of vacancies at equilibrium. Thus interstitial defects are not expected to participate in processes that occur near equilibrium. However, they may play an important role in crystals that are far from equilibrium. For example, in neutron irradiation, collisions of high-energy neutrons with atoms on normal sites may displace atoms over large distances. These displaced atoms come to rest in regions in which the normal sites are all occupied and thus these atoms reside in interstitial sites. The empty sites they leave behind are vacancies. During subsequent annealing these collections of equal numbers of vacancies and interstitial defects may recombine to form normal, occupied lattice sites. However, interstitial defects do not play an important role in most applications.

Defects may occur in combinations in an elemental crystal. The most common of these is the divacancy, which is a pair of adjacent vacant lattice sites. Analysis shows that such a defect would be significantly more mobile than a single vacancy. However, the equilibrium concentration of such defects in a crystal will be much smaller than that of single vacancies. Adapt the result in Equation 13.13 to this case; at equilibrium,

$$X_{vv} = e^{\bar{S}_{vv}^{xs}/k} e^{-\bar{H}_{vv}/kT} \tag{13.15}$$

where the subscript vv denotes properties of divacancies.

It is useful to visualize the formation of a divacancy in two steps: (1) the separate formation of two vacancies from a perfect crystal; and (2) the formation of the divacancy configuration from two separated single vacancies. The enthalpy change associated with the first process is simple $2\bar{H}_v$. Let $\Delta \bar{H}_{int}$, the interaction enthalpy, be the change in enthalpy associated with the second process. Thus, the enthalpy of formation of a divacancy may be written:

$$\Delta \bar{H}_{vv} = 2\bar{H}_v + \Delta \bar{H}_{int} \tag{13.16}$$

The same argument may be used to write the excess entropy

$$\Delta \bar{S}_{vv} = 2\bar{S}_v + \Delta \bar{S}_{int} \tag{13.17}$$

$\Delta \bar{H}_{int}$ and $\Delta \bar{S}_{int}$ are the interaction enthalpy and entropy for the pair of vacancies, both of which may be expected to be negative since the property values for a divacancy may be expected to be smaller than for a pair of isolated vacancies. Equation 13.15 may be written

$$X_{vv} = e^{(1/k)[2\bar{S}_v^{xs}+\Delta \bar{S}_{int}]}e^{-(1/kT)[2\bar{H}_v+\Delta \bar{H}_{int}]}$$

$$X_{vv} = \left[e^{(1/k)[\bar{S}_v^{xs}]}e^{-(1/kT)[\bar{H}_v]}\right]^2 e^{\Delta \bar{S}_{int}/k}e^{-(\Delta \bar{H}_{int}/kT)} \tag{13.18}$$

$$X_{vv} = (X_v)^2 e^{\Delta \bar{S}_{int}/k}e^{-(\Delta \bar{H}_{int}/kT)} \tag{13.19}$$

It is concluded that at equilibrium the concentration of divacanies will be somewhat larger than the square of the concentration of single vacancies. Since the square of a small fraction is a much smaller fraction, equilibrium concentrations of divacancies probably do not play a significant role in most processes. However, it is possible to introduce a supersaturated concentration of vacancies into a crystal by quenching from a high temperature where vacancy concentrations are high, by neutron irradiation or by ion bombardment, such as is routinely done in ion implantation of thin films for microelectronic devices or in surface treatments of bulk materials. Under these conditions, far from equilibrium, significant divacancy concentrations may be developed and may play a role in enhancing diffusion rates and in electronic processes.

13.2 POINT DEFECTS IN STOICHIOMETRIC COMPOUND CRYSTALS

Chapter 8 dealt with the thermodynamic description of binary and higher order solutions. In the context of that chapter the crystal structure of solid phases was not explicitly treated because the phenomenological approach did not require it. It was assumed more or less explicitly that such a crystal consisted of a single class of lattice sites, which could be readily occupied by atoms of any of the components in the system. Many intermediate phases and line compounds have crystal structures that have two (or more) distinct classes of lattice sites (Figure 13.4) called sublattices. In ionic crystals one set of sites typically contains the cations (positively charged ions) and the other the anions (negatively charged ions). In such crystals one set of sites is termed the cation sites and the other the anion sites. This characteristic holds even if the bonding in the crystal is not purely ionic; in this case the more electronegative atom occupies the anion sites and the less electronegative atom the cation sites.

The anion (or more electronegative) component in a compound crystal is typically (though not always) a nonmetallic element, such as oxygen, nitrogen, carbon, sulfur, chlorine, etc. In the following description this element will be

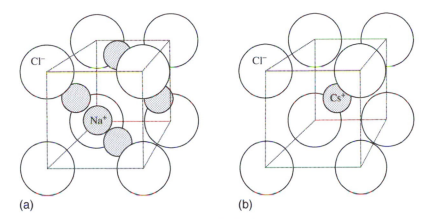

(a) (b)

FIGURE 13.4 Common crystal structures with two classes of lattice sites (sublattices): (a) NaCl and (b) CsCl.

designated generically as **X**. The other more metallic (cation) element in the system will be designated with an **M**. A vacant lattice site will be designated with a **V**. These components (atoms or ions) may occupy cation (M) sites, anion (X) sites or interstitial (i) sites. Finally, each entity visualized has an associated electronic charge.

A broad variety of defects may be visualized to exist, even in a simple crystal with only two types of normal lattice sites. To facilitate the explicit description of this variety of entities a widely accepted notation was devised by Kröger and Vink.[4] This notation exhibits three important elements in the identification of a particular defect:

a. The entity occupying the defect site (M, X, V or substitutional elements).
b. A subscript for the type of site occupied (M, X or i).
c. A superscript for the excess charge associated with the site, $(^X)$,$(\cdot)$ or $(')$.

Designations (a) and (b) are self-evident. The notation used to describe electrical charge associated with the defect requires some discussion. It is convenient to describe the charge in comparison with that which is normally associated with a particular site, i.e., to describe the local excess charge. If the entity occupying the site carries the charge of the specie normally occupying that site, then a superscript $(^X)$ is used. Thus in an alumina crystal (Al_2O_3) normal sites would be designated Al_{Al}^X and O_O^X. Similarly, a trivalent chromium ion on a cation site in alumina would be designated Cr_{Al}^X because it carries the same charge as the aluminum ions in the structure. A superscript $(\cdot)$ designates one unit of excess positive charge. Thus $Al_i^{\cdots}$ describes a trivalent aluminum ion in an interstitial site where its three positive charges are all excess charge; $Ca_K^{\cdot}$ represents a divalent calcium ion on a normally monovalent potassium site in KCl. A superscript $(')$ represents an excess negative charge. An oxygen ion in an interstitial site is represented by O_i''; Mg_{Al}' describes a divalent magnesium ion on a (normally trivalent) aluminum cation site. Some generic examples of this notation are reviewed in Table 13.2.

TABLE 13.2
Defect Designations Using the Kroger–Vink[4] Notation Applied to a Compound with Nominal Composition MX and Normal Valence of M as + 2, X as − 2

Defect	Excess Charge	Symbol
Vacancy on M sublattice	− 2	V_M''
Vacancy on X sublattice	+2	$V_X^{\cdot\cdot}$
M atom on an interstitial site	+2	$M_i^{\cdot\cdot}$
X atom on an interstitial site	− 2	X_i''
M atom on an X site	+4	$M_X^{\cdot\cdot\cdot\cdot}$
X atom on an M site	− 4	X_M''''
Divacancy on M and X sites	0	$(V_M V_X)$
M interstitial paired with M on an X site	+6	$(M_i M_X)^{\cdot\cdot\cdot\cdot\cdot\cdot}$
Solute cation L with +3 charge on M site	+1	$L_M^{\cdot}$
Solute anion Y with − 1 charge on X site	+1	$Y_X^{\cdot}$
Free (unattached) electron	− 1	e'
Electron hole	+1	$h^{\cdot}$

Vacant lattice sites on either sublattice have an associated excess charge of equal magnitude but opposite sign to the ion that normally occupies the site. For example, the removal of a cation from a cation site to produce a cation vacancy leaves an excess negative charge associated with the surrounding anions, which is no longer balanced. Thus a cation vacancy in potassium chloride is designated V_K'; in alumina, the designation is V_{Al}''' An anion vacancy in MgO is represented by $V_O^{\cdot\cdot}$. Cation vacancies carry an excess negative charge and anion vacancies an excess positive charge.

13.2.1 FRENKEL DEFECTS

The equilibrium concentration of defects in a compound crystal may be derived from the general strategy with careful attention paid to the conservation equations that form the isolation constraints. The results take a form, which is analogous to the conditions for equilibrium in a multivariate reacting system, derived in Section 11.4 of Chapter 11. Indeed, these conditions may be formulated in terms of defect reactions, affinities, formation energies of the defects and corresponding equilibrium constants. These results are illustrated in this section for the simplest form of defect in a binary compound, known as a Frenkel defect. A Frenkel defect is formed on the cation sublattice by removing an M ion from a normal M site and placing it in an interstitial site (Figure 13.5). It is also possible to form a Frenkel defect on the anion sublattice. A Frenkel defect is called an intrinsic defect because it may be formed without any interaction with the surroundings of the crystal.

Consider a crystal MX in which the normal valance of M is +2 and that of X is − 2. If this crystal contains Frenkel defects derived from cation sites, four distinct entities exist in such a crystal: M_M^X, X_X^X, V_M'' and $M_i^{\cdot\cdot}$ (see Figure 13.5). In words this

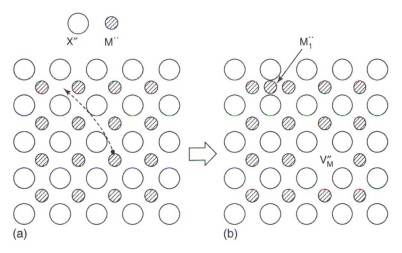

FIGURE 13.5 A Frenkel defect is formed on the cation sublattice by removing an M atom from a normal cation lattice site and placing it in an interstitial site.

notation describes: M ions on cation (M) sites; X ions on anion (X) sites; vacant sites on the cation (M) sublattice; and M ions in an interstitial (i) site. The number of each of these entities may be varied in the crystal; however, these variations are not independent. It is possible to define a chemical potential for each of these entities as the rate of change of the Gibbs free energy of the crystal with the number of each particular entity at constant temperature and pressure.

To find the conditions for equilibrium in such a system, apply the familiar strategy. First, write an expression for the entropy of this homogeneous crystalline phase incorporating changes in the numbers of each possible entity:

$$dS'_{sys} = \frac{1}{T}dU' + \frac{P}{T}dV' - \frac{1}{T}[\mu_{M_M}dn_{M_M} + \mu_{X_X}dn_{X_X} + \mu_{V_X}dn_{V_X} + \mu_{M_i}dn_{M_i}]$$

(13.20)

In an isolated system, $dU' = 0$ and $dV' = 0$. Conservation of atoms of M and X requires

$$dm_X = dn_{X_X} = 0$$

(13.21)

$$dm_M = 0 = dn_{M_M} + dn_{M_i} \Rightarrow dn_{M_M} = -dn_{M_i}$$

(13.22)

since all X atoms remain on anion sites while the M atoms are distributed over cation and interstitial sites. Further, each Frenkel defect consists of one vacancy and one interstitial atom. In the process of creating Frenkel defects the change in the number of vacancies and interstitials therefore must be the same:

$$dn_{V_M} = dn_{M_i}$$

(13.23)

Insertion of these constraints into Equation 13.20 simplifies to

$$dS'_{\text{sys,iso}} = -\frac{1}{T}\left[\mu_{M_M}(-dn_{M_i}) + \mu_{X_X}(0) + \mu_{V_M}(dn_{M_i}) + \mu_{M_i}dn_{M_i}\right]$$

$$dS'_{\text{sys,iso}} = -\frac{1}{T}\left[\mu_{M_i} + \mu_{V_M} - \mu_{M_M}\right]dn_{M_i} \tag{13.24}$$

The quantity in brackets may be thought of as the affinity for the defect reaction

$$M_M = V_M + M_i \tag{13.1}$$

which represents the unit process in the formation of a Frenkel pair from an M atom on a normal M site.

The condition for equilibrium in the crystal may be obtained by setting the affinity equal to zero:

$$\mu_{M_i} + \mu_{V_M} - \mu_{M_M} = 0 \tag{13.25}$$

Each of the chemical potentials in this expression may be written in the form

$$\mu_k = G_k^0 + kT \ln a_k \qquad (k = M_i, V_M, M_M) \tag{13.26}$$

Substitute these expressions into the condition for equilibrium, Equation 13.25:

$$\lfloor G_{V_M}^0 + G_{M_i}^0 - G_{M_M}^0 \rfloor + kT \lfloor \ln a_{V_M} + \ln a_{M_i} - \ln a_{M_M} \rfloor = 0 \tag{13.27}$$

This result may be cast in a form that is very similar to the law of mass action for a reacting system, Equation 11.34.

$$\Delta G_{\text{fd}}^0 = -kT \ln K_{\text{fd}} \tag{13.28}$$

where

$$\Delta G_{\text{fd}}^0 \equiv G_{V_M}^0 + G_{M_i}^0 - G_{M_M}^0 \tag{13.29}$$

and

$$K_{\text{fd}} \equiv \frac{a_{V_M} a_{M_i}}{a_{M_M}} \tag{13.30}$$

Since the solution is very dilute, Raoult's law applies to the solute, M_M:

$$a_{M_M} = X_{M_M} \approx 1 \tag{13.31}$$

Equation 13.30 may be rewritten:

$$K_{\text{fd}} = \frac{1}{a_{V_M} a_{M_i}} = \frac{1}{\gamma_{V_M} \gamma_{M_i} X_{V_M} X_{M_i}} \approx \frac{1}{X_{V_M} X_{M_i}} \tag{13.32}$$

The approximation is necessary; information about the activity coefficients of defects is generally unavailable. Precise calculations of the defect concentrations are thus not accessible. The results of defect equilibrium analyses therefore provide trends, such as the dependence of the equilibrium upon temperature and relations among the concentrations of defects. Insert this evaluation into Equation 13.28:

$$K_{fd} = \frac{1}{X_{M_v} X_{M_i}} = e^{-(\Delta G^0_{fd}/kT)} = e^{\Delta S^{\circ}_{fd}/k} e^{-(\Delta H^0_{fd}/kT)} \tag{13.33}$$

where ΔH^0_{fd} and ΔS°_{fd} are implicitly defined in this equation. The condition for equilibrium may now be expressed as

$$X_{V_M} X_{M_i} = e^{-(\Delta S^{\circ}_{fd}/k)} e^{\Delta H^0_{fd}/kT} \tag{13.34}$$

Finally, since the number of vacancies and interstitials are constrained to be the same, Equation 13.23, $X_{Vm} = X_{Mi}$ so that $(X_{Vm})(X_{Mi}) = (X_{Vm})^2 = (X_{fd})^2$, where X_{fd} is the atom fraction of Frenkel defects in the structure. Thus,

$$X_{V_M} X_{M_i} = (X_{V_M})^2 = (X_{fd})^2 = e^{-(\Delta S^{xs}_{fd}/k)} e^{\Delta H_{fd}/kT}$$

or

$$X_{fd} = e^{\Delta S^{\circ}_{fd}/2k} e^{-(\Delta H^0_{fd}/2kT)} \tag{13.35}$$

If ΔH^0_{fd} and ΔS°_{fd} are insensitive to temperature, the equilibrium concentration of Frenkel defects is expected to increase with temperature with the typical Arrhenius functional form.

13.2.2 SCHOTTKY DEFECTS

In the simplest case of a crystal with formula MX the anions and cations carry the same number of charges. In an MX crystal a Schottky defect consists of a vacant cation site and a vacant anion site (Figure 13.6); in this simple case the formation of Schottky defects does not disturb the electrical neutrality of the crystal. Similar to the Frenkel defect, this structural imperfection is intrinsic since it may be formed in a perfect crystal without adding or subtracting atoms or charges to the crystal.

The equilibrium concentration of Schottky defects in a crystal with formula MX begins with an expression for the change in entropy allowing in this case for changes in the number of cation and anion vacancies:

$$dS'_{sys} = \frac{1}{T} dU' + \frac{P}{T} dV' - \frac{1}{T}\left[\mu_{M_M} dn_{M_M} + \mu_{X_X} dn_{X_X} + \mu_{V_M} dn_{V_M} + \mu_{V_X} dn_{V_X}\right] \tag{13.36}$$

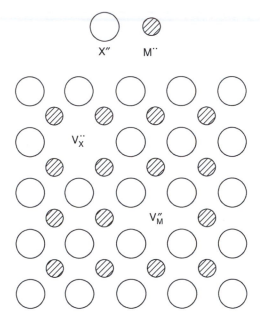

FIGURE 13.6 A Schottky defect consists of a vacant cation site and a vacant anion site.

The isolation constraints in this case are:

$$dU' = 0; \qquad dV' = 0$$
$$dm_M = dn_{M_M} = 0 \qquad\qquad (13.37)$$
$$dm_X = dn_{X_X} = 0$$

The latter two equations report the conservation of atoms of M and X, respectively. The condition that the ratio of anion to cation sites is constrained to be 1/1 requires that the equal numbers of each type of vacancies be formed:

$$dn_{V_X} = dn_{V_M} \qquad\qquad (13.38)$$

Put the constraints into the expression for the entropy:

$$dS'_{sys,iso} = -\frac{1}{T}\left[\mu_{V_X} dn_{V_X} + \mu_{V_M} dn_{V_M}\right] = -\frac{1}{T}\left[\mu_{V_X} + \mu_{V_M}\right]dn_{V_M} \qquad (13.39)$$

The coefficient of dn_{V_M} in this expression may be thought of as the affinity for the defect reaction:

$$\text{null} = V_X = V_M \qquad\qquad [13.2]$$

which describes the formation of two vacancies in a region that is initially a perfect crystal; the notation, "null", in this context means the initially defect-free crystal. The condition for equilibrium is obtained by setting the coefficient in Equation 13.39

equal to zero:

$$\mu_{V_X} + \mu_{V_M} = 0 \tag{13.40}$$

Following an argument that parallels that connecting Equation 13.25 with Equation 13.35 yields an expression for the equilibrium concentration of Schottky defects at any temperature T:

$$X_{sd} = e^{\Delta S_{sd}^{\circ xs}/2k} e^{-(\Delta H_{sd}^0/2kT)} \tag{13.41}$$

where ΔS_{sd}° is the excess entropy associated with the formation of the pair of vacancies from a perfect crystal and ΔH_{sd}^0 is the corresponding enthalpy change.

In a crystal with formula $M_u X_v$, in which the ratio of M to X sites is u/v, conservation of the ratio of anion to cation sites dictated by the geometry of the crystalline arrangement requires that a Schottky defect be made up of u cation vacancies and v anion vacancies. This condition also preserves charge neutrality with the formation of a Schottky defect. The condition for equilibrium reflects the constraint on the ratio of lattice sites and is described by the defect equation:

$$null = uV_M + vV_M \tag{13.3}$$

13.2.3 COMBINED DEFECTS IN BINARY COMPOUNDS

A cation site is normally occupied by an M ion carrying a positive charge ez_M, where z_M is the valence of the cation and e is the magnitude of the charge carried by an electron. If the site is vacant, i.e., if there is no M ion occupying it, then there exists an uncompensated negative charge equal to $(-ez_M)$ associated with surrounding anions. Similarly, a vacant anion site carries a charge of $(-ez_X)$ where z_X is the normal valence of the anion. Since z_X is negative, this excess charge is positive. These oppositely charged entities in a crystal may be expected to attract each other to form a cation–anion vacancy pair (Figure 13.7). If z_X and z_M are equal, this vacancy complex has zero excess charge; for the more general case for which z_M and z_X are not equal, the complex carries a net charge. This relationship may be represented by the defect reaction,

$$V_M + V_X = (V_M V_X) \tag{13.4}$$

where the notation $(V_M V_X)$ represents the complex defect consisting of the oppositely charged pair of vacancies. The condition for equilibrium with respect to this interaction is determined by setting the corresponding affinity equal to zero:

$$\mu_{(V_M V_X)} - (\mu_{V_M} + \mu_{V_X}) = 0 \tag{13.42}$$

so that

$$\frac{X_{(V_M V_X)}}{X_{V_M} X_{V_X}} = e^{\Delta S_{mX}^{\circ}/k} e^{(\Delta H_{mX}^0/kT)} \tag{13.43}$$

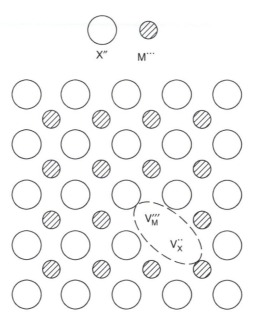

FIGURE 13.7 Vacancies on the two sublattices tend to attract each other to form a complex defect, $V_M V_X$. The net charge on this complex is $e(z_X - z_M)$.

where ΔS_{mX}° is the difference in excess entropy and ΔH_{mX}^0 is the enthalpy difference for the process expressed by Equation 13.4. The entropy difference is arguably small. The enthalpy difference is negative since the oppositely charged vacancies are spontaneously attracted. Thus the concentration of complexes is larger than the product of the two single vacancy concentrations by an amount that depends upon the energy of the attraction. The enthalpy difference may be estimated as the difference in energy between two charged particles at the equilibrium separation distance in the vacancy pair and that of a pair of infinitely separated vacancies. This estimate gives reasonable accuracy for ionic crystals in which the excess charges are strongly associated with the sites. More sophisticated models are required to describe the behavior of compound crystal structures in which other types of bonding play a significant role.

13.2.4 Multivariate Equilibrium among Defects in a Stoichiometric Compound Crystal

This section introduces the strategy for determining defect concentrations in a crystal of the compound $M_u X_v$ that is isolated from its surroundings. Thus the collections of defects that may form may be considered to be intrinsic, since no interaction with the surroundings is involved in their formation. Section 13.3.1 and Section 13.4 treat more general problems in which interactions with surroundings are explicitly included. The entities that may exist in this crystal when it comes to equilibrium include M_M, M_i, V_M, X_X, X_i and V_X. Defect complexes may also exist but are neglected in this development. In applying the general strategy for finding conditions for equilibrium, the expression for the change in entropy will contain a term of the form $\mu_k dn_k$ for each of the six entities listed above.

$$dS'_{sys} = \frac{1}{T}dU' + \frac{P}{T}dV' - \frac{1}{T}\Big[\mu_{M_M}dn_{M_M} + \mu_{M_i}dn_{M_i} + \mu_{V_M}dn_{V_M}$$
$$+ \mu_{X_X}dn_{X_X} + \mu_{X_i}dn_{X_i} + \mu_{V_X}dn_{V_X}\Big] \quad (13.44)$$

In addition to the usual isolation constraints on changes in internal energy and volume, three constraining equations operate on the changes in the numbers of each entity in this isolated crystal:

a. Conservation of M atoms.

$$dm_M = dn_{M_M} + dn_{M_i} = 0 \Rightarrow dn_{M_M} = -dn_{M_i} \quad (13.45)$$

b. Conservation of X atoms:

$$dm_X = dn_{X_X} + dn_{X_i} = 0 \Rightarrow dn_{X_X} = -dn_{X_i} \quad (13.46)$$

c. Conservation of the ratio of sites in the two sublattices:

$$v\,dn_{S_M} = u\,dn_{S_X} \quad (13.47)$$

where n_S denotes the number of each kind of site in the lattice. Since sites are either occupied or vacant,

$$v\lfloor dn_{M_M} + dn_{V_M}\rfloor = u\lfloor dn_{X_X} + dn_{V_X}\rfloor \quad (13.48)$$

Thus three of the six dn_k's are dependent upon the other three. Choose dn_{Mi}, dn_{Xi} and dn_{VX} as independent variables. The third equation permits expression of dn_{VM} in terms of these variables:

$$dn_{V_M} = \frac{u}{v}\Big[dn_{X_X} + dn_{V_X}\Big] - dn_{M_M} \quad (13.49)$$

Substitute these constraints into Equation 13.44 and collect terms:

$$dS'_{sys,iso} = -\frac{1}{T}\Big[(\mu_{M_i} + \mu_{V_M} - \mu_{M_M})dn_{M_i} + (\mu_{X_i} + \mu_{V_X} - \mu_{X_X})dn_{X_i}$$
$$+ \Big(\frac{u}{v}\mu_{V_M} + \mu_{V_X}\Big)dn_{V_X}\Big] \quad (13.50)$$

To find the conditions for equilibrium, set the coefficients equal to zero. These three coefficients are the affinities for the three defect chemistry equations:

$$M_M = M_i + V_M \quad [13.5]$$

$$X_X = X_i + V_X \qquad [13.6]$$

$$\text{null} = uV_M + vV_X \qquad [13.7]$$

The first two equations correspond to the formation of Frenkel defects on the cation and anion sublattices, respectively; the last equation is the Schottky reaction for an M_uX_v crystal.

It is customary to describe the concentrations of each of the entities as a fraction of the sites in the corresponding sublattice. These measures of composition are designated with brackets. Thus $[V_M]$ is the ratio of the number of vacant cation sites to the total number of cation sites: n_{VM}/n_{SM}. Similarly, $[X_i]$ is the ratio of the number of interstitial X ions to the number of normal anion sites in the crystal: n_{Xi}/n_{SX}. Focus upon 1 mole of the compound M_uX_v; more explicitly, focus upon a quantity of the crystal that contains $(u + v)N_0$ lattice sites, where N_0 is Avagadro's number. The number of M sites is $u(N_0)$; and the number of X sites is $v(N_0)$. Thus the total number of cation vacancies in 1 mole of compound may be written $n_{VM} = uN_0[V_M]$, the number of cation interstitials is $n_{Mi} = uN_0[M_i]$, the number of anion vacancies is $n_{VX} = vN_0[V_X]$ and the number of anion interstitials may be written $n_{Xi} = vN_0[X_i]$.

The condition for equilibrium corresponding to the formation of a Frenkel defect on the cation lattice represented in Equation 13.5 may be written:

$$\frac{[M_i][V_M]}{[M_M]} = K_{fd,c} = K^0_{fd,c}e^{-(\Delta H^0_{fd,c}/kT)}$$

where $K^0_{fd,c}$ is an entropy term and $\Delta H^0_{fd,c}$ is the enthalpy of formation of a Frenkel defect on the cation lattice. Since the defect concentrations are generally very small, $[M_M]$ may be taken to be 1. A similar approximation is also valid for $[X_X]$. The conditions for equilibrium corresponding to the defect Equation 13.5 to Equation 13.7 are:

$$[M_i][V_M] = K^0_{fd,c}e^{-(\Delta H^0_{fd,c}/kT)} \qquad (13.51)$$

$$[X_i][V_X] = K^0_{fd,a}e^{-(\Delta H^0_{fd,a}/kT)} \qquad (13.52)$$

$$[V_M]^u[V_X]^v = K^0_{sd}e^{-(\Delta H^0_{sd}/kT)} \qquad (13.53)$$

The K^0_r coefficients contain the entropy factors in these equations. These results represent three equations among the four variables, $[V_M]$, $[M_i]$, $[V_X]$ and $[X_i]$. A fourth equation is supplied by the condition for charge neutrality in the crystal.

Each of the four defect entities has an excess charge associated with it. If z is the normal charge on an occupied cation site, then $-(u/v)z$ is the charge on a normal anion site. For example, for the crystal Al_2O_3, $z = +3$ and the charge on the anion site is $-(2/3)(+3) = -2$. The excess charge associated with each entity is thus:

a. Cation vacancy, $V_M: -z$.
b. Cation interstitial, $M_i: +z$.

    c. Anion vacancy, V_X:$+(u/v)z$.
    d. Anion interstitial, X_i:$-(u/v)z$.

Charge neutrality requires that the total excess charge sum to zero. Mathematically

$$-zn_{V_M} + zn_{M_i} + \frac{u}{v}zn_{V_X} - \frac{u}{v}zn_{X_i} = 0$$

Substitute the expressions for the defect concentrations and simplify:

$$-[V_M] + [M_i] + \frac{u}{v}[V_X] - \frac{u}{v}[X_i] = 0 \qquad (13.54)$$

If the entropy and energy factors are known, then Equation 13.51 to Equation 13.54 provide four equations among four variables and the concentrations of all four defects may be computed for the crystal at any given temperature. Unfortunately, well-documented and consistent information of this type is not widely available even for thoroughly studied ceramic systems.[2] A critical obstacle to the development of such information is the crucial role associated with impurity atoms in compound crystals (see Section 13.4). However, these equations may provide reasonable relative values for the four types of defects. For example, in sapphire (undoped alumina) in the range from 1200 to 1500°C the dominant defect is V_{Al}''' with the concentration of anion defects about ten orders of magnitude lower and aluminum interstitials more than 20 orders of magnitude smaller.[2]

13.3   NONSTOICHIOMETRIC COMPOUND CRYSTALS

The notion that chemical substances are stoichiometric, i.e., that the numbers of atoms of the elements that compose them occur in ratios of small whole numbers, may be traced to the discovery that many substances exist as molecules. H_2O, CO_2 and CH_4 are familiar examples of stoichiometric compounds. When elements M and X combine to form a crystalline compound M_uX_v, the resulting structure forms in two sublattices, one normally occupied by M atoms and the other by X atoms. The geometry of the crystal structure dictates that the number of the two kinds of sites occurs in the ratio (v/u); the ratio of sites is stoichiometric. This is a hard, fast requirement that comes from the geometry of the crystal structure. In a perfect crystal, with all M sites occupied by M atoms and X sites by X atoms, the composition of the system will also be stoichiometric, i.e., the ratio of X to M atoms is (v/u). If the crystal contains only intrinsic defects, such as those described in Section 13.2, the composition will still be stoichiometric.

The defect structure of real crystals is not confined to intrinsic defects. Interactions with the surroundings may produce defects that are not simply equivalent to an internal rearrangement of atoms on sites in the crystal. In an oxidizing atmosphere the compound M_uO_v may dissolve more oxygen atoms than the number given by the stoichiometric ratio (v/u). The composition of the compound departs from its stoichiometric ratio and becomes oxygen rich. However, the geometric requirements of the crystal lattice for the compound require that the ratio of sites remains at (u/v). Thus the excess oxygen atoms must be accommodated either by:

a. Placing oxygen ions in interstitial positions.
b. Placing the oxygen ions on normal anion sublattice sites and simultaneously creating vacancies in the cation sublattice.

In the first case the compound is said to be oxygen excess; this case may be represented by the formula, $M_u O_{v+\delta}$, where δ reports the quantity of excess oxygen in the system. The second case is said to be metal deficient and may be represented: $M_{u-\delta} O_v$. Both of these conditions describe a system in which the ratio of oxygen to metal is greater than (v/u).

For example, an oxide with the stoichiometric formula MO_2 may analyze at the composition $MO_{2.05}$. This departure from stoichiometry may correspond to either an oxygen excess compound, $MO_{2+0.05}$, or a metal deficient compound, $M_{1-0.024}O_2$. Note that the value of δ corresponding to the O/M ratio 2.05/1 is not the same in these two cases. In either case the corresponding number of defects may be computed from the measured composition. In the oxygen excess presumption δ is the ratio of interstitial oxygen atoms to anion sites in the system. For the metal deficient model, δ is the fraction of cation sites that are vacant.

If in contrast the ratio of metal to oxygen atoms in the system is less than (v/u), the compound is metal rich. This excess of metal relative to oxygen may be accommodated by:

c. Placing metal ions in interstitial sites (metal excess).
d. Placing metal ions on normal cation sites and creating an equal number of vacancies on the anion sublattice (oxygen deficient).

The metal excess case may be represented by $M_{u+\delta} O_v$ and alternately the oxygen deficient case by $M_u O_{v-\delta}$. In the first case δ is the ratio of interstitial metal atoms to cation sites while for the oxygen deficient description, δ is the fraction of anion sites that are vacant.

In a real $M_u O_v$ crystal all four of the defects listed above may be expected to exist simultaneously in the system. Indeed, in principle, all of the possible defects and defect combinations that can be visualized for a system may be expected to exist at thermodynamic equilibrium. However, the relative concentrations of defect types may vary by many orders of magnitude so that some defects will be present in concentrations that are negligibly small while others dominate. The identification of which defects are important in a given crystal under a given set of conditions depends upon their thermodynamic properties.

13.3.1 EQUILIBRIUM IN COMPOUND CRYSTALS WITH A VARIETY OF DEFECTS

The problem of identifying those defects that are present in sufficient concentrations to play a role in the behavior of the system is closely analogous to the description of homogeneous multivariant reacting systems developed in Section 11.1.2 of Chapter 11. All of the defects that can be visualized may be expected to exist in any real compound crystal in equilibrium with its surrounding atmosphere. The internal energy and entropy of a crystal are a function of the numbers of each of these entities; each has a chemical potential derived from coefficient relations in the expression for dU' or dG'. The change in entropy of such a crystal for an arbitrary change in its state may be written:

$$dS'^\alpha = \frac{1}{T^\alpha}dU'^\alpha + \frac{P^\alpha}{T^\alpha}dV'^\alpha - \frac{1}{T^\alpha}\sum_{k=1}^{c^\alpha}\mu_k^\alpha dn_k^\alpha \qquad (13.55)$$

Here the summation covers the complete list of separately countable entities presumed to exist in the crystal, including:

a. Ions or atoms on their normal sublattice sites.
b. Vacant sites on each sublattice.
c. Ions or atoms on interstitial sites.
d. Impurity ions or atoms on each type of sublattice site.
e. Unassociated electrons and holes in the system.
f. Combinations of these entities.

If the total number of distinguishable types of these enumerated atoms, ions and defects is c^α, then there are $[c^\alpha + 2]$ terms in this expression for the entropy.

Since departure from stoichiometry requires exchange of components with the surroundings, in deriving the conditions that define the final state for a such a crystal in equilibrium with its surrounding atmosphere it is also necessary to compute the change in entropy of the gas phase with its components.

$$dS'^g = \frac{1}{T^g}dU'^g + \frac{P^g}{T^g}dV'^g - \frac{1}{T^g}\sum_{k=1}^{c^g}\mu_k^g dn_k^g \qquad (13.56)$$

where c^g is the number of components in the gas phase. There are $[c^g + 2]$ terms in this equation. The change in entropy of the system is the sum of the two expressions given in Equation 13.55 and Equation 13.56; it contains $[c^\alpha + c^g + 4]$ terms.

The general strategy for finding the conditions for equilibrium focuses on an isolated system. Some of the isolation constraints are familiar in this two-phase system:

$$dU'_{sys} = dU'^\alpha + dU'^g = 0 \Rightarrow dU'^g = -dU'^\alpha \qquad (13.57)$$

$$dV'_{sys} = dV'^\alpha + dV'^g = 0 \Rightarrow dV'^g = -dV'^\alpha \qquad (13.58)$$

The number of gram atoms of each of the elements in the system is conserved:

$$dm_j = 0 = \sum_{k=1}^{c^\alpha} v_{kj}dn_k^\alpha + \sum_{k=1}^{c^g} v_{kj}dn_k^g \qquad (13.59)$$

Here, as in Section 11.1.2 of Chapter 11, the coefficients v_{kj} represent the number of atoms of the element j contained in each enumerated entity, k, including normally occupied sites, defects and components in both phases. Many of these coefficients may be expected to be zero; most entities contain either M but not X or vice versa. The numbers of vacant sites also have zero coefficients in these equations since they do not contain atoms.

Two new constraints arise from the nature of the system. The ratio of the number n_{S_X} of X sites to that M sites, n_{S_M}, is a characteristic of the crystal lattice structure and must be rigidly maintained. If the compound has the nominal formula $M_u X_v$, this ratio is (v/u). Thus,

$$vn_{S_M} = un_{S_X} \Rightarrow vdn_{S_M} = udn_{S_X} \tag{13.60}$$

Here n_{S_M} includes all of the entities in the list that occupy a site on the M sublattice and n_{S_X} incorporates those on the X sublattice. This includes atoms or ions and vacancies on such sites, but does not include interstitials, electrons or holes, which do not reside on sublattice sites.

Yet another constraint arises because most of the entities enumerated carry an electric charge. Since the crystal is electrically neutral at equilibrium, whatever changes in the numbers of these entities occur, the change in total electrical charge in the crystal is zero. This charge balance relationship is conveniently expressed in terms of the excess charges on each of the entities present, assuming an M site normally carries a charge of $-e(z_M)$ charge and an X site carries a charge of $e(z_X)$:

$$dq = 0 = \sum_{k=1}^{c^\alpha} z_k dn_k^\alpha \tag{13.61}$$

where z_k is the excess number of charge units carried by entity k in the crystal. The value of z_k may be positive, negative or zero, in agreement with the defined nature of the entity, k.

Examination of the constraining equations shows that there are $[2 + e + 1 + 1] = [4 + e]$ equations relating the $[c^\alpha + c^g + 4]$ variables in the expression for the entropy. Thus $[4 + e]$ variables may be eliminated, leaving $[(c^\alpha + c^g + 4) - (4 + e)]$ or $[c^\alpha + c^g - e]$ independent variables. The conditions for equilibrium are obtained by setting the coefficients of each of these independent variables in the expression for the change in entropy of the system equal to zero.

Two of the resulting equations are the familiar conditions for thermal and mechanical equilibrium. The remaining $[c^\alpha + c^g - e - 2]$ equations are linear combinations of the chemical potentials of the entities involved. Each of these independent equations has the form of an affinity, as was found to be characteristic of multivariate reacting systems in Section 11.1.2 of Chapter 11. Each affinity expression may be represented by a chemical equation with reactants and products and appropriate coefficients to balance the equation (see, e.g., Equation 13.1). However, in addition to balancing the number of atoms of each element on both sides of the equation, these defect chemistry reactions must also balance the electric charge in order to satisfy Equation 13.61 and the ratio of X to M sites required to satisfy Equation 13.60.

The construction of balanced defect chemistry equations is a nontrivial exercise because all three of the above conditions must be satisfied. In simple cases it is possible to balance such an equation by inspection or by trial and error. However, a proper set of coefficients may be deduced for any assumed reaction by applying the associated conservation equations. To illustrate this procedure, again focus on a crystal with the nominal formula M_uX_v. The normal valence of M is z and that of X is $-(u/v)z$. Consider the reaction with its surroundings that adds M to a normal site on the cation lattice. This incorporation reaction must be accompanied by the formation of vacant sites on the anion lattice, each of which carries a charge of $+(u/v)z$, and may also require formation of unassociated electrons to balance the charge. The corresponding incorporation reaction might be written

$$aM(g) \xrightarrow{M_uX_v} bM_M + cV_X{}^{(u/v)z} + de^-$$
[13.8]

To find the coefficients, apply the conservation equations:

   a. Conservation of M: $a = b$.
   b. X/M site ratio is v/u: $c/b = v/u$.
   c. Balance of excess charge: $b(0) + c[(u/v)z] - d = 0$.

These three conditions yield values for the coefficients in Equation 13.8:

$$b = a; \quad c = (v/u)a; \quad d = az$$

Write the reaction for one atom of M; thus $a = b = 1$. The defect Equation 13.8 may be written:

$$M(g) \xrightarrow{M_uX_v} M_M + \frac{v}{u}V_X + ze^-$$
[13.9]

EXAMPLE 13.1
Consider the incorporation of Si into SiO_2. For this case: $u = 1$, $v = 2$, and $z = 4$. Take a to be 1. Then Equation 13.9 may be adapted to read:

$$Si(g) \xrightarrow{SiO_2} Si_{Si} + 2V_O^{\cdot\cdot} + 4e'$$
[13.10]

It may be verified by inspection that all three conditions are satisfied.

EXAMPLE 13.2
Write a balanced defect equation for the incorporation of aluminum into alumina. The constants are: $u = 2$; $v = 3$; and $z = 3$. Apply Equation 13.9, again taking $a = 1$

$$Al(g) \xrightarrow{Al_2O_3} Al_{Al} + \frac{3}{2}V_O^{\cdot\cdot} + 3e'$$

$$Al(g) \xrightarrow{Al_2O_3} 2Al_{Al} + 3V_O^{\cdot\cdot} + 6e'$$
[13.11]

Alternate reactions may be written for the incorporation of M into the crystal. For example, M may enter as an interstitial. A generic equation analogous to Equation 13.9 could also be derived for this case by incorporating the balance equations as above.

In further analogy with the results for ordinary chemical equilibria obtained in Chapter 11, it may also be concluded that every other reaction (balanced equation) that may be written among the enumerated entities will be a linear combination of these independent equations. Thus at equilibrium, the affinity will be equal to zero for every defect reaction that can be written among the participants.

Finally, by following steps similar to those connecting Equation 13.25 with Equation 13.30 for the simple case of the Frenkel defect, each of these affinity expressions may be converted to a form reminiscent of the law of mass action

$$\Delta \bar{G}_r^0 = \Delta H_r^0 - T\Delta S_r^0 = -kT \ln K_r \tag{13.62}$$

where ΔH_r^0 is the change in enthalpy associated with the defect reaction and ΔS_r^0 is the corresponding entropy change. The equilibrium constant is the ratio of atom fractions of the entities on the product side of the reaction to those on the reactant side. This completes the connection between the thermodynamics of the structure and the compositional distribution of defects at equilibrium. The variation of defect concentrations with departures from stoichiometry, which is usually controlled by the composition of the gas atmosphere in equilibrium with the crystal, may be conveniently displayed in equilibrium defect diagrams such as that shown in Figure 1.8.

13.3.2 ILLUSTRATION OF THE CONDITIONS FOR EQUILIBRIUM FOR ALUMINA

Alumina has the nominal formula Al_2O_3, with $z_{Al} = +3$ and $z_O = -2$. In this example assume that an alumina crystal is in equilibrium with its vapor; neglect the presence of impurities. The components in the gas phase are $Al(g)$ and $O_2(g)$. The list of entities that may exist in the alumina crystal includes Al_{Al}^X, O_O^X, $Al_i^{\cdots}$, O_i'', V_{Al}''', and $V_O^{\cdots}$. Neglect combinations of defects. The expression for the change in entropy that this two-phase system may exhibit contains terms for each of this list of components. The internal energy and volume terms give rise to the usual conditions for thermal and mechanical equilibrium. This treatment focuses upon the defect chemistry and other chemical changes for the system. Thus Equation 13.55 and Equation 13.56 may be combined to read

$$\begin{aligned} dS'_{sys} = &[[\ ]] - \frac{1}{T^g}\left[\mu_{Al}^g dn_{Al}^g + \mu_{O_2}^g dn_{O_2}^g\right] \\ &- \frac{1}{T^\alpha}\left[\mu_{Al_{Al}} dn_{Al_{Al}} + \mu_{Al_i} dn_{Al_i} + \mu_{V_{Al}} dn_{V_{Al}}\right] \\ &- \frac{1}{T^\alpha}\left[\mu_{O_O} dn_{O_O} + \mu_{O_i} dn_{O_i} + \mu_{V_O} dn_{V_O} + \mu_e dn_e\right] \end{aligned} \tag{13.63}$$

Conservation of aluminum and oxygen atoms in this isolated system requires

$$dm_{Al} = 0 = dn_{Al}^g + dn_{Al_{Al}} + dn_{Al_i} \Rightarrow dn_{Al}^g = -[dn_{Al_{Al}} + dn_{Al_i}] \tag{13.64}$$

$$dm_O = 0 = 2dn_{O_2}^g + dn_{O_O} + dn_{O_i} \Rightarrow dn_{O_2}^g = -\frac{1}{2}\left[dn_{O_O} + dn_{O_i}\right] \qquad (13.65)$$

Maintenance of the ratio of cation to anion sites at 2/3 requires

$$3\lfloor dn_{Al_{Al}} + dn_{V_{Al}}\rfloor = 2\lfloor dn_{O_O} + dn_{V_O}\rfloor \qquad (13.66)$$

The requirement for charge balance may be written:

$$\delta q^{xs} = +3dn_{Al_i} - 3dn_{V_{Al}} - 2dn_{O_i} + 2dn_{V_O} - dn_e \qquad (13.67)$$

Substitute these constraints into Equation 13.63 and collect terms:

$$\begin{aligned}
dS'_{sys,iso} =[[\ ]] &- \frac{1}{T}\left[\mu_{Al_{Al}} + 3\mu_e - \mu_{Al}^g - \mu_{V_{Al}}\right]dn_{Al_{Al}} \\
&- \frac{1}{T}\left[\mu_{Al_i} + 3\mu_e - \mu_{Al}^g\right]dn_{Al_i} \\
&- \frac{1}{T}\left[\mu_{O_O} + \frac{2}{3}\mu_{V_{Al}} - \frac{1}{2}\mu_{O_2}^g - 2\mu_e\right]dn_{O_O} \qquad (13.68) \\
&- \frac{1}{T}\left[\mu_{O_i} - \frac{1}{2}\mu_{O_2}^g - 2\mu_e\right]dn_{O_i} \\
&- \frac{1}{T}\left[\frac{2}{3}\mu_{V_{Al}} + \mu_{V_O}\right]dn_{V_O}
\end{aligned}$$

The coefficients in these equations are the affinities for the following balanced defect reactions:

$$Al(g) + V_{Al}''' = Al_{Al}^x + 3e' \qquad [13.12]$$

$$Al(g) = Al_i^{\cdots} + 3e' \qquad [13.13]$$

$$3O_2(g) + 12e' = 6O_0^x + 4V_{Al}''' \qquad [13.14]$$

$$O_2(g) + 4e' = 2O_i'' \qquad [13.15]$$

$$null = 2V_{Al}''' + 3V_O^{\cdots} \qquad [13.16]$$

Other reactions that may be written for this system may all be expressed as linear combinations of these five independent reactions. For example, the Frenkel reaction,

$$Al_{Al}^x = Al_i''' + V_{Al}'''$$

may be obtained by subtracting reaction 13.12 from 13.13.

13.4 IMPURITIES IN NONSTOICHIOMETRIC COMPOUNDS

Up to this point the description of the defect structure in a compound M_uX_v has been limited to systems in which the only elements existing are M and X. No real crystalline compound is pure; additional elements are always incorporated to some extent. Indeed, in some systems it is possible to replace one element, say M, completely with another, L; the crystalline compound phase is capable of forming a complete series of solid solutions from M_uX_v to L_uX_v; the phase diagram is isomorphous. The system NiO–CoO is an example of such an isomorphous system (see Figure 9.15 in Chapter 9). In order for significant solubility to exist in such cases the elements involved must have the same valence, nearly the same atom size and very similar electronegativity values. Where these three conditions are not met, which corresponds to the vast majority of combinations of the elements, the solubility of L in M_uX_v will be limited to impurity levels.

As might be expected, the conditions for equilibrium in these more complex cases are derived from the general strategy and may be represented by defect energetics and equilibrium constants that correspond to defect chemistry equations. The construction of balanced reaction equations must conserve the atoms of all of the elements involved, preserve the ratio of anion to cation sites and maintain electrical neutrality. Some examples follow.

Magnesium oxide (MgO) is widely used as a sintering aid in the consolidation of alumina powders. MgO may be incorporated into the alumina lattice in a variety of reactions.[2]

$$2MgO \xrightarrow{Al_2O_3} 2Mg'_{Al} + 2O^x_O + V^{\cdot\cdot}_O \qquad [13.17]$$

Addition of Mg (which is divalent) at a trivalent cation sublattice site leaves an uncompensated charge of (-1). In addition to adding oxygen ions at anion sites, it is necessary to create an anion vacancy in order to preserve the ratio of anion to cation sites of 3/2. The charges are thus balanced. Alternatively,

$$3MgO + Al^x_{Al} \xrightarrow{Al_2O_3} 3Mg'_{Al} + 3O^x_O + Al^{\cdot\cdot\cdot}_i \qquad [13.18]$$

has the net effect of increasing the number of cation sites by two while creating three anion sites and an aluminum atom in an interstitial site. A third possibility,

$$3MgO \xrightarrow{Al_2O_3} 2Mg'_{Al} + 3O^x_O + Mg^{\cdot\cdot}_i \qquad [13.19]$$

produces a magnesium interstitial while creating no vacancies on either sublattice. Estimates of the energetics of these reactions remain speculative. Calculations based upon such estimates show that the dominant defect in MgO doped sapphire is $V^{\cdot\cdot}_O$, i.e., vacancies on the oxygen sublattice. Equilibrium concentrations of V'''_{Al}, estimated to be the dominant defect in undoped sapphire, are about ten orders of magnitude lower in sapphire doped with 500 ppm of MgO.[2] Clearly, impurities may have a dominant effect upon the distribution of defects in compound crystals.

These results demonstrate the complexity of thermodynamic relationships involved in determining the distribution of defects in compound crystals. Most undoped ceramic materials contain very significant concentrations of impurities. Impurity additions may completely determine the equilibrium distribution of defects. The thermodynamic treatment of defect chemistry presented in this chapter provides a foundation for understanding physical phenomena, such as ionic conductivity, diffusion, oxide layer growth, dielectric and optical behavior in crystalline compounds. The application of these principles to the interpretation of experimental observations in real ceramic structures constitutes a challenging area of study for the foreseeable future.

13.5 SUMMARY

When it attains its equilibrium state a crystal contains defects in its lattice structure. These defects usually take the form of vacant lattice sites or atoms in interstitial sites. Associations of defects are also common in crystals.

Application of the general strategy for finding conditions for equilibrium in crystals with defects yields results that are formally identical to those obtained for multicomponent reacting systems. Interactions of defects with normal lattice sites and with the surroundings may be described in terms of defect chemistry equations, with accompanying enthalpies and excess entropies of reaction and equilibrium constants.

In compound crystals balanced defect chemistry reactions must conserve the elements involved, the ratio of lattice sites and charge neutrality.

Intrinsic defects form in stoichiometric crystals as Frenkel pairs or as Schottky clusters. Departures from stoichiometry always involve the formation of extrinsic defects.

Impurity atoms may be incorporated into crystals in a variety of ways, each requiring the formation of a specific set of defects. Since the number of defects may be simply related to the number of impurity atoms, small concentrations of impurity atoms may produce defect concentrations far in excess of those that form intrinsically. It is thus essential to treat explicitly impurity effects in considering aspects of the behavior of compound crystals that are governed by their defect structure.

HOMEWORK PROBLEMS

Problem 13.1. Careful dilatometric and high temperature diffraction measurements of pure face centered cubic A give the following information:

	a_0 (cm)	V (cm^3/mole)
At $T_1 = 1040$ K	3.7021×10^{-8}	7.6404
At $T_2 = 1340$ K	3.7113×10^{-8}	7.6992

Derive a relationship for the temperature dependence of the vacancy concentration in pure A.

Problem 13.2. The enthalpy of formation of an interstitial in an FCC material is found to be 188 (KJ/mole); the excess entropy is 7.6 (J/mole K):
   a. Calculate and plot the equilibrium concentration of interstitials as a function of temperature in the range from 600 to 1300 K.
   b. Compare this result to that found for vacancies in this material in Problem 13.1.

Problem 13.3. Suppose that the interaction parameters for divacancies are about 10% of the values of the corresponding single defect parameters; i.e., suppose

$$\Delta \bar{H}_{int} = -0.1 \Delta \bar{H}_v$$

$$\Delta \bar{S}_{int} = -0.1 \Delta \bar{S}_v$$

Using the results of Problem 13.1, compute and plot the mole fraction of divacancies as a function of temperature for pure A in the range from 600 to 1300 K.

Problem 13.4. Enthalpies of formation of Frenkel defects range from 350 to 550 (KJ/mole) and entropies from 1 to 10 (J/mole K). Use these ranges to compute and plot limits of the defect concentrations expected as a function of T. Could very careful measurements of volume be used to estimate values of formation energies of Frenkel defects?

Problem 13.5. Frenkel and Schottky defects are intrinsic to the crystal in which they occur. While both of these classes of defects may form in compound crystals, only one may form in a pure elemental crystal. Identify which of these defects may form in an elemental crystal and explain why this is not possible for the other defect.

Problem 13.6. Consider the stoichiometric crystal, silicon nitride: Si_3N_4. The following defects may form in this system: V_{Si}; Si_i; V_N; N_i.
   a. Write out the conditions for equilibrium that permit the computation of these defect concentrations at any temperature.
   b. What information would be required to permit actual equilibrium defect concentrations as a function of temperature?

Problem 13.7. The system M–O forms the oxides M_2O^γ, $M_2O_3^\varepsilon$ and MO_2^η at 1600 K. The composition of the ε phase (M_2O_3) in equilibrium with γ (M_2O) is $X_O = 0.5994$. At the other phase boundary of ε, the composition in equilibrium with η (MO_2) is $X_O = 0.6003$. Describe the composition of the ε phase when equilibrated with γ
   a. As a metal excess oxide.
   b. As an oxygen deficient oxide.
Describe the composition of the ε phase when equilibrated with η
   c. As a metal deficient oxide.
   d. As an oxygen excess oxide.

Explain briefly what each of these descriptions implies in terms of the defect structure.

Problem 13.8. Write a balanced equation that represents the incorporation of silicon into a silicon sublattice site in silicon nitride, Si_3N_4.

Problem 13.9. Develop a series of balanced defect equations necessary to relate defect concentrations in boron nitride (BN) to the composition of the atmosphere in which it is equilibrated.

Problem 13.10. Construct a set of equations that describe the incorporation of yttria (Y_2O_3) into zirconia (ZrO_2). Describe in words what defects are created in each case.

REFERENCES

1. Simmons, R. and Balluffi, R., *Phys. Rev.*, 117, 52, 1960.
2. Lagerlof, K.P.D., Mitchell, T.E., and Heuer, A.H., Lattice diffusion kinetics in undoped and impurity-doped sapphire (α-Al_2O_3): a dislocation loop annealing study, *J. Am. Ceram. Soc.*, 72, 2159, 1989.
3. Mukherjee, K., *Trans. AIME*, 6, 1324, 1966.
4. Kröger, F.A., *The Chemistry of Imperfect Crystals*, North-Holland Publishing Co., Amsterdam, 1964.

14 Equilibrium in Continuous Systems: Thermodynamic Effects of External Fields

CONTENTS

In treating equilibria between phases in earlier chapters it has been tacitly assumed that the intensive properties within a phase are uniform. In assigning a value T^{α} to the α phase, it is assumed that the α phase can be described by a single value of the temperature which is characteristic of that part of the system. This same assumption has been made for pressure, P^{α}, composition, X_k^{α}, chemical potential, μ_k^{α}, and all of the partial molal properties. In an equilibrium between two phases some of these properties have different values in the two phases; for example, if the phases are separated by a curved interface, the pressures in the two phases are different. However, it has been assumed that the pressure within each phase is uniform.

If external fields act on a system the intensive properties are found to vary with position within the system when it comes to equilibrium. For example, in the gravitational field of the Earth, at equilibrium the pressure varies with height. In a centrifugal field, such as may be maintained in an ultracentrifuge rotating at constant velocity, the composition varies along the radius. This nonuniform distribution may be used to separate components, or isotopes, as was done in the Manhattan Project during World War II, within a single phase. In an electrostatic field it will be shown that the chemical potential of charged components varies with position when the system attains equilibrium.

In order to describe the variation of thermodynamic properties in such homogeneous but nonuniform systems it is necessary to define and derive densities of thermodynamic extensive properties, such as the entropy density or Gibbs free energy density. Densities are local intensive properties that may be associated with each point in the system. This notion was introduced in Section 12.2 in Chapter 12 in defining surface excess properties. Formulation of the combined statement of the first and second laws of thermodynamics for a volume element in terms of the local densities lays the foundation for applying the general strategy, Figure 1.4, for finding conditions for equilibrium in nonuniform systems. This strategy will first be used for a system without external fields to demonstrate the approach and to show that the assumption that the intensive properties are uniform in such a system is indeed valid. Subsequently, conditions for equilibrium in the presence of external fields will be developed. The presentation in Sections 14.1 to 14.3 rests heavily upon the foundation laid down by Haase[1] in his brief but comprehensive treatment.

It is also important to explore the thermodynamic description of nonuniform systems as a basis for describing the behavior of systems that are not at equilibrium. Much of materials science examines the processes that occur in materials as they experience spontaneous changes on their path toward equilibrium. The evolution of the system is generally viewed as being driven by precisely the nonuniformities that represent departures from equilibrium. Heat flow is a response to temperature gradients; diffusion responds to concentration or chemical potential gradients; electrical flow arises from electrical potential gradients. The description of such processes cannot begin without an apparatus for describing nonuniformities in the properties of the system.

In the description of ordinary nonuniform systems, it is generally assumed that the local properties at a point depend upon each other through conventional thermodynamic relationships. For example, the chemical potentials of the components at a point are assumed to be determined by the temperature, pressure, and composition at that point with the same functional relationship that would exist for a uniform large system with the same intensive properties. It has been found that this simple picture is inadequate for the description of nonuniform systems in which the intensive properties vary significantly over short distances. In very fine structures, for example, one in which the composition varies periodically with position and the wavelength of the variation is of the order of a few nanometers, the Gibbs free energy is larger than the value that would be calculated from the local intensive properties. This extra energy, associated with property variations as opposed to property values, has been called the gradient energy of the system. The concept of the gradient energy and some of its applications are presented in Section 14.4. The classic papers introducing this concept are due to Cahn[2] and Cahn and Hilliard.[3] Hilliard[4] provides an excellent review of the concepts and pertinent experimental observations.

14.1 THERMODYNAMIC DENSITIES AND THE DESCRIPTION OF NONUNIFORM SYSTEMS

Consider any extensive thermodynamic property, B'; B' may thus be S', V', U', H', F', G', or n_k. Each of these is a property associated with the system as a whole. A local density of each of these properties may be defined as in Chapter 12, Section 12.2, Equation 12.8:

$$b_v \equiv \lim_{V' \to 0} \frac{B'}{V'} \tag{14.1}$$

Familiar examples of this concept include the local mass density, in which B' is m, the mass of the volume V' and the limiting process yields ρ_m, the mass density, which may vary from point to point in the system. Alternatively, if B' is taken to be n_k, the number of moles of component k, then the limit of the ratio of n_k/V' is the local concentration of component k, c_k.

Focus on a volume element δv in the vicinity of a point P located at coordinates (x, y, z) in the system. If $b_v(x, y, z)$ is the local density of B at P, then $b_v(x, y, z)\delta v$ is the total value of the extensive property B' in the volume element. The total value B' for the system is obtained by summing over all of the volume elements:

$$B' = \iiint_{V'} b_v(x, y, z)\delta v \tag{14.2}$$

The change in B' when the system is taken through an infinitesimal process is obtained by taking the differential of both sides of Equation 14.2:

$$dB' = d \iiint_{V'} b_v(x, y, z)\delta v = \iiint_{V'} db_v \delta v \tag{14.3}$$

The order of the differential and integral operations can be interchanged because the volume elements may be considered to be fixed during the changes in local density values. The differential of the density of any extensive property may be obtained by taking the differential of the definition, Equation 14.1:

$$db_v = d\left(\frac{B'}{V'}\right) = \frac{1}{V'} dB' - \frac{B'}{V^2} dV' \tag{14.4}$$

It is possible to formulate a local version of the combined statement of the first and second laws in terms of the corresponding density functions. Apply Equation 14.4 to describe the change in the internal energy density:

$$du_v = d\left(\frac{U'}{V'}\right) = \frac{1}{V'} dU' - \frac{U'}{V'^2} dV' \tag{14.5}$$

The first term can be evaluated from the combined statement for the first and second laws for the macroscopic system since:

$$dU' = TdS' - PdV' + \sum_{k=1}^{c} \mu_k dn_k \tag{14.6}$$

Integration of this equation at constant intensive properties (T, P and μ_k) gives an expression for the internal energy:

$$U' = TS' - PV' + \sum_{k=1}^{c} \mu_k n_k \tag{14.7}$$

Substitute Equation 14.6 and Equation 14.7 into Equation 14.5:

$$du_v = \frac{1}{V'}\left[TdS' - PdV' + \sum_{k=1}^{c} \mu_k dn_k \right]$$

$$- \frac{1}{V'^2}\left[TS' - PV' + \sum_{k=1}^{c} \mu_k n_k \right] dV'$$

Group like terms:

$$du_v = T\left[\frac{1}{V'} dS' - \frac{S'}{V'^2} dV' \right] - P\left[\frac{1}{V'} dV' - \frac{V'}{V'^2} dV' \right]$$

$$+ \sum_{k=1}^{c} \mu_k \left[\frac{1}{V'} dn_k - \frac{n_k}{V'^2} dV' \right] \tag{14.8}$$

Apply Equation 14.4 the first term in brackets is precisely the differential of the entropy density, ds_v. The second term in brackets is zero: this result arises because V' is the normalization factor in the definition of the density. The term in brackets inside the summation is the differential of the density of number of moles of component k, generally referred to as the molar concentration of component k, dc_k. Thus Equation 14.8 may be written:

$$du_v = Tds_v + \sum_{k=1}^{c} \mu_k dc_k \tag{14.9}$$

This is the combined statement of the first and second laws applied to the densities of thermodynamic functions in a volume element. Rearrangement of this equation gives the corresponding expression for the change in entropy density:

$$ds_v = \frac{1}{T}du_v - \frac{1}{T}\sum_{k=1}^{c}\mu_k dc_k \tag{14.10}$$

This expression is required in the application of the general strategy in order to find the conditions for equilibrium within a single phase continuous system.

14.2 CONDITIONS FOR EQUILIBRIUM IN THE ABSENCE OF EXTERNAL FIELDS

Consider a nonuniform multicomponent, single phase, nonreacting otherwise simple system. The steps in deriving the conditions for equilibrium are familiar. To evaluate the change in entropy for an infinitesimal alteration of the distribution of densities in this nonuniform system, insert Equation 14.10 into the generic Equation 14.3:

$$dS' = \iiint_{V'}\left[\frac{1}{T}du_v - \frac{1}{T}\sum_{k=1}^{c}\mu_k dc_k\right]\delta v \tag{14.11}$$

If during its displacement the system is isolated from its surroundings, then its internal energy cannot change:

$$dU' = \iiint_{V'}du_v\delta v = 0 \tag{14.12}$$

Isolation prevents exchange of components with the surroundings. Since the system being considered is also nonreacting, the number of moles of each component is conserved:

$$dn_k = \iiint_{V'}dc_k\delta v = 0 \qquad (k = 1, 2, ..., c) \tag{14.13}$$

The constant volume isolation constraint is implicit in this treatment because volume is used as the normalization factor.

Equilibrium is attained in an isolated system when its entropy reaches a maximum. To find this extremum in this case it is necessary to use the method of Lagrange multipliers applied in Chapter 6, Section 6.2.3, to derive the Boltzmann distribution function in statistical thermodynamics. Multiply each of the $(c + 1)$ constraining equations by an arbitrary constant and add the result to the expression for the function to be maximized, Equation 14.11:

$$dS' + \alpha dU' + \sum_{k=1}^{c}\beta_k dn_k = 0 \tag{14.14}$$

where α and the set of β_k values are constants in the system, which means explicitly that they are not functions of position. Write this equation in terms of the corresponding integrals,

$$\iiint_{V'} \left[\frac{1}{T} du_v - \frac{1}{T} \sum_{k=1}^{c} \mu_k dc_k \right] \delta v + \alpha \iiint_{V'} du_v \, \delta v + \sum_{k=1}^{c} \beta_k \iiint_{V'} dc_k \delta v = 0$$

Collect terms:

$$\iiint_{V'} \left[\left(\frac{1}{T} + \alpha \right) du_v - \frac{1}{T} \sum_{k=1}^{c} (\mu_k - \beta_k) dc_k \right] \delta v = 0 \qquad (14.15)$$

In order for this integral to equal zero for arbitrary changes in the differentials of the densities, the coefficients of these differentials must simultaneously vanish. Thus, for the extremum,

$$\frac{1}{T} = -\alpha \Rightarrow T = \text{constant} \qquad (14.16)$$

$$\mu_k = \beta_k \Rightarrow \mu_k = \text{constant} \qquad (k = 1, 2, ..., c) \qquad (14.17)$$

where the notation " $=$ constant" in this context means "is not a function of position" in the system. A more explicit statement of these conditions uses the concept of the gradient developed in vector calculus. For any function of position in space, $f(x, y, z)$, the gradient is a vector given by:

$$\nabla f \equiv \left(\frac{\partial f}{\partial x} \right) \hat{i} + \left(\frac{\partial f}{\partial y} \right) \hat{j} + \left(\frac{\partial f}{\partial z} \right) \hat{k} \qquad (14.18)$$

where $\mathbf{i}$, $\mathbf{j}$, and $\mathbf{k}$ are unit vectors in the x, y, and z directions. The gradient vector at any position (x, y, z) gives the magnitude and direction of the most rapid change in the function f at that point. Expressed in terms of this concept the conditions for equilibrium in a continuous system may be stated:

$$\text{grad } T = 0 \qquad \text{(Thermal equilibrium)} \qquad (14.19)$$

$$\text{grad } \mu_k = 0 \qquad (k = 1, 2, ...c) \qquad \text{(Chemical equilibrium)} \qquad (14.20)$$

The condition for mechanical equilibrium may be derived from these results. Recall the variation of the chemical potential of a component with temperature, pressure, and composition, Equation 12.72:

$$d\mu_k = -\overline{S_k} dT + \overline{V_k} dP + \sum_{j=2}^{c} \mu_{kj} dX_j \qquad (14.21)$$

where the summation is over $(c - 1)$ independent mole fractions. In the context of a nonuniform system, the variations being considered in this equation are variations with position. Accordingly, each of the differentials may be replaced by the corresponding gradient of the intensive property:

$$\text{grad } \mu_k = -\overline{S_k}\text{grad } T + \overline{V_k}\text{grad } P + \sum_{j=2}^{c} \mu_{kj}\text{grad } X_j \qquad (14.22)$$

The conditions for thermal and chemical equilibrium, Equation 14.19 and Equation 14.20, eliminate two of the terms in this expression:

$$0 = \overline{V_k}\text{grad } P + \sum_{j=2}^{c} \mu_{kj}\text{grad } X_j$$

Multiply by X_k and sum over all of the components:

$$0 = \sum_{k=1}^{c} X_k \overline{V_k}\text{grad } P + \sum_{k=1}^{c} X_k \sum_{j=2}^{c} \mu_{kj}\text{grad } X_j$$

Note that the summation in the first term is the molar volume. Interchange the order of the summations in the second term:

$$0 = V \text{ grad } P + \sum_{k=1}^{c} X_k \sum_{j=2}^{c} \mu_{kj} \text{ grad } X_j$$

The complicated second term in this expression can be shown to be equal to zero by writing the sums out explicitly and repeatedly applying the Gibbs–Duhem equation. Thus the condition for mechanical equilibrium is as expected:

$$\text{grad } P = 0 \qquad (14.23)$$

This result completes the demonstration that temperature, pressure chemical potential and hence composition are uniform within any single-phase region provided there are no external fields acting on the system. This presumption, used extensively if implicitly in earlier chapters, is thus justified.

14.3 CONDITIONS FOR EQUILIBRIUM IN THE PRESENCE OF EXTERNAL FIELDS

This section considers the influence that time invariant gravitational, electrical, and centrifugal fields have upon the conditions for equilibrium in a multicomponent single-phase system. It is essential to limit consideration to time invariant fields because if the fields and their influence on the properties of the system were changing with time, then it would not be possible for the system to come to equilibrium, which is a time invariant condition.

In classical physics the influence that these phenomena have upon the behavior of a system can be formulated in terms of the concept of potential energy. Each is visualized to have a corresponding force field, the strength of which varies from point to point in space. The force field at some point P in space is a representation of the force that would act upon a particle placed at P. Gravitational and centrifugal fields interact with the mass of the particle while an electrical field interacts with its charge.

14.3.1 POTENTIAL ENERGY OF A CONTINUOUS SYSTEM

An alternate formulation of the same phenomena is based upon the idea of the potential field, which describes the potential energy of a particle in a force field. Introduce the concept of the specific potential function $\psi(x, y, z)$ defined to be the potential energy per unit mass at the point (x, y, z) derived from a gravitational or centrifugal field. The force acting on a particle depends upon its mass, m, and the gradient of the potential function:

$$\vec{F} = -m \text{ grad } \psi \tag{14.24}$$

The work done on the particle in moving it through some infinitesimal displacement dx is

$$\delta W_{PF} = \vec{F} \cdot d\vec{x}$$

$$\delta W_{PF} = -m\left[\left(\frac{\partial \psi}{\partial x}\right)\hat{i} + \left(\frac{\partial \psi}{\partial y}\right)\hat{j} + \left(\frac{\partial \psi}{\partial z}\hat{k}\right)\right][d\vec{x}\hat{i} + d\vec{y}\hat{j} + d\vec{z}\hat{k}]$$

$$\delta W_{PF} = -m\left[\left(\frac{\partial \psi}{\partial x}\right)dx + \left(\frac{\partial \psi}{\partial y}\right)dy + \left(\frac{\partial \psi}{\partial z}\right)dz\right] = -md\psi \tag{14.25}$$

since the quantity in brackets in the last equation is precisely the differential of the specific potential function ψ. For any finite change in position of the particle in the field from point a to point b, the work done on the particle is

$$W_{PF} = \int_a^b \vec{F} \cdot d\vec{x} = \int_a^b -md\psi = -m[\psi(b) - \psi(a)] \tag{14.26}$$

In a conservative field, the work done depends only upon the potential at the initial and final positions in the field and is independent of the path by which the particle moves between these points. Since gravitational, centrifugal, and electrostatic fields are conservative, the specific potential function gives the potential energy per unit mass of a particle situated in the field.

In applying these concepts to the thermodynamic description of these systems the focus is upon changes that are associated, not with displacements of a particle in the field, but with the influence of the potential field upon an element of volume in a system that may be experiencing a variety of internal changes that may affect its mass. Focus upon an element of volume δv in a multicomponent single-phase system. If c_k is the molar concentration of component k in the element, then the number of moles of k is $c_k \delta v$; the mass of component k in the element is $M_k c_k \delta v$ where M_k is the molecular weight (gram atomic weight for an element) of component k. The total mass δm contained in the volume element δv is given by

$$\delta m = \sum_{k=1}^{c} M_k c_k \delta v \tag{14.27}$$

Since ψ is the potential energy per unit mass, the potential energy of this volume element is $\psi \delta m$. The total potential energy of the system is obtained by summing the contributions from each volume element:

$$E'_{\text{pot}} = \int \int \int_{V'} \psi \sum_{k=1}^{c} M_k c_k \delta v \tag{14.28}$$

Suppose the system experiences an infinitesimal alteration of its internal state in which all of its intensive properties are allowed to vary. Since ψ is fixed in time (the field is assumed to be time invariant) and the molecular weight of each component is constant, the change in potential energy of each volume element can only arise from changes in the concentrations of the components in that element. The change in potential energy of the system is thus given by

$$dE'_{\text{pot}} = \int \int \int_{V'} \psi \sum_{k=1}^{c} M_k dc_k \delta v \tag{14.29}$$

The total energy of the system is the sum of its internal energy, derived from its temperature, pressure, composition, and the potential energy associated with the external field:

$$E'_{\text{tot}} = U' + E'_{\text{pot}} = \int \int \int_{V'} u_v \delta v + \int \int \int_{V'} \psi \sum_{k=1}^{c} M_k dc_k \delta v \tag{14.30}$$

The change in total energy of the system accompanying an arbitrary change in its state is given by

$$dE'_{\text{tot}} = \int \int \int_{V'} \left[du_v + \psi \sum_{k=1}^{c} M_k dc_k \right] \delta v \tag{14.31}$$

All of the effects of external force fields upon the conditions for equilibrium arise from this relationship.

14.3.2 CONDITIONS FOR EQUILIBRIUM

The entropy of the system is not influenced by the external field. In applying the general strategy, the change in entropy experienced by the system is given by Equation 14.11. The influence of the external field operates through the isolation constraints. In an isolated system, which cannot exchange energy with its surroundings, the total energy of the system cannot change. Equation 14.12 is replaced by the more general condition,

$$dE'_{tot} = 0 = \iiint_{V'} \left[du_v + \psi \sum_{k=1}^{c} M_k dc_k \right] \delta v \qquad (14.32)$$

Conservation of the components, described by Equation 14.13, is unaffected by the external field. To find the extremum in the entropy, apply the method of Lagrange multipliers, slightly generalized from Equation 14.14:

$$dS' + \alpha dE'_{tot} + \sum_{k=1}^{c} \beta_k dn_k = 0 \qquad (14.33)$$

Substitute from Equation 14.11, Equation 14.13, and Equation 14.32:

$$\iiint_{V'} \left[\frac{1}{T} du_v - \frac{1}{T} \sum_{k=1}^{c} \mu_k dc_k \right] \delta v + \alpha \iiint_{V'} \left[du_v + \psi \sum_{k=1}^{c} M_k dc_k \right] \delta v$$
$$+ \sum_{k=1}^{c} \beta_k \iiint_{V'} dc_k \delta v = 0$$

Combine like terms:

$$\iiint_{V'} \left[\left[\frac{1}{T} + \alpha \right] du_v + \sum_{k=1}^{c} \left[-\frac{1}{T} \mu_k + \alpha \psi M_k + \beta_k \right] dc_k \right] \delta v = 0 \qquad (14.34)$$

Set the coefficients of each of the differentials in the integrand equal to zero to find the conditions for equilibrium. Keep in mind that α and the set of β_k values are constants, independent of position in the system. The coefficient of du_v gives the usual condition for thermal equilibrium:

$$\frac{1}{T} = -\alpha \Rightarrow \text{grad } T = 0 \qquad (14.35)$$

Use this result to evaluate α in the coefficients of each of the dc_k's:

$$-\frac{1}{T}\mu_k - \frac{1}{T}\psi M_k = \beta_k \Rightarrow \mu_k + \psi M_k = \text{constant} \qquad (14.36)$$

Express this result in terms of the gradients of the variables:

$$\text{grad}(\mu_k + \psi M_k) = 0$$

Since the gradient of a sum is the sum of the gradients, this result may also be written:

$$\text{grad } \mu_k = -M_k \text{ grad } \psi \qquad (k = 1, 2, ..., c) \qquad (14.37)$$

Note that M_k, the molecular weight of k is a constant. Thus, in the presence of an external force field, the chemical potential of each of the components in a homogeneous system is not uniform a^e equilibrium.

The condition for mechanical equilibrium is derived by paraphrasing Equation 14.21 to Equation 14.23. The left-hand side of Equation 14.22 is not zero in the presence of an electrical field but is given by Equation 14.37:

$$-M_k \text{ grad } \psi = \overline{V_k} \text{ grad } P + \sum_{j=2}^{c} \mu_{kj} \text{ grad } X_j \qquad (14.38)$$

Multiply by X_k and sum:

$$-\sum_{k=1}^{c} X_k M_k \text{ grad } \psi = \sum_{k=1}^{c} X_k \overline{V_k} \text{ grad } P + \sum_{k=1}^{c} X_k \sum_{j=2}^{c} \mu_{kj} \text{ grad } X_j$$

The Gibbs–Duhem relation demonstrates that the double summation term is zero. The coefficient of grad ψ may be recognized as M, the mass per mole of solution at the position (x, y, z), while the coefficient of grad P is V, the volume per mole. Their ratio $M/V = \rho$, the mass density (gm/cc) at the point P. The condition for mechanical equilibrium is therefore:

$$\text{grad } P = -\rho \text{ grad } \psi \qquad (14.39)$$

It is concluded that when a system comes to equilibrium in the presence of an external force field, the temperature does not vary with position. In contrast, both the pressure and the chemical potential of each component vary with position in a manner that is determined by the positional variation of the potential field.

14.3.3 EQUILIBRIUM IN A GRAVITATIONAL FIELD

In a gravitational field the force acting on a mass m is given by Newton's second law:

$$\vec{F} = m\vec{a} = m\vec{g} \tag{14.40}$$

where $\vec{g}$ is the acceleration engendered by the field. Comparison of this equation with the general definition of the specific potential, Equation 14.24, yields:

$$\vec{g} = -\text{grad } \psi \tag{14.41}$$

The conditions for equilibrium stated in Equation 14.37 and Equation 14.39 for the case of a gravitational field are:

$$\text{grad } \mu_k = M_k \vec{g} \qquad (k = 1, 2, ..., c) \tag{14.42}$$

$$\text{grad } P = \rho \vec{g} \tag{14.43}$$

The latter equation is a general form of the familiar barometric equation relating pressure to altitude in the Earth's atmosphere. The following examples illustrate these results.

EXAMPLE 14.1

Derive an expression for the variation of pressure with altitude for a unary system.
The condition for mechanical equilibrium, Equation 14.43, provides the basis for the relationship. In a Cartesian coordinate system with $\mathbf{k}$ the unit vector in the vertical direction, $\mathbf{g} = -g\mathbf{k}$, where g is the magnitude of the gravitational acceleration. The negative sign is required because the acceleration vector is directed downward while $\mathbf{k}$ by this convention points upward. The vector Equation 14.43 may be written explicitly as

$$\left(\frac{\partial P}{\partial x}\right)\hat{i} + \left(\frac{\partial P}{\partial y}\right)\hat{j} + \left(\frac{\partial P}{\partial z}\right)\hat{k} = \rho(-g\hat{k})$$

A vector equation represents three equations: corresponding coefficients of $\mathbf{i}$, $\mathbf{j}$, and $\mathbf{k}$ on the two sides of the equation must be separately equal. Since the $\mathbf{i}$ and $\mathbf{j}$ components on the right side are zero, the variation of P in the horizontal directions is zero. Equating coefficients in the $\mathbf{k}$ direction yields

$$\frac{dP}{dz} = -\rho g \tag{14.44}$$

It will be recalled that ρ is the mass density of the system, which for a unary system is the ratio of the molecular weight, M_1 (mass/mol), to the molar volume, V:

$$\rho = \frac{M_1}{V} = \frac{M_1}{(RT/P)}$$

assuming the atmosphere is an ideal gas. The mass density thus varies with height, z, since the pressure varies with height. Insert this result into the condition for mechanical equilibrium, Equation 14.44:

$$\frac{dP}{dz} = -M_1\left(\frac{P}{RT}\right)g$$

Separate the variables:

$$\frac{dP}{P} = -\frac{M_1 g}{RT}dz$$

Integrate:

$$\ln\left(\frac{P(z)}{P(z_0)}\right) = -\frac{M_1 g}{RT}(z - z_0) = \frac{M_1 g}{RT}h$$

$$P(h) = P(z_0)e^{-\frac{M_1 g}{RT}h} \qquad\qquad (14.45)$$

This is the barometric equation, which demonstrates that at equilibrium in a unary system the pressure decreases exponentially with altitude. The analogous result is more complex for a multicomponent system because the molar mass varies with composition, which also varies with altitude.

EXAMPLE 14.2
Planet X has an atmosphere that consists of a binary mixture of hydrogen and nitrogen. Derive an expression for the variation of composition of this atmosphere with altitude. Assume the atmosphere is an ideal gas mixture. At the planetary surface mole fraction of H_2 is 0.35. Planet X has a radius of 10,000 km and a mass that yields a gravitational acceleration of 1000 cm/sec^2. Take the temperature of the atmosphere to be 800 K. Compute the composition of the atmosphere at 100 km above the surface.
For this one-dimensional field the vector equation describing the condition for chemical equilibrium in this atmosphere Equation 14.42 simplifies to

$$\frac{d\mu_k}{dz} = -M_k g$$

again noting that the vector $\mathbf{g}$ is $-g\mathbf{k}$. The chemical potential change for the mixing process for an ideal gas mixture is

$$\mu_k - \mu_k^\circ = RT \ln\frac{P_k}{P^\circ} = RT \ln P_k$$

Since P° is one atmosphere. The partial pressure of component k in the mixture, P_k, is given by

$$P_k = X_k P(z)$$

where $P(z)$ is the total pressure in the gas mixture, which, in this application, depends upon height, z, see Equation 14.45. The variation of chemical potential with height is the derivative

$$\frac{d\mu_k}{dz} = RT\frac{d\ln(X_k P(z))}{dz} = RT\frac{d\ln X_k}{dz} + RT\frac{d\ln P(z)}{dz}$$

The condition for mechanical equilibrium, Equation 14.44, for an ideal gas gives

$$\frac{dP}{dz} = -\rho g = -\frac{M}{V}g = -\frac{M}{\left(\dfrac{RT}{P}\right)}g$$

So that

$$\frac{d\ln P}{dz} = -\frac{\langle M\rangle}{RT}g$$

where $\langle M\rangle$ is the molecular weight of the gas mixture,

$$\langle M\rangle = \sum_{k=1}^{c} X_k M_k$$

Thus, the variation with chemical potential with height at equilibrium gives

$$\frac{d\mu_k}{dz} = RT\frac{\partial\ln X_k}{\partial z} + RT\frac{\partial\ln P(z)}{\partial z} = RT\frac{\partial\ln X_k}{\partial z} + \langle M\rangle g = -M_k g$$

Rearrange

$$\frac{d\ln X_k(z)}{dz} = \frac{g}{RT}[\langle M\rangle - M_k] \tag{14.46}$$

Thus, if the molecular weight of component k is larger than the average, the right side of the equation is negative and the mole fraction of that component *decreases* with altitude. If M_k is smaller than the average, the right side is positive and the mole fraction of that component *increases* with altitude. These results are in agreement with intuition which suggests heavier components will be enriched at lower altitudes.

For a binary system the average molar mass is

$$\sum_{j=1}^{c} X_j M_j = X_1 M_1 + X_2 M_2 = (1 - X_2)M_1 + X_2 M_2$$

Substitute this result into Equation 14.46:

$$\frac{d\ln X_2(z)}{dz} = -M_2 g + g[(1 - X_2)M_1 + X_2 M_2] = g(1 - X_2)(M_1 - M_2)$$

Separate the variables:

$$\frac{d \ln X_2}{1 - X_2} = \frac{g(M_1 - M_2)}{RT} dz$$

Integrate:

$$\frac{X_2(z)}{1 - X_2(z)} = \frac{X_2(z_0)}{1 - X_2(z_0)} e^{\frac{g}{RT}(M_1 - M_2)h} \tag{14.47}$$

Choose component 2 to be H_2 and insert the values given in the problem:

$$\frac{X_2(z)}{1 - X_2(z)} = \frac{0.35}{0.65}$$

$$\times \exp\left[\frac{1000(cm/sec^2)}{8.314 \times 10^7 (ergs/mole\,K)800\,K} (28 - 2) gm/mole\ 100km \times 10^5 (cm/km) \right]$$

Convert gm (cm^2/sec^2) to joules to make the units compatible:

$$\frac{X_2(z)}{1 - X_2(z)} = \frac{0.35}{0.65} e^{2.11} = 4.42$$

Solve for the composition of hydrogen at 100 km:

$$X_{H_2}(100km) = 0.82$$

The atmosphere is significantly enriched in the lighter element at 100 km.

14.3.4 Equilibrium in a Centrifugal Field

Consider now a system that is rotating with an angular velocity ω(rad/sec). The acceleration vector for an element of volume contained a distance r from the axis of rotation is given by

$$\vec{a} = \omega^2 \vec{r} \tag{14.48}$$

where the vector **r** is directed toward the center of rotation. The acceleration is directed inward. The force acting upon the mass dm contained in the volume element is

$$\overrightarrow{dF} = dm\,\vec{a} = dm\,\omega^2 \vec{r}$$

Comparison with Equation 14.24 yields the specific potential field for a centrifugal force field:

$$\text{grad } \psi = -\omega^2 \vec{r} \tag{14.49}$$

For this case the conditions for chemical equilibrium [Equation 14.37] and mechanical equilibrium [Equation 14.39] become

$$\text{grad } \mu_k = M_k \omega^2 \vec{r} \tag{14.50}$$

$$\text{grad } P = \rho \omega^2 \vec{r} \tag{14.51}$$

EXAMPLE 14.3

Estimate the maximum pressure achieved in a turbine blade in a jet turbine engine rotating at 22,000 rpm. The root of the blade is 3 cm from the axis of rotation and the blade is 8 cm long. Take the density of the blade to be 9 gm/cc. Assume the state of stress is a simple hydrostatic pressure.
Compute the angular velocity in radians per second:

$$\omega = \frac{2\pi \cdot 22,000}{60} = 2300 \; \frac{\text{rad}}{\text{sec}}$$

The condition for mechanical equilibrium for a radial centrifugal field is

$$\frac{dP}{dr} = \rho \omega^2 r$$

Integrate:

$$P(r) = \frac{1}{2} \rho \omega^2 r^2$$

Insert the properties stated in the problem

$$P(r) = \frac{1}{2} \left(9 \, \frac{\text{gm}}{\text{cc}} \right) \left(2300 \, \frac{\text{rad}}{\text{sec}} \right) ((8+3)\text{cm})^2 \left(\frac{1 \, J}{10^7 \, \text{ergs}} \right) \left(\frac{82.06 \, \text{cc} - \text{atm}}{8.314 \, J} \right)$$

Calculate:

$$P(r) = 2850 \text{ atm} \cong 42,000 \text{ psi}$$

Thus, centrifugal stresses of the order of two kilobars are not uncommon in an operating turbine blade. Actual stresses are not hydrostatic; indeed, the deviations from this simple equilibrium condition play a much more significant role in determining the performance of a turbine blade.

14.3.5   EQUILIBRIUM IN AN ELECTROSTATIC FIELD

The description of forces that operate in an electrostatic field is formulated in terms that are slightly different from those used for gravitational and centrifugal fields. An electrical potential field $\phi(x, y, z)$ is defined such that the force acting on a particle carrying a *charge q* is given by

$$\vec{F} = -q \text{ grad } \phi \qquad (14.52)$$

In this view, $-$ grad ϕ is the force per unit charge. This contrasts with the definition of the specific potential field, ψ, defined in Equation 14.24 such that $-$ grad ψ is the force per unit mass.

These two potentials are related through the charge per unit mass that a system may possess. Focus on a volume element δv. Within that volume element focus upon the atoms of component k. Let z_k be the charge number associated with a unit of component k. z_k is a small whole number; it may be zero, positive, or negative depending upon the nature of the component. The mass of the atoms of component k in δv,

$$\delta m_k = M_k c_k \delta v \qquad (14.53)$$

The charge carried by these atoms is

$$\delta q_k = z_k(N_0 e)c_k \delta v \qquad (14.54)$$

where e is the magnitude of the charge carried by an electron and N_0 is Avogadro's number. The product $(N_0 e)$ is the magnitude of the electric charge carried by one mole of electrons; this quantity is called the Faraday, written $\mathscr{F}$, after the genius who pioneered the study of electrochemical effects early in the nineteenth century. The force acting upon this subset of atoms in the volume element δv may be formulated in terms of either Equation 14.24 or Equation 14.52:

$$\delta\vec{F}_k = -\delta m_k \text{ grad } \psi = -\delta q_k \text{ grad } \phi - (M_k c_k \delta v)\text{grad } \psi = -(z_k\mathscr{F}c_k\delta v)\text{grad } \phi$$

$$M_k \text{ grad } \psi = z_k\mathscr{F} \text{ grad } \phi \qquad (14.55)$$

This establishes the relationship between the specific potential ψ and the electric potential, ϕ. The conditions for equilibrium, Equation 14.37 and Equation 14.39, may now be written in terms of the electrical potential. In particular, for chemical equilibrium,

$$\text{grad } \mu_k = -z_k\mathscr{F}\text{grad } \phi \qquad (k = 1, 2, ...c) \qquad (14.56)$$

Introduce the following definition:

$$\eta_k \equiv \mu_k + z_k\mathscr{F}\phi \qquad (14.57)$$

The property, η_k is called the electrochemical potential of component k. With this definition, the condition for equilibrium in the presence of an electrostatic field may be written

$$\text{grad } \eta_k = 0 \qquad (k = 1, 2, \dots c) \tag{14.58}$$

Thus, in an electrostatic field the electrochemical potential is uniform at equilibrium. Note that the gradient in chemical potential, and hence qualitatively the gradient in composition, depends upon the sign of z_k. For positively charged components the chemical potential decreases with increasing electrical potential; for negatively charged components the chemical potential and electric field gradients are in the same direction.

For mechanical equilibrium,

$$\text{grad } P = -\rho \text{ grad } \psi = -\frac{1}{V}\left(\sum_{k=1}^{c} M_k X_k\right)\text{grad } \psi$$

Distribute grad ψ throughout the sum:

$$\text{grad } P = -\frac{1}{V}\sum_{k=1}^{c} M_k X_k \text{grad } \psi$$

Substitute in each term from Equation 14.55:

$$\text{grad } P = -\frac{1}{V}\sum_{k=1}^{c} X_k(z_k \mathscr{F} \text{ grad } \phi)$$

Note that in each term $X_k/V = c_k$, the molar concentration. Finally,

$$\text{grad } P = -\mathscr{F} \text{ grad } \phi \sum_{k=1}^{c} z_k c_k \tag{14.59}$$

The quantity in the summation may be thought of as the average charge density at some position (x, y, z) in the system. Since in general the right side of Equation 14.59 differs from zero for a system with charged particles in an electrical field, a pressure gradient is expected to develop in such a system when it comes to equilibrium.

EXAMPLE 14.4

Given an electrostatic field that decays exponentially from the surface of the system toward its interior, describe the variation of composition of the components with position when this system comes to equilibrium. Let $X_k(\infty)$ be the composition well away from the surface of the system corresponding to the chemical potential, $\mu_k(\infty)$.

Since at equilibrium the electrochemical potential [Equation 14.57] is constant [Equation 14.58]

$$\mu_k + z_k \mathcal{F} \phi = \text{const} = \mu_k(\infty)$$

$$\mu_k - \mu_k(\infty) = -z_k \mathcal{F} \phi_0 e^{-ax}$$

Components for which z_k is negative exhibit a chemical potential distribution that mimics the electric field function, with μ_k and thus X_k high near the surface and decaying down to $X_k(\infty)$. The distribution for positively charged components is opposite, rising from the lowest value of concentration adjacent to the surface toward their far field value. This result is consistent with the notion that negatively charged particles move toward regions of high electrical potential while positively charged ions are repelled from those regions.

14.4 THE GRADIENT ENERGY IN NONUNIFORM SYSTEMS

An important trend in materials science is the preparation of microstructures that have extremely fine characteristic length scales. Microelectronic devices are prepared with controlled composition distributions that span a few nanometers. Sol gel materials have pore structures in the nanometer size range. Rapidly solidified materials yield very fine scale microstructures. Nanocomposite materials are being designed and fabricated. Solute rich clusters that form prior to precipitation in age hardening materials have composition gradients on a fine scale. Spinodal decomposition and ordering transformations also occur at this level of microstructural scale. Nucleation processes also may involve large changes in intensive properties over short distances.

The description of the thermodynamics of nonuniform systems presented in Section 14.1 to Section 14.3 is founded on the concept of the local density of thermodynamic properties, Equation 14.1. The total value of any extensive property is then taken to be the integral of the local density function over the volume of the system, Equation 14.2. It has been found that this description is inadequate if the properties of the system vary significantly over very short distances. Such a system is characterized as having extremely large values of gradients of its intensive properties.

To illustrate this idea consider the concentration gradient in a typical interdiffusion study. In an experimental diffusion couple joining A to B the atom fraction of B may change from 0 to 1 over a range of perhaps 10 microns (10^{-6} m). The average gradient in atom fraction in this structure is thus $1/10^{-6}$ or 10^6 m^{-1}. In a microstructure undergoing spinodal decomposition the atom fraction of B may change from, say, 0.8 to 0.2 in a few nanometers. The gradient in this system is thus ($0.6/3 \times 10^{-9}$ m^{-1} = 2×10^9 m^{-1}) or about three orders of magnitude larger than in the diffusion couple.

In the presence of large gradients it is found that the value of the local density of a thermodynamic function, b_v, depends not only upon the values of the intensive properties in the volume element but also upon their rates of change, i.e., their local gradients. A rigorous formulation of this contribution to the properties of systems was first presented by Cahn and Hilliard in 1959.[3]

Consider as an example the Helmholtz free energy of a binary system, F'. In the traditional description the local density of Helmholtz free energy is a function of the local temperature, pressure (in solids, the state of stress), and composition:

$$f_v = f_v(T, P, X_2) = f_v(T, P, c) \tag{14.60}$$

To simplify the problem suppose the temperature and pressure are uniform; the system only exhibits a nonuniformity in composition. Then, according to Equation 14.2, the Helmholtz free energy of the system is the integral:

$$F' = \int\int\int_{V'} f_v[c(x, y, z)]\delta v \tag{14.61}$$

If the variation of composition with position is known, together with a solution model that permits calculation of the Helmholtz free energy from the composition, then F' may be computed.

Cahn and Hilliard assumed that f_v depends not only upon c_2 but also on all measures of the variation of c_2 with position including c_2', $(c_2')^2$, $(c_2')^3 \ldots$, and c_2'', $(c_2'')^2$, etc. Here, the ($'$) and ($''$) denote the first and second derivatives with respect to position. Thus, it is assumed that f_v depends upon the composition, the gradient, the square of the gradient, the second derivative, and so on. A generalization of the Taylor expansion of a function about a point permits this dependence to be expressed:

$$f_v = f_v{}^0[c] + L[c'] + K_1[c''] + K_2[c'^2] + \cdots \tag{14.62}$$

where the coefficients have explicit but unfamiliar meaning:

$$L = \left(\frac{\partial f_v}{\partial c'}\right) \tag{14.63}$$

$$K_1 = \left(\frac{\partial f_v}{\partial c''}\right) \tag{14.64}$$

Terms beyond these second order quantities are neglected in the development.

In crystals with a center of symmetry in the unit cell, the value of f_v must be independent of the choice of direction for the coordinate system describing the variation of composition. Since altering the direction of the x axis changes the sign of the first derivative, c_2', the coefficient of this term, L, must be zero. Further mathematical manipulation demonstrates that the remaining two terms may be combined:

$$K_1 \frac{d^2c}{dx^2} + K_2 \left(\frac{dc}{dx} \right)^2 = K \left(\frac{dc}{dx} \right)^2$$

where

$$K = K_2 - \frac{dK_1}{dc} \qquad (14.65)$$

Thus, the local free energy density may be written:

$$f_v = f_v^{\,0}(c) + K \left(\frac{dc}{dx} \right)^2 \qquad (14.66)$$

The total Helmholtz free energy of the system is then

$$F' = \int \int_{V'} \int [f_v^{\,0}(c) + K(\nabla c)^2] \delta v \qquad (14.67)$$

where the symbol ∇ is shorthand notation for the *gradient* of the composition function (see Equation 14.18). Comparison of this result with Equation 14.61 identifies the additional term in the Helmholtz free energy which is proportional to the *square of the gradient* of the composition distribution. Since it can be shown that the coefficient K must be positive, it follows that this gradient energy contribution is always positive.

It may be useful to view the gradient energy as a distributed or smeared out surface energy. In Gibbs' development of the thermodynamics of surfaces the effects of the nonuniformity associated with the transition from the properties of one phase to those of its neighbor are assigned to a discontinuity located at the dividing surface, Section 12.2 in Chapter 12. The gradient energy formalism recognizes the diffuse nature of this distribution in many real interfaces, characterizing them with a smooth variation in intensive properties, i.e., a gradient, yet realizing an excess energy in the system that is associated with such nonuniformities.

Similar to surface energy, the gradient energy may provide a barrier to be overcome by a process, as in nucleation, or it may contribute to the driving force, as in coarsening processes. Consider, for example, the spinodal decomposition process described in connection with miscibility gaps in Section 10.1.4 in Chapter 10. If a uniform solution is quenched into the spinodal region it may spontaneously unmix. In more detail, the as-quenched structure will contain an array of random statistical fluctuations in composition (Figure 14.1a). The spontaneous unmixing process develops as an increasing variation in the composition range in the system (Figure 14.1b). This may be viewed as an increase in the amplitude of components of the composition wave. The process evidently requires the motion of atoms of the components through the solution, i.e., by diffusion. Components of the composition fluctuation that have the shortest wavelength will have the shortest diffusion distance

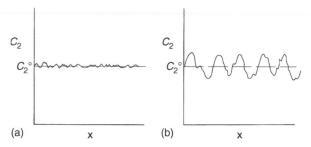

FIGURE 14.1 A random composition fluctuation about the average composition c_2^0 in a solution quenched into a miscibility gap (a) begins to unmix spontaneously (b) by amplifying some of the harmonic components of the compositonal wave form.

and largest concentration gradient. Amplitudes of these shortest waves will therefore grow most rapidly, ultimately resulting in a modulated structure with a wavelength equal to the distance between atoms.

This is not observed experimentally. The structure that results has a characteristic wavelength in the nanometer range. The value of this characteristic wavelength varies in a predictable way with temperature and the starting composition of the solution. The atomically fine compositional variation does not form because the energy associated with the extremely large gradients in this system prevents it. Rather, the system chooses a length scale which optimizes the competition between shortened diffusion distances and increased gradient energy. These observations could not be explained without invoking the gradient energy concept.

The gradient energy can add to the driving force for diffusion for a process that requires the spontaneous elimination of gradients artificially introduced into a system. Hilliard and Philofsky[5] prepared a series of diffusion couples by alternately vapor depositing thin layers of gold and palladium on a substrate (Figure 14.2).

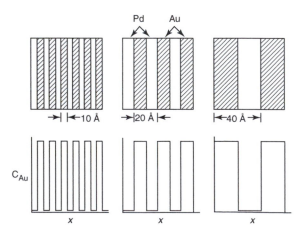

FIGURE 14.2 Thin film diffusion couple formed by alternately vapor depositing layers of gold and palladium.

They were able to prepare these layered structures with different layer thicknesses ranging from about one to four nanometers. The gradients in the one-nanometer composite were thus about four times larger than those in the coarser, four-nanometer structure. Since the gradient energy term is proportional to the square of the gradient, the ratio of gradient energies in the two extreme cases was about 16.

In spinodal decomposition the spontaneous process is unmixing, i.e., the spontaneous amplification of compositional fluctuations. In these initially layered structures the spontaneous process is the elimination of these compositional variations with the equilibrium structure exhibiting a uniform composition. In this homogenization process the system has gradients initially; in its final state the gradients are eliminated. Thus, in this case, the gradient energy provides an additional driving force that increases the diffusion rate. In the experiments of Hilliard and Philofsky this showed up as a variation of the diffusion coefficient with the initial spacing between the layers. Their one-nanometer structure with the largest gradient energy gave the lowest diffusion coefficient (Figure 14.3). The diffusion coefficient reports the proportionality between the flux and the driving force. Since the driving force is increased by the gradient energy term for the finer structures, a lower value of the diffusion coefficient is required to obtain the same flux.

The sharp drop in diffusion coefficient that appears near three nanometers in Figure 14.3 is associated with a loss of mechanical coherency in these modulated structures. Elastic effects are found to play an important role in ultrafine structures, and may even eclipse the gradient energy effects in the system. For example, if the elastic strains associated with the compositional variations that develop in spinodal decomposition are sufficiently large, the process may be completely suppressed.

For an excellent review of the theory and associated experimental observations see Hilliard.[4]

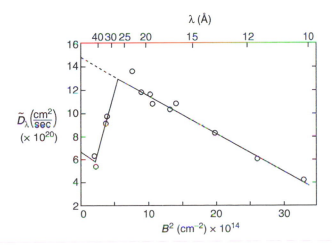

FIGURE 14.3 Experimentally determined variation of the diffusion coefficient with wavelength of the starting layered structure demonstrates that the gradient energy contribution is real [14.5]. *Source:* Philofsky, E.M. and Hilliard, J.E., J. Appl. Phys., 40, 2198, 1969.

A major application of the gradient energy formalism lies in the development of *phase field models* for evolving microstructures. The traditional description of the evolution of microstructures describes the motion of the interphase interfaces that define the geometry of the particles of the individual phases in the microstructure. In the traditional view discontinuities in properties (composition, chemical potential, pressure) occur at these interfaces as required in ordinary thermodynamics. If, on the other hand, the interfaces are modeled as local gradients in these properties, then the discontinuities are removed from the mathematical description and the spatial distribution of individual properties can be described by continuous functions that vary smoothly through all of the phases and their boundaries. Such functions are much more amenable to insertion into computer models that include all of the thermodynamics variables and kinetic formalism in generating images of evolving microstructures. Phase field modeling of microstructural evolution is finding growing application in materials science, particularly in the nanometer range of microstructural scales.

14.5 SUMMARY

In systems of ordinary dimensions, the thermodynamics of nonuniform systems may be described in terms of the variation of the local density of any of its extensive properties, b_v, with position.

The global value of a property is the integral over the volume of the system of its local density.

Application of the general strategy for finding conditions for equilibrium in continuous systems showed that:

a. In the absence of external fields, at equilibrium,

$$\text{grad } T = 0 \tag{14.19}$$

$$\text{grad } P = 0 \tag{14.23}$$

$$\text{grad } \mu_k = 0 \quad (k = 1, 2, ...c) \tag{14.20}$$

b. In the presence of external fields, described by the specific potential energy function ψ, at equilibrium,

$$\text{grad } T = 0 \tag{14.35}$$

$$\text{grad } P = -\rho \, \text{grad } \psi \tag{14.39}$$

$$\text{grad } \mu_k = -M_k \text{grad } \psi \quad (k = 1, 2, ...c) \tag{14.37}$$

The specific potential energy function is related to familiar physical quantities:

a. In a gravitational field, grad $\psi = -\mathbf{g}$, the gravitational acceleration vector.

b. In a centrifugal field, grad $\psi = -\omega^2 \mathbf{r}$, where T is the angular velocity and $\mathbf{r}$ is the distance from the axis of rotation.

c. In an electrostatic field, M_k grad $\psi = z_k \mathscr{F}$ grad ϕ, where M_k is the molecular weight of component k, z_k is its charge number, $\mathscr{F}$ is Faraday's constant, and ϕ is the electric potential.

If the length scales of variation of properties is in the nanometer range, then it is necessary to include a gradient energy term in the description of the thermodynamic state of each volume element. For example, the density of the Helmholtz free energy must be written

$$F' = \int\int\int_{V'} [f_v^0(c) + K(\nabla c)^2] \delta v \tag{14.67}$$

where $f_v(c)$ is the free energy density associated with an element of the same temperature, pressure, and composition in a homogeneous system. Phenomena that occur in fine scale structures cannot be adequately explained without including this gradient energy contribution.

HOMEWORK PROBLEMS

Problem 14.1.   A bar of copper is insulated along its length L. One end is held at 498 K and the other end fixed at 298 K until a steady state temperature profile

$$T(x) = T_2 - \frac{T_2 - T_1}{L} x$$

is reached. The ends of the bar are then insulated from the heat source and sink (the system is isolated) If $L = 2$ cm:

a. Find the equilibrium temperature distribution.

b. Compute the change in entropy for the system during its transition to equilibrium.

Begin by writing an expression for the change in entropy density of the system as a function of position.

Problem 14.2.   Explain why there is no PdV type term in the local formulation of the combined statement of the first and second laws.

Problem 14.3.   Use the conditions for equilibrium in a single-phase system to prove that, in the absence of external fields, the *composition*, X_k, of each component is uniform in the system.

Problem 14.4.   Potential field effects are incorporated into the conditions for equilibrium through Equation 14.31, which expresses one of the constraints that operates in an isolated system. Justify the contention that the energy function,

$E'_{Tot} = U' + E'_{pot}$, is constrained to be constant, rather than just the internal energy, U', in developing the description of an isolated system.

Problem 14.5. The mixing behavior of the A–B system is described by the familiar equation

$$\Delta G_{mix} = a_0 X_A X_B + \Delta G^{id}_{mix}$$

where a_0 is independent of temperature and pressure.
  a. Write an expression for the chemical potential of component B as a function of T, P, and X_B.
  b. Derive an expression for the variation of composition with altitude for this system.
  c. Compare this result with that obtained for an ideal solution.

Problem 14.6. One strategy for separating isotopes of uranium (U^{235} and U^{238}) in the Manhattan Project was to form a dilute solution containing these isotopes and centrifuge it. Examine the thermodynamics of this situation and assess its feasibility.

Problem 14.7. Describe an electrical field that will generate the same pressure distribution as a centrifugal field developed in an ultracentrifuge rotating at 20,000 r/min. State all of your assumptions. (Several essential elements have been left out of the statement of this problem.)

Problem 14.8. Develop an expression for the variation of composition with distance from the surface of a system for a dilute NaCl aqueous solution given that the electrical potential decays exponentially from the surface as in Example 14.4.

Problem 14.9. Consider a nonuniform single-phase system in which the composition varies periodically in one direction:

$$X_2(x) = 0.50 + 0.05 \cos\frac{2\pi x}{\lambda} \qquad 0 \le x \le L$$

where $\lambda = 5$ nm (nanometers). Assume

$$\Delta G_{mix} = \Delta F_{mix} = 12,200 X_A X_B + \Delta G^{id}_{mix}$$

and $V = 7.1$ (cc/mol), independent of composition.
  a. Calculate the Gibbs free energy of formation of this periodic system from an initially uniform system for a cube 1 cm on a side without the gradient energy contribution.
  b. Make the same calculation with the gradient energy assuming $\kappa = 6 \times 10^{-11}$ J/cm.
  c. Compare the contribution of the gradient energy to the total energy of mixing.

REFERENCES

1. Haase, R., *Thermodynamics of Irreversible Processes,* Addison-Wesley, Reading, MA, pp. 64–69, 1969.
2. Cahn, J.W. and Hilliard, J.E., *J. Chem. Phys.*, 28, 258, 1958.
3. Cahn, J.W. and Hilliard, J.E., *J. Chem. Phys.*, 31, 688, 1959.
4. Hilliard, J.E., Spinodal decomposition, in *Phase Transformations*, Aaronson, H.I., Ed., ASM, Materials Park, OH, p. 497, 1970.
5. Philofsky, E.M. and Hilliard, J.E., *J. Appl. Phys.*, 40, 2198, 1969.

CONTENTS

Many of the processes that sustain life combine chemical change with electrical exchanges. This complex combination of influences may degrade materials in corrosion, or may be used to refine them, as in the electrolysis of aluminum, titanium, copper, and other metals, and in the preparation of ceramic powders and polymer blends. The subset of science that deals with phenomena that combine chemical and electrical effects, electrochemistry, is exceedingly broad.[1-3] In this introductory text the focus is limited to the description of the conditions for equilibrium in such systems and applications derived from these conditions.

The ionization energy of an element is the difference in energy between an atom of that element in its vapor state and an ion of the same element. When that element is present as a component in a solution its ionization energy may be greatly reduced as a result of its interactions with the surrounding solvent molecules. A salt dissolved in water dissociates to a greater or lesser extent to form cations and anions. The resulting solution is called an electrolyte because it is found that a current of charge is transported through the solution by the ions when it is subject to an electric field.[1]

---

[1] The term electrolyte is also used to describe the compound that forms ions in solution. NaCl is an electrolyte; a solution of NaCl in water is also called an electrolyte.

Water is the most familiar and important solvent for electrolytes; life itself depends upon this property of water. However, many other so-called polar solvents, such as acetone or glycerol, also form electrolyte solutions. Solid electrolytes, in which a charge current is carried by diffusion of the ions in the system, also exist and find useful applications. The conditions for equilibrium in an electrolyte solution are derived in Section 15.1.

In addition to the intensive properties dealt with in earlier chapters, every phase has an internal electric potential. If no external electrostatic field is applied, this potential is uniform within the phase at equilibrium. In considering the equilibrium between phases, it is not necessary to consider their differing electric potentials unless charged particles are transferred between the phases in the process being considered. Section 15.2 presents the conditions for equilibrium between phases and electrolytes that may communicate electrical charges.

Absolute values of the electric potentials of phases cannot be measured nor can the potential difference between two phases. Since the measurement of an electrical potential requires a closed circuit, the minimum configuration of system that may be devised in which an electric potential may be measured is the galvanic cell. Such a system consists of at least four phases: two electrodes inserted into an electrolyte and externally connected by a wire (which also provides connections to the electric measuring devices that monitor the system), Figure 15.1. The properties of galvanic cells and associated conditions for equilibrium are presented in Section 15.3.

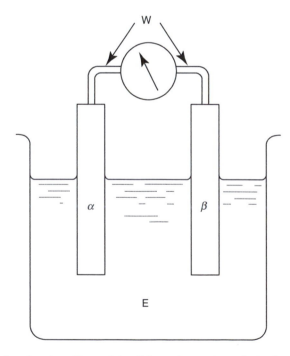

FIGURE 15.1 A galvanic cell consists of four phases: two electrodes (α) and (β), an electrolyte (E), and the external wire (W) connecting them with a measuring apparatus.

Practical visualization of the complexities of the conditions for equilibrium in cells is conveniently represented in potential-pH predominance diagrams, commonly called Pourbaix diagrams after their inventor. An example of such an equilibrium map was shown in Figure 1.9. The strategy for deriving and interpreting Pourbaix diagrams is laid out in Section 15.4, together with applications to corrosion and related phenomena.

15.1 EQUILIBRIUM WITHIN AN ELECTROLYTE SOLUTION

The compound M_uX_v is an electrolyte if it dissociates, either partially or completely, into its cations and anions when dissolved in water or some other appropriate solvent. This compound may be an **acid** (in which case M is hydrogen), a base (X is the hydroxyl group OH) or a **salt** (M is usually a metal and X a nonmetal as in NaCl). Depending upon the level of detail required in a given application, this solution may be considered to consist of two components: H_2O and M_uX_v; or alternatively, six components: undissociated solvent (H_2O) and its associated ions, H^+ [2] and OH^-, and undissociated compound (M_uX_v) and its ions, M^{z+} and X^{z-}. Complexes involving other combinations of these elements, e.g., $MX_2O_4^z$, may also exist in the system and thereby further increase the number of components that may be considered in its behavior.

It is awkward to designate the chemical potentials, number of moles, molar concentrations, activities and activity coefficients of all of these components with self-explanatory subscripts because these designations themselves have superscripts. For example, M^{z+} describes the cation obtained from the compound electrolyte; the number of moles of this ionic component could be written $n_{M^{z+}}$. In the descriptions that follow the notation can be simplified without loss of clarity with the designations shown in Table 15.1.

The solution is formed by adding n_c moles of the compound M_uX_v to n_w moles of water. In the solution the compound dissociates to form n_+ moles of the cation M^{z+} and n_- moles of the anion X^{z-}, leaving n_U moles of the compound undissociated. Introduce the concept of the degree of dissociation of the electrolyte, α, such that αn_c is the number of moles of the electrolyte that dissociates into ions. Since each molecule of the compound M_uX_v contains u cations and v anions, $u(\alpha n_c) = n_+$ moles of the cation M^{z+} and $v(\alpha n_c) = n_-$ moles of the anion, X^{z-}, are formed. The number of moles of undissociated electrolyte in the solution, $n_U = (1 - \alpha)n_c$. This latter relation emphasizes the distinction between n_c, the total number of moles of the compound added to the system and n_U, the number of moles of the compound that remain undissociated in the solution after it comes to equilibrium.

---

[2] Since the hydrogen atom consists of a single proton and an electron, the hydrogen ion, H^+, is a proton. Experimental studies of kinetic phenomena demonstrate that free protons do not generally exist in aqueous solution but are associated with a water molecule, forming the hydronium ion, H_3O^+. However, this distinction, while critical for explaining kinetic phenomena, is not important in rationalizing thermodynamic behavior. Accordingly, in this chapter, H^+ will be used to designate the positive ion derived from the dissociation of H_2O.

TABLE 15.1
Notation Used to Designate the Number of Moles
of Components in Electrolytes

Component	Symbol
Preparation of the solution	
Numbers of moles of water added	n_w
Number of moles of compound M_uX_v added	n_c
Components in the solution	
Number of moles of undissociated M_uX_v	n_u
Number of moles of cation formed	n^+
Number of moles of anion formed	n^-

There exists a class of compounds called strong electrolytes which dissociate completely into ions in the solution, i.e., for which $\alpha = 1$ so that n_U is zero. **Weak electrolytes** are partially dissociated, with α values ranging between 0 and 1. The behavior of strong electrolytes and weak electrolytes is discretely different and is treated separately in the thermodynamics of electrolytes.

15.1.1 EQUILIBRIUM IN WEAK ELECTROLYTES

The change in Gibbs free energy associated with a change in composition at constant temperature and pressure of an electrolyte solution may be written in either of two forms:

$$dG'_{T,P} = \mu_W dn_W + \mu_C dn_C \tag{15.1}$$

$$dG'_{T,P} = \mu_{H_2O}\, dn_{H_2O} + \mu_{H^+}\, dn_{H^+} + \mu_{OH^-}\, dn_{OH^-} + \mu_U\, dn_U + \mu_+\, dn_+ + \mu_-\, dn_- \tag{15.2}$$

A chemical potential may be defined for each of the components. For example,

$$\mu_+ \equiv \left(\frac{\partial G'}{\partial n_+}\right)_{T,P,n_{H^+},n_{OH^-},n_U,n_-} \tag{15.3}$$

Guggenheim[4] has observed that the chemical potential of ionic components is of academic interest only since it is not possible to concoct a process in which one ionic component, e.g., n_+ is added to a system while keeping the number of moles of the other ionic components constant; such a process would violate the condition of charge neutrality. In practice, the composition of the electrolyte may be changed only by adding the compound, M_uX_v (increasing n_c) or the solvent, H_2O (increasing n_w). Thus, the chemical potentials (and the related activities and activity coefficients introduced below), of the ionic components are conceptually useful, but cannot be measured.

The activity corresponding to the chemical potential defined in Equation 15.3 is given by:

$$\mu_+ = \mu_+^0 + RT \ln a_+ \tag{15.4}$$

where μ_+^0 is the chemical potential of the cation in its standard state. An activity coefficient for the ionic component, γ_+, may be defined in the usual way:

$$a_+ \equiv \gamma_+ X_+ \tag{15.5}$$

Similar definitions may be applied to the other ionic components.

In the description of the composition of electrolytes the molar concentration or molarity of a component is usually used in the literature in place of the mole fraction. The molarity of a component is defined to be the number of moles of the component per liter of solution. The molarity of component k is usually written as [k]; it is identical to the molar concentration, c_k, defined and used in Chapter 14. The molarity of component k is simply related to its mole fraction:

$$c_k = [k] = \frac{X_k}{V} \tag{15.6}$$

where V is the molar volume of the solution. This quantity is a convenient measure of composition because it is easy to prepare a liquid solution of a desired molarity in the laboratory. A known number of moles of solute is weighed and added to the solvent and dissolved. Solvent is then added to bring the total volume of the solution to one liter, which then has the desired number of moles of solute per liter of solution.

In this context it is useful to introduce an alternate definition of the activity coefficient such that

$$a_k = f_k c_k = f_k [k] \tag{15.7}$$

Note that, since a_k is unitless, f_k must have units of (liters/mole of k). Evidently,

$$f_k = \gamma_k V \tag{15.8}$$

The strategy for finding conditions for equilibrium within an electrolyte yields the usual conditions for thermal and mechanical equilibrium. Applying the isolation constraints that conserve H, O, M, and X atoms yields combinations of the chemical potentials of the components that arrange themselves into affinities, each implying its corresponding reaction:

$$A_w = (\mu_{H^+} + \mu_{OH^-}) - \mu_{H_2O} \tag{15.9}$$

$$H_2O = H^+ + OH^- \tag{15.1}$$

$$A_c = (u\mu_+ + v\mu_-) - \mu_U \qquad (15.10)$$

$$M_uX_v = uM^{z+} + vX^{z-} \qquad [15.2]$$

Additional independent equations will appear if complex ionic components are assumed to be present.

At equilibrium these affinities are each separately equal to zero. Substitution of the definitions of the corresponding activities and standard free energy changes leads to the familiar working equations for chemical equilibrium:

$$\Delta G_W^0 = -RT \ln K_W \quad \text{where} \quad K_W = \frac{a_{H^+}\, a_{OH^-}}{a_{H_2O}} \qquad (15.11)$$

The equilibrium constants, K_w and K_c are called dissociation constants in this application. The ΔG^0 terms are differences in free energies between reactants and products when they are in their reference states.

The equilibrium constant for the dissociation reaction 15.2 may be expressed in terms of the activity coefficients and molarities:

$$K_C = \frac{a_+^u a_-^v}{a_U} = \frac{f_+^u}{f_U} \frac{[M^{z+}]^u [X^{z-}]^v}{[M_uX_v]}$$

$$K_C = K_f \frac{[M^{z+}]^u [X^{z-}]^v}{[M_uX_v]} \qquad (15.12)$$

Reference states for the components in an electrolyte solution are chosen so that in dilute solutions the activity coefficients, f_k, approach 1. Values of the dissociation constants that are tabulated are based upon this choice of reference state. Thus, the ratio of activity coefficients, K_f, is approximately 1 for dilute solutions and the dissociation constant is expressible in terms of the concentrations:

$$K_C \cong \frac{[M^{z+}]^u [X^{z-}]^v}{[M_uX_v]} \qquad (15.13)$$

This assumption, i.e., that $K_f = 1$, is made in most practical applications and will be made in the applications presented in this chapter. Evidently, use of this assumption in concentrated solutions may lead to significant errors.

The concentration ratio in Equation 15.13 may be expressed in terms of c_c, the total concentration of electrolyte compound added to make the solution and the degree of dissociation, α.

$$[M^{z+}] = u\alpha c_C; \quad [X^{z-}] = v\alpha c_C : \quad [M_uX - v] = (1 - \alpha)c_C \qquad (15.14)$$

$$K_C = \frac{(u\alpha c_C)^u (v\alpha c_C)^v}{(1 - \alpha)c_C} = u^u v^v \frac{\alpha^{(u+v)} c_C^{(u+v-1)}}{(1 - \alpha)} \qquad (15.15)$$

Let $\nu = u + v$ be the total number of ions formed when a molecule of $M_u X_v$ dissociates. Equation 15.15 may be written:

$$K_C = (u^u v^v) \frac{\alpha^\nu c_C^{\nu-1}}{(1-\alpha)} \tag{15.16}$$

For example, for the dissociation of $CaCl_2$, $u = 1$, $v = 2$, and $\nu = 3$:

$$K_{CaCl_2} = (1^1 \cdot 2^2) \frac{\alpha^3}{1-\alpha} c_{CaCl_2}^{(3-1)} = 4 \frac{\alpha^3}{1-\alpha} c_{CaCl_2}^2 \tag{15.17}$$

With a value of the dissociation constant, for a given total concentration of electrolyte the value of α may be computed and the equilibrium concentrations of cations, anions, and undissociated electrolyte may be determined with Equation 15.14.

For many dissociation reactions K_c is many orders of magnitude smaller than one and over a broad concentration range the resulting degree of dissociation $\alpha \ll 1$. In this case $(1-\alpha)$ may be replaced safely by 1 and Equation 15.16 takes on a simpler form:

$$K_C \cong (u^u v^v) \alpha^\nu c_C^{\nu-1} \tag{15.18}$$

This equation may be solved explicitly for the degree of dissociation in terms of the concentration of electrolyte added to the solution:

$$\alpha = \left(\frac{K_C}{u^u v^v}\right)^{\frac{1}{\nu}} \cdot c_C^{\left(\frac{1}{\nu}-1\right)} \tag{15.19}$$

Thus, for most weak electrolytes the degree of dissociation increases as the electrolyte concentration decreases.

EXAMPLE 15.1

The dissociation constant of pure water at 25 °C is very close to $K_w = 10^{-14}$. Compute the hydrogen and hydroxyl concentrations in one liter of water at 25 °C. For the dissociation of water,

$$H_2O = H^+ + OH^- \qquad K_w = 10^{-14}$$

Since water is the only component, the reaction requires that equal numbers of H^+ and OH^- ions be formed: $[H^+] = [OH^-]$. The concentration of undissociated H_2O will be negligibly different from 1. Assuming K_f for this reaction is 1, the dissociation constant may be written:

$$K_w = 10^{-14} = \frac{[H^+][OH^-]}{[H_2O]} = \frac{[H^+][H^+]}{1} = [H^+]^2$$

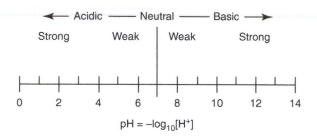

FIGURE 15.2 The scale of pH values in electrolyte solutions.

so that

$$[H^+] = 10^{-7}\left(\frac{moles}{liter}\right)$$

Thus, in pure water the activities of hydrogen and hydroxyl ions are equal to their concentrations, which are each about 10^{-7} moles per liter.

One of the most important characteristics of aqueous solutions is the hydrogen ion concentration contained. Since values of this quantity are found to vary over a range of several orders of magnitude in ordinary solutions, a convenient measure of this quantity has been defined as the pH of an electrolyte solution:[3]

$$pH \equiv -\log_{10}(a_{H^+}) = -\log_{10}[H^+] \qquad (15.20)$$

The value of the pH for pure water is $-\log(10^{-7}) = -(-7) = +7$. The hydrogen ion activity is higher than 7 in acids, which contribute hydrogen ions to the solution, and lower than 7 in basic solutions, which contribute hydroxyl ions. (Since H_2O is the solvent, $a_{H_2O} = 1$; thus, $a_{H^+}a_{OH^-} = K_w = 10^{-14}$. A high activity of OH^- implies a low H^+ activity and vice versa.) The scale of pH values normally obtained in the laboratory is illustrated in Figure 15.2. The pH ranges from 2 for strong acids to 12 for strong base solutions. Since in most applications the concentrations of H^+ and OH^- are in the dilute range, the activities may be replaced by the molar concentrations.

EXAMPLE 15.2
Compute the pH of a solution containing 0.05 M concentration of acetic acid, HAc (where Ac is shorthand for CH_3COO^-). The dissociation constant for acetic acid at 25 °C is 1.8×10^{-5}.
The dissociation reaction for acetic acid is:

$$HAc = H^+ + Ac^- \qquad [15.3]$$

---

[3] Since the chemical potential of hydrogen is proportional to its activity at a fixed temperature, pH represents the potential of hydrogen in the electrolyte. Since the activity of H^+ is normally a small fraction, its logarithm is negative: inclusion of the minus sign in the definition makes the pH a positive number for the dilute solutions of H^+ ions ordinarily encountered.

In writing the relation between the dissociation constant and the degree of dissociation, Equation 15.15, set $u = v = 1$:

$$K_{HAc} = (1^1 \cdot 1^1) \frac{\alpha^2}{1 - \alpha} c_{HAc}$$

Substitute for K_{HAc} and c_c from the information given:

$$1.8 \times 10^{-5} = \frac{\alpha^2}{1 - \alpha}(0.05)$$

Solve for the degree of dissociation:

$$\alpha = 0.019$$

The hydrogen ion concentration is

$$[H^+] = \alpha[HAc] = 0.019(0.05) = 9.4 \times 10^{-4}$$

Set this concentration equal to the activity in the definition of pH, Equation 15.20:

$$pH = -\log_{10}(9.4 \times 10^{-4}) = 3.0$$

In a multicomponent electrolyte containing a variety of cations and anions, the conditions for equilibrium may be written

$$\Delta G_W^0 = -RT \ln K_W \tag{15.21}$$

$$\Delta G_r^0 = -RT \ln K_r \quad (r = 1, 2, ...) \tag{15.22}$$

where Equation 15.21 applies to the solvent and Equation 15.22 to all of the other ionic components. Each of the latter equilibrium constants may be written

$$K_r = K_{f,r} \cdot K_{c,r} = K_{f,r} \frac{[+]_r^u [-]_r^v}{[U]_r} \quad (r = 1, 2, ...) \tag{15.23}$$

where $K_{f,r}$ is the proper quotient of activity coefficients and $K_{c,r}$ is the corresponding ratio of the concentrations. Values for the dissociation constants are tabulated for many electrolyte compounds including salts, acids, and bases; examples of some common compounds are given in Table 15.2.

The problem of the determination of the equilibrium concentrations of all of the ions in the system is mathematically identical to that treated in Section 11.1.2 in Chapter 11 for chemical reactions in a multicomponent ideal gas mixture. Indeed, provided the dissociation constants are known and the $K_{f,r}$ terms are taken to be unity, the same computer program, SOLGASMIX, that provides solutions for a multivariate reacting mixture of gases will compute the concentrations of ions in an equilibrated multicomponent electrolyte solution.

Example 15.3

1.5 g of phosphoric acid (H_3PO_4) is added to a beaker; water is added to fill the beaker to the 200 ml level. Compute the concentrations of all the components of this solution at equilibrium.

Similar to most acids that contain more than one hydrogen atom, phosphoric acid dissociates in stages, one hydrogen atom at a time. Thus, the components existing in this system may be considered to be: H^+, OH^-, H_2O, H_3PO_4, $H_2PO_3^-$, HPO_4^{2-} and

TABLE 15.2
Dissociation Constant for Common Inorganic Acids and Bases

Compound	Dissociating into	K
NH_4OH	$NH_4 + OH^-$	1.79×10^{-5}
$Ca(OH)_2$	$Ca(OH)^+ + OH^-$	3.74×10^{-3}
$Ca(OH)^+$	$Ca^{++} + OH^-$	4.0×10^{-2}
$AgOH$	$Ag^+ + OH^-$	1.1×10^{-4}
H_2CO_3	$H^+ + HCO_3^-$	4.3×10^{-7}
HCO_3^-	$H^+ + CO_3^-$	5.61×10^{-11}
HI	$H^+ + I^-$	1.69×10^{-1}
CH_2O_2	$H^+ + CHO_2^-$	1.77×10^{-4}

Source: Weast, R.C., Ed., *Handbook of Chemistry and Physics*, 57th ed., CRC Press, Boca Raton, FL, pp. D-149 and D-151, 1976.

PO_4^{3-}. Conservation equations for the elements H, O, and P yield three equations among their concentrations. The four independent reaction equations required to provide seven relations among the concentrations of the seven components are presented with their dissociation constants.

$$H_2O = H^+ + OH^- \qquad K_W = 1 \times 10^{-14} = \frac{[H^+][OH^-]}{[H_2O]}$$

$$H_3PO_4 = H^+ + H_2PO_4^- \qquad K_{a1} = 7.1 \times 10^{-3} = \frac{[H^+][H_2PO_4^-]}{[H_3PO_4]}$$

$$H_2PO_4^- = H^+ + HPO_4^{2-} \qquad K_{a2} = 6.3 \times 10^{-8} = \frac{[H^+][HPO_4^{2-}]}{[H_2PO_4^-]}$$

$$HPO_4^{2-} = H^+ + PO_4^{3-} \qquad K_{a3} = 4.7 \times 10^{-13} = \frac{[H^+][PO_4^{3-}]}{[HPO_4^{2-}]}$$

The atomic weights of the elements involved are: $H = 1$; $O = 16$; and $P = 31$ g. The molecular weight of H_3PO_4 is thus $3(1) + 31 + 4(16) = 98$ g. 1.5 g of H_3PO_4 correspond to $1.5/98 = 0.015$ mol. 0.015 mol dissolved to form 200 ml of solution corresponds to a concentration of 0.015 mol/$(0.2 \, l) = 0.077$ mol/l. Since this solution is very dilute, the 200 ml of solution corresponds closely to 200 g of water. (Recall that the specific gravity of water is 1 g/cc.) The molecular weight of H_2O is $2(1) + 16 = 18$ g/mol. The number of moles of water in the solution is 200 g/18 g/mol $= 11.11$ mol. The number of gram atoms of each of the elements in the system may be computed from the initial compositions:

$$m_H = 2n_{H_2O} + 3n_{H_3PO_4} = 2(11.11) + 3(0.015) = 22.27 \text{ gm atoms of } H$$

$$m_O = n_{H_2O} + 4n_{H_3PO_4} = 11.11 + 4(0.015) = 11.17 \text{ gm atoms of O}$$

$$m_P = n_{H_3PO_4} = 0.015 \text{ gm atoms of P}$$

The corresponding molar concentrations of each of the elements may be obtained by dividing by the volume of the solution. The conservation of the elements may be expressed in terms of molar concentrations:

$$\frac{m_H}{V} = \frac{22.27 \text{ gm atoms}}{0.2 \text{ l}} = 111.4\left(\frac{\text{gm atoms}}{\text{liter}}\right)$$

$$= 2[H_2O] + [H^+] + [OH^-] + 3[H_3PO_4] + 2[H_2PO_4^-] + [HPO_4^{2-}]$$

$$\frac{m_O}{V} = \frac{11.17 \text{ gm atoms}}{0.2 \text{ l}} = 55.85\left(\frac{\text{gm atoms}}{\text{liter}}\right)$$

$$= [H_2O] + [OH^-] + 4([H_3PO_4] + [H_2PO_4^-] + [HPO_4^{2-}] + [PO_4^{3-}])$$

$$\frac{m_P}{V} = \frac{0.015 \text{ gm atoms}}{0.2 \text{ l}} = 0.075\left(\frac{\text{gm atoms}}{\text{liter}}\right)$$

$$= [H_3PO_4] + [H_2PO_4^-] + [HPO_4^{2-}] + [PO_4^{3-}]$$

These three conservation equations combine with the four relations derived from the dissociations constants to provide seven equations in seven unknowns. The equilibrium composition is the solution to these equations, obtained by using a solver program in a math applications software package.

15.1.2 EQUILIBRIUM IN A STRONG ELECTROLYTE

By definition a strong electrolyte is a compound that dissociates completely into its ionic components when added to a solvent. Thus, n_U, the number of moles of undissociated molecules in the solution is zero and the degree of dissociation, α, is one for a strong electrolyte. Some compounds are essentially completely dissociated (are strong electrolytes) in dilute solutions but become weak electrolytes with increasing concentration; indeed, this behavior is true of essentially all electrolytes at sufficiently high concentrations.

The thermodynamic behavior of weak electrolytes is based upon the condition for equilibrium expressed in terms of dissociation constants, Equation 15.12 or Equation 15.15. These equations may be used to compute the degree of dissociation and then the pH and other properties of the system. For strong electrolytes it is not necessary to invoke this calculation since, by definition, the degree of dissociation, $\alpha = 1$. Thus, if a strong acid such as HCl is dissolved in a known quantity of water, the concentration of H^+ ions is evident by inspection, being equal to the concentration of HCl added; every HCl molecule dissociates and contributes one hydrogen ion.

EXAMPLE 15.4

Compute the pH of a 0.08 M solution of HCl in water.

The number of moles of HCl added per liter of solution is 0.08. Since every HCl molecule dissociates forming one hydrogen ion, the concentration of H^+ resulting is 0.08 mol/l. The corresponding pH of this solution is

$$pH = -\log_{10}(0.08) = 1.09$$

EXAMPLE 15.5

Evaluate the pH of a solution formed by mixing equal quantities of 0.04 M HCl with 0.06 M NaOH, both of which are strong electrolytes.

The [H^+] contributed from HCl is 0.04 mol/l of HCl added, the sodium hydroxide contributes 0.06 mol of [OH^-] per liter of NaOH added. Each of these is diluted by a factor of two when the two solutions are mixed so that the concentrations contributed are, respectively, 0.02 and 0.03 mol/l. Essentially all 0.02 mol/l of the hydrogen ions contributed will react with 0.02 mol/l of hydroxyl ions to form 0.02 mol/l of water. The unreacted concentration of OH^- ions (0.03 − 0.02 = 0.01 mol/l) remain in solution. The corresponding hydrogen ion concentration may be determined from the dissociation constant for water:

$$K_W = 10^{-14} = [H^+][OH^-] = [H^+][0.01]$$

Solve for

$$[H^+] = 10^{-14}(0.01) = 10^{-12}$$

The corresponding pH is

$$pH = -\log_{10}(10^{-12}) = 12$$

which represents a strongly basic solution.

Manipulation of compositions of solutions involving strong electrolytes is evidently significantly simplified in comparison with the analysis of weak electrolytes.

15.2 EQUILIBRIUM IN TWO-PHASE SYSTEMS INVOLVING AN ELECTROLYTE

The general strategy for finding conditions for equilibrium may be applied to a system consisting of two phases that are capable of exchanging electrical charge. A simple system of this class is shown in Figure 15.3. A copper rod is immersed in an aqueous solution of copper chloride, $CuCl_2$. The copper rod is called an electrode and the solution an electrolyte.

If the concentration of $CuCl_2$ in the electrolyte is dilute, copper atoms will dissolve from the rod and form cupric ions in the solution. Copper atoms are said to be oxidized to cupric ions. For each copper atom dissolved two electrons are released and remain in the metal rod. The rod becomes negatively charged as electrons accumulate while the solution develops a positive charge. Both of these accumulations alter the total energy of the system by changing the electrostatic potential energy of the phases involved.

On the other hand if the electrolyte is sufficiently concentrated, cupric ions will plate out on the copper rod, each ion capturing two electrons from the metal. Cupric ions in the solution are reduced to copper atoms on the metal surface. The number of free electrons in the copper rod is reduced and it develops a positive charge. In the solution the number of cupric ions is reduced relative to the negatively charged chloride ions and the solution develops a negative charge. Both effects alter the electrical potential energy of the two-phase system.

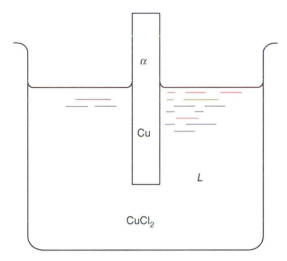

FIGURE 15.3 A copper rod (the α phase) is immersed in a solution of copper chloride in water (the, ε phase).

These exchanges of atoms and charges may be succinctly represented by the reaction

$$Cu^\alpha = Cu^{L2+} + 2e^{\alpha-} \qquad [15.4]$$

If this reaction proceeds in the direction as written it is an oxidation reaction; the reverse is a reduction reaction.

Eventually, in each phase after sufficient charge is transferred an equilibrium condition is attained. The general strategy for finding conditions for equilibrium may be applied to determine the relations that govern this equilibrium state.

Figure 15.4 shows a rod of the metal M (the α phase) immersed in an aqueous solution (the ε phase) containing the electrolyte M_uX_v. Let z^+ be the valence of the cations (M^{z+}) and z^- that of the anions (X^{z-}) in the electrolyte. The components in the α phase are atoms of the metal, M, and electrons, e. Components in the electrolyte are M^{z+}, X^{z-}, M_uX_v, H^+, OH^- and H_2O. Let φ^α be the electrical potential of the metallic phase (defined as in Chapter 14 as the potential energy per unit charge) and φ^ε that of the electrolyte solution.[4] The exchange of charged particles between the α and ε phases contributes to the change in the total energy of the system and must be incorporated into its thermodynamic description.

Equilibria established within the electrolyte will be determined by application of the conditions for equilibrium within an electrolyte derived in Section 15.2. The dissociation constants of the electrolyte and of water establish the relationships

---

[4] φ^α may be thought of as the change in potential energy when unit charge is transferred from infinity to the α phase. φ^ε may be similarly defined for the ε phase. It is not practical to determine the absolute values of these potentials; only differences in potentials appear in the working equations.

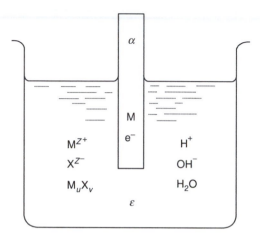

FIGURE 15.4 A metal rod M (the α phase) is immersed in a solution of the electrolyte M_uX_v in water (the ε, phase).

between the electrolyte components. The following development focuses upon the equilibrium established between the electrode and the electrolyte. The components involved in this exchange are the metal M and the electrons e^- in the electrode and the metal ions M^{z+} in the electrolyte.

The change in entropy of the metallic phase α that constitutes the electrode is

$$dS'^{\alpha}_{sys} = \frac{1}{T^{\alpha}}dU'^{\alpha} + \frac{P^{\alpha}}{T^{\alpha}}dV'^{\alpha} - \frac{1}{T^{\alpha}}[\mu^{\alpha}_{M}dn^{\alpha}_{M} - \mu^{\alpha}_{e}dn^{\alpha}_{e}] \tag{15.24}$$

In the electrolyte phase

$$dS'^{\varepsilon}_{sys} = \frac{1}{T^{\varepsilon}}dU'^{\varepsilon} + \frac{P^{\varepsilon}}{T^{\varepsilon}}dV'^{\varepsilon} - \frac{1}{T^{\varepsilon}}\mu^{\varepsilon}_{M^{z+}}dn^{\varepsilon}_{M^{z+}} \tag{15.25}$$

The change in entropy of the system is the sum of these two expressions.

$$dS'_{sys} = dS'^{\alpha} + dS'^{\varepsilon}$$

The isolation constraints must be carefully and completely formulated for this case. Conservation of the element M may be simply stated:

$$dm_M = 0 = dn^{\alpha}_{M} + dn^{\varepsilon}_{M^{z+}} \Rightarrow dn^{\varepsilon}_{M^{z+}} = -dn^{\alpha}_{M} \tag{15.26}$$

Each electron that is added to the α phase adds a charge of (-1) e, where e is the magnitude of the charge on an electron. If dn^{α}_{e} moles of electrons are added to the α phase, the change in the charge contained in that phase is

$$dq^{\alpha} = (-1)\mathscr{F}\,dn^{\alpha}_{e} = -\mathscr{F}\,dn^{\alpha}_{e} \tag{15.27}$$

where $\mathscr{F} = N_0 e$ is Faraday's constant (see Section 14.3.5 in Chapter 14). Each metal ion M^{z+} carries a charge of z^+. If $dn^\varepsilon_{M^{z+}}$ moles of these ions are added to the electrolyte, the associated change in charge on the ε phase is

$$dq^\varepsilon = (+z^+)\mathscr{F}\, dn^\varepsilon_{M^{z+}} \tag{15.28}$$

The total charge accumulated in the two-phase system must be zero:

$$dq_{Tot} = 0 = dq^\alpha + dq^\varepsilon = -\mathscr{F}\, dn^\alpha_e + (z^+)\mathscr{F} dn^\varepsilon_{M^{z+}} \tag{15.29}$$

So that

$$dn^\alpha_e = (z^+)dn^\varepsilon_{M^{z+}} \tag{15.30}$$

From Equation 15.26,

$$dn^\alpha_e = z^+(-dn^\alpha_M) = -z^+ dn^\alpha_M \tag{15.31}$$

Equation 15.26 and Equation 15.31 permit expression of the three compositional variables in Equation 15.24 and Equation 15.25 in terms of dn^α_M.

Additional isolation constraints derive from the condition of constant volume in an isolated system:

$$dV'_{sys} = 0 = dV'^\alpha + dV'^\varepsilon \Rightarrow dV'^\varepsilon = -dV'^\alpha \tag{15.32}$$

and the requirement that the total energy of an isolated system cannot change. In formulating the total energy constraint it is necessary to include changes in the electrical energies of the phases due to the exchange of charged particles.

$$dE'_{Tot} = dU'^\alpha + dU'^\varepsilon + \varphi^\alpha dq^\alpha + \varphi^\varepsilon dq^\varepsilon \tag{15.33}$$

Equation 15.27 and Equation 15.28 may be used to evaluate dq^α and dq^e:

$$dE'_{Tot} = dU'^\alpha + dU'^\varepsilon + \varphi^\alpha(-\mathscr{F}\, dn^\alpha_e) + \varphi^\varepsilon(z^+\mathscr{F}\, dn^\varepsilon_{M^{z+}}).$$

Both dn^α_e and dn^ε_{M+} may be expressed in terms of dn^α_M, see Equation 15.26 and Equation 15.31:

$$dE'_{Tot} = dU'^\alpha + dU'^\varepsilon + z^+\mathscr{F}(\varphi^\alpha - \varphi^\varepsilon)dn^\alpha_M$$

In an isolated system, $dE'_{Tot} = 0$; this isolation constraint therefore requires that

$$dU'^\varepsilon = -[(dU'^\alpha + z^+\mathscr{F}(\varphi^\alpha - \varphi^\varepsilon)dn^\alpha_M)] \tag{15.34}$$

The change in the total entropy of the system obtained by adding expressions for dS' in Equation 15.24 and Equation 15.25 is written in terms of seven variables. The isolation constraints provide four relations among these seven variables. Use Equation 15.26 and Equation 15.31 to replace dn^α_e and $dn^\varepsilon_{M^{z+}}$ with dn^α_M.

Use Equation 15.32 to eliminate dV'^{ε} and Equation 15.34 to substitute for dU'^{ε}.

$$dS'_{\text{sys,iso}} = \frac{1}{T^{\alpha}} dU'^{\alpha} + \frac{P^{\alpha}}{T^{\alpha}} dV'^{\alpha} - \frac{1}{T^{\alpha}} [\mu_M^{\alpha} dn_M^{\alpha} + \mu_e^{\alpha}(-z^+ dn_M^{\alpha})]$$
$$+ \frac{1}{T^{\varepsilon}}(-1)[dU'^{\alpha} + z^+ \mathscr{F}(\varphi^{\alpha} - \varphi^{\varepsilon}) dn_M^{\alpha}] + \frac{P^{\varepsilon}}{T^{\varepsilon}}(-dV^{\alpha})$$
$$- \frac{1}{T^{\varepsilon}} \mu_{M^{z+}}^{\varepsilon}(-dn_M^{\alpha})$$

Group like terms:

$$dS'_{\text{sys,iso}} = \left(\frac{1}{T^{\alpha}} - \frac{1}{T^{\varepsilon}}\right) dU'^{\alpha} + \left(\frac{P^{\alpha}}{T^{\alpha}} - \frac{P^{\varepsilon}}{T^{\varepsilon}}\right) dV'^{\alpha}$$
$$- \frac{1}{T^{\alpha}} [\mu_M^{\alpha} - z^+ \mu_e^{\alpha} - \mu_{M^{z+}}^{\alpha} + z^+ \mathscr{F}(\varphi^{\alpha} - \varphi^{\varepsilon})] \, dn_M^{\alpha} \qquad (15.35)$$

At equilibrium all of the coefficients in this equation are zero. This condition for the extremum in entropy in an isolated system yields the following conditions for equilibrium:

$$\frac{1}{T^{\alpha}} - \frac{1}{T^{\varepsilon}} = 0 \Rightarrow T^{\alpha} = T^{\varepsilon} \qquad \text{[Thermal Equilibrium]} \qquad (15.36)$$

$$\frac{P^{\alpha}}{T^{\alpha}} - \frac{P^{\varepsilon}}{T^{\varepsilon}} = 0 \Rightarrow P^{\alpha} = P^{\varepsilon} \qquad \text{[Mechanical Equilibrium]} \qquad (15.37)$$

$$\mu_M^{\alpha} - (z^+ \mu_e^{\alpha} + \mu_{M^{z+}}^{\varepsilon}) + z^+ \mathscr{F}(\varphi^{\alpha} - \varphi^{\varepsilon}) = 0 \quad \text{[Electrochemical Equilibrium]}$$
$$(15.38)$$

These relations are familiar except Equation 15.38, which is the condition for electrochemical equilibrium in the system. Rewrite this expression as

$$\mu_M^{\alpha} - (z^+ \mu_e^{\alpha} + \mu_{M^{z+}}^{\varepsilon}) = -z^+ \mathscr{F}(\varphi^{\alpha} - \varphi^{\varepsilon}) \qquad (15.39)$$

The left side of this equation may be recognized as the affinity for the reduction reaction

$$(M^{z+})^{\varepsilon} + z^+ e^- = M^{\alpha} \qquad [15.5]$$

Evidently equilibrium is reached when the affinity for this reaction equals the change in potential energy of the system associated with transferring z^+ gram atoms of positive charge from the metal electrode to the electrolyte.

Unfortunately, the potential difference between two phases in equilibrium cannot be measured. A probe inserted into the electrolyte and connected through external instrumentation to the electrode in order to measure this potential would itself constitute an electrode/electrolyte system with its own equilibrium conditions. Such a system with two electrodes immersed in an electrolyte is called an electrochemical cell. It is the simplest configuration that can be assembled to explore the relation between electrical and chemical effects in such systems.

15.3   EQUILIBRIUM IN AN ELECTROCHEMICAL CELL

Figure 15.5 shows an electrochemical cell consisting of two dissimilar electrodes, α and β, immersed in an appropriate electrolyte solution ε. To make the illustration concrete, suppose one electrode is copper immersed in a $CuCl_2$ solution, and the other is zinc in $ZnCl_2$. Each electrode forms divalent ions in its electrolyte solution. If each electrode reaches equilibrium with its electrolyte separately, as in Figure 15.5a, the condition for equilibrium in each case is given by the appropriate adaptation of Equation 15.39. For the copper electrode (α) the reaction

$$(Cu^{2+})^{\varepsilon} + 2(e^-)^{\alpha} = Cu^{\alpha} \qquad [15.6]$$

The condition for equilibrium for this electrode is:

$$\mu^{\alpha}_{Cu} - (2\mu^{\alpha}_e + \mu^{\varepsilon}_{Cu^{2+}}) = -2\mathscr{F}(\varphi^{\alpha} - \varphi^{\varepsilon}) \qquad (15.40)$$

At the zinc electrode (β) the reaction is

$$(Zn^{2+})^{\varepsilon} + 2(e^-)^{\beta} = Zn^{\beta} \qquad [15.7]$$

and the condition for equilibrium gives

$$\mu^{\beta}_{Zn} - (2\mu^{\beta}_e + \mu^{\varepsilon}_{Zn^{2+}}) = -2\mathscr{F}(\varphi^{\beta} - \varphi^{\varepsilon}) \qquad (15.41)$$

Let the two electrolyte baths be connected by a salt bridge, Figure 15.5b. The salt bridge is a tube filled with an electrolyte that permits transfer of charge between the electrolytes without disturbing the equilibrium between Zn and its electrolyte and Cu and its electrolyte. As a consequence, within the continuous electrolyte phase ε, the potential is uniform with the value φ^{ε}. The chemical potentials of the electrons in the metallic electrodes may also be taken to be the same. Subtraction of Equation 15.40 from Equation 15.41 yields an expression for the potential difference

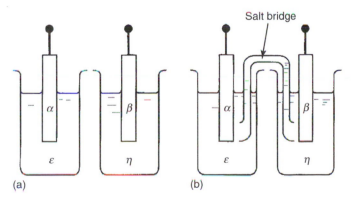

FIGURE 15.5 (a) Two separate single electrode systems achieve their own internal equilibrium. (b) They are then connected in the electrolyte with a salt bridge. The electrical potential difference between the electrodes can now be measured.

between the electrodes.

$$\lfloor\mu_{Zn}^{\beta} - (2\mu_e^{\beta} + \mu_{Zn^{2+}}^{\varepsilon})\rfloor - \lfloor\mu_{Cu}^{\alpha} - (2\mu_e^{\alpha} + \mu_{Cu^{2+}}^{\varepsilon})\rfloor = -2\mathscr{F}(\varphi^{\beta} - \varphi^{\alpha})$$

(15.42)

The first term in brackets on the left side of this equation corresponds to the affinity, $A_{15.7}$, for the electrode reaction 15.7 for the zinc (β) electrode and the second term is the affinity, $A_{15.6}$ for the copper (α) reaction 15.6. Equation 15.42 demonstrates that when a galvanic cell reaches equilibrium the external potential, which can be measured experimentally with a potentiometer or sensitive voltmeter, is determined by the difference in affinities for the two electrode reactions:

$$A^{\beta} - A^{\alpha} = -2\mathscr{F}(\varphi^{\beta} - \varphi^{\alpha})$$

(15.43)

Since $\mu_e^{\alpha} = \mu_e^{\beta}$, the left side of Equation 15.43 may also be written

$$\lfloor\mu_{Zn}^{\beta} - \mu_{Zn^{2+}}^{\varepsilon}\rfloor - \lfloor\mu_{Cu}^{\alpha} - \mu_{Cu^{2+}}^{\varepsilon}\rfloor = -2\mathscr{F}(\varphi^{\beta} - \varphi^{\alpha})$$

(15.44)

The left side now corresponds to the affinity for the overall cell reaction

$$Zn^{\beta} + (Cu^{2+})^{\varepsilon} = Cu^{\alpha} + (Zn^{2+})^{\varepsilon}$$

[15.8]

15.3.1   CONDITIONS FOR EQUILIBRIUM IN A GENERAL GALVANIC CELL

Equation 15.44 can be written for a general galvanic cell. In order to avoid confusion with respect to the meaning of the sign of the cell potential, it is useful to establish the following conventions for the representation of the cell:

(a) The affinities and the electrode reactions are written for reduction reactions, i.e., so that electrons are consumed as reactants.
(b) In visualizing the cell geometry the electromotive force (emf) that is reported for the cell is the potential for the right electrode minus that for the left.[5]

This convention (right minus left) may be conveniently remembered by recognizing that it is the same as the convention for interpreting chemical reactions: (products−reactants) is (right side of reaction−left side) by convention.

In order to help visualize the cell under consideration it is useful to adopt a simple notation for describing its configuration. The generic cell representation:

Left Electrode|Electrolyte 1‖Electrolyte 2|Right Electrode

---

[5] In a significant subset of the literature the opposite convention is adopted, and the emf is written as that of the left electrode minus that of the right. As a consequence, for the same cell the reported emf would be the negative of that obtained for the convention adopted here.

The single vertical dash represents the interface between electrode and electrolyte on each side. The double vertical line represents an interface between the two electrolytes. The cell discussed in the last section may be written

$$Zn^{\alpha}|ZnCl_2\|CuCl_2|Cu^{\beta}$$

More explicit information may be supplied in representing the system:

$$Zn^{\alpha}(\text{pure})|Zn^{++}(a_{Zn++} = 0.005)\|Cu^{++}(a_{Cu++} = 0.002)|Cu(\text{pure})$$

All of the electrode reactions written so far in this section are examples of oxidation reactions in which electrons are generated. The reverse of these reactions are reduction reactions. All subsequent reactions will be written as reduction reactions according to the convention just adopted. For a set of examples of reduction reactions see Table 15.3.

Consider the cell shown in Figure 15.6. The α electrode is visualized as on the left and β on the right. By the convention chosen, the emf of this cell would be reported as $\varphi^{\beta} - \varphi^{\alpha}$. Suppose the α phase is composed of metal M_1 which forms ions with charge z_1 in the electrolyte solution, and electrode β is the metal M_2 with ions of charge z_2. The electrode reactions are:

$$(M_1^{z_1+})^{\varepsilon} + z_1(e^-)^{\alpha} = M_1^{\alpha} \qquad [15.9]$$

and

$$(M_2^{z_2+})^{\varepsilon} + z_2(e^-)^{\beta} = M_2^{\beta} \qquad [15.10]$$

TABLE 15.3
Selected Values from the Electromotive Series

Reduction Reaction	Emf (V)
$Ca^+ + e^- = Ca$	-3.80
$Na^+ + e^- = Na$	-2.71
$Al^{+++} + 3\,e^- = Al$	-1.662
$Fe^{++} + 2\,e^- = Fe$	-0.447
$Ni^{++} + 2\,e^- = Ni$	-0.257
$Fe^{+++} + 3\,e^- = Fe$	-0.037
$2\,H^+ + 2\,e^- = H_2$	0.000
$Cu^{++} + e^- = Cu^+$	0.153
$Cu^+ + e^- = Cu$	0.521
$Fe^{+++} + e^- = Fe^{++}$	0.771
$O_2 + 4\,H^+ + 4\,e^- = 2\,H_2O$	1.229
$Au^+ + e^- = Au$	1.692
$Ag^{++} + e^- = Ag^+$	1.980

Source: D.R. Lide, Ed., *CRC Handbook of Chemistry and Physics*, CRC Press, Boca Raton, FL, 1991; compare with Appendix G.

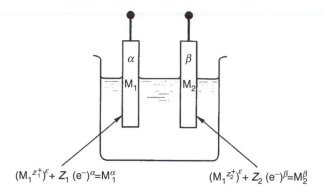

$(M_1^{z_1^+})^\varepsilon + Z_1\,(e^-)^\alpha = M_1^\alpha$

$(M_1^{z_2^+})^\varepsilon + Z_2\,(e^-)^\beta = M_2^\beta$

FIGURE 15.6 A general galvanic cell with two electrodes; the phase α on the left is the metal M_1; and the β phase on the right is the metal M_2. The electrode reactions, written as reduction reactions are indicated.

Apply the condition for equilibrium, Equation 15.39, to each of these electrode/electrolyte systems:

$$A^\alpha = \mu^\alpha_{M_1} - (z_1 \mu^\alpha_e + \mu^\varepsilon_{M_1^{z_1+}}) = -z_1\mathscr{F}(\varphi^\alpha - \varphi^\varepsilon) \qquad (15.45)$$

$$A^\alpha = \mu^\beta_{M_2} - (z_2 \mu^\beta_e + \mu^\varepsilon_{M_2^{z_2+}}) = -z_2\mathscr{F}(\varphi^\beta - \varphi^\varepsilon) \qquad (15.46)$$

Multiply Equation 15.45 by (z_2/z_1) and subtract the result from Equation 15.46:

$$\left[\mu^\beta_{M_2} - \left(z_2\mu^\beta_e + \mu^\varepsilon_{M_2^{z_2+}}\right)\right] - \frac{z_2}{z_1}\left[\mu^\alpha_{M_1} - \left(z_1\mu^\alpha_e + \mu^\varepsilon_{M_2^{z_1+}}\right)\right]$$
$$= -z_2\mathscr{F}(\varphi^\beta - \varphi^\alpha) \qquad (15.47)$$

The left-hand side is again seen to correspond to the difference in the affinities for the two electrode reactions, balanced so that each transfers the same number of electrons.

$$A^\beta - \frac{z_2}{z_1}A^\alpha = -z_2\mathscr{F}(\varphi^\beta - \varphi^\alpha) \qquad (15.48)$$

Equation 15.48 demonstrates that in a galvanic cell at equilibrium the difference in affinity for the two electrode reactions is proportional to the externally measured difference in potential between the electrodes.

Because the chemical potentials of the electrons in the electrodes are equal, Equation 15.47 may also be written

$$A_{cell} = \left[\mu^\beta_{M_2} + \frac{z_2}{z_1}\mu^\varepsilon_{M_1^{z_1+}}\right] - \left[\mu^\varepsilon_{M_2^{z_2+}} + \frac{z_2}{z_1}\mu^\alpha_{M_1}\right] = -z_2\mathscr{F}(\varphi^\beta - \varphi^\alpha)$$
$$(15.49)$$

The left side of this equation corresponds to the affinity for the overall cell reaction

$$(M_2^{z_{2+}})^\varepsilon + \frac{z_2}{z_1} M_1^\alpha = M_2^\beta + \frac{z_2}{z_1} (M_1^{z_{1+}})^\varepsilon \qquad [15.11]$$

To illustrate this general result, suppose α is zinc for which $z_1 = 2$ and β is aluminum for which z_2 is 3. This result then reads

$$\left[\mu_{Al}^\beta + \frac{3}{2} \mu_{Zn^{2+}}^\varepsilon \right] - \left[\mu_{Al^{3+}}^\varepsilon + \frac{3}{2} \mu_{Zn}^\alpha \right] = -3\mathscr{F}(\varphi^\beta - \varphi^\alpha) \qquad (15.50)$$

The left side is the affinity for the cell reaction

$$(Al^{3+})^\varepsilon + \frac{3}{2} Zn^\alpha = Al^\beta + \frac{3}{2}(Zn^{2+})^\varepsilon \qquad [15.12]$$

Recall from Section 11.1, Equation 11.32 in Chapter 11, that the affinity for a reaction may be written in terms of the standard free energy change for the reaction and the proper quotient of activities:

$$A = \Delta G^0 + RT \ln Q \qquad (11.32)$$

Substitute this expression into Equation 15.49,

$$A_{cell} = \Delta G_{cell}^0 + RT \ln Q_{cell} = -z\mathscr{F}\mathscr{E}_{cell} \qquad (15.51)$$

where $\mathscr{E}$ is the potential difference between the electrodes, $\mathscr{E} = (\varphi^\beta - \varphi^\alpha)$, reported, in accordance with the convention adopted, as the potential of the right electrode minus the left. Note that the order of the potentials in this definition ($\beta - \alpha$) corresponds to the order of the electrode reactions, written as reduction reactions, implied by the affinity on the right-hand side of Equation 15.51.

If a cell is constructed so that the components in the electrode reactions are in their standard states, then for that cell $Q = 1$ and $\ln Q = 0$. As an example, for the $Cu/CuCl_2//ZnCl_2/Zn$ cell considered previously the electrodes would be pure copper and pure zinc and the two electrolytes would be prepared with a molality that yields an activity of 1. Equation 15.51 becomes:

$$\Delta G_{cell}^0 = -z\mathscr{F}\mathscr{E}_{cell}^0 \qquad (15.52)$$

The measured emf, $\mathscr{E}^0$, is the standard electrode potential for the cell reaction. Substitution of this result into Equation 15.51 yields, after minor rearrangement,

$$\mathscr{E} = \mathscr{E}^0 - \frac{RT}{z\mathscr{F}} \ln Q \qquad (15.53)$$

This relation is known as the Nernst equation; it is the working equation of electrochemistry, most often applied in the solution of practical problems.

Most applications of electrochemistry to aqueous solutions take place near room temperature, taken to be 25 °C. It is also convenient in most applications to replace the natural logarithm, in this expression with base ten logarithms, where $\ln(x) = 2.303 \log_{10}(x)$. With these constraints, the Nernst equation may be written

$$\mathscr{E} = \mathscr{E}^0 - \frac{8314\left(\dfrac{J}{\text{mole K}}\right)(298.17 \text{ K})(2.303)}{z \times 96,512\left(\dfrac{J}{\text{volt} - \text{mole}}\right)} \ln Q$$

$$\mathscr{E} = \mathscr{E}^0 - \frac{0.05915}{z} \ln Q \tag{15.54}$$

where the unit of all terms is volts. While this is the most widely used form of the Nernst equation, it is emphasized that the relationship in Equation 15.53 is general and is also frequently applied to cells at elevated temperatures in which the electrolyte is a molten salt or even a solid ionic conductor.

15.3.2 Temperature Dependence of the Electromotive Force of a Cell

Electrochemical cells designed to study a given chemical reaction may be used to measure the standard entropy change for the reaction and the heat of reaction. This may be achieved by constructing a cell with electrodes and their solutions in their standard states and measuring their reversible emf at a series of temperatures. The computation of these properties is then straightforward.

Recall that the change in entropy for a reaction is the negative of the temperature derivative of its Gibbs free energy change. Apply this coefficient relation to Equation 15.52:

$$\Delta S^\circ = -\left(\frac{\partial \Delta G^0}{\partial T}\right)_{P,n_k} = +z\mathscr{F}\left(\frac{\partial \mathscr{E}^0}{\partial T}\right)_{P,n_k} \tag{15.55}$$

The enthalpy change for the reaction may be computed from the definitional relationship:

$$\Delta H^0 = \Delta G^0 + T\Delta S^\circ = z\mathscr{F}\left(-\mathscr{E}^0 + T\left(\frac{\partial \mathscr{E}^0}{\partial T}\right)_{P,n_k}\right) \tag{15.56}$$

If experiments are carried out carefully, the resulting values for ΔS° and ΔH^0 compare well with those obtained calorimetrically. The electrochemical cell is evidently an extremely useful tool in the experimental evaluation of thermodynamic properties.

15.3.3 The Standard Hydrogen Electrode

It would be very convenient if it were possible to measure and tabulate the electrical potentials of single electrodes, e.g., the left electrode in Figure 15.5a. The equilibrium electrical potential of any galvanic cell could then be computed as the difference between two "single electrode potentials." A simple strategy has been devised that essentially achieves this level of convenience without the necessity of measuring true "single electrode potentials." Experimentalists have agreed to report equilibrium emf measurements for cells in which one electrode is chosen because it bears upon the practical problem at hand and the other electrode is always the same half cell configuration. If all potential differences are recorded relative to this standard electrode as $(\varphi^\alpha - \varphi^{STD})$, $(\varphi^\beta - \varphi^{STD})$, etc., then potential differences for cells with electrodes α and β can be determined by subtraction:

$$\mathscr{E}^{\alpha\beta} = \varphi^\beta - \varphi^\alpha = (\varphi^\beta - \varphi^{STD}) - (\varphi^\alpha - \varphi^{STD}) \tag{15.57}$$

In practice it is not even necessary to make all measurements on cells that contain the standard electrode. It is only necessary to measure the emf of a system in which one of the electrodes has previously been calibrated against the standard electrode. Suppose the emf of a cell has been measured with an electrode β and the standard electrode: $(\varphi^\beta - \varphi^{STD})$. Then the potential of a new electrode material α against the standard electrode may be determined from a cell with α and β as electrodes: $(\varphi^\beta - \varphi^\alpha)$,

$$\varphi^\alpha - \varphi^{STD} = (\varphi^\beta - \varphi^{STD}) - (\varphi^\beta - \varphi^\alpha) \tag{15.58}$$

Since the potential of the standard electrode cancels in the computation of cell potentials, it is convenient to choose the potential φ^{STD} to be at zero volts. With this choice, any potential measured relative to the standard electrode configuration

$$\mathscr{E}^\beta = \varphi^\beta - \varphi^{STD} = \varphi^\beta - 0 = \varphi^\beta \tag{15.59}$$

may be thought of as a "half cell potential" for that electrode.

The choice of electrode configuration for the standard is based on considerations that are analogous to those that determined the choice of the 0° and 100 °C benchmarks for the centigrade temperature scale. The ice point and boiling point of pure water were chosen because water is available in reasonable purity in every laboratory and the benchmarks (freezing and boiling at one atmosphere pressure) are reproducibly obtained. The standard electrode to which all electrochemical emfs are referred is the standard hydrogen electrode (SHE) shown in Figure 15.7. An inert platinum electrode is immersed in an acid solution and purified hydrogen gas at one atmosphere pressure is bubbled over the platinum/electrolyte interface with the system maintained at 25 °C. The reaction at the electrode is

$$2(H^+)^\varepsilon + 2(e^-)^{Pt} = H_2(g) \tag{15.13}$$

A variety of electrodes have been designed that may be conveniently used as reference electrodes in cell measurements in the laboratory.[5] All have been carefully calibrated with respect to the SHE so that potentials determined with their use may be computed relative to the agreed standard, as in Equation 15.58, and results compared with compiled potential measurements.

Application of this strategy for determining half cell potentials or single electrode potentials to cells in which the components are in their standard states generates $\mathscr{E}^0$ values, standard single electrode potentials, for the corresponding

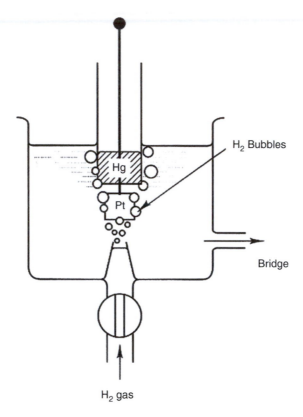

FIGURE 15.7 One form of standard hydrogen electrode used as a reference electrode in reporting cell potentials.

single electrode reactions. A collection of standard single electrode potentials (measured relative to the SHE and reported for the reduction reactions) is presented in Table 15.3. Some compilations report these electrode potentials for the oxidation reactions. In this case, all signs would be reversed and the order of the reactions in the table inverted. This ordering of the elements according to their single electrode cell emf is called the electromotive series of the elements.

The standard free energy change for a reaction is related to its equilibrium constant K. From Equation 15.52,

$$\Delta G^0 = -RT \ln K = -z\mathscr{F}\mathscr{E}^0$$

or

$$\mathscr{E}^0 = \frac{RT}{z\mathscr{F}} \ln K \tag{15.60}$$

Reactions near the top of Table 15.3, with relatively large positive values of $\mathscr{E}^0$ correspond to the condition $K > 1$. Thus, for these reactions at equilibrium the activity of the products (the metal) is large in comparison with the reactants (the ion in solution). Elements at this end of the series are said to be noble because they resist dissolution to form their ions. Near the bottom of the series in Table 15.3 $\mathcal{E}^0$ values are negative, implying $K < 1$; the reactant (in this case the ion in solution) dominates at equilibrium. Elements at the bottom end of the series are said to be reactive. The behavior changes smoothly from noble through neutral to reactive as the electromotive series is traversed from top to bottom.

EXAMPLE 15.6

The measured reversible emf for the following cell

$$Pt/H_2|H^+\|Cu^{++}|Cu$$

at 25 °C is found to be +0.295 V. Compute the activity of Cu^{++}.
The standard electrode potential for the reduction of cupric ion is found in Appendix G to be +0.342 V. The electrode reaction is

$$Cu^{2+} + 2e^- = Cu$$

for which $Q = a_{Cu}/a_{Cu2+} = 1/a_{Cu^{2+}}$. Apply the Nernst Equation 15.54:

$$0.295\ v = 0.342\ v - \frac{0.05915\ v}{2}\log_{10}\frac{1}{a_{Cu^{2+}}}$$

$$\log_{10}\frac{1}{a_{Cu^{2+}}} = (0.295\ v - 0.337\ v)\frac{2}{0.05915\ v} = -1.59$$

$$a_{Cu^{2+}} = 10^{+1.59} = 39$$

15.4 POURBAIX DIAGRAMS

Predominance diagrams, which are similar in concept to those devised in Section 11.4 in Chapter 11, to describe the behavior of multivariate reacting systems may be constructed to represent electrochemical equilibria. Variables chosen to describe the system are cell emf and electrolyte pH, which form the axes of Pourbaix diagrams. Construction of these plots is based upon repeated application of the Nernst Equation 15.54 to all of competing reactions that may occur within a given cell. Lines derived from these conditions divide the $(pH, \mathcal{E})$ space into regions of predominance for each of the components known to exist in the system.

15.4.1 THE STABILITY OF WATER

The simplest Pourbaix diagram is for pure water. The components assumed to exist in such a system are H_2O, H^+, OH^-, $H_2(g)$ and $O_2(g)$. The activity of H^+, and implicitly that of OH^-, is represented on the pH axis of the diagram. Regions of predominance of the remaining components, H_2, O_2, and H_2O, may be computed by considering two electrode reactions.

For the formation of hydrogen gas consider the reaction

$$2H^+ + 2e^- = H_2(g) \qquad \mathscr{E}^0 \equiv 0.000 \, v \qquad\qquad [15.14]$$

For this half cell reaction,

$$\log_{10}Q = \log_{10}\left[\frac{a_{H_2}^g}{a_{H^+}^2 + a_{e^-}^2}\right] = \log_{10}\left[\frac{P_{H_2}}{a_{H^+}^2}\right] = \log_{10}P_{H_2} - 2\log_{10}a_{H^+}$$

$$\log_{10}Q = \log_{10}P_{H_2} + pH \qquad\qquad (15.61)$$

Let $b = 2.303 \, RT/\mathscr{F} = 0.05915$ V, since the constant appears repeatedly in these calculations. The Nernst equation for the given value of $\mathscr{E}^0$ and Q is

$$\mathscr{E} = 0.000 - \frac{b}{2}[\log_{10}P_{H_2} + 2pH]$$

$$\mathscr{E} = -bpH - \frac{b}{2}\log_{10}P_{H_2} \qquad\qquad (15.62)$$

For a given value of P_{H_2} this relationship plots as a straight line in (pH, $\mathscr{E}$) space. The slope of the line is $(-b) = -0.05915$ V, and the intercept on the vertical line at pH = 0 is equal to $[-(b/2)\log P_{H_2}]$. Values of intercept for $P_{H_2} = 10^{-2}, 10^{-1}, 10^0, 10^1$ and 10^2 are respectively $+0.05915$, $+0.02958$, 0.000, -0.02958, and -0.05915 V. This set of five lines is plotted in Figure 15.8. As the emf of the cell decreases in the region of these lines the equilibrium hydrogen pressure increases from 10^{-2} atm to 10^2 atm. Thus, in the (pH, $\mathscr{E}$) domain below the line representing $P_{H_2} = 1$ atm, water decomposes to form H_2 gas. Above this line the equilibrium hydrogen pressure decreases very rapidly as the emf increases and H_2O

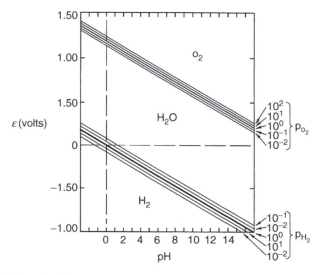

FIGURE 15.8 Domain of stability of water in $\mathscr{E}$-pH space.

is stable. The line represented by the equation

$$\mathscr{E} = -b\text{pH} = -0.05915 \text{ pH} \tag{15.63}$$

may be thought of the limit of predominance between H_2 gas and H_2O.

The competition between H_2O and O_2 may be examined by considering the electrode reaction

$$O_2 + 4H^+ + 4e^- = 2H_2O \qquad \mathscr{E}^0 = +1.229 \text{ V} \tag{15.15}$$

For this reaction,

$$\log_{10}Q = \log_{10}\frac{a_{H_2O}^2}{a_{O_2}^g a_{H^+}^4 a_{e^-}^4} = \log_{10}\frac{1}{P_{O_2} a_{H^+}^4}$$

$$= -\log_{10}P_{O_2} - 4\log_{10}a_{H^+}$$

$$\log_{10}Q = -\log_{10}P_{O_2} + 4\text{pH} \tag{15.64}$$

The condition for equilibrium expressed in the Nernst equation is

$$\mathscr{E} = +1.229 - \frac{1}{4}b[-\log_{10}P_{O_2} + 4\text{pH}] \tag{15.65}$$

$$\mathscr{E} = \left[1.229 + \frac{1}{4}b\,\log_{10}P_{O_2}\right] - b\text{pH} \tag{15.66}$$

For a fixed value of P_{O_2}, this equation represents a straight line with slope equal to $-b = -0.05915$ and intercept $[1.229 + (b/4)\log P_{O_2}]$. As P_{O_2} varies from 10^{-2} to 10^2 the intercept varies from 1.199 to 1.259. These lines have the same slope as do the hydrogen reaction lines. They are closer together because the coefficient of $\log P_{O_2}$ is $(b/4)$ as compared to $(b/2)$ for the hydrogen case. Also, the order is reversed because the sign of the $\log P_{O_2}$ term is opposite to that for the H_2 lines. Thus as the cell emf increases, the oxygen pressure changes from 10^{-2} to 10^2 atm. The region of predominance of oxygen gas thus lies above the line corresponding to $P_{O_2} = 1$ atm; H_2O is the predominant component below that line.

Figure 15.8 shows that the Pourbaix diagram for pure water has three regions of predominance. At high cell emf water decomposes to form oxygen gas; at intermediate emf water is stable; and in a specific range of negative emf, water decomposes to form hydrogen gas. From a strictly theoretical point of view, the consideration of all electrochemical reactions that presume the presence of liquid water must be limited to the $(\text{pH}, \mathscr{E})$ range corresponding to the stability of H_2O in Figure 15.8. However, from a practical point of view the rate of evolution of hydrogen at low (negative) emfs and oxygen at high positive values may be very slow in comparison to other electrochemical reactions that may be of interest in the system. Accordingly, the practical application of these thermodynamic considerations span a range that is significantly broader than the region of stability of H_2O.

15.4.2 POURBAIX DIAGRAM FOR COPPER

The general strategy for computing Pourbaix diagrams is illustrated for the copper–water system in this section. In addition to the components normally present in water, it is necessary to enumerate the ionic and nonionic components containing copper atoms that are known to exist in the system. The seven copper containing components considered in calculating this diagram may be divided into two categories: three solid substances, Cu, CuO, and Cu_2O, and four dissolved ionic species, Cu^+, Cu^{++}, $HCuO_2^-$ and CuO_2^-. It is further assumed that the electrolyte contains additional components, anions, or cations, necessary to vary the pH of the solution, which do not participate in the reactions involving copper.

When H^+, H_2O, and e^- are included, there are ten components and three elements (Cu, H, O) participating in this system. Thus, there are $r = (c - e) = (10 - 3) = 7$ independent reactions. Standard electrode potentials must be obtained for each of these reactions.

Each line on the diagram represents a zone of transition in predominance between two competing components. The equation describing a particular line is obtained by applying the condition for equilibrium, the Nernst equation, to a cell in which the competing pair of components are involved in the electrode reactions. The number of reactions that must be considered is the number of ways the seven copper-containing components can be placed in competition two at a time: $[7!/(2!)(5!)] = 21$. Thus, the Pourbaix diagram for copper consists at most of 21 lines (more precisely, 21 zones of transition). Some of these lines may not appear on the diagram. For example, the line calculated representing the competition between components X and Y may lie within a region in which a third component, Z, predominates over both X and Y. In this case, the XY line will be absent from the diagram.

The 21 reactions that must be considered are listed in Table 15.4. According to the convention established, reactions that involve the transfer of electrons are written as reduction reactions. Also, when a reaction may be written involving either H^+ or OH^- ions, the chosen reaction is expressed in terms of the H^+ ion. This choice is made because the abscissa on the Pourbaix diagram is expressed in terms of the activity of the hydrogen ion and OH^- and H^+ activities are related through the water reaction. Thus, for example, reaction [H] describing the competition between Cu and CuO could be written

$$Cu + 2OH^- = CuO + H_2O + 2e^-$$

However, the Nernst equation for this reaction involves $\log a_{OH^-}$, whereas the Pourbaix diagram describes this aspect of the behavior of the system with $pH = -\log a_{H^+}$. Subtraction of the water reaction from the above equation yields

$$Cu + 2OH^- + 2H_2O = CuO + H_2O + 2e^- + 2H^+ + 2OH^-$$

which simplifies to

$$Cu + H_2O = CuO + 2H^+ + 2e^-$$

TABLE 15.4
Reactions Considered in Computing the Pourbaix Diagram for Copper

(A) $Cu^{++} + 2 H_2O = HCuO_2^- + 3 H^+$
(B) $Cu^{++} + 2 H_2O = CuO_2^{2-} + 4 H^+$
(C) $HCuO_2^- = CuO^{2-} + H^+$
(D) $Cu^+ = Cu^{++} + e^-$
(E) $Cu^+ + 2 H_2O = HCuO_2^- + 3 H^+ + e^-$
(F) $Cu^+ + 2 H_2O = CuO_2^{2-} + 4 H^+ + e^-$
(G) $2 Cu + H_2O = Cu_2O + 2 H^+ + 2 e^-$
(H) $Cu + H_2O = CuO + 2 H^+ + 2 e^-$
(I) $Cu_2O + H_2O = 2 CuO + 2 H^+ + 2 e^-$
(J) $2 Cu^+ + H_2O = Cu_2O + 2 H^+$
(K) $Cu^{++} + H_2O = CuO + 2 H^+$
(L) $CuO + H_2O = HCuO_2^- + H^+$
(M) $CuO + H_2O = CuO_2^{2-} + 2 H^+$
(N) $Cu = Cu^+ + e^-$
(O) $Cu = Cu^{++} + 2 e^-$
(P) $Cu + 2 H_2O = HCuO_2^- + 3 H^+ + 2 e^-$
(Q) $Cu + 2 H_2O = CuO_2^{2-} + 4 H^+ + 2 e^-$
(R) $Cu_2O + 2 H^+ = 2 Cu^{++} + H_2O + 2 e^-$
(S) $Cu_2O + 3 H_2O = 2 HCuO_2^- + 4 H^+ + 2 e^-$
(T) $Cu_2O + 3 H_2O = 2 CuO_2^{2-} + 6 H^+ + 2 e^-$
(U) $Cu^+ + H_2O = CuO + 2 H^+ + e^-$

The inverse reaction is the proper reduction reaction for computation of the competition for the Pourbaix diagram.

$$CuO + 2e^- + 2H^+ = Cu + H_2O \qquad (H)$$

This simple strategy permits any reaction involving OH^- ions to be rewritten in terms of H^+ ions so that the corresponding condition for equilibrium is expressed directly in terms of the pH of the solution.

The reactions summarized in Table 15.4 may be divided into three classes:

(a) Reactions in which no electrons are transferred to the electrode include [A], [B], [C], [J], [K], [L] and [M]. These reactions rearrange charge over the ions in the electrolyte. As no charge is transferred to the electrode, the condition for equilibrium is independent of the external emf of the system. The corresponding limits of predominance plot as vertical lines at a fixed pH on the Pourbaix diagram.

(b) Reaction [D], [N] and [O] do not depend upon the hydrogen ion concentration; the limit of predominance for this competition plots as a horizontal line on the diagram.

(c) The remaining reactions involve both electrons and hydrogen ions. These are represented through the Nernst equation by sets of lines that have a slope that is a simple multiple of 0.05915; the multiplying factor depends upon the stoichiometric coefficients of H^+ and e^- in the reaction.

Sets of lines corresponding to these 21 reactions may be combined to generate the Pourbaix diagram for the copper-water system. It remains to determine which sets of lines do not appear on the diagram and to identify the regions of predominance that are outlined by each set. Automatic algorithms based on linear programming theory that make these decisions have been adapted in the development of computer programs to calculate Pourbaix diagrams.

The completed Pourbaix diagram for the copper water system is presented in Figure 15.9. From a practical point of view the variety of regions shown can be grouped into three classes:

1. A region of immunity in which the copper metal is the predominant component.
2. Regions of corrosion in which the predominant component is one of the copper-containing ionic species.
3. Regions of passivation in which the predominant component is one of the solid copper-containing species.

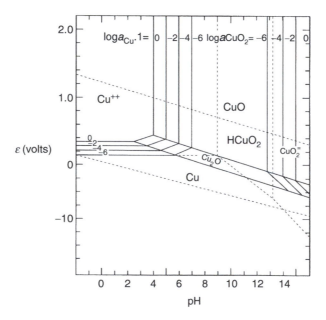

FIGURE 15.9 Complete Pourbaix diagram for copper in water. *Source:* Potter, E.C., *Thermodynamics*, 5th ed., North-Holland, Amsterdam, pp. 101–123, 1967.

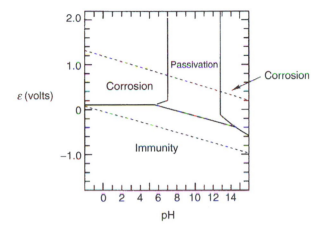

FIGURE 15.10 Simplified Pourbaix diagram for copper in water. *Source:* Potter, E.C., *Thermodynamics*, 5th ed., North-Holland, Amsterdam, pp. 101–123, 1967.

The latter condition will be protective of the metal with respect to corrosion if the solid component reaction product forms a coherent layer on the metal, isolating it from the electrolyte solution and rendering it passive to corrosion. Figure 15.10 shows a simplified version of the Pourbaix diagram for copper in which only these three classes of behavior are identified.

Pourbaix diagrams may be computed for any metal or alloy for which the required thermodynamic information is available. The atlas of Pourbaix diagrams[6] provides a continuing compilation of this information. A computer program for generating such diagrams from the input thermodynamic data is available for use on personal computers.[7] With limited additional information it is possible to explore the effects of additional variables upon these regions of predominance. For example, it is relatively straightforward to introduce a third variable, such as chloride ion activity into the analysis and generate three-dimensional ($\mathscr{E}$, pH, $-\log a_{Cl^-}$) diagrams or to plot sections at selected chloride activities.

As is the case for phase diagrams, knowledge of Pourbaix potential-pH diagrams is just the starting point for developing an understanding of corrosion phenomena. These diagrams are generated by repeated application of the conditions for equilibrium in electrochemical systems. Thus the domains of predominance they identify describe the predominant specie for a given (pH, $\mathscr{E}$) at equilibrium. The diagram describes the ultimate condition of the system left indefinitely under fixed conditions. How the system approaches its ultimate destination may be of paramount importance in practical corrosion. Also, the systems treated in Pourbaix diagrams are generally greatly simplified in comparison to the electrolyte and electrode combinations that are of practical interest. Real-world corrosion problems are inherently dirty, experience changing conditions with time, and may involve human factors that are unrelated to the physics, chemistry, and thermodynamics of the system. Pourbaix diagrams provide a point of departure in the solution of corrosion problems.

15.5 SUMMARY

When dissolved in an appropriate solvent an electrolyte dissociates into ionic components. Equilibrium within an electrolyte solution is determined by the dissociation constant through a condition for equilibrium that is analogous to that obtained for reacting systems. Weak electrolytes partially dissociate into ions with some of the compound remaining undissociated in solution. The degree of dissociation, α, for an electrolyte, M_uX_v, is governed by the dissociation constant, K, according to the equation

$$K_c = (u^u v^v)\frac{\alpha^\nu}{1 - \alpha}[k]^{\nu-1} \tag{15.16}$$

where $\nu = u + v$ and $[k]$ is the molarity (concentration in moles per liter) of the electrolyte in the solvent. Strong electrolytes, acids, bases, and salts dissociate completely into ions in solution.

Equilibrium between an electrode and its enclosing electrolyte solution gives rise to an electrostatic potential difference between the two phases. The system attains equilibrium when the affinity for the reaction that governs transfer of charge between the phases is proportional to the potential difference between electrode and electrolyte.

$$A = \mu_M^\alpha - (z^+\mu_{e^-}^\alpha + \mu_{M^{z+}}^\varepsilon) = -z^+\mathscr{F}(\varphi^\alpha - \varphi^\varepsilon) \tag{15.38a}$$

Application of this condition for equilibrium to both electrodes in an electrochemical cell yields the Nernst equation

$$\mathscr{E} = \mathscr{E}^0 - \frac{RT}{z\mathscr{F}}\ln Q \tag{15.53}$$

where $\mathscr{E}^0$ is the standard electrode potential for the cell and Q is the proper quotient of activities for the overall cell reaction.

A compilation of standard electrode potentials for a collection of reduction reactions involving elemental electrodes, called the electromotive series, provides a basis for ordering the elements in terms of their tendency to dissolve in water.

The conditions for electrochemical equilibrium of materials may be usefully visualized with potential-pH predominance diagrams devised by Pourbaix. Regions of immunity, corrosion and passivation of materials may be mapped out in potential-pH space.

HOMEWORK PROBLEMS

Problem 15.1. The dissociation constant for aluminum phosphate in aqueous solution at 298 K is 9.8×10^{-21}. Assume $K_f = 1$. Calculate the degree of dissociation of $AlPO_4$ as a function of concentration. Plot the result.

Problem 15.2. The dissociation constant of hydrocyanic acid (HCN) at 25 °C is 7×10^{-10}. Compute the pH of a solution prepared with 0.05 M concentration of HCN.

Problem 15.3. Beryllium hydroxide is a weak base with a dissociation constant given by 5×10^{-11} at 25 °C. How many grams of $Be(OH)_2$ must be added to a liter of water in order to obtain a pH of 8.5?

Problem 15.4. What concentration of sodium hydroxide must be added to water to adjust the pH to 9.5?

Problem 15.5. Calcium hydroxide and nitric acid are strong electrolytes. Two stock solutions are available: one with 0.02 M $Ca(OH)_2$ and the other with 0.05 M HNO_3. How much of the nitric acid solution must be added to one liter of the calcium hydroxide solution to make a solution with a pH = 8.2?

Problem 15.6. Explain why the potential difference computed in Equation 15.39 for a single electrode cannot be measured.

Problem 15.7. Describe all of the subsystems in a galvanic cell including the phases and the interfaces. Describe the conditions that must be met by each of these subsystems in order for the cell to operate reversibly.

Problem 15.8. Sketch a galvanic cell represented by the following notation.

$$Ni^{\alpha}(a_{Ni} = 0.23)|NiCl_2(a_{Ni} = 0.03)\|CuCl_2(a_{Cu} = 0.02)|Cu^{\alpha}(a_{Cu} = 1)$$

Label all of the parts of the cell.

Problem 15.9. Given that it is not possible to measure the potential difference between an electrode and an electrolyte in which it is immersed, how is it possible to evaluate half cell potentials for the subsystems in an electrochemical cell?

Problem 15.10. Reversible emfs are measured for the following cell at 1073 K: Mg (pure, liquid)|$MgCl_2$–$CaCl_2$ liquid eutectic|Mg (in Al) liquid for a sequence of compositions:

$X_{Mg} = 0.0447$	0.1130	0.1905	0.3400	0.5825	0.7490
ξ (mv) = 152.840	109.868	82.501	54.195	30.446	20.871

Source: From Basant L. Tiwari, *Metall. Trans.*, 18A, 1987, p. 1645.

Compute and plot the activity of lithium in liquid Li–Sn alloys as a function of composition.

Problem 15.11. Review the concept of the limit of predominance as it applies to the construction of Pourbaix pH-emf diagrams.

Problem 15.12. Construct a potential-pH diagram for the Ni–water system at 25 °C.
   a. Determine the ionic species that are pertinent.
   b. List the pertinent reactions.
   c. Look up the essential electrochemical information.
   d. Calculate and plot the diagram.

Problem 15.13. A sample of copper immersed in water at 25 °C will dissolve. Use the Pourbaix diagram in Figure 15.10 to suggest three strategies that could be applied to avoid the corrosion of copper in water.

REFERENCES

1. Robinson, R.A. and Stokes, R.H., *Electrolyte Solutions*, Butterworths, London, 1965.
2. Potter, E.C., *Electrochemistry Principles and Applications*, Cleaver-Hume Press, London, 1961.
3. Devereaux, O.F., *Topics in Metallurgical Thermodynamics*, Wiley, New York, 1983, Chapter 6.
4. Guggenheim, E.A., *Thermodynamics*, 5th ed., North-Holland, Amsterdam, 1967.
5. Potter, E.C., *Thermodynamics*, 5th ed., North-Holland, Amsterdam, pp. 101–123, 1967.
6. Pourbaix, M., *Atlas of Electrochemical Equilibria in Aqueous Solutions*, 2nd ed., National Association of Corrosion Engineers, Houston, TX, 1974.
7. Froning, M.H., Shanley, M.E., and Verink, E.D. Jr., *Corros. Sci.*, 16, 371, 1976.

Appendices

A Fundamental Physical Constants and Conversion Factors

Physical Constants

Avogadro's number	$N_0 = 6.0221415 \times 10^{23}$
Rest mass of the electron	$m_e = 9.1093826 \times 10^{-31}$ (kg)
Elementary charge	$e = 1.60217653$ C
Plank's constant	$H = 6.6260693$ (J s)
Gas constant	$R = 8.314313$ (J/mol K)
	$R = 1.98727$ (cal/mol K)
	$R = 82.0589$ (cc atm/mol K)
	$R = 0.0820589$ (l atm/mol K)
Boltzmann's constant ($k = R/N_0$)	$k = 1.3806505 \times 10^{-23}$ (J/atm K)

Conversion Factors

Length	$1 \text{ m} = 100 \text{ cm} = 1000 \text{ mm} = 10^6 \ \mu\text{m} = 10^9 \text{ nm}$
Volume	$1 \text{ l} = 1000 \text{ ml} = 1000 \text{ cm}^3 = 10^{-3} \text{ m}^3$
Temperature	$T°C = T K - 273.15 = (5/9)(T°F - 32)$
Pressure	$1 \text{ atm} = 101325 \text{ Pa} = 0.101325 \text{ MPa} = 0.06805 \text{ psi}$
Energy	$1 \text{ J} = 1 \text{ N m} = 4.184 \text{ cal} = 9.8699 \text{ cc atm} = 0.0098699 \text{ l atm}$

Hint: A useful method for converting energy units is to multiply by the ratio of two *R*-values that contain the units to be converted.

Source: From Lide, D.R., Ed., *CRC Handbook of Chemistry and Physics*, 71st ed., Boca Raton, FL, 1991.

B Properties of Selected Elements

		A Wt.	$\alpha \times 10^6$	$\beta \times 10^7$	V^S	V^L	S°_{298}
				Properties of the Stable Form at 298 K			
Element	AN	(gm/mol)	(K^{-1})	(atm^{-1})	(cc/mol)	(cc/mol)	(J/mol K)
Ag	47	107.868	57	9.0	10.27	11.54	42.6
Al	13	26.98	70.5	12	9.99	11.31	28.3
Au	79	196.97	43	—	10.21	11.35	47.5
C(d)	6	12.011	21.3	—	3.42	—	2.4
Co	27	58.94	37	5.6	6.62	7.59	30.0
Cu	29	63.546	51	6.6	7.09	7.94	33.1
Fe	26	55.85	36	5.9	7.10	7.96	27.3
H$_2$	1	1.009	Id.Gas	Id.Gas	—	—	130.6
Hf	72	178.58	18	89	13.61	16.07	43.5
Mg	12	24.32	78	—	13.97	15.29	32.7
N$_2$	7	14.008	Id.Gas	Id.Gas	—	—	191.5
Ni	28	58.71	40	26	6.60	7.43	29.9
O$_2$	8	16.000	Id.Gas	Id.Gas	—	—	205.1
Pb	82	207.2	87	23	17.74	19.40	64.8
Pt	78	159.09	27	3.5	9.10	10.3	41.5
Pu	94	(244)	165		12.3	14.7	
Si	14	28.086	23	—	12.00	—	18.8
Ti	22	47.90	27	—	10.64	11.65	30.7
U	92	238.03	—	—	12.49	13.30	50.2
Zn	30	65.38	93	16	9.16	9.94	41.6
Zr	40	91.22	18	—	14.06	(15.7)	39.0

Continued

Element	AN	A Wt. (gm/mol)	$C_P(T) = a + bT + \dfrac{c}{T^2} + dT^2$ (J/mol K)			
			a	$b \times 10^3$	$c \times 10^{-5}$	$d \times 10^6$
Ag	47	107.868	21.30	8.54	1.51	—
Al	13	26.98	31.38	− 16.40	− 3.60	20.75
Au	79	196.97	31.46	−13.47	− 2.89	10.96
C(d)	6	12.011	9.12	13.22	− 6.19	—
Co	27	58.94	18.12	23.14	− 0.42	− 0.08
Cu	29	63.546	30.29	− 10.71	− 3.22	9.47
Fe	26	55.85	28.18	− 7.32	− 2.90	25.04
H_2	1	1.009	27.37	3.33	—	—
Hf	72	178.58	23.44	7.63	—	—
Mg	12	24.32	21.13	11.92	0.15	—
N_2	7	14.008	30.42	2.54	− 2.37	—
Ni	28	58.71	11.17	37.78	3.18	—
O_2	8	16.000	29.96	4.18	− 1.67	—
Pb	82	207.2	24.23	8.70	—	—
Pt	78	159.09	24.69	5.02	− 0.50	—
Pu	94	(244)	24.7	24		—
Si	14	28.086	23.93	2.22	− 4.14	—
Ti	22	47.90	24.94	6.57	− 1.63	1.34
U	92	238.03	25.10	2.38	—	23.68
Zn	30	65.38	22.38	10.04	—	—
Zr	40	91.22	22.84	8.95	− 0.67	—

AN, atomic number; A Wt., atomic weight (gm/mol); α, volume coefficient of thermal expansion $(K)^{-1}$; β, volume coefficient of compressibility $(atm)^{-1}$; V^S, molar volume of the solid at 298 K and 1 atm (cc/mol); V^L, molar volume of the liquid at 298 K and 1 atm (cc/mol); S°_{298}, absolute entropy of the element in its stable phase form at 298 K and 1 atm (J/mol K); $C_P(T)$, heat capacity at constant pressure (J/mol K).

Notes: The value of α is the number in the table $\times 10^{-6}$. The value of β is the number in the table $\times 10^{-7}$.

Sources: From Kubaschewski, O., Alcock, C.B., and Spencer, R.J., *Materials Thermochemistry*, 6th ed., Pergamon Press, Oxford, 1993; Reynolds, C.L., Faugham, K.A., and Barker, R.E., *J. Chem. Phys.*, 59, 2943, 1973; Reynolds, C.L. and Barker, R.E., *J. Chem. Phys.*, 61, 2564, 1974.

C Phase Transformations for the Elements

Element	Transformation	T_T	ΔS_T°	ΔH_T°
Ag	fcc → l	1235	9.2	11.3
	l → g	2470	104.2	257.4
Al	fcc → l	934	11.4	10.7
	l → g	2793	104	290.5
Au	fcc → l	1338	9.4	12.6
	l → g	3130	109.4	342.4
B	hex → l	2450	9.2	22.5
	l → g	4100	141	
Cd	hex → l	594	10.4	6.2
	l → g	1040	95.8	
Co	hcp → fcc	700	0.7	0.5
	fcc → l	1768	9.2	16.2
	l → g	3203	(130)	416.4
Cr	bcc → l	2130	8.1	17.234
	l → g	2945	116	1.6
Cu	fcc → l	1358	9.8	13.3
	l → g	2830	108	305.6
Fe	bcc → fcc	1184	0.8	0.9
	fcc → bcc	1665	0.5	0.8
	bcc → l	1809	7.6	13.8
	l → g	3130	109	341.2
Ga	orth → l	303	18.4	5.6
	l → g	2690	100	269.0
Hf	hex → bcc	2016	2.9	5.9
	bcc → l	2504	10.9	27.2
	l → g	4870	117	569.8
Ir	fcc → l	2720	9.6	26.1
	l → g	4700	130	611.0
Mg	hcp → l	923	9.2	8.5
	l → g	1360	93.7	127.4
Mo	bcc → l	2896	13.5	39.1
	l → g	1880	121	227.5
Ni	fcc → l	1728	10.0	17.2
	l → g	3180	118	375.2

Continued

563

Element	Transformation	T_T	ΔS_T°	ΔH_T^0
Pb	fcc → l	601	8.0	4.8
	l → g	2020	88.4	178.6
Pt	fcc → l	2042	9.6	19.7
	l → g	4400	107	470.8
Si	dia → l	1685	29.8	50.2
	l → g	2873	109	385.9
Ti	hex → bcc	1155	3.6	4.2
	bcc → l	1943	8.6	16.7
	l → g	3558	120	427.0
U	orth → tetrag	941	3.0	2.8
	tetrag → cubic	1049	4.6	4.8
	cubic → l	1408	6.5	9.2
	l → g	4700	89	418.3
V	bcc → l	2183	10.5	23.0
	l → g	3680	124	456.3
W	bcc → l	3693	13.5	50.0
	l → g	5828	(126)	734
Zn	hcp → l	693	10.5	7.3
	l → g	1180	96.9	114.3
Zr	hcp → bcc	1136	3.4	3.9
	bcc → l	2128	8.8	18.8
	l → g	4700	123	578.1

T_T, transformation temperature (K); ΔS_T°, entropy of the transformation (J/mol K); ΔH_T^0, enthalpy of the transformation (kJ/mol).

Source: Kubaschewki, O., Alcock, C.B., and Spencer, R.J., *Materials Thermochemistry*, 6th ed., Pergamon Press, Oxford, 1993.

D Properties of Some Random Solutions

The thermochemical literature contains a wide variety of models for the mixing behavior of solutions. In the CALPHAD approach, the Gibbs free energy of any phase, p, is computed relative to the reference states for the elements that form the mixtures assumed to be the stable phase form of the element at 298 K and ambient pressure:

$$G^p_{\text{mix}}(X^p_k, T) = G^{0,p} + G^{i,\text{dp}} + G^{\text{xs},p}$$

$$G^p_{\text{mix}}(X^p_k, T) = \sum_{k=1}^{c} X^p_k G^{0,p}_k(T) + RT \sum_{k=1}^{c} X^p_k \ln X^p_k + G^{\text{xs},p}_{\text{mix}}(X^p_k, T) \qquad (D.1)$$

where $G^{0,p}_k$ is the stability of phase p for the element k at the temperature T and $G^{\text{xs},p}_{\text{mix}}$ is the excess-free energy of mixing. The focus of the description of mixing behavior is upon the excess Gibbs free energy term. Depending upon the nature of the phase (stoichiometric compound, ordered or random solution) a variety of mixing models exist in the literature. The most widely used model is based upon the Redlich–Kister power series is introduced in Section 8.7.2:

$$G^{\text{xs}}_{\text{mix}} = \sum_{i=1}^{c} \sum_{j>i}^{c} X_i X_j \sum_{v=0}^{n} \Omega^v_{ij}(X_i - X_j)^v \qquad (8.117)$$

For a binary system, this equation simplifies to

$$\Delta G^{\text{xs}}_{\text{mix}} = X_1 X_2 \sum_{v=0}^{n} \Omega^v(X_1 - X_2)^v \qquad (D.2)$$

The database parameters are contained in the set of coefficients, Ω^v which generally have the form $\Omega^v = a_v + b_v T$. The first few terms may be written explicitly as

$$\Delta G^{\text{xs}}_{\text{mix}} = X_1 X_2 [(a_0 + b_0 T) + (a_1 + b_1 T)(X_1 - X_2) + (a_2 + b_2 T)(X_1 - X_2)^2 + ...] \qquad (D.3)$$

This table presents values for these parameters for a selected set of random solutions. These data may be used in solving problems dealing with mixing behavior that may be assigned.

System	Phase	a_0 (J/mol)	b_0 (J/mol K)	a_1 (J/mol)	b_1 (J/mol K)	a_2 (J/mol)	b_2 (J/mol K)
Ag–Au	FCC	− 15599	—	—	—	—	—
	Liq	− 16402	1.14	—	—	—	—
Ag–Cu	FCC	33819	− 8.12	− 5602	1.33	—	—
	Liq	17535	− 4.54	2251	− 2.67	493	—
Au–Cu	FCC	− 168314	16	34311	0	4162	27.3
	Liq	− 197088	30.4	5450	0	54624	− 11.4
Cr–Fe	BCC	20500	− 9.7	—	—	—	—
	FCC	10833	− 7.5	1410	0	—	—
Cu–Ni	FCC	8048	3.4	− 2041	1.0	—	—
	Liq	12049	1.3	− 1862	0.9	—	—
Cu–Zn	FCC	− 42804	10.0	2936	− 3.1	9034	− 5.4
	Hex	− 35422	5.2	25277	− 10.0	—	—
Mo–W	BCC	2000	—	—	—	—	—
	Liq	0	—	—	—	—	—
Zr–Nb	BCC	15911	3.4	3919	− 1.1	—	—
	Hex	24411	—	—	—	—	—

Source: Chang, A., *WinPhad Software*, Computherm, Madison, WI, 1998. With permission.

E Properties of Selected Compounds

Compound	S°_{298} (J/mol K)	$-\Delta H^0_{298}$ (KJ/mol)	$C_P(T) = a + bT + \dfrac{c}{T^2} + dT^2$ (J/mol K)			
			a	$b \times 10^3$	$c \times 10^{-5}$	$d \times 10^6$
Carbides						
B_4C	27.1	71.5	96.52	21.92	-44.98	—
$Cr_{23}C_6$	612.1	328.4	683.25	209.20	-104.6	—
Cr_7C_3	201.0	160.7	233.89	62.34	-38.07	—
Cr_3C_2	85.4	85.4	123.26	25.90	-28.24	—
Fe_3C	104.6	-25.1	82.01	83.68	—	—
HfC	39.5	225.9	42.34	12.13	-7.36	-2.43
SiC	16.5	66.9	42.59	8.37	-16.61	-1.26
Ta_2C	81.6	208.4	66.44	13.93	-8.58	—
TiC	42.3	142.7	43.30	8.16	-7.95	—
UC	59.0	97.5	143.59	-1.26	-8.70	4.39
U_2C_3	134.8	181.6	125.10	12.80	-15.52	—
UC_2	68.8	88.3	69.04	8.54	-9.41	—
W_2C	81.6	26.4	89.75	10.88	-14.56	—
WC	34.7	40.6	43.49	8.62	-9.33	-1.03
Hydrides						
AlH_2	30.0	11.4	45.19	—	—	—
$CH_4(g)$	186.3	74.8	12.45	76.69	1.45	-17.99
$C_2H_2(g)$	200.8	-226.8	43.63	31.65	-7.51	-6.32
$C_2H_4(g)$	219.2	-52.5	32.64	59.83	—	—
MgH_2	31.0	76.1	27.20	49.37	-5.86	—
$NH_3(g)$	192.7	49.5	37.32	18.66	-6.49	—
UH_3	63.7	127.0	30.38	42.34	5.65	—
Nitrides						
AlN	20.2	318.4	32.26	22.68	-7.91	—
BN	14.8	252.3	41.21	9.41	-21.76	—
Fe_4N	155.6	11.1	110.79	34.14	—	—
GaN	29.7	109.6	38.07	9.00	—	—
Nb_2N	79.5	253.1	62.38	17.11	—	—
NbN	43.9	236.4	36.36	22.59	—	—
Ta_2N	74.5	272.4	70.50	17.66	-7.11	—

Continued

Compound	S°_{298} (J/mol K)	$-\Delta H^{0}_{298}$ (KJ/mol)	$C_P(T) = a + bT + \dfrac{c}{T^2} + dT^2$ (J/mol K)			
			a	$b \times 10^3$	$c \times 10^{-5}$	$d \times 10^6$
TaN	41.8	252.3	55.27	2.72	-12.64	—
UN	62.5	294.6	55.73	4.98	-8.79	—
Oxides						
Al_2O_3	50.9	1675.7	117.49	10.38	-37.11	—
CO	197.5	110.5	28.41	4.10	-0.46	—
CO_2	213.7	393.5	44.14	9.04	-8.54	—
CoO	53.0	237.7	55.40	-6.44	—	7.11
Co_3O_4	114.3	910.0	140.75	17.28	-24.35	53.97
Cu_2O	92.9	173.2	58.20	23.97	-1.59	—
CuO	42.6	161.9	46.44	11.55	-7.11	2.59
$Fe_{0.945}O$	60.1	263.0	48.79	8.37	-2.80	—
Fe_2O_3	87.4	823.4	98.28	77.82	-14.85	—
Fe_2O_3	150.9	1108.8	91.55	202.00	—	—
$H_2O(l)$	60.9	285.8	75.44	—	—	—
(g)	188.7	241.8	30.0	10.71	0.33	—
MgO	26.9	601.6	48.99	3.43	-11.34	—
NbO	46.0	419.7	42.97	8.87	-4.02	—
NbO_2	54.5	795.0	61.30	25.77	-4.56	—
Nb_2O_5	137.3	1899.5	162.17	14.81	-30.63	—
NiO	38.0	239.7	20.88	157.23	16.28	—
SiO_2	43.4	908.3	46.90	31.51	-10.08	—
TiO	34.7	542.7	44.22	15.06	-7.78	—
Ti_2O_3	77.2	1520.9	31.80	213.38	—	—
Ti_3O_5	129.4	2459.1	231.04	-24.07	-61.25	—
TiO_2	50.6	944.0	73.35	3.05	-17.03	—
UO_2	77.0	1085.0	80.33	6.78	-16.57	—
U_4O_9	344.1	4510.4	356.27	35.44	-66.40	—
U_3O_8	282.5	3574.8	282.42	36.94	-49.96	—
UO_3	96.2	1223.8	93.30	10.88	-10.88	—
Y_2O_3	99.2	1905.0	109.62	20.08	-11.30	—
ZnO	43.6	350.5	48.99	5.10	-9.12	—
ZrO_2	50.4	1100.8	60.62	7.53	-14.06	—

S°_{298}, absolute entropy of compound at 298 K (J/mol K); ΔH^{0}_{298}, enthalpy of formation of compound at 298 K (KJ/mol); a, first heat capacity coefficient (J/mol K); b, second heat capacity coefficient (J/mol K^2); c, third heat capacity coefficient (J K/mol); d, fourth heat capacity coefficient (J/mol K^3).

Notes: The negative of ΔH^{0}_{298} is reported in the table. The value of b is table value $\times 10^{-3}$. The value of c is table value $\times 10^5$. The value of d is table value $\times 10^{-6}$.

Source: From Kubaschewski, O., Alcock, C.B., and Spencer, R.J., *Materials Thermochemistry*, 6th ed., Pergamon Press, Oxford, 1993.

F Interfacial Energies of Selected Elements

Element	σ^{LV}	γ^{SV}	γ^{SL}
Ag	926	1100	—
Al	865	(910)	93
Au	(731)	(1370)	132
Bi	376	(550)	61
Cs	68.6	—	—
Cu	1360	(1700)	177
Fe	1880	(1870)	204
Ga	708	—	—
Ni	1822	(2040)	255
Pt	1865	(1900)	240
Si	720	—	—
W	2500	(2500)	—

σ^{LV}, surface tension of liquid–vapor interface at the melting point (mN/m); γ^{SV}, surface free energy of solid–vapor interface at the melting point (mJ/m^2 = erg/cm^2); γ^{SL}, surface free energy of solid–liquid interface at the melting point (mJ/m^2 = erg/cm^2).

Sources: From Lide, D.R., Ed., *CRC Handbook of Chemistry and Physics*, 71st ed., Boca Raton, FL, 1991; Murr, L.E., *Interfacial Phenomena in Metals and Alloys*, Addison-Wesley, Reading, MA, 1975.

G Electrochemical Series

Electrode potentials relative to the standard hydrogen potential arranged in descending order.

Reduction Reaction	Emf (volts)	Reduction Reaction	Emf (volts)
$Sr^+ + e^- = Sr$	− 4.10	$2\,H^+ + 2\,e^- = H_2$	0.000
$Ca^+ + e^- = Ca$	− 3.80	$Cu^{++} + e^- = Cu^+$	0.153
$Li^+ + e^- = Li$	− 3.0401	$Re^{+++} + 3\,e^- = Re$	0.300
$Rb^+ + e^- = Rb$	− 2.98	$Cu^{++} + 2\,e^- = Cu$	0.3419
$K^+ + e^- = K$	− 2.931	$Ru^{++} + 2\,e^- = Ru$	0.455
$Cs^+ + e^- = Cs$	− 2.92	$Cu^+ + e^- = Cu$	0.521
$Ba^{++} + 2\,e^- = Ba$	− 2.912	$Rh^{++} + 2\,e^- = Rh$	0.600
$Sr^{++} + 2\,e^- = Sr$	− 2.89	$Rh^+ + e^- = Rh$	0.600
$Ca^{++} + 2\,e^- = Ca$	− 2.868	$Rh^{+++} + 3\,e^- = Rh$	0.758
$Na^+ + e^- = Na$	− 2.71	$Fe^{+++} + e^- = Fe^{++}$	0.771
$Mg^+ + e^- = Mg$	− 2.70	$Ag^+ + e^- = Ag$	0.7996
$Mg^{++} + 2\,e^- = Mg$	− 2.372	$Hg^{++} + 2\,e^- = Hg$	0.851
$Sc^{+++} + 3\,e^- = Sc$	− 2.077	$Pd^{++} + 2\,e^- = Pd$	0.951
$Be^{++} + 2\,e^- = Be$	− 1.847	$Pt^{++} + 2\,e^- = Pt$	1.118
$Al^{+++} + 3\,e^- = Al$	− 1.662	$Ir^{+++} + 3\,e^- = Ir$	1.156
$Ti^{++} + 2\,e^- = Ti$	− 1.630	$O_2 + 4\,H^+ + 4\,e^- = 2\,H_2O$	1.229
$Mn^{++} + 2\,e^- = Mn$	− 1.185	$Au^{+++} + e^- = Au^{++}$	1.401
$V^{++} + 2\,e^- = Vi$	− 1.175	$Au^{+++} + 3\,e^- = Au$	1.498
$Nb^{+++} + 3\,e^- = Nb$	− 1.099	$Mn^{+++} + e^- = Mn^{++}$	1.5415
$Cr^{+++} + 3\,e^- = Cr$	− 0.744	$Au^+ + e^- = Au$	1.692
$Ga^{+++} + 3\,e^- = Ga$	− 0.560	$Ag^{++} + e^- = Ag^+$	1.980
$Fe^{++} + 2\,e^- = Fe$	− 0.447		
$Cd^{++} + 2\,e^- = Cd$	− 0.352		
$Co^{++} + 2\,e^- = Co$	− 0.28		
$Ni^{++} + 2\,e^- = Ni$	− 0.257		
$Mo^{+++} + 3\,e^- = Mo$	− 0.200		
$Pb^{++} + 2\,e^- = Pb$	− 1205		
$Fe^{+++} + 3\,e^- = Fe$	− 0.037		
$2\,H^+ + 2\,e^- = H_2$	0.000		

Source: From Lide, D.R., Ed., *CRC Handbook of Chemistry and Physics*, CRC Press, Boca Raton, FL, 1991.

H The Carnot Cycle

The connection between the entropy change and the reversible heat absorbed for an arbitrary process can be traced to a development first presented by Sadi Carnot early in the 19th century. Carnot visualized a thermodynamic system containing an ideal gas taken through a specific process sequence that returns it to its original state. This process is called the Carnot cycle and the sequence of states of the Carnot engine as its cycle is traversed is represented in Figure H.1. A pictorial representation of the steps in the process is shown in (P, V) coordinates in Figure H.2. All steps in the process are assumed to be carried out reversibly.

Process I. The state of the system is initially given by (T_1, P_0), with a corresponding volume, V_0. It is in contact with a heat reservoir at the same temperature T_1. Except for its area of contact with the heat reservoir, the boundary is thermally insulated. The pressure is reduced to P_1, allowing the gas to expand to a volume V_1 isothermally at T_1. A quantity of heat Q_I flows from the system into the reservoir (Q_I is negative). When the system reaches the state (P_1, V_1) a thermally insulating barrier is inserted to separate it from the reservoir.

Process II. The pressure is further reduced to P_2. Since the entire system is thermally insulated from its surroundings, process II is adiabatic and the heat absorbed Q_{II} is equal to zero. It is presently shown that the path in (P, V) space can be computed for this process so that the final volume (V_2) and temperature (T_{III}) can be calculated given P_2.

Process III. The system is brought into contact with an external heat reservoir with a temperature T_{III} matching the value attained at the end of step II. The insulating partition is removed and the system is isothermally compressed at T_{III} to the pressure P_3. A quantity of heat is exchanged with the reservoir, and since heat flows from the surroundings into the system, Q_{III} is positive. The final state is represented by the condition (P_3, V_3). An insulating barrier is again inserted between the reservoir and the system in preparation for the next step.

Process IV. The final pressure P_3 at the end of step III is not chosen arbitrarily. It is the unique pressure at T_{III} that takes the system along an adiabatic path back to the starting state at (P_0, V_0) completing the cycle. The heat absorbed in this process, Q_{IV}, is again equal to zero.

The property changes associated with each of these processes can be computed for one mole of an ideal gas without invoking the second law of thermodynamics. This observation is crucial, since the argument is intended to establish a key element

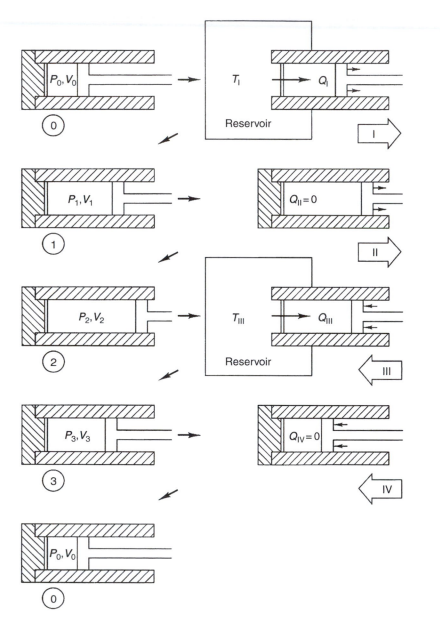

FIGURE H.1 States and processes that connect them in the Carnot cycle.

of the second law, namely, the relation between the change in a state function and the heat absorbed during an arbitrary reversible process. Later the state function is identified as the entropy.

Process I is an isothermal expansion. Since the internal energy of an ideal gas depends only upon its temperature, $dU = 0$ for an isothermal process.

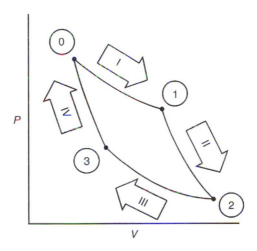

FIGURE H.2 Sequence of states in (V, P) space traversed by a system during a Carnot cycle.

From the first law

$$dU = \delta Q + \delta W = 0 \Rightarrow \delta Q = -\delta W \tag{H.1}$$

Since, for any reversible process $\delta W = -PdV$,

$$\delta Q = -(-PdV) = PdV$$

For the isothermal process at T_{I},

$$dV_T = d\left(\frac{RT_{\mathrm{I}}}{P}\right)_T = -\frac{RT_{\mathrm{I}}}{P^2}dP_T \Rightarrow PdV_T = -\frac{RT_{\mathrm{I}}}{P}dP_T$$

Integration gives the heat absorbed:

$$Q_{\mathrm{I}} = \int_{P_0}^{P_1} \frac{RT_{\mathrm{I}}}{P}dP = RT_{\mathrm{I}}\ln\frac{P_1}{P_0} \tag{H.2}$$

Similarly, for process III carried out at T_{III} from P_2 to P_3:

$$Q_{\mathrm{III}} = \int_{P_2}^{P_3} \frac{RT_{\mathrm{III}}}{P}dP = RT_{\mathrm{III}}\ln\frac{P_3}{P_2} \tag{H.3}$$

For an adiabatic process, $\delta Q = 0$, and the first law becomes

$$dU = C_V dT = \delta W = -PdV = -\frac{RT}{V}dV$$

Separate the variables:

$$\frac{dT}{T} = -\frac{R}{C_V}\frac{dV}{V}$$

In process II the temperature changes from T_I to T_{III} and the volume changes from V_1 to V_2. Integration gives

$$\ln\frac{T_{III}}{T_I} = \ln\left(\frac{V_1}{V_2}\right)^{R/C_V}$$

Note that the minus sign in the previous expression causes the ratio of the volumes to be inverted in the integrated expression. Thus, for an adiabatic process for an ideal gas

$$\frac{T_{III}}{T_I} = \left(\frac{V_1}{V_2}\right)^{R/C_V} \tag{H.4}$$

Since the states are described in terms of the pressures involved, the ideal gas law can be used to convert V to corresponding (T, P) values.

$$\frac{T_{III}}{T_I} = \left(\frac{RT_I/P_1}{RT_{III}/P_2}\right)^{R/C_V}$$

Algebraic manipulation of this result gives the relation between T and P along an adiabatic path as

$$\frac{P_2}{P_1} = \left(\frac{T_{III}}{T_I}\right)^{(C_V+R)/C_V} \tag{H.5}$$

The analogous relation holds for process IV in which the system changes adiabatically from (T_{III}, P_3) to (T_I, P_0):

$$\frac{P_0}{P_3} = \left(\frac{T_I}{T_{III}}\right)^{(C_V+R)/C_V} \tag{H.6}$$

Next, examine the hypothesis that, for any reversible process

$$\int\frac{\delta Q_{rev}}{T} \tag{H.7}$$

is a state function with a value independent of the path followed during the process and thus depends only upon the initial and final states. The crucial test for this hypothesis rests on the proof that, for any cyclic process, that is, one that returns the system to its original state, this integral is zero. This integral can be easily evaluated for the Carnot cycle:

Process I:

$$\int \frac{\delta Q_{rev}}{T} = \frac{1}{T_I} \int \delta Q_{rev} = \frac{Q_1}{T_I} = \frac{1}{T_I} RT_1 \ln P_1 OVERP_0 = R \ln \frac{P_1}{P_0}$$

Process II:

$$\int \frac{\delta Q_{rev}}{T} = 0$$

Process III:

$$\int \frac{\delta Q_{rev}}{T} = \frac{1}{T_{III}} \int \delta Q_{rev} = \frac{Q_{III}}{T_{III}} = 1OVERT_{III}RT_{III} \ln \frac{P_3}{P_2} = R \ln \frac{P_3}{P_2}$$

Process IV:

$$\int \frac{\delta Q_{rev}}{T} = 0$$

The result for the system is the sum of these four contributions:

$$\int \frac{\delta Q_{rev}}{T} = R \ln \frac{P_1}{P_0} + 0 + R \ln \frac{P_3}{P_2} + 0$$

which can be written

$$\int \frac{\delta Q_{rev}}{T} = R \ln\left(\frac{P_1}{P_0} \frac{P_3}{P_2}\right) \tag{H.8}$$

The pressure ratios in this equation, (P_2/P_1) and (P_0/P_3), can be expressed simply in terms of the temperatures of the two isothermal steps by applying the adiabatic path Equation H.5 and Equation H.6:

$$\int \frac{\delta Q_{rev}}{T} = R \ln\left(\frac{P_1}{P_0}\right)\left(\frac{P_3}{P_2}\right) = R \ln\left(\frac{T_1}{T_{III}}\right)^{(C_V+R)/R}\left(\frac{T_{III}}{T_1}\right)^{(C_V+R)/R} \tag{H.9}$$

or

$$\int \frac{\delta Q_{rev}}{T} = R \ln(1) = 0 \tag{H.10}$$

Thus it is demonstrated that, for any Carnot cycle using an ideal gas operating between any pair of temperatures T_I and T_{III} and any choice of independent pressures

P_0 and P_1 (the other two pressures are determined by the definition of the cycle)

$$\int \frac{\delta Q_{rev}}{T} = 0$$

Accordingly, within these restrictions, the integral is a state function and its integrand is an exact differential of a state function.

Carnot and others went on to demonstrate that every reversible cyclic path that could be devised can be decomposed into an infinite sequence of connected infinitesimal Carnot cycles. Since the integral in question is zero for every Carnot cycle it totals to zero for any sum of Carnot cycles. Thus, the integral has the character of a state function for every cyclic process that can be imagined for an ideal gas.

The result was further extended to systems containing arbitrary working substance (i.e., other than an ideal gas). This advancement was achieved by imagining the system with the real working substance coupled to one with an ideal gas working substance so that the energy, heat and work exchanges accompanying the Carnot cycle for each of these two systems are equal and opposite at each step. It is concluded that

$$\int \frac{\delta Q_{rev}}{T} = 0$$

for any real system taken along an arbitrary path. Thus, the result is completely general and this integral is demonstrated to be a state function of general application. The symbol widely adopted for this state function is S: the origins of the name, entropy, are obscure.

For any arbitrary system taken along any reversible path

$$\Delta S = \int dS = \int \frac{\delta Q_{rev}}{T} \tag{H.11}$$

It follows that, for each infinitesimal step along the reversible path

$$dS = \frac{\delta Q_{rev}}{T} \tag{H.12}$$

and the general relation between the entropy and the reversible heat absorbed is established.

The development of the concept of entropy did not end with the Carnot cycle argument. The statistical interpretation of the entropy, introduced in Chapter 6, evolved half a century later. Notions of configurational entropy represented, for example, by the entropy of mixing of solutions, expanded the concept. The quantification of the role of entropy production in irreversible processes took another half a century, and continues to evolve. The Carnot cycle lies at the foundation of the development of thermodynamics; however, its role is largely hidden in the treatment of complex multicomponent, multiphase systems with capillarity and electrochemical effects, which are the bread and butter of materials science.

Answers to Homework Problems

CONTENTS

NUMERICAL ANSWERS TO HOMEWORK PROBLEMS

Homework problems in the text that have mathematical answers (as opposed to text answers) use as a database either: (1) data that are given the statement of the problem, or (2) data obtained from the Appendices of this text. These data are intended for use as exercises for homework assignments.

Thermochemical data evolve with time as new experimental results appear, and new optimizations are applied to existing data. For applications in the real world the reader is directed to a variety of databases (some free, some expensive) available through the Internet, which will give results that are not necessarily identical to those in the Appendices, because they are up to date.

CHAPTER 2

Problem 2.5

$$dz = 36u^2v\cos(x)du + 12u^3\cos(x)dv - 12u^3v\sin(x)dz$$

CHAPTER 3

Problem 3.16

Reaction	ΔS_{298} (J/mol K)
$C + O_2 = CO_2$	2.92
$2Al + \frac{3}{2}O_2 = Al_2O_3$	-313.04
$Si + C = SiC$	-7.98
$C + \frac{1}{2}O_2 = CO$	89.70
$Si + O_2 = SiO_2$	-182.36
$Si + CO_2 = SiC + O_2$	-10.93
$Al_2O_3 + \frac{3}{2}Si = \frac{3}{2}SiO_2 + 2Al$	42.56
$2Al + 3CO_2 = Al_2O_3 + 3CO$	-54.3
$SiO_2 + 2C = CO_2 + SiC$	191.6
$CO + \frac{1}{2}O_2 = CO_2$	-86.35
...	

CHAPTER 4

Problem 4.3

After the isobaric step, $V = 25.97$ (cc/mol)
After the isothermal step, $V = 25.76$ (cc/mol)

Problem 4.4

(a) 55.83 (J/mol K)
(b) -2.68 (J/mol K)
(c) 96.42 (J/mol K)
(d) -2.23 (J/mol K)
(e) 42.67 (J/mol K)
(f) -95.72 (J/mol K)

Problem 4.5

32.14 (J/mol K); -0.74 (J/mol K)

Problem 4.6

(a) $n = 0.536$ (moles); $\Delta U = 81 - 9 = 72$ (l-atm)
(b) $T_f = 1365$ K; $\Delta U = 72$ (l-atm)

Problem 4.7

Function to be plotted:

$$\Delta S(T,P) = C_p \ln\left(\frac{T}{300}\right) - R \ln\left(\frac{P}{1}\right)$$

Problem 4.8

$$\Delta U(P, V) = C_V(PV - P_0V_0)$$
$$\Delta H(P, V) = C_P(PV - P_0V_0)$$

Problem 4.9

$$\left(\frac{\partial H}{\partial G}\right)_S = \frac{C_P}{C_P - TS\alpha}$$

Problem 4.10

$$dF = -\left[S + PV\alpha - \frac{P\beta}{T\alpha}C_P\right]dT - \frac{P\beta}{\alpha}dS$$

Problem 4.12

(a) $T_i = 219.4$ K, $T_f = 292.5$ K
(b) $Q = -33.77$ (J/mol)
(c) $W = 945.62$ (J/mol)
(d) $\Delta U = 911.84$ (J/mol); $\Delta H = 1520$ (J/mol); $\Delta S = 0.217$ (J/mol K);
$\Delta F = -7902$ (J/mol); $\Delta G = -7294$ (J/mol)

Problem 4.13

(a) For silver, $P = 9.6 \times 10^6$ (atm)
(b) For alumina, $P = 978$ (atm)

Problem 4.14

Benchmark values:
ΔG (298 K, 10^{-10} atm) $= -57,050$ (J/mol)
ΔG (1000 K, 100 atm) $= -95,050$ (J/mol)

Problem 4.15

For a combination of the maximum values for α, β and V, $C_P - C_V = 0.435$.
For a combination of the minimum values, $C_P - C_V = 0.011$

CHAPTER 5

Problem 5.7

Unconstrained, $z_{max} = 4$ at $(x = 2, y = 2)$;
Constrained by $y = 1 - x$, $z'_{max} = 2.4$ at $(x = 6/5, y = 12/5)$

CHAPTER 6

Problem 6.2

List the macrostates states systematically starting with (10, 0, 0), (9, 0, 1), (9, 1, 0), (8, 0, 2), (8, 1, 1), (8, 2, 0), etc.

Problem 6.3
(a) $4^4 = 16$ microstates
(b) Enumerate systematically. There are 10 macrostates

Problem 6.4
(a) $4^3 = 64$
(b) $4^{15} = 1.074 \times 10^9$
(c) $15^4 = 50,625$
(d) $30^{50} = 7.18 \times 10^{73}$
(e) $100^{1000} = (10^2)^{1000} = 10^{2000}$

Problem 6.5
(a) $\Delta U = 0$ (J/mol)
(b) $\Omega_{II}/\Omega_I = 12$
(c) State II is more likely

Problem 6.6
(a) Calculator:
      $10! = 3.629 \times 10^6$, $30! = 2.653 \times 10^{32}$, $60! = 8.321 \times 10^{81}$
(b) Stirling's Formula: 4.54×10^5, 1.927×10^{31}, 4.28×10^{80}
(c) Errors in $\ln x!$: 0.138, 0.035, 0.016

Problem 6.7
$\Delta S = -13.05$ (J/mol K) $= -2.17 \times 10^{-23}$ (J/atom-K) (Applying Equation 6.9)

Problem 6.9
$\Delta U = 2311$ (J/mol)

Problem 6.10
(a) $\Delta S = 2.162$ (J/mol K)
(b) $\Delta S = 2.162$ (J/mol K)

CHAPTER 7

Problem 7.3

$$\mu(T, P) - \mu(298, 1) = -20.79T \ln T + 13.16T + 31,370 + RT \ln P$$

Problem 7.6
The error in $\log P$ is about 1.5% at the melting point (over a range of about 1500 K)

Problem 7.7
$\Delta S = 18.6$ (J/mol K)

Problem 7.8
  $T^{\varepsilon L} = 1710$ K

Problem 7.9
  $\Delta H_v = 269,000$ (J/mol)

Problem 7.10
  $P_{\alpha\gamma G} = 2.6 \times 10^{-10}$ (atm); $P_{\gamma\delta G} = 9.0 \times 10^{-6}$ (atm); $P_{\delta LG} = 7.0 \times 10^{-5}$ (atm)

Problem 7.11
  $(T,P)_{\beta LG} = (576\,\text{K}, 6.38 \times 10^{-10}(\text{atm}))$; $(T,P)_{\varepsilon\beta G} = (500\,\text{K}, 4.47 \times 10^{-12}\,\text{atm})$

Problem 7.12
  Metastable triple point: $(T,P)_{\varepsilon LG} = (569$ K, 4.34×10^{-10} atm$)$

CHAPTER 8

Problem 8.1

$$(Wt.\%O) = 0.044\left(\frac{\text{gm of O}}{\text{gm of soln}}\right); \quad c_O = 0.011\left(\frac{\text{mol of O}}{\text{cc of soln}}\right);$$

$$\rho_O = 0.18\left(\frac{\text{gm of O}}{\text{cc of soln}}\right)$$

Problem 8.3
  Benchmark:

$$\Delta\overline{V}_2 = 54X_1^2X_2; \quad \Delta\overline{V}_1 = 27X_1X_2^2$$

Problem 8.5
  Plot the results of a Gibbs–Duhem integration:

$$\Delta\overline{H_{Pn}} = 12,500\,X_{Cu}{}^2\left(X_{Cu} - \frac{1}{2}\right); \quad \Delta H_{mix} = 6250X_{Cu}^2X_{Pn}$$

Problem 8.9
  Plot the activity of zinc obtained from

$$RT \ln a_{Zn} = \left(1 - \frac{T}{4000}\right)X_{Al}^2[13,200(X_{Al} - X_{Zn}) + 19,200X_{Zn}]$$

$$+ RT \ln X_{Zn}$$

Problem 8.10

	Total Properties	PMP of A	PMP of B
ΔG^{xs}_{mix} (J/mol)	2513	1353	4667
ΔS^{xs}_{mix} (J/mol K)	0.565	0.304	1.049
ΔV_{mix} (cc/mol)	-1.749	-0.942	-3.247
ΔH_{mix} (J/mol)	2824	1521	5244
ΔU_{mix} (J/mol)	2842	1511	5211
ΔF^{xs}_{mix} (J/mol)	2531	1363	4700

Problem 8.13

Plot as a function of temperature

$$\gamma^0_A = \gamma^0_B = e^{-13,500/RT}$$

Problem 8.14

$$\varepsilon_{AB} = -6.111 \times 10^{-20} \text{ (J/bond)}$$

Problem 8.15

Plot $f_{AB} = [2X_A X_B](1.3)$; then $f_{AA} = X_A - f_{AB}/2$, $f_{BB} = X_B - f_{AB}/2$

Note that f_{AA} and f_{BB} go negative at compositions not too far from $X_B = 1/2$; the model loses physical significance for compositions far from $X_B = 1/2$.

Problem 8.16

Plot

$$f_{AB} = \frac{\Delta H_{mix}}{\left[\frac{1}{2}N_0 z\left(\varepsilon_{AB} - \frac{1}{2}(\varepsilon_{AA} + \varepsilon_{BB})\right)\right]}$$

$$f_{AA} = X_A - \frac{1}{2}f_{AB}; \qquad f_{BB} = X_B - \frac{1}{2}f_{AB}$$

CHAPTER 9

No numerical problems.

CHAPTER 10

Problem 10.1

Given:

$$\Delta G^\alpha_{mix}\{\alpha;\alpha\} = 8400X^\alpha_A X^\alpha_B + RT(X^\alpha_A \ln X^\alpha_A + X^\alpha_B \ln X^\alpha_B)$$

$$\Delta G^L_{mix}\{L;L\} = 10,500X^L_A X^L_B + RT(X^L_A \ln X^L_A + X^L_B \ln X^L_B)$$

(a) Compute and plot for the {L;L} reference state:

$$\Delta G^{\alpha}_{mix}\{L;L\} = \Delta G^{\alpha}_{mix}\{\alpha;\alpha\} - X^{\alpha}_A \Delta G^{0\,\alpha \to L}_A - X^{\alpha}_B \Delta G^{0\,\alpha \to L}_B$$

$$\Delta G^{L}_{mix}\{L;L\} = \Delta G^{L}_{mix}\{L;L\}$$

(b) Compute and plot for the {L;α} reference state:

$$\Delta G^{\alpha}_{mix}\{L;\alpha\} = \Delta G^{\alpha}_{mix}\{\alpha;\alpha\} - X^{\alpha}_A \Delta G^{0\,\alpha \to L}_A$$

$$\Delta G^{L}_{mix}\{L;\alpha\} = \Delta G^{L}_{mix}\{L;L\} + X^{L}_B \Delta G^{0\,\alpha \to L}_B$$

(c) Compute and plot for the {$\alpha;\alpha$} reference state:

$$\Delta G^{\alpha}_{mix}\{\alpha;\alpha\} = \Delta G^{\alpha}_{mix}\{\alpha;\alpha\}$$

$$\Delta G^{L}_{mix}\{\alpha;\alpha\} = \Delta G^{L}_{mix}\{L;L\} + X^{L}_A \Delta G^{0\,\alpha \to L}_A + X^{L}_B \Delta G^{0\,\alpha \to L}_B$$

Problem 10.2

(a) Compute and plot:

$$\Delta G^{\beta}_{mix}\{\beta;\beta\} = -8200 X^{\beta}_A X^{\beta}_B + RT(X^{\beta}_A \ln X^{\beta}_A + X^{\beta}_B \ln X^{\beta}_B)$$

$$\Delta G^{L}_{mix}\{L;L\} = -10,500 X^{L}_A X^{L}_B + RT(X^{L}_A \ln X^{L}_A + X^{L}_B \ln X^{L}_B)$$

(b) Compute and plot:

$$\Delta G^{\beta}_{mix}\{\beta;\lambda\} = \Delta G^{\beta}_{mix}\{\beta;\beta\} - X^{\beta}_B \frac{6800}{660}(660 - T)$$

$$\Delta G^{L}_{mix}\{\beta;L\} = \Delta G^{L}_{mix}\{L;L\} + X^{L}_A \frac{8200}{1050}(1050 - T)$$

Problem 10.3

(a) Compute and plot:

$$a_B\{L;L\} = X_B e^{(8400/RT)X^2_A}$$

(b) Compute and plot:

$$a_B\{L;\alpha\} = a_B\{L;L\}e^{-1200/RT}$$

Problem 10.9

(a) $X_2 = 0.307$ to $X_2 = 0.693$

(b) Compute and plot:

$$\Delta \mu_2^\alpha = 12,700(1 - X_2)^2 + RT \ln X_2$$

(c) Compute and plot:

$$\left(\frac{d\Delta \mu_2^\alpha}{dX_2}\right) = \frac{RT}{X_2} - 2 \times 12,700(1 - X_2)$$

Problem 10.16

Solve the following equations simultaneously:

$$317 - 5600X_2^{L2} + 11,400X_2^{\alpha2} + R(1300)\ln\left[\frac{1 - X_2^L}{1 - X_2^\alpha}\right] = 0$$

$$- 1246 - 5600(1 - X_2^L)^2 + 11,400(1 - X_2^\alpha)^2 + R(1300)\ln\left[\frac{X_2^L}{X_2^\alpha}\right] = 0$$

At 1300 K $X_2^\alpha = 0.763$; $X_2^L = 0.824$

Problem 10.17

$T_c = 637.5$ K

Spinodal boundary is given by

$$X_{2,\text{spin}}(T) = \frac{1}{2}\left[1 \pm \sqrt{1 - \frac{T}{T_c}}\right]$$

Problem 10.18

$X_{2\text{max}} = 0.649$; $T_c = 652.3$ K

Problem 10.19

$$T(X^*) = \frac{-9380X^*(1 - X^*) + 1283(8.8)(1 - X^*) + 942(6.3)X^*}{8.8(1 - X^*) + 6.3X^*}$$

$$T(X^*) = \frac{10,000X^*(1 - X^*) + 1283(8.8)(1 - X^*) + 942(6.3)X^*}{8.8(1 - X^*) + 6.3X^*}$$

Problem 10.20

$\gamma_0^0 = 370$

Problem 10.21

18,930 K

CHAPTER 11

Problem 11.2

$A = -161{,}200$ (J); products form.

Problem 11.3

$$H_2 + \tfrac{1}{2}O_2 = H_2O \qquad A = -97{,}700 \text{ (J)}$$

$$CO + \tfrac{1}{2}O_2 = CO_2 \qquad A = -145{,}000 \text{ (J)}$$

$$CH_4 + O_2 = CO_2 + 2H_2 \qquad A = -328{,}000 \text{ (J)}$$

Oxygen is consumed, oxides form.

Problem 11.4

$X_{CO} = 0.295,\ X_{CO_2} = 0.705,\ X_{O_2} = 1.8 \times 10^{-20}$

Problem 11.6

(a) $\Delta G_f(T) = -239{,}700 + 94.5T$ (J/mol)
(b) $K(T) = \exp[-\Delta G_f^0(T)/RT]$
(c) $P_{O_2} = 6.9 \times 10^{-17}$ (atm)
(d) $A = -136{,}400$ (J/mol)

Problem 11.7

(a) $\Delta G_f(SiO_2) = -714{,}600$ (J); ΔG_f $(H_2O) = -194{,}100$ (J)
(b) $\Delta G_r = -326{,}400$ (J)
(c) $K_r = 7.78 \times 10^{15}$, $(H_2/H_2O) = 8.76 \times 10^7$
(d) $P_{O_2} = 1.6 \times 10^{-35}$ (atm)
(e) $P_{O_2 2} = 1.6 \times 10^{-35}$ (atm)

Problem 11.9

(a) $P_{O_2} = 2 \times 10^{-12}$ (atm)
(b) $K = 3 \times 10^{33}$
(c) $(CO/CO_2) = 4 \times 10^5$
(d) Yes
(e) $(H_2/H_2O) = 6 \times 10^2$
(f) $P_{O_2} 10^{-27}$

Problem 11.10

(a) Yes
(b) No
(c) No
(d) Yes

Problem 11.11

Cu_2O: $A = -254,800$ (J); CaO: $A = -516,000$ (J)

NiS: $A = 120,900$ (J); $CaCl_2$: $A = -564,000$ (J)

Problem 11.13

$$Na_2O + O_2 + \tfrac{1}{2}S_2 = Na_2SO_3$$

$$2/5Nb_2O_5 + O_2 + 4/5S_2 = 4/5NbS_2O_5$$

$$2/7Cu_2O + O_2 + 2/7S_2 = 4/7CuSO_4$$

$$1/4Cu_2S + O_2 + 1/8S_2 = \tfrac{1}{2}CuSO_4$$

Problem 11.16

If total pressure is 1 atm, for $P_{CO} > 0.447$ oxidation will be prevented. Carburization requires $P_{CO} > 0.96$. Thus, if $P_{CO} > 0.96$, both conditions are satisfied.

Problem 11.17

Reactions: $2CO + O_2 = 2CO_2$ and $2H_2 + O_2 = 2H_2O$

(a) equilibrium mole fractions: $CO = 0.40$; $CO_2 = 0.09$; $H_2 = 0.377$; $H_2O = 0.133$; $O_2 = 1.6 \times 10^{-20}$

(b) Affinities: $A_{[2H_2O]} = -378,000$ (J); $A_{[2CO_2]} = -374,400$ (J)

CHAPTER 12

Problem 12.6

Equation of the surface:

$$P(H, T) = P(H = 0, T)e^{(2\gamma V^L/RT)H}$$

where $P(H = 0, T)$ is given by Equation 7.39.

Problem 12.7

$P(H = 0) = 1.728 \times 10^{-6}$ (atm); $P(r = -0.5\mu) = 1.722 \times 10^{-6}$ (atm). Note that the vapor pressure is lowered because the curvature is negative.

Problem 12.9

(a) $\lambda^\alpha = 1.98 \times 10^{-7}$ (cm); $\lambda^\beta = 1.68 \times 10^{-7}$ (cm)

Problem 12.10

(a) $X_B^\varepsilon = 0.109$; $X_B^\beta = 0.127$

(b) $\lambda^\beta = 1.16 \times 10^{-7}$ (cm); $\lambda^\varepsilon = 1.18 \times 10^{-7}$ (cm)

Problem 12.12
  0.866

Problem 12.14
  550 (ergs/cm^2)

Problem 12.15
  Yes, it will wet the grain boundaries completely.

Problem 12.16
  (a) $\phi = 163°$
  (b) $s_c = 4.1(\mu)$

CHAPTER 13

Problem 13.1
  (a) $\Delta S_v = 8.85$ (J/mol K); $\Delta H_v = 95,800$ (J/mol)

Problem 13.2
  (a) $X_i = \exp[7.6/R]\exp[-188,000/RT]$

Problem 13.3
  $X_{vv} = X_v{}^2\exp[0.89/R]\exp[-9580/RT]$

Problem 13.4
  High range: $X_f = \exp[10/R]\exp[-550,000/RT]$
  Low range: $X_f = \exp[1/R]\exp[-350,000/RT]$

Problem 13.7
  (a) $M_{2+0.00501}O_3$
  (b) $M_2O_{3-0.00749}$
  (c) $M_{2-0.0025}O_3$
  (d) $M_2O_{3+0.00375}$

CHAPTER 14

Problem 14.1
  $T = 398.38$ K at all x; (b) $\Delta S = 0.266$ (J/mol K)

Problem 14.6
  Doubling the isotope ratio requires a centrifuge with 50 cm radius rotating at 20,000 rpm.

Problem 14.9
  (a) $\Delta G = -219.59$ (J)
  (b) $\Delta G_{grad} = 0.235$ (J); $\Delta G_{Tot} = -219.35$ (J)

CHAPTER 15

Problem 15.1

Compute and plot:

$$\alpha(c) = \frac{-K \pm \sqrt{K^2 - 4Kc}}{2c} \qquad \text{Use} + \text{sign}$$

Problem 15.2

pH $= 5.72$

Problem 15.3

5.2×10^{-6} (gm)

Problem 15.4

$C_{\text{NaOH}} = 3.162 \times 10^{-5}$ (mol/liter)

Problem 15.5

1.26×10^{-7} (liters) $= 0.126$ microliters

Problem 15.10

For the compositions given,

$a_{\text{Mg}} = 0.006,\ 0.024,\ 0.061,\ 0.160,\ 0.357,\ 0.494$

Problem 15.12

Formation energies needed as input (J/mol):
NiO $= -215{,}900$; NiO$_2$ $= -215{,}100$; Ni^{++} $= -48{,}200$;
HNiO$_2^{--}$ $= -349{,}200$; H$_2$O $= -273{,}200$.

Triple points (E, pH) occur as follows:
(Ni, Ni^{++}, NiO) at $(-0.250,\ 6.09)$;
(Ni^{++}, NiO, NiO$_2$) at $(0.863,\ 6.09)$;
(NiO, NiO$_2$, HNiO$_2^{-}$) at $(0.146,\ 18.21)$;
(Ni, NiO, HNiO$_2^{-}$) at $(-0.967,\ 18.21)$

Index